Plant Specialized Metabolism

Genomics, Biochemistry,
and Biological Functions

Plant Specialized Metabolism

Genomics, Biochemistry, and Biological Functions

Edited by
Gen-ichiro Arimura
Tokyo University of Science
Tokyo, Japan

Massimo Maffei
University of Turin
Torino, Italy

CRC Press
Taylor & Francis Group
Boca Raton London New York

CRC Press is an imprint of the
Taylor & Francis Group, an **informa** business

CRC Press
Taylor & Francis Group
6000 Broken Sound Parkway NW, Suite 300
Boca Raton, FL 33487-2742

© 2017 by Taylor & Francis Group, LLC
CRC Press is an imprint of Taylor & Francis Group, an Informa business

No claim to original U.S. Government works

Printed on acid-free paper
Version Date: 20160414

International Standard Book Number-13: 978-1-4987-2628-3 (Hardback)

This book contains information obtained from authentic and highly regarded sources. Reasonable efforts have been made to publish reliable data and information, but the author and publisher cannot assume responsibility for the validity of all materials or the consequences of their use. The authors and publishers have attempted to trace the copyright holders of all material reproduced in this publication and apologize to copyright holders if permission to publish in this form has not been obtained. If any copyright material has not been acknowledged please write and let us know so we may rectify in any future reprint.

Except as permitted under U.S. Copyright Law, no part of this book may be reprinted, reproduced, transmitted, or utilized in any form by any electronic, mechanical, or other means, now known or hereafter invented, including photocopying, microfilming, and recording, or in any information storage or retrieval system, without written permission from the publishers.

For permission to photocopy or use material electronically from this work, please access www.copyright.com (http://www.copyright.com/) or contact the Copyright Clearance Center, Inc. (CCC), 222 Rosewood Drive, Danvers, MA 01923, 978-750-8400. CCC is a not-for-profit organization that provides licenses and registration for a variety of users. For organizations that have been granted a photocopy license by the CCC, a separate system of payment has been arranged.

Trademark Notice: Product or corporate names may be trademarks or registered trademarks, and are used only for identification and explanation without intent to infringe.

Library of Congress Cataloging-in-Publication Data

Names: Arimura, Gen-ichiro, editor. | Maffei, Massimo, editor.
Title: Plant specialized metabolism : genomics, biochemistry, and biological functions / [editors]: Gen-ichiro Arimura and Massimo Maffei.
Description: Boca Raton : Taylor & Francis, 2017. | Includes bibliographical references and index.
Identifiers: LCCN 2016015579 | ISBN 9781498726283 (alk. paper)
Subjects: LCSH: Plants--Metabolism. | Plant metabolites--Synthesis.
Classification: LCC QK881 .P553 2017 | DDC 572/.42--dc23
LC record available at https://lccn.loc.gov/2016015579

Visit the Taylor & Francis Web site at
http://www.taylorandfrancis.com

and the CRC Press Web site at
http://www.crcpress.com

Printed and bound in the United States of America by Publishers Graphics, LLC on sustainably sourced paper.

Contents

Preface vii
Editors ix
Contributors xi

Chapter 1   Introduction to plant specialized metabolism 1
            Gen-ichiro Arimura and Massimo Maffei

Chapter 2   Site of synthesis of specialized metabolites 9
            Massimo Maffei

Chapter 3   Biodiversity and chemotaxonomic significance
            of specialized metabolites 23
            Francesca Barbero and Massimo Maffei

Chapter 4   Biosynthesis and roles of Salicaceae salicylates 65
            Riitta Julkunen-Tiitto and Virpi Virjamo

Chapter 5   Alkaloid biosynthesis and regulation in plants 85
            Tsubasa Shoji

Chapter 6   Pinaceae alkaloids 119
            Virpi Virjamo and Riitta Julkunen-Tiitto

Chapter 7   Biosynthesis, regulation, and significance
            of cyanogenic glucosides 131
            *Lasse Janniche Nielsen, Nanna Bjarnholt, Cecilia Blomstedt,
            Roslyn M. Gleadow, and Birger Lindberg Møller*

Chapter 8   Glucosinolate biosynthesis and functional roles 157
            Tomohiro Kakizaki

Chapter 9 Biosynthesis and regulation of plant volatiles and their functional roles in ecosystem interactions and global environmental changes 185
Gen-ichiro Arimura, Kenji Matsui, Takao Koeduka, and Jarmo K. Holopainen

Chapter 10 Microbial volatiles and their biotechnological applications 239
Birgit Piechulla and Marie Chantal Lemfack

Chapter 11 Volatile glycosylation—A story of glycosyltransferase for volatiles: Glycosylation determining the boundary of volatile and nonvolatile specialized metabolites 257
Eiichiro Ono and Toshiyuki Ohnishi

Chapter 12 Plant secondary metabolites as an information channel mediating community-wide interactions 283
André Kessler and Kaori Shiojiri

Chapter 13 Metabolic engineering and synthetic biology of plant secondary metabolism 315
Dae-Kyun Ro, Yang Qu, and Moonhyuk Kwon

Index 361

Preface

Recent advances in science have clarified the role of plant specialized metabolites (classically known as plant secondary metabolites), which should not only be considered as bioactive molecules used for human health but must also be considered as pivotal factors for the global ecosystem. They play a major role in plant life, evolution, and mutualism. To provide the reader with a general view of plant specialized metabolites, it is important to consider both the biochemistry and the functional and ecological roles of these important compounds.

Readers will be fascinated by the fact that ~200,000 specialized metabolites are formed by a wide array of plant metabolic pathways from numerous plant taxa and by learning how other species (including human beings) rely on them. Accordingly, this book provides a wealth of both fundamental knowledge and up-to-date research mostly for undergraduate and graduate students, and scientists in the fields of botany, plant physiology, ecology, pharmacology, and biotechnology.

To meet this goal, we invited several experts to provide the most up-to-date contributions in several fields of study of specialized metabolites. After a brief introductory chapter, a thorough description of the sites of synthesis provides the reader with the basics for understanding the diversity of secretory structures and the biodiversity and the chemotaxonomic significance of specialized metabolites. The biochemistry and the functional roles of specialized metabolites are covered by several chapters, including those on salicylates, alkaloids, cyanogenic glycosides, glucosinolates, glycosylation, and the production of plant volatiles. Biotechnology of specialized metabolites is covered by two chapters, which describe the frontiers of synthetic biology and the production of volatiles by microorganisms. Finally, the role of specialized metabolites in plant communication is described by illustrating models of plant interaction.

The chapters are written in a contemporary fashion, in which functions, metabolism, and physiology are merged to show the reader both the complexity and the comprehensiveness of the topics explored. We are confident that all of these topics will provide the reader with special insights into the sophisticated nature of the metabolites and their various valuable uses, based on the most recent findings in science.

Gen-ichiro Arimura
Massimo Maffei

Editors

Gen-ichiro Arimura, PhD, received a PhD in science from Hiroshima University (Japan). He began his career in 1998 at Kyoto University, working in the field of molecular ecology of plant communications mediated by herbivore-induced plant volatiles. Sponsored by a fellowship from the Japan Society for the Promotion of Science (JSPS), he spent 2 years (2002–2004) at the University of British Columbia, Vancouver, Canada, working on terpene biosynthesis. Beginning in 2004, Dr. Arimura pursued his interest in the biosynthesis and regulation of plant terpene biosynthesis at the Max Planck Institute for Chemical Ecology, Jena, Germany. He has worked mainly on molecular and chemical ecology of plant and arthropod mutualisms since joining the faculties at Kyoto University (2008–2013) and Tokyo University of Science (from 2013 to present). Dr. Arimura is one of the editorial board members of *Scientific Reports* and *Applied Entomology and Zoology*.

Massimo Maffei, PhD, received a PhD in plant biology from the University of Turin (Italy) in 1981. In 1984, he became a research associate for the Istituto di Botanica Speciale Veterinaria (University of Turin); from 1989 to 1992 he was an assistant professor; from 1992 to September 2000 he was an associate professor of plant morphology and physiology; and since October 2000, he has been a professor of plant physiology in the Department of Life Sciences and Systems Biology of the University of Turin. He is the coordinator of the PhD School in Pharmaceutical and Biomolecular Sciences. He was also the director of the Department of Plant Biology and vice dean of the Faculty of Sciences from 2000 to 2006. From October 2003 to December 2006, Dr. Maffei was the coordinator of the Centre of Excellence for Plant and Microbial Biosensing (CEBIOVEM). From July 2012 to February 2014, he was the vice director of research in the Department of Life Sciences and Systems Biology. From 1987 to 1988, he was a postdoctoral research associate at the Institute of Biological Chemistry at Washington State University under the guidance of Professor Rodney Croteau. From 1990 to 1993, he was a visiting professor for the Mediterranean Agronomic Institute of Chania, Greece

(MAICH–CHIEAM), teaching secondary metabolism of Mediterranean medicinal and aromatic plants.

Dr. Maffei is a member of the Academy of Agriculture and of the Academy of Sciences of the University of Turin. He is also a member of a number of international and national scientific associations; a member of the editorial board of the international publication, the *Journal of Essential Oil Research*; and is a reviewer for several other journals. He is the editor in chief of the Open Access JCR *Journal of Plant Interactions* (published by Taylor & Francis). Dr. Maffei's research is dedicated to the study of secondary plant metabolites, with a particular reference to terpenoids, wax constituents, and phenolic compounds, as well as primary/secondary metabolic interactions. Recently, electrophysiological, transmission electron microscopical, and laser confocal scanning microscopical methods have been used to detect the involvement of cytological, membrane-related, and nuclear factors in the transduction of signals arising from plant–plant, plant–herbivore, and plant–pathogen interactions.

Contributors

Francesca Barbero
Department of Life Sciences and
  Systems Biology
University of Turin
Turin, Italy

Nanna Bjarnholt
Plant Biochemistry Laboratory
Department of Plant and
  Environmental Sciences
University of Copenhagen
and
VILLUM Research Center "Plant
  Plasticity"
Frederiksberg C, Denmark

Cecilia Blomstedt
School of Biological Sciences
Monash University
Victoria, Australia

Roslyn M. Gleadow
School of Biological Sciences
Monash University
Victoria, Australia

Jarmo K. Holopainen
Department of Environmental
  Science
University of Eastern Finland
Kuopio, Finland

Riitta Julkunen-Tiitto
Department of Environmental and
  Biological Sciences
University of Eastern Finland
Joensuu, Finland

Tomohiro Kakizaki
NARO Institute of Vegetable and
  Tea Science
Mie, Japan

André Kessler
Department of Ecology and
  Evolutionary Biology
Cornell University
Ithaca, New York

Takao Koeduka
Department of Biological
  Chemistry
and
Department of Applied Molecular
  Bioscience
Graduate School of Medicine
Yamaguchi University
Yamaguchi, Japan

Moonhyuk Kwon
Department of Biological Sciences
University of Calgary
Calgary, Alberta, Canada

Marie Chantal Lemfack
Institute for Biological Sciences
University of Rostock
Mecklenburg-Vorpommern,
  Germany

Kenji Matsui
Department of Biological Chemistry
and
Department of Applied Molecular
  Bioscience
Graduate School of Medicine
Yamaguchi University
Yamaguchi, Japan

Birger Lindberg Møller
Plant Biochemistry Laboratory
Department of Plant and
  Environmental Sciences
University of Copenhagen
and
VILLUM Research Center "Plant
  Plasticity"
and
Center for Synthetic Biology
  "bioSYNergy"
Frederiksberg C, Denmark
and
Carlsberg Research Laboratory
Copenhagen V, Denmark

Lasse Janniche Nielsen
Plant Biochemistry Laboratory
Department of Plant and
  Environmental Sciences
University of Copenhagen
and
VILLUM Research Center "Plant
  Plasticity"
and
Center for Synthetic Biology
  "bioSYNergy"
Frederiksberg C, Denmark

Toshiyuki Ohnishi
College of Agriculture
Academic Institute
Shizuoka University
Shizuoka, Japan

Eiichiro Ono
Research Institute
Suntory Global Innovation Center
  (SIC) Ltd.
Kyoto, Japan

Birgit Piechulla
Institute for Biological Sciences
University of Rostock
Rostock, Germany

Yang Qu
Department of Biological Sciences
Brock University
St. Catharines, Ontario, Canada

Dae-Kyun Ro
Department of Biological Sciences
University of Calgary
Calgary, Alberta, Canada

Kaori Shiojiri
Faculty of Agriculture
Ryukoku University
Otsu, Japan

Tsubasa Shoji
Graduate School of Biological
  Sciences
Nara Institute of Science and
  Technology
Nara, Japan

Virpi Virjamo
Department of Environmental and
  Biological Sciences
University of Eastern Finland
Joensuu, Finland

chapter one

Introduction to plant specialized metabolism

Gen-ichiro Arimura and Massimo Maffei

Contents

1.1 Overview 1
1.2 Molecular evolution 3
1.3 Human health needs 5
References 6

1.1 Overview

Before focusing on the topic of plant specialized metabolites (PSMs, traditionally referred to as plant secondary metabolites), we need to make a general distinction between metabolic pathways essential for growth and development (i.e., primary metabolism) and those not essential for such purposes (specialized metabolism). The availability of carbon, nitrogen, and sulfur resources has a major impact on the production of specific classes of primary metabolites and consequently on the levels and composition of PSMs derived from these primary metabolites (Aharoni and Galili 2011). The focal point in framing the PSMs in a metabolic and functional context is to define their "dispensability" or "indispensability." The dispensability of these molecules in the face of growth and development has led in the past to consider these molecules as "the ebbs and flows on the metabolism beach" or even waste and/or detoxification products. Today we know that many PSMs are used as chemical signals in the ecosystem and as tools of metabolic defense. PSMs may therefore be regarded as molecules that are "dispensable" for growth and development (with the exception of those metabolic pathways involved in the synthesis of plant hormones), but "indispensable" for survival of the species. Many times the difference between the primary and specialized metabolism is best expressed in functional rather than structural terms, since the same compound can exhibit qualities of both primary and specialized metabolites (Harborne 1993). The most evident qualities of PSMs are definitely their enormous structural diversity, the restriction of their presence to certain families or genera, and their high interspecific and intraspecific variability.

In the genomics age, various genomic tools have transformed the study of PSMs. The field of metabolomics is continually adding novelty and complexity to our information on the chemistry of PSMs, and the availability of whole-genome sequences for an ever-increasing list of plants enables us to examine the genomic basis of PSM production (Zhao et al. 2013). Now just beyond the genomics age, it has become the post-genome era. In recent years, a huge number of genes and gene products involved in PSM metabolism (~200,000 metabolites) have been discovered and characterized. PSMs that are known terpenoids, phenols, alkaloids, fatty acids, and so forth, play several important roles in plant growth or development, environmental adaptation, and trade and defense toward other surrounding organisms (Figure 1.1). Fraenkel (1959) classically stated, "The occurrence

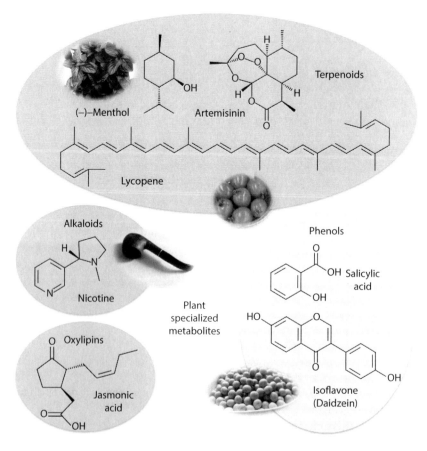

Figure 1.1 (See color insert.) Representative plant specialized metabolites.

of PSMs is sporadic but may be specific for a narrow set of species within a phylogenetic group. It is almost inconceivable that PSMs play a function in the basic metabolism of plants but they significantly function in repelling and attracting insects." Notably, recent biological and biochemical research involving metabolomic analysis have awakened us to the importance of PSMs, while also disclosing the sophisticated mechanisms underlying plant life cycles and interactions. For instance, plant-derived volatile organic chemicals (e.g., terpenoids and green leaf volatiles) mediate a web of communications between plants and their animal partners, including pollinators, herbivores, and herbivore enemies, therefore contributing to the configuration of biological habitats, that is, ecosystems, and those eventually lead to the diversity and coevolution of organisms. Therefore, the old concept that PSMs have minor functions as "secondary metabolites" is no longer accepted.

1.2 Molecular evolution

The great breadth of PSMs is not a coincidence, rather it is the result of a harmonious adjustment of various biogenetic pathways that have been fully integrated into the primary metabolism during molecular evolution. According to several authors of this book, it appears evident as to how plants use the limited numbers of basic metabolic pathways from which diverge an infinite number of variations that lead to hundreds of thousands of molecules. One possible explanation of the origin of PSMs can be sought right in the primary metabolism: some intermediates or side compounds of a primary metabolic pathway are accidently produced due to metabolic impairments caused by environmental pressure. The biosynthesis of such primary metabolites can create additive metabolic flow(s) by a chain of "assembly" that leads to catabolic or anabolic end products. Sometimes such a new metabolic flow triggers a feedback or competitive reaction that acts on the origin of primary metabolism either positively or negatively. However, if excess primary metabolite is available for a parallel metabolic pathway to form a characteristic PSM(s), then the pathway may play an important role in the homeostasis of the primary metabolism. Similarly, several other "parallel" metabolic pathways produce arrays of specialized metabolites that eventually give rise to the hundreds of thousands of PSMs.

One of the characteristics of PSMs is the high "degree of freedom" of compounds, which can vary in amount (concentration) or in quality (chemical structure) irrespective of the development and growth of the organ where they are produced. This is different from the case of primary metabolism, in which all the components required for plant growth and development processes must be sustained at the appropriate cellular levels for the organism's structural and functional integrity. An important consequence of the high intraspecific variation of PSMs is the capacity to evolve rapidly. It is becoming increasingly clear that trait variance linked to both macro- and microenvironmental variations can also evolve and respond more strongly to selection than mean trait values. Multiple selection forces acting across many spatial and temporal scales probably maintain PSM polymorphisms, but convincing examples that recognize the diversity of plant population structures are limited (Moore et al. 2014).

Undoubtedly, the transition of plants to the land (terrestrialization) was one of the most significant evolutionary events in the history of life on Earth. Plants made their appearance ~400 million years ago, during the Silurian age. At that time, there were already scorpions and centipedes, and scientists have not ruled out the possibility that during the Ordovician (Early Silurian) the first organisms appeared on land, including arthropods feeding on algae and cyanophytes that were predominant even before the Cambrian. Namely, at that time, plant feeders (ancient herbivore arthropods) may have coexisted with plants. During the Permian, flora and fauna changed slowly, and gymnosperms and cycads took the place of Licopodales and tree ferns. At that time, many insects, probably 50% of the species, fed on plants. Butterflies appeared at the beginning of the Cretaceous period and contributed to the process of pollination of flowering plants, which gradually evolved. Herbivorous insects were dominantly present in the Paleozoic Era, as well as during the Mesozoic era. It is likely that the algae Charoficeae gave rise to the terrestrial plants that later evolved into woody plants. In fact, predation of leaves and seeds had not occurred until the Carboniferous, given that the first documentation of feeding dates back to the Mississippian period and predation of seeds in the early Pennsylvanian. Fossil evidence of other feeding methods in the Paleozoic Era, such as the formation of galls and root herbivory, appears even later, from the middle to the end of the Pennsylvanian (Labandeira and Sepkoski 1993). Clearly, PSMs helped ancient plants to withstand attacks from herbivores, as seen in contemporary ecosystems (see Chapters 9 and 13). According to Theis and Lerdau (2003), the evolution of PSM function in such plant defense against herbivores can occur as a result of change at any of several scales of biological organizations. Changes in the distribution of PSMs within plant taxa and in the structures used for their storage can be involved in functional shifts. Moreover, as described in Chapter 9, in some PSMs such as volatile compounds, evolution of function can occur through changes in biological communications using a vapor of PSMs as airborne signals. A better understanding of the evolution of PSMs requires studying the regulation of function across all of these scales.

The broad range of the chemical diversity and biological activities of PSMs raises many questions about their roles in nature and the specific traits leading to their evolution. The answers to these questions would not only be of fundamental interest but may also provide lessons that could help to improve the screening protocols of pharmaceutical companies and strategies for rational PSM engineering (Jenke-Kodama et al. 2008). Since a given structural type of PSM has almost invariably arisen on a number of occasions in different plant taxa, the co-occurrence of a structural class in two taxa is a possible, but not a conclusive indication of a monophyletic relationship of the need for such a PSM. The co-occurrence of a structural class of PSM in two taxa could be due to either convergent evolution of the responsible gene(s) or variable regulation of the genes: in some cases, the genes encoding the enzymes for the production of PSMs with a given structure or structural skeleton evolved at early stages of plant evolution. Even if these genes are not structurally lost during evolution, their expression may cease suddenly because of some accident in, for example, genetic and epigenetic modification. In some cases, however, those genes can be "switched on" again at some later point, leading to a gap in the molecular evolution of PSMs (Wink 2003).

1.3 Human health needs

PSMs are also highly valuable for human life, being used as medicines, flavorings, self-medication drugs such as dietary supplements, and industrial products such as rubbers made from long carbon chains (Figure 1.1). For instance, isoflavones (flavonoid family), produced by most members of the Fabaceae family, are believed to reduce the risk of hormone-dependent cancers, including breast cancer and endometrial cancer, due to their estrogen-like activity (Eden 2012; Varinska et al. 2015). Soy isoflavones are also widely acknowledged to exert various favorable effects against diseases, especially including prevention of cardiovascular disease, osteoporosis, and adverse postmenopausal symptoms, and improvement of physiological conditions, such as maintaining cognitive function (Ko 2014), although soy products have been reported not to relieve menopausal vasomotor symptoms any more effectively than placebo (Eden 2012). Similarly, a wide array of PSMs has been characterized as healthy compounds. On the other hand, nicotine (alkaloid family) is an addictive substance that can cause not only lung cancer, but also have deleterious effects on pregnancy outcome and the reproductive health of offspring (Wong et al. 2015). Similarly, a broad range of medical, pharmacological, and biochemical studies have clarified various traits of several PSMs (valuable or not and, if valuable, their solo and additive functions), and recent advances in biotechnology enables the production of PSMs in various more efficient, more economical, and more ecologically friendly ways, in addition to the classically used methods for extraction and purification from either cultivated and natural plants, or for production by industrial chemical reactions.

Today, despite the radical changes enabled by chemical synthesis, large numbers and amounts of drugs used for treating various diseases are extracted from plant resources directly or indirectly. Substances that were once extracted from plants have in many cases given way to synthetic molecules, which can be obtained with more effective and definitely cheaper procedures, but this is not true in all cases. Recently, a number of novel molecules discovered in plants have appeared on the market as medicinal products and established a wave of novel pharmacological and clinical studies and research. In particular, there has been significant progress in using these molecules for treating diseases such as cancer or diseases caused by infectious agents such as malaria. Aside from the pharmaceutical industry, which extracts and purifies bioactive substances from plants, there is a popular industry, less standardized and controlled, which uses medicinal plants in the form of herbs, extracts, and powders for the treatment of a large variety of health disorders (Astin et al. 1998).

Notably, the discovery of natural substances is frequently made based on knowledge of customs and traditions, through ethnopharmacologic and ethnopharmacognostic studies, contributing substantially to innovation of medical products by providing novel chemical structures and mechanisms of action.

References

Aharoni, A. and G. Galili. 2011. Metabolic engineering of the plant primary-secondary metabolism interface. *Curr. Opin. Biotechnol.* 22: 239–244.

Astin, J.A., A. Marie, K.R. Pelletier, E. Hansen, and W.L. Haskell. 1998. A review of the incorporation of complementary and alternative medicine by mainstream physicians. *Arch. Intern. Med.* 158: 2303–2310.

Eden, J.A. 2012. Phytoestrogens for menopausal symptoms: A review. *Maturitas.* 72: 157–159.

Fraenkel, G.S. 1959. The raison d'etre of secondary plant substances; these odd chemicals arose as a means of protecting plants from insects and now guide insects to food. *Science.* 129: 1466–1470.

Harborne, J.B. 1993. *Introduction to Ecological Biochemistry*, 4th edition. San Diego, CA: Academic Press.

Jenke-Kodama, H., R. Müller, and E. Dittmann. 2008. Evolutionary mechanisms underlying secondary metabolite diversity. *Prog. Drug Res.* 65: 119, 121–140.

Ko, K.P. 2014. Isoflavones: Chemistry, analysis, functions and effects on health and cancer. *Asian Pac. J. Cancer Prev.* 15: 7001–7010.

Labandeira, C.C. and J.J. Sepkoski, Jr. 1993. Insect diversity in the fossil record. *Science.* 261: 310–315.

Moore, B.D., R.L. Andrew, C. Kulheim, and W.J. Foley. 2014. Explaining intraspecific diversity in plant secondary metabolites in an ecological context. *New Phytol.* 201: 733–750.

Theis, N. and M. Lerdau. 2003. The evolution of function in plant secondary metabolites. *Int. J. Plant Sci.* 164: S93–S102.

Varinska, L., P. Gal, G. Mojzisova, L. Mirossay, and J. Mojzis. 2015. Soy and breast cancer: Focus on angiogenesis. *Int. J. Mol. Sci.* 16: 11728–11749.

Wink, M. 2003. Evolution of secondary metabolites from an ecological and molecular phylogenetic perspective. *Phytochemistry.* 64: 3–19.

Wong, M.K., N.G. Barra, N. Alfaidy, D.B. Hardy, and A.C. Holloway. 2015. Adverse effects of perinatal nicotine exposure on reproductive outcomes. *Reproduction.* 150: R185–R193.

Zhao, N., G.D. Wang, A. Norris, X.L. Chen, and F. Chen. 2013. Studying plant secondary metabolism in the age of genomics. *Crit. Rev. Plant Sci.* 32: 369–382.

chapter two

Site of synthesis of specialized metabolites

Massimo Maffei

Contents

2.1 Introduction 9
2.2 Secretion 9
2.3 Glandular trichomes 12
2.4 Secretory cavities and resin ducts 16
2.5 Lysigenous cavities 17
2.6 Oil-bearing cells and secretory cells associated with bacteria 17
2.7 Laticifers 18
References 19

2.1 Introduction

Photosynthesis, the process by which plants build organic molecules from carbon dioxide and water, provides the basic molecules for the synthesis of many metabolites, which are essential for plant life. Among these are specialized metabolites, which are synthesized from sugars, amino acids, and numerous intermediates of the primary metabolism. Some plants have evolved the ability to synthesize and store in large quantities specialized metabolites, through specific structures called secretory tissues (Maffei 2010). Specialized metabolites are synthetized and stored, sometimes in considerable amounts, by these tissues.

This chapter describes the main sites of synthesis and storage of the major specialized metabolites, with particular reference to the secretory structures that produce lipophilic substances.

2.2 Secretion

In general terms, secretion is the passage of substances produced by the cell from the inside to the outside of the plasma membrane; secreted substances (the products of secretion) play a special function in organs or tissues where they are produced. The term secretion is often used together with the term excretion and the literature reports numerous attempts to define the two terms. In animal biology, excretion means the process related to substances that no longer have any use for the body and therefore are excluded from cells. On the other hand, secretion is the process of elimination of molecules that have a particular function in the organism's growth and development processes (Fahn 1988).

Secretion is usually the result of the metabolic activity of almost all cells but is particularly evident in secretory cells. The process usually takes place by molecular extrusion, but it may also involve the destruction of the entire cell. The definition of secretory (or glandular) cells depends on the degree of cytological specialization that usually occurs in the presence of a dense cytoplasm with an elevated number of active membranes and organelles (Fahn 1988). A certain degree of physiological specialization is also required, which depends more or less on the specific elimination processes of substances. The concept of secretion is opposed to diffusion because it involves work to move the substance through the plasma membrane against the concentration gradient.

Plant cells have the ability to extrude the substances secreted either out of the protoplasm or inside the vacuoles and the ability to fulfill both metabolic processes often depends on the chemical nature of the molecules produced. Secretion can take place by intracellular storage, as occurring during the storage of many substances inside the cell wall (e.g., lignin, cutin, waxes, suberin, etc.), by secretion of intracellular substances secreted into membrane-delimited compartments or by extracellular secretion, through the delivery of secreted substances outside the plasma membrane. In the latter case, an intracellular accumulation may be present or not (Maffei 2015).

The main mechanisms through which substances are eliminated are shown in Figure 2.1, and can be classified into:

- Substances secreted as a result of disintegration of the cells that produced them, in this case, the secretion is defined as *holocrine*.
- Substances secreted from the cytoplasm of cells that remain intact, in this case, the secretion is defined as *merocrine*.

Merocrine secretion can be further subdivided into:

- *Eccrine secretion*, where secreted substances pass through the plasma membrane or across the tonoplast by concentration gradients or through active transport processes.
- *Granulocrine secretion*, where the produced substances are accumulated in vesicles bounded by a membrane which may join with the plasma membrane or with the tonoplast. Otherwise, vesicles may undergo phagocytosis by the plasma membrane that surrounds them, thus eliminating their content outside the protoplast.

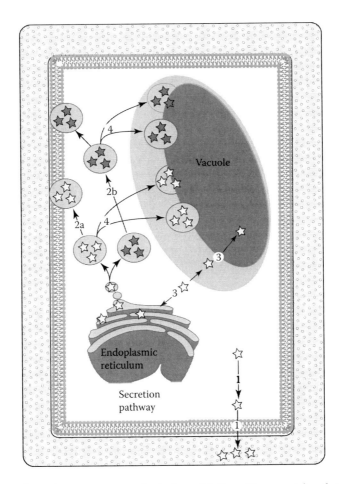

Figure 2.1 Secretion process: cytological model. Secretion can take place via the transmembrane transport of molecules by "eccrine" secretion (1) or through the transport of molecules into vesicles, by "granulocrine" secretion (2). The latter can take place as such (2a) or upon product modification (2b) that can be stored in vesicles or in cellular compartments. The storage in compartments such as the vacuole can take place via molecular (3) or vesicular (4) transport. Lysis of the cell eventually leads to the loss of cellular integrity.

We can divide secreting cells into two major categories, depending on the nature of the secreted compound: secretory cells that produce hydrophilic substances and secretory cells whose secretion is of a lipophilic nature.

The first group of secretory cells characterizes tissues that produce a mixture of various substances, mainly attributable to the primary metabolism or related to metabolic processes of basal type. Hydrophilic tissues include mucilage glands, which secrete mucopolysaccharides and are typical of carnivorous plants such as *Drosophyllum*, *Drosera*, and *Pinguicola*. They are used to catch small insects that remain glued to the secretion. Very similar to these glands are digestive glands that, in addition to producing polysaccharides, secrete proteolytic enzymes. In some cases (as in *Nepenthes*), modified leaves contain a fair amount of a "digestive" liquid while in other cases (as in *Drosophyllum*, *Drosera*, and *Pinguicola* mentioned above), secretion begins after insect contact and capture (Maffei 2015).

Nectaries are secretory tissues releasing a fluid with a high sugar content (known as nectar, a sugary fluid secreted by plants to encourage pollination by insects and other animals). They are located principally in the flowers, but there are cases of extrafloral nectaries (Kost and Heil 2008). Other secretory tissues that produce hydrophilic substances are hydathodes, which may or may not be arranged in glandular tissues. Hydathodes extrude water and other substances in the external environment. Hydathodes are very similar to salt glands, secretory structures able to extrude large quantities of mineral salts (particularly sodium chloride) with active transport mechanisms (Fahn 1988).

Particularly involved in specialized metabolism are secretory tissues that produce lipophilic molecules. Their importance is both morpho-physiological, for their functional role provided by their high bioactivity, and economical, since they represent a significant part of the world economy. Unlike hydrophilic secretion, which is mainly extruded with a merocrine secretion, most lipophilic compounds are released through holocrine secretion. In some cases, the compound is released in a first step with merocrine secretion and then is eventually expelled through holocrine secretion (Maffei 2015).

2.3   Glandular trichomes

Lipophilic molecules are secreted by a wide variety of anatomical structures, from simple cells such as epidermal idioblasts to aggregates, to structures with a large degree of specialization present on the plant surface (such as glandular trichomes) or inside roots, shoots, leaves, and fruits (like resin ducts and lysigenous cavities). Volatile organic compounds (VOCs) are emitted in the atmosphere and most of them are of biogenic origin. Several plant species store VOCs in specialized glandular trichomes (Gershenzon et al. 2000), which release their contents in response to tissue damage, thus deterring herbivores or inhibiting microbial growth (Langenheim 1994; Zebelo et al. 2011). Over 90% of VOC emissions are produced from natural forests around the world; the most important among them is the Amazonian rainforest. Up to 36% of the carbon taken up by plants is released as a complex mixture of VOCs (Kesselmeier 2001;

Kesselmeier and Hubert 2002; Kesselmeier and Staudt 1999). Unlike methane, VOCs produced by plants in the troposphere are extremely reactive, with lifetimes ranging from minutes to hours, contributing to the formation of an aerosol that diffuses the light to produce the blue sky.

VOCs are also released into the atmosphere from leaf and flower trichomes (Dudareva et al. 2006). The main features of VOCs are released into the air for plant defense against herbivores and pathogens, to attract pollinators, seed dispersers, and other animals and beneficial microorganisms, and as plant–plant communication signals (Pearse and Karban 2013). In some plants, the release of VOCs can also act as a sealing of wounds (Hematy et al. 2009).

Some VOCs can be hazardous to people's health when present in high concentrations and are also important precursors of tropospheric phytotoxic compounds (Jahodar and Klecakova 1999). Because some VOCs may serve as precursors of photochemical smog, control of their concentration is one of the key parameters for evaluating air quality (Ulman and Chilmonczyk 2007). The VOC can adjust the oxidative capacity of the troposphere, carbon monoxide, O_3 (Vuorinen et al. 2005).

The pubescence is typical of the surfaces of many plant species and such a term indicates the presence of structures emerging from the epidermis known by the name of glandular trichomes or glandular hairs. The morphology of these structures is extremely variable and more than 300 morphological types have been described thus far (Maffei 2010). These structures develop from protodermal cells formed because of anticlinal and periclinal divisions. Being protodermal extrusions, glandular trichomes are present on plant surfaces, with particular reference to leaf blades, flowers and, in some cases, seeds. In *Arabidopsis thaliana* mutants unable to produce leaf trichomes continue to produce root hairs, clearly indicating that different genes control the formation of trichomes and root hairs. Regarding secretion, glandular trichomes isolation techniques allowed to demonstrate that these structures are the only site of synthesis of certain substances, such as, for example, mono- and sesquiterpenes (Bertea et al. 2006; Gershenzon et al. 1989).

The most common form of glandular trichomes is of the peltate type, characterized by a globular dome formed by the detachment of the cuticle from the cell wall. Other types of structures include monoseriate and biseriate trichomes with one (capitate or sessile trichomes) or more (peltate trichomes) secretory cells. In any case, a basal cell always serves as a physical connection between the secreting cells and the rest of the plant. Sometimes the metabolite flow passes through one or more connecting cells placed between the basal cell and the secretory cells. These cells are called stem cells.

Biochemistry of specialized metabolites produced by glandular trichomes is important because of the interaction that many VOCs have with the external environment, with herbivores, and with predators and pollinators (Cseke et al. 2007). Understanding the mechanisms of formation and differentiation of trichomes and assessing the gene expression of the enzymes that produce many substances of practical interest are of fundamental importance for the sustainable use and biological control strategies (Maffei 1999, 2015; Maffei et al. 2011).

One of the most studied plants is *Mentha piperita*. In this hybrid (as well as in other species like *M. spicata*), there are two types of glandular trichomes: peltate, containing eight secretory cells, a stalk cell, and a basal cell that, because of their size and number, are expected to contain the bulk of the oil; and capitate, with a secretory cell, a stalk cell, and a basal cell (Gershenzon et al. 1989; Maffei et al. 1989). Histochemical and biochemical studies have shown that in this plant the secretory cells are the only site of synthesis of products secreted lipophilic (mainly monoterpenes and sesquiterpenes) (Caissard et al. 2012; Gershenzon et al. 1989, 2000; Turner et al. 2000). Maffei et al. (1989) showed that a larger number of peltate trichomes was initiated on adaxial epidermises than abaxial epidermises, and that during leaf development a higher gland number was produced on abaxial epidermises. Moreover, it was found that the trichome number was not fixed at the time of leaf emergence, and that trichomes grow dynamically during the ontogenetic development producing a VOC composition that is different, both from a quantitative and qualitative point of view. With the use of morphometric techniques, microfluorimetric and electron microscopy (TEM and SEM) the dimension, the level of ploidy and morphology of the nucleus of different cell types (basal, stem, and secretory) was investigated. The results showed a nuclear hypertrophy in the secretory cell of both types, which was explained by polyploidization and by variations in chromatin structure that may be related to increased transcriptional activity (Berta et al. 1993). Glandular trichomes are not restricted to leaves. In the genus *Mentha*, flowers are especially rich in glandular trichomes of the peltate and capitate type in the green parts of the flower, the calix, while the density of the trichomes is extremely low on the petals (Maffei and Sacco 1987) (Figure 2.2a–e).

The Asteraceae (Compositae) family is composed of numerous genera producing glandular trichomes that secrete substances of lipophilic nature. In some cases, such as in the genus *Inula*, mixtures of hydrophilic and lipophilic compounds of different chemical nature are secreted. One of the most studied genera is *Artemisia* because of the economic importance of terpenoids used both for flavoring and for the treatment of infectious diseases, such as malaria (Brown 2010). In this genus, the trichomes are of the biseriate type and are formed by two basal cells, two stem cells, and by three pairs of secretory cells. During early phases of cell division, plastids are present as proplastids with few thylakoids, while when

Chapter two: Site of synthesis of specialized metabolites 15

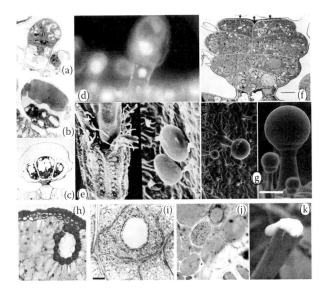

Figure 2.2 Secretory tissues in plants. (a–e) The first divisions of secretory cells and a clear distinction of basal and stem cells in peppermint (*Mentha x piperita*) (a). The volatile terpenes accumulate in a subcuticular space (b) where the staining with *p*-phenylenediamine highlights its lipophilic nature (dark color). In old, mature peppermint trichomes, secretory cells degenerate and form large vacuoles, while the contents of the subcuticular space appear empty (c). DNA staining in peppermint capitate trichomes with DAPI allows quantifying the DNA content in the nuclei of different cells. The fluorescence microscopy image (d) shows that the nucleus of the secretory cell has a higher ploidy (larger) than the nuclei of other cells. Flowers of the genus *Mentha* show many glandular trichomes, especially in the calix (e). (Photo by M. Maffei.) (f) The last stage of development of an *Artemisia annua* biseriate galdular trichome; apical secretory cells, and cells containing chloroplasts are clearly visible. Two stem cells and two basal cells are present below the secretory cells. Arrows indicate a dense osmiophilic deposit in direct contact with the secretory cells. (Reprinted with permission from Duke, S.O. and R.N. Paul. 1993. *International Journal of Plant Sciences* 154: 107–118.) (g) *Salvia sclarea* glandular trichomes: ESEM view of a calyx epidermis showing the diversity of trichomes (left); ESEM view of a capitate gland (right) (metric bar 50 µm). (Reprinted with permission from Caissard, J.C., T. Olivier, C. Delbecque et al. 2012. *Plos One* 7: e48253.) (h) Gymnosperm needles contain numerous resin ducts. A cross section of a pine (*Pinus* sp.) leaf shows a resin duct surrounded by a sheath of cells with thickenings of suberin. Inside the duct, secretory cells align the duct walls. (Photo by M. Maffei.) (i) In *Hymenaea stigonocarpa*, a paradermal of an adult leaf shows the secretory cavity in the mesophyll; note the equidistance of the secretory cavities compared to vascular tissues (metric bar 5 µm). (Reprinted with permission from Paiva, E.A.S. and S.R. Machado. 2004. *Nord. J. Bot.* 24: 423–431.) (j) Secretory cells and the presence of bacteria in the root of *Vetiveria zizanioides*. (Photo by M. Maffei.) (k) Great profusion of latex from the cut stem of *Munnozia hastifolia*. (Reprinted with permission from Gutierrez, D.G. and M.L. Luna. 2013. *Flora* 208: 33–44.)

divisions are completed, numerous chloroplasts differentiate in all cell types (Figure 2.2f).

The leaves of hops (*Humulus lupulus*) have two types of glandular trichomes: peltate and bulbous. Peltate trichomes are formed from a protodermal cell following two anticlinal divisions followed by two periclinal divisions, leading to the formation of the initial cells of the secretory head, basal cell, and those of the stem. On development, peltate glandular trichomes consist of a head made up of 30–72 cells, stem cells, and four basal cells. Bulbous trichomes are formed by a head with four (rarely eight) secretory cells, two stem cells, and two basal cells (Nagel et al. 2008) (Figure 2.2g).

In *Cannabis sativa*, cannabinoids are synthesized in glandular trichomes present mainly on female flowers. The main product of the secretion in *C. sativa* is Δ^9-tetrahydrocannabinol (THC). Recent advances on the cellular localization of cannabinoids biosynthesis focused on capitate trichomes as the main site of their storage. It was also confirmed that the head of the secretory trichomes is the main production site of cannabinoids demonstrating the presence of cDNAs encoding three polyketides, the MEP pathway, and tetrahydrocannabinolate (TCHA) synthase (Stout et al. 2012).

Tobacco (*Nicotiana tabacum*) leaves are covered with glandular trichomes that produce sucrose esters and diterpenoids in varying amounts, depending on the type of cultivar. In *N. tabacum*, *N. sylvestris*, and *N. rustica*, there are two types of glandular trichomes: long and short. The long trichomes consist of 4–6 stem cells and 1–6 secretory cells that increase in number from young to adult trichomes. Both stem and secretory contain chloroplasts, while secretory cells show druse crystals associated to the nuclei. The short trichomes consist of 8–16 secretory cells and 1 stem cell. Unlike the long trichomes, secretory cells of these trichomes contain neither druse crystals nor chloroplasts (Glas et al. 2012).

2.4 Secretory cavities and resin ducts

Lipophilic secretion also occurs in secretory tissues that are located within the body of the plant. These tissues are capable of secreting their products in intercellular spaces that develop by either schizogeny (i.e., a spreading apart) or lysigeny (i.e., programmed death and dissolution) of cells, sometimes these structures form by a mechanism that involves both stages of cellular development. In some cases, as we shall see in laticifers, the secreted substances accumulate within the cells (Pickard 2008).

Secretory ducts that produce lipophilic substances are found in several plant families, especially in the Pinaceae, Anacardiaceae, Asteraceae, Hypericaceae, Leguminosae, and Apiaceae.

In conifers, secretory ducts produce a resin formed by mono-, sesqui-, and diterpenes and for this reason they are also known as resin ducts. The development of the duct takes place via schizogeny through the dissolution of the middle lamella between the duct initials and the formation of an intercellular space. The ducts are normally oriented parallel to the longitudinal axis of the organ containing them, but they may anastomose tangentially (Pickard 2008). In conifers, resin ducts are found throughout the body of the plant and consist of elongated structures wrapped in epithelial cells surrounding an interior space. These cells are in turn surrounded by one or more layers of cells with relatively thick cell walls rich in pectic substances (Figure 2.2h) (Fahn 1988; Maffei 2010).

2.5   Lysigenous cavities

Lysigenous cavities are present in many families such as Myrtaceae, Rutaceae, Myoporaceae, Leguminosae, and Hypericaceae. These secreting structures are formed by a single epidermal cell. Some species show the formation of a meristemoid tissue characterized by cells with a dense cytoplasm and a large nucleus. The first two divisions of this structure form two bicellular layers, an upper one and a lower one. From the latter, secretory cells are formed providing the epithelium surrounding the cavity inner space (Vieira et al. 2001). In some families (as in the Myrtaceae) the formation of the inner cavity is the result from schizogeny of an initial group of cells, while in others (as in the Rutaceae), both schizogeny and lysigeny may occur. In any case, it cannot be excluded, also in the case of schizogeny the formation of the cavity may result from an eventual lysigeny (Fahn 1988).

In legume leaves, secretory resin cavities are referred to as translucent glands; they are used in taxonomic descriptions as a diagnostic character in identification keys. In *Hymenaea stigonocarpa*, the resin content in the leaves occurs in the epithelial cells of the secretory cavities that are distributed randomly (Paiva and Machado 2004). These clearly defined cavities are covered with a single layer secretory epithelium. In the leaf blade, the cavities are located mainly near the adaxial surface (Figure 2.2i).

2.6   Oil-bearing cells and secretory cells associated with bacteria

Other tissues able to produce lipophilic substances are represented by secretory cells that accumulate the secreted products inside their vacuoles. This is the case of VOCs produced by the odorous roots of the grass *Vetiveria zizanioides* Nash (vetiver). Vetiver VOCs are produced in secretory cells localized in the first cortical layer outside the endodermis of mature vetiver roots (Akhila and Rani 2002; Viano et al. 1991a,b). By using culture-based and culture-independent approaches to analyze the microbial community of the vetiver root, Del Giudice and co-workers (2008) demonstrated the presence of a broad phylogenetic spectrum of bacteria, including α-, β-, and γ-proteobacteria, high-G + C-content gram-positive bacteria, and microbes belonging to the Fibrobacteres/Acidobacteria group. The same group isolated root-associated bacteria and showed that most of them were able to grow by using vetiver sesquiterpenes as a carbon source and to metabolize them releasing into the medium a large number of compounds typically found in commercial vetiver oils. Several of these bacteria were also able to induce gene expression of a vetiver sesquiterpene synthase (Del Giudice et al. 2008). Figure 2.2j shows a cross section of the vetiver root, where the essential oil-producing cells are evidenced along with the associated bacteria (Maffei 2002). These results support the intriguing hypothesis that bacteria may have a role in essential oil biosynthesis, opening the possibility for using them to maneuver the vetiver oil molecular structure (Alifano et al. 2010). These results are in accordance with those obtained by Viano et al. (1991a,b) who analyzed the vetiver root ultrastructure using electron transmission microscopy and detected essential oil crystals in the inner cortical layer close to the endodermis. According to these authors, the secretion of the essential oil occurs in this region and successively reaches the whole cortex.

VOCs can be synthesized by a variety of other anatomical structures such as solitary cells and areas of epidermal cells. The typical fragrance of flowers results from VOCs occurring in the form of small droplets in the cytoplasm of the epidermal and neighboring mesophyll cells of sepals (Fahn 1988). In flowers, the biosynthesis of VOCs usually occurs in epidermal cells, allowing an easy escape of VOCs into the atmosphere (Kolosova et al. 2001). Flowers usually produce their attractive fragrance in osmophores or in conical cells located on the petals. These cells do not stock VOCs but release them into the air (Caissard et al. 2004; Ibanez et al. 2010). In species belonging to the Orchidaceae and Araceae, VOCs produced by osmophores also produce amines and ammonia (Pridgeon and Stern 1983). Although these stored and induced VOCs have useful roles, non-terpene-emitting species also survive the onslaught of herbivores and competition, and can set seed (Owen and Penuelas 2005). In fact, lack of specific anatomical structures for VOC storage does not imply negligible internal VOC concentrations (Niinemets et al. 2004).

2.7 Laticifers

The latex is a suspension or in some cases an emulsion of small particles in a liquid with a particular index of refraction. Although mainly of milky color, this sap can also be yellow, orange, red, brown, or even colorless.

The chemical composition of the latex ranges from polyisoprenic hydrocarbons to triterpenols and from sterols to fatty acids and aromatic compounds, and can also contain carotenes, phospholipids, proteins, and inorganic compounds. Latex can also contain protein crystals, starch granules, tannins, alkaloids, proteolytic enzymes (papain), vitamins, calcium oxalate, and malate (Figure 2.2k) (Pickard 2008).

The tissues that contain latex are defined laticifers and are present in about 12,500 species belonging to 900 genera and 20 families, especially dicotyledonous (especially Apocynaceae, Asclepiadaceae, Asteraceae, Euphorbiaceae, Papaveraceae, and Sapotaceae) and less in monocots (Araceae, Liliacae, and Musaceae) (Maffei 2015). However, latex can accumulate in tissues that are not laticifers, as found in *Parthenium argentatum*.

Laticifers are divided into nonarticulated and articulated. The former (also known as laticifers cells) are multinucleated and are derived from a single cell that extends enormously during the plant development. In some species, laticifers grow in the form of elongated tubes, in others branch and are defined as not articulated branched laticifers. Articulated laticifers (also known as laticifer vessels) consist of a series of usually elongated simple or branched cells. Laticifers are also divided into articulated not anastomosed laticifers (not branched) and articulated anastomosed laticifers (branched). Not articulated and unbranched laticifers are located in the stem outside of the primary phloem, being absent in the roots and secondary tissues (Fahn 1988).

The species *Camptotheca acuminata* (Nyssaceae) is a tree native to southern China and Tibet. Interest in *C. acuminata* resides in the importance of certain specialized products, in particular camptothecin (a pentacyclic quinoline alkaloid) and some of its derivatives. Camptothecin is known for its remarkable inhibitory activity against cancer cells and the human immunodeficiency virus (HIV). The distribution of laticifers and ultrastructure of secretory cells were analyzed in this species by light and electron microscopy and histochemical analysis to identify the major components of the latex. Histological analysis revealed that primary laticifers are already present in the leaf primordia. The vacuolar content of laticifers in the proximal area of the leaf is more intensely colored with histochemical reagents. In some laticifers, the latex does not accumulate evenly and the histological analysis of the roots has not revealed the presence of laticifers in both primary and secondary tissues (Monacelli et al. 2005).

References

Akhila, A. and K. Rani. 2002. Chemical constituents and essential oil biogenesis in *Vetiveria zizanioides*. In *Vetiveria, the Genus Vetiveria*, ed. M. Maffei, 73–109. London: Taylor & Francis.

Alifano, P., L. Del Giudice, A. Tala, M. De Stefano, and M.E. Maffei. 2010. Microbes at work in perfumery: The microbial community of vetiver root and its involvement in essential oil biogenesis. *Flavour Frag. J.* 25: 121–122.

Berta, G., M. Delapierre, and M. Maffei. 1993. Nuclear morphology and DNA content in the glandular trichomes of peppermint (*Mentha* X *piperita* L). *Protoplasma* 175: 85–92.

Bertea, C.M., A. Voster, F.W.A. Verstappen, M. Maffei, J. Beekwilder, and H.J. Bouwmeester. 2006. Isoprenoid biosynthesis in *Artemisia annua*: Cloning and heterologous expression of a germacrene A synthase from a glandular trichome cDNA library. *Arch. Biochem. Biophys.* 448: 3–12.

Brown, G.D. 2010. The Biosynthesis of Artemisinin (Qinghaosu) and the Phytochemistry of *Artemisia annua* L. (Qinghao). *Molecules* 15: 7603–7698.

Caissard, J.C., A. Meekijjironenroj, S. Baudino, and M.C. Anstett. 2004. Localization of production and emission of pollinator attractant on whole leaves of *Chamaerops humilis* (Arecaceae). *Am. J. Bot.* 91: 1190–1199.

Caissard, J.C., T. Olivier, C. Delbecque et al. 2012. Extracellular localization of the diterpene sclareol in clary sage (*Salvia sclarea* L., Lamiaceae). *Plos One* 7: e48253.

Cseke, L.J., P.B. Kaufman, and A. Kirakosyan. 2007. The biology of essential oils in the pollination of flowers. *Nat. Prod. Commun.* 2: 1317–1336.

Del Giudice, L., D.R. Massardo, P. Pontieri et al. 2008. The microbial community of vetiver root and its involvement into essential oil biogenesis. *Environ. Microbiol.* 10: 2824–2841.

Dudareva, N., F. Negre, D.A. Nagegowda, and I. Orlova. 2006. Plant volatiles: Recent advances and future perspectives. *Crit. Rev. Plant Sci.* 25: 417–440.

Duke, S.O. and R.N. Paul. 1993. Development and fine-structure of the glandular trichomes of *Artemisia annua* L. *International Journal of Plant Sciences* 154: 107–118.

Fahn, A. 1988. Secretory tissues in vascular plants. *New Phytol.* 108: 229–257.

Gershenzon, J., M. Maffei, and R. Croteau. 1989. Biochemical and histochemical localization of monoterpene biosynthesis in the glandular trichomes of spearmint (*Mentha spicata*). *Plant Physiol.* 89: 1351–1357.

Gershenzon, J., M.E. McConkey, and R.B. Croteau. 2000. Regulation of monoterpene accumulation in leaves of peppermint. *Plant Physiol.* 122: 205–213.

Glas, J.J., B.C.J. Schimmel, J.M. Alba, R. Escobar-Bravo, R.C. Schuurink, and M.R. Kant. 2012. Plant glandular trichomes as targets for breeding or engineering of resistance to herbivores. *Int. J. Mol. Sci.* 13: 17077–17103.

Gutierrez, D.G. and M.L. Luna. 2013. A comparative study of latex-producing tissues in genera of Liabeae (Asteraceae). *Flora* 208: 33–44.

Hematy, K., C. Cherk, and S. Somerville. 2009. Host-pathogen warfare at the plant cell wall. *Curr. Opin. Plant Biol.* 12: 406–413.

Ibanez, S., S. Dotterl, M.C. Anstett et al. 2010. The role of volatile organic compounds, morphology and pigments of globeflowers in the attraction of their specific pollinating flies. *New Phytol.* 188: 451–463.

Jahodar, L. and J. Klecakova. 1999. Toxicity with respect to pharmaceutically important species in the family Asteraceae. *Chem. Listy* 93: 320–326.

Kesselmeier, J. 2001. Exchange of short-chain oxygenated volatile organic compounds (VOCs) between plants and the atmosphere: A compilation of field and laboratory studies. *J. Atmos. Chem.* 39: 219–233.

Kesselmeier, J. and A. Hubert. 2002. Exchange of reduced volatile sulfur compounds between leaf litter and the atmosphere. *Atmos. Environ.* 36: 4679–4686.

Kesselmeier, J. and M. Staudt. 1999. Biogenic volatile organic compounds (VOC): An overview on emission, physiology and ecology. *J. Atmos. Chem.* 33: 23–88.

Kolosova, N., D. Sherman, D. Karlson, and N. Dudareva. 2001. Cellular and subcellular localization of S-adenosyl-L-methionine: Benzoic acid carboxyl methyltransferase, the enzyme responsible for biosynthesis of the volatile ester methylbenzoate in snapdragon flowers. *Plant Physiol.* 126: 956–964.

Kost, C. and M. Heil. 2008. The defensive role of volatile emission and extrafloral nectar secretion for lima bean in nature. *J. Chem. Ecol.* 34: 2–13.

Langenheim, J.H. 1994. Higher-plant terpenoids—A phytocentric overview of their ecological roles. *J. Chem. Ecol.* 20: 1223–1280.

Maffei, M. 1999. Sustainable methods for a sustainable production of peppermint (*Mentha x piperita* L.) essential oil. *J. Essent. Oil Res.* 11: 267–282.

Maffei, M. 2002. *Vetiveria, the Genus Vetiveria*. London: Taylor & Francis.

Maffei, M. 2010. Sites of synthesis, biochemistry and functional role of plant volatiles. *S. Afr. J. Bot.* 76: 612–631.

Maffei, M.E. 2015. *Molecole Bioattive delle Plante*. Rome: Gruppo Editoriale l'Espresso.

Maffei, M., F. Chialva, and T. Sacco. 1989. Glandular trichomes and essential oils in developing peppermint leaves. I. Variation of peltate trichome number and terpene distribution within leaves. *New Phytol.* 111: 707–716.

Maffei, M.E., J. Gertsch, and G. Appendino. 2011. Plant volatiles: Production, function and pharmacology. *Nat. Prod. Rep.* 28: 1359–1380.

Maffei, M. and T. Sacco. 1987. Chemical and morphometrical comparison between two peppermint notomorphs. *Planta Med.* 53: 214–216.

Monacelli, B., A. Valletta, N. Rascio, I. Moro, and G. Pasqua. 2005. Laticifers in *Camptotheca acuminata* Decne: Distribution and structure. *Protoplasma* 226: 155–161.

Nagel, J., L.K. Culley, Y.P. Lu et al. 2008. EST analysis of hop glandular trichomes identifies an O-methyltransferase that catalyzes the biosynthesis of xanthohumol. *Plant Cell* 20: 186–200.

Niinemets, U., F. Loreto, and M. Reichstein. 2004. Physiological and physicochemical controls on foliar volatile organic compound emissions. *Trends Plant Sci.* 9: 180–186.

Owen, S.M. and J. Penuelas. 2005. Opportunistic emissions of volatile isoprenoids. *Trends Plant Sci.* 10: 420–426.

Paiva, E.A.S. and S.R. Machado. 2004. Structural and ultrastructural aspects of ontogenesis and differentiation of resin secretory cavities in *Hymenaea stigonocarpa* (Fabaceae-Caesalpinioideae) leaves. *Nord. J. Bot.* 24: 423–431.

Pearse, I.S. and R. Karban. 2013. Do plant-plant signals mediate herbivory consistently in multiple taxa and ecological contexts? *J. Plant Interact* 8: 203–206.

Pickard, W.F. 2008. Laticifers and secretory ducts: Two other tube systems in plants. *New Phytol.* 177: 877–887.

Pridgeon, A.M. and W.L. Stern. 1983. Ultrastructure of smophores in *Restrepia* (Orchidaceae). *American Journal of Botany* 70: 1233–1243.

Stout, J.M., Z. Boubakir, S.J. Ambrose, R.W. Purves, and J.E. Page. 2012. The hexanoyl-CoA precursor for cannabinoid biosynthesis is formed by an acyl-activating enzyme in *Cannabis sativa* trichomes. *Plant J.* 71: 353–365.

Turner, G.W., J. Gershenzon, and R.B. Croteau. 2000. Development of peltate glandular trichomes of peppermint. *Plant Physiol.* 124: 665–679.

Ulman, M. and Z. Chilmonczyk. 2007. Volatile organic compounds—Components, sources, determination. A review. *Chem. Analit.* 52: 173–200.
Viano, J., E. Gaydou, and J. Smadja. 1991a. Sur la presence des bacteries intracellulaires dans les racines du *Vetiveria zizanioides* (L.) Staph. *Rev. Cytol. Biol. Veget. Bot.* 14: 65–70.
Viano, J., J. Smadja, J.Y. Conan, and E. Gaydou. 1991b. Ultrastructure des racines de *Vetiveria zizanioides* (L.) Staph (Gramineae). *Bull. Mus. Natl. Hist. Nat, Paris, 4e sér., 13, section B* 1–2: 61–69.
Vieira, R.C., P.G. Delprete, G.G. Leitao, and S.G. Leitao. 2001. Anatomical and chemical analyses of leaf secretory cavities of *Rustia formosa* (Rubiaceae). *Am. J. Bot.* 88: 2151–2156.
Vuorinen, T., A.M. Nerg, E. Vapaavuori, and J.K. Holopainen. 2005. Emission of volatile organic compounds from two silver birch (Betula pendula Roth) clones grown under ambient and elevated CO_2 and different O-3 concentrations. *Atmos. Environ.* 39: 1185–1197.
Zebelo, S.A., C.M. Bertea, S. Bossi, A. Occhipinti, G. Gnavi, and M.E. Maffei. 2011. *Chrysolina herbacea* modulates terpenoid biosynthesis of *Mentha aquatica* L. *PLoS One* 6: e17195.

chapter three

Biodiversity and chemotaxonomic significance of specialized metabolites

Francesca Barbero and Massimo Maffei

Contents

- 3.1 Introduction 24
- 3.2 Distribution of biodiversity 25
- 3.3 Chemotaxonomy of specialized compounds 26
  - 3.3.1 Chemotaxonomy of phenolic compounds 27
    - 3.3.1.1 Asteraceae 28
    - 3.3.1.2 Lamiaceae 29
    - 3.3.1.3 Leguminosae 30
    - 3.3.1.4 Other plant families 32
  - 3.3.2 Chemotaxonomy of terpenoids 32
    - 3.3.2.1 Monoterpenes 33
    - 3.3.2.2 Sesquiterpenes 36
    - 3.3.2.3 Diterpenes 38
    - 3.3.2.4 Triterpenes 40
    - 3.3.2.5 Tetraterpenes 41
    - 3.3.2.6 Polyterpenes 42
  - 3.3.3 Chemotaxonomy of specialized products containing nitrogen 42
    - 3.3.3.1 Alkaloids 42
    - 3.3.3.2 Glucosinolates 46
    - 3.3.3.3 Cyanogenic glycosides 46
    - 3.3.3.4 Nonprotein amino acids 47
  - 3.3.4 Chemotaxonomic significance of fatty acids and surface alkanes 47
- References 50

3.1 Introduction

Biodiversity, or biological diversity, is a global concept of biology that includes the analysis and description of life variability, whether of microbes, plants, or animals belonging to aquatic and terrestrial ecosystems. The concept of biodiversity can also be extended to molecules produced by living organisms, regardless of their function or biosynthetic pathways. In the biosphere, there are several biodiversity areas, among the largest are insects, which may include 2–5 million species, angiosperms, with more than 275,000 known species, and specialized metabolites, which exceed 100,000 molecular structures recognized. Biodiversity can be settled to other ecosystems, but also reduced to minor scales, referring to the biodiversity of a particular nation, park, and plant group or even in one single species (Buchs 2003).

The term "biodiversity," became popular after the signing of the "Convention on Biological Diversity" by 168 countries. More recent interpretations of the term biodiversity are not restricted to the concept of "species richness," but are also linked to a variety of races, life forms, and genotypes, as well as landscape types, habitat, and structural elements (e.g., shrubs, stonewalls, hedges, ponds). Molecular plant biodiversity is one the most intriguing subjects because it reflects the huge diversity in chemical structures produced by individual species (Cropper 1993).

Biodiversity is assessed using the same criteria employed for taxonomical classification. The biosystematics, including taxonomy, is a powerful tool for the study of biodiversity that relies on many biological disciplines such as evolutionary biology, phylogeny, population genetics, and phytogeography (Crawford et al. 2005). Another fundamental component in the evaluation of biodiversity is the study of genetic diversity, which, within species, allows a specific individual to evolve under environmental pressures and natural selection. The variability that we observe between individuals (the phenotype) is in part the result of the interaction between genetic differences (genotype) with the environment that surrounds them (Ilnicki 2014). In the specific case of many specialized metabolites, such as the monoterpenes, the genotypic expression can be influenced by many factors, both biotic (such as the attack of herbivores) and abiotic (such as environmental variations). In this case, the biodiversity is found in phenotypic plasticity, that is, in the ability of genes to be differently expressed even within the same population, or by the same genotypes (Maffei 1988, 1990). Thus, evolution is not necessarily the only way a species reacts to the surrounding environment.

The incredible diversity of plant chemicals, most of which belong to a specialized metabolite category, would require a vast number of enzymes, but plant genome encodes only a small fraction of all the protein needed for their synthesis. However, there are several substrates and products in plant specialized metabolism and only a few types of reactions. In this respect, Pichersky and Gang (2000) argue that parallel to the evolution of new genes for novel specialized compounds, what often occurs is a special form of convergent evolution in which new enzymes with the same function evolve independently in separate plant lineages from a shared pool of related enzymes with similar but not identical functions.

The analysis of chemical data in taxonomy allows the study of wide arrays of gene expressions and represents an extremely powerful tool for an extensive range of research. Pushed to their extreme limit, chemical data should be able to go further beyond the cytology and genetics, allowing direct comparison of DNA sequences (Gnavi et al. 2010a).

3.2 Distribution of biodiversity

Medicinal plants are undoubtedly one of the most fascinating categories of plants both because their use dates back to ancient times in humankind history and because they are sources of bioactive molecules. About 80% of the nearly 30,000 known natural products are derived from plants (Balandrin et al. 1988; Maffei et al. 2011). The number of known chemical structures is estimated to be almost four times greater than that in the microbial kingdom and, in addition, some specialized metabolites are unique to the plant kingdom, not being produced by microbes or animals. In different parts of the world, about 20,000 species of plants with medicinal properties are used and their distribution encompasses all land where a plant can survive, from the steppes close to polar circles to tropical forests, to the semidesert areas bordering the great deserts. The distribution of medicinal plants can be classified based on the separation in the centers of origin of some species:

- North American: *Echinacea angustifolia, Hamamelis virginiana, Sassafras officinale, Lobelia inflata, Hydrastis canadensis,* and *Podophyllum peltatum*
- South and Central American: *Vanilla planifolia, Carica papaya, Aloe vera, Erythroxylon coca, Ilex paraguariensis, Theobroma cacao, Dioscorea composita,* and *Echinocactus williamsii*
- Mediterranean: Most of Lamiaceae, *Valeriana officinalis, Digitalis purpurea, Crocus sativus, Laurus nobilis, Foeniculum vulgare, Glycyrrhiza glabra, Colchicum autumnale,* and *Atropa belladonna*
- African: *Acacia senegal, Ricinus communis, Cassia acutifolia, Datura stramonium, Rauwolfia vomitori,* and *Physostigma vevenosum*
- Madagascar: *Eugenia caryophyllata, Catharanthus roseus,* and *Piper nigrum*
- Indian: *Rauwolfia serpentina, Datura* spp., *Cannabis sativa* var. *indica, Curcuma longa, Strychnos nux-vomica, Cinnamomum zeilanicum, Cassia angustifolia, Zingiber officinale,* and *Dioscorea* spp.

- Asiatic: *Papaver somniferum, Panax ginseng, Rheum* spp., *Cinnamomum camphora, Thea sinensis,* and *Vaccinium myrtillus*
- Indonesia: *Myristica fragrans, Illicium verum, Eugenia caryophyllata,* and *Piper methysticum*
- Australia: *Eucalyptus* spp. and *Duboisia myoporoides*

Certainly one of the most important centers of origin of aromatic and medicinal plants is the Mediterranean Basin. This area is difficult to define because it does not coincide with any political boundaries and is characterized by nations of different ethnic groups and climates.

According to Heywood (1995), the importance of the Mediterranean region is derived from a number of considerations, including the high variability of soil and climatic conditions which favor the high biodiversity of plant species and the fact that it contains high proportions of annual species belonging particularly to the Caryophyllaceae, Brassicaceae, Asteraceae, and Apiaceae families.

3.3 Chemotaxonomy of specialized compounds

The use of chemicals to distinguish plants is lost in the midst of time. In fact, humans have learned to distinguish plants primarily for their medicinal properties. Albeit unconsciously, the first distinctions were chemical, linked to the effect that the substances contained in a particular plant practiced on humans or due to the presence of compounds with smells or tastes. Associating an effect to a form of the plant was the first attempt to combine the presence of a chemical compound with the morphology of a plant. Egyptians, Babylonians, Greeks, and Romans knew the medicinal properties of plants but only in the twentieth century the technology to define the chemical compounds responsible for certain medicinal effects has evolved. This event established the basis for a branch of taxonomy that has developed as chemotaxonomy (Maffei 2015).

Chemical data are useful at any level of taxonomic characterization and the hierarchical level depends on the type of molecule or class of compound examined. In general, many authors consider micromolecular data the most useful for classifications at the specific or subspecific level, while macromolecular data are more effective at higher taxonomic levels (e.g., genus, subfamily, family) (Hadacek 2002).

The importance of chemotaxonomy has grown in recent years thanks to the introduction of molecular biology techniques, using macromolecular data. The biosystematics utilizes DNA sequences that represent one of the most powerful tools of investigation (Gnavi et al. 2010a).

The chemical data have often been used to overcome the difficulties encountered in the classification of species based solely on morphological data. Sometimes chemical data allow getting to where the morphological data stops and, in other cases, the results obtained are consistent with those collected with the classical taxonomy. Occasionally, the extreme variability of the classes of compounds used for chemotaxonomic purposes creates contrasting results. For example, in the Leguminosae, phylogeny obtained by biomolecular data agrees with that obtained by morphological data, but disagrees with that resulting from alkaloids or other natural substance analysis. One of the main reasons for these discrepancies is that specialized metabolites are molecules produced in response to external agents, both biotic and abiotic. In particular, the micromolecular data play an important role in plant adaptation to various stress conditions, especially to changing environmental conditions (Maffei 2015). Thus, the use of a particular metabolite for taxonomic purposes, without taking into account any stress that the species in question is suffering, can lead to underestimation or overestimation of the chemotaxonomic meaning of that particular metabolite. This does not mean that micromolecular data are unreliable, but require the rigorous evaluation of these molecules. By contrast, data obtained from DNA analysis are insensitive to environmental changes or the attack of pathogens, although some authors speculate that the repeated fraction of the plant genome is susceptible to some environmental stresses that would cause DNA qualitative and quantitative changes (Bassi et al. 1995).

For the purposes of this chapter, we will focus on the main classes of compounds used for micromolecular analysis; however, we will also consider the correlation between micromolecular and macromolecular data.

The most common micromolecular data include specialized phenolic compounds such as phenylpropanoids, benzoic acids, coumarins and furanocoumarins, stilbenes, flavonoids, anthocyanidins, catechins, procyanidins, and polymeric forms as hydrolysable and condensed tannins. In addition, terpenoids may be listed as micromolecular compounds, especially monoterpenes and sesquiterpenes. Low-molecular-weight alkaloids, betalains and glucosinolates, correlate well with morphological and genetic data, providing additional useful information for classification criteria. Macromolecular compounds are often used to solve taxonomic doubts at higher hierarchical levels. For instance, electrophoresis of proteins contained in the seed storage tissues shows a number of bands that can be compared between different taxa and can be considered phenetically as many other data (Maffei 2015). In order to give an overview of the chemotaxonomic significance of specialized metabolites, we will present data on three major plant families, namely, Leguminosae, Lamiaceae, and Asteraceae, and we will add some information from other plant families.

3.3.1 *Chemotaxonomy of phenolic compounds*

Phenolic compounds are among the most frequently used molecules in chemotaxonomic studies. They are divided into different classes, from simple phenols such as benzoic and cinnamic acids, stilbenes and coumarins to more complex structures such as flavonoids and anthocyanidins and their polymers. The extensive use of these compounds is principally due to the ease of their extraction and separation, together with their stability even at room temperature.

The benzoic acids are the simplest molecules, made by a benzoic ring with an acid group in position 1. They may bear substituents of various natures, but the most common groups are –OH and –OCH$_3$. Cinnamic acid derivatives, in particular *trans*-cinnamic acid (**1**), are basic building molecules for complex phenolic compounds and the constituents of one of the most widespread polymers in the plant kingdom, lignin. Flavonoids consist of three aromatic rings, two of which are found in all plants with the exception of algae and are stored predominantly in the vacuole. By far the most used compounds in chemotaxonomy of phenolic compounds are flavonoids and anthocyanidins.

Among the main families used for chemotaxonomic surveys are the Asteraceae, Lamiaceae, and Leguminosae. We will give detailed information about these families, while other families will be treated briefly.

3.3.1.1 Asteraceae

The effectiveness of flavonoids as molecular markers for phylogenetic and evolutionary studies in angiosperms has been widely accepted by the majority of botanists (Bates-Smith 1962; Crawford 1978; Giannasi 1978). In very large families, like the Asteraceae, the flavonoid distribution is not able to reflect the phylogenetic relationships among genera and even less explains the evolutionary steps of distinct tribes. However, flavonoids can be very useful, also in the case of Asteraceae, to overcome problems related to taxonomic revisions beyond the tribe level. From an evolutionary point of view, flavonoids can confirm the reticulate evolution of the tribes in this family (Emerenciano et al. 2007; Harborne 1994; Seeligmabb 1994).

Flavonoids exist in plant cells in two main forms: aglycones and glycosides. The most detailed studies have been made on aglycones. Species showing high production of flavonoid exudates as well as those with the largest number of flavonoids among plants belong to the Asteroideae subfamily (Seitz et al. 2015; Wollenweber et al. 1997, 1998). Aglycones of flavones and flavonols, with substitutions in 6- and 8-positions, which represent the basic structures of Asteraceae flavonoids, are also widespread. While aglycones of this subfamily are useful taxonomic markers for species identification, they prove to be useless at tribe or genus levels.

In the Asteraceae, the majority of flavonoids are found in leaves, stems, and flowers as water-soluble glycosides. The plant surfaces can also be rich in fat-soluble aglycone flavonoids in the form of either powdery deposits or mixed with resins and waxes. Usually in the latter case, they are produced by glandular trichomes, secreted outward with other substances and laid on the epicuticular layers (Maffei 2015).

A comprehensive study on anthocyanidins of the Asteraceae has shown that most of these molecules are acetylated with organic dicarboxylic acids. The succinic acid is exclusively found in the tribe of Cynareae while the acetylating acid in other families is the malonic acid. The family comprises other rare anthocyanidins which are acetylated by (di)malonylate derivatives and glycosylated by glucuronic acid. In the Asteraceae there are three most common basic structures of anthocyanidins, that is, cyanidin (**2**), delphinidin (**3**), and pelargonidin (**4**), with sugar molecules linked mostly in the positions 3 or in 3 and 5. Unlike other flavonoids, anthocyanidins are useful chemotaxonomic markers for tribe identification. For example, in the Astereae 3-malonylglucoside derivatives are found, while in the Heliantheae there is a second acyl group. In the Senecioneae substituents of malonic acid and caffeic acid (**5**) exist, while in the Anthemideae there are molecules with two units of malonic acid linked to glucose in position 3. Finally, the acetylating compound in the Cynareae is succinic acid (Harborne 1994).

Anthocyanidins are mainly responsible for flower color, but for certain hues more than one pigment is needed. The dark blue of some species of Asteraceae, for instance, is made of at least three metabolic pathways (Mishio et al. 2015; Park et al. 2015; Tanaka and Brugliera 2013). Usually, the anthocyanidin is linked to a flavone and to one or more aromatic rings; in addition, metals in ionic form, as magnesium and iron, are essential for color stability (Tanaka and Brugliera 2013).

3.3.1.2 Lamiaceae

Several phenolic compounds are present in the Lamiaceae family, and in this case as well, the most studied structures are flavonoids and anthocyanidins. There is a general correlation between the chemistry of flavonoids and the systematics of Lamiaceae, both within the family and at the genus level. Again, these molecules are valuable chemotaxonomic tools. One of the family features is to accumulate 5,6-dihydroxy-7,8-dimetoxyflavones, except for the Lamioideae subfamily, which is particularly rich in 8-hydroxyflavone-7-allosyl glucosides or *p*-coumaryl glycosides, and Nepetoideae, which pile up 6-hydroxy flavones (Tomas-Barberan and Gil 1992). 5,7-Dihydroxy-6-methoxyflavone and 5,6-dihydroxy-7-methoxyflavone with a substituted B-ring and 5,6-dihydroxy-7,8-dimethoxyflavone are characteristic flavonoid constituents of the subfamily Nepetoideae and of the tribe Saturejeae (Zaidi et al. 1998).

One of the most studied genera is *Salvia*. The chemotaxonomic diversity of *Salvia* species shows a consistent intraspecific diversification based on different quantity and quality of flavones, flavonols, flavanones, isoflavones, dihydroflavonols, and chalcones (Kharazian 2014; Nikolova et al.

2006; Wollenweber et al. 1992). Only about 40% of *Plectranthus* species were found to produce exudate flavonoids, which were mainly flavones. Flavanones were restricted to five species of the genus *Plectranthus*, whereas flavonols were only found in two species of *Coleus*. The most common flavones, occurring in both genera, were cirsimaritin (**6**) and salvigenin (**7**), which are methoxylated at the 6- and 7-positions. 6-Hydroxylated flavones such as scutellarein (**8**) and ladanein (**9**) were restricted to *Plectranthus* species (Grayer et al. 2010). A significant intraspecific variation also occurs in the genus *Origanum*. Taxa in subgeneric populations accumulated flavonoids with methoxyl groups at both C-6 and C-4'; however, taxa in other subgeneric groups did not accumulate 4'- or 6-methoxylated compounds (Skoula et al. 2008). The widespread genus *Stachys* comprises about 300 species and is considered to be one of the largest genera of the Lamiaceae. Several flavonoids, among others apigenin (**10**), chrysoeriol (**11**), penduletin (**12**), luteolin-7-O-β-D-glucoside (**13**), and stachyspinoside (**14**) are characteristic of this genus and have been used for chemotaxonomic purposes (Skaltsa et al. 2007; Tomas-Barberan and Wollenweber 1990; Valant-Vetschera et al. 2003; Wollenweber et al. 1989).

As regards the anthocyanidins, a direct comparison with other families indicates that a distinctive feature of the Lamiaceae is the greater frequency of aromatic and aliphatic acylations. A cyanidin derivative present in several members of the Lamiaceae has been characterized as the related 3-p-coumarylglucoside-5-malonylglucoside (**15**) (Takeda et al. 1986). In the blue-purple flowers of *Triteleia* species, one of the main anthocyanidins is delphinidin 3-*trans*-p-coumaroylglucoside-5-malonylglucosides (**16**) (Toki et al. 1998). A good correlation was also found between flower color and anthocyanidin type in several Lamiaceae species (Saito and Harborne 1992).

3.3.1.3 Leguminosae

Anthocyanins and flavonoids have also been used as chemotaxonomic markers in the large family of the Leguminosae (Hegnauer and Hegnauer 2001). Despite the massive amount of information that has accumulated, much remains to be done to obtain good quality comparative data that will allow for sensible chemosystematic interpretations at higher taxonomic levels (van Wyk 2003). In the Cesalpinioideae subfamily, numerous anthocyanidins are in the form of 3-glucosides and 6-alkylated-3-soforoside of cyanidin. In the Lotoideae subfamily, the tribe of the Genisteae is characterized by 3,5-diglucosides of pelargonidin, the Trifolieae by 3-glucosides of delphinidin, the Galegeae by 3,5-diglucosides of delphinidin, the Hedysareae by 3,5-diglucosides of malvidin and the Phaseoleae from 3-soforosides of pelargonidin and cyanidin (Harborne 1971; Hegnauer and Grayerbarkmeijer 1993; van Wyk 2003; Wink 2013). Flavonoids bearing 6,7-(dimethylpyran) and 8-(γ,γ-dimethyl allyl) substituents such as mundulin (**17**) and minimiflorin (**18**) are characteristic for some *Lonchocarpus* species (Alavez-Solano et al. 2000), whereas in the genus *Sophora* the presence or absence of oligostilbenes and prenylated flavonoids was found to represent an important chemotaxonomic character (Ohyama et al. 1995).

Common in this family are flavonols and flavones. Among flavonols, those with the highest chemotaxonomic value are the kaempferol (**19**) and the quercetin (**20**). Quercetagetin (**21**), gossypetin (**22**), and their derivatives are primarily found in the subfamilies Mimosoideae, Caesalpinioideae, and Loteae, whereas chalcones are distributed mainly in the Lotoideae (Harborne 1971). A survey of foliar flavonoids in the swartzioid legume genus *Cordyla s.l.* revealed that three species, *C. haraka*, *C. pinnata*, and *C. richardii*, were rich in the flavonol pentaglycosides 3-*O*-α-L-rhamnopyranosyl (1 → 3)-α-L-rhamnopyranosyl (1 → 2) [α-L-rhamnopyranosyl (1 → 6)]-β-D-galactopyranoside-7-*O*-α-L-rhamnopyranosides of quercetin and kaempferol [cordylasins A (**23**) and B (**24**), respectively] (Veitch et al. 2008). Flavone glycosides have also been demonstrated to possess chemotaxonomic character (Veitch et al. 2010).

Esterification of anthocyanins appears to be a unique apomorphy for the tribe Podalyrieae (including Liparieae). When present, these compounds are esterified with coumaric or acetic acid, while species from other tribes have glycosides only. In the Podalyrieae and in the related Crotalarieae, only hydroxylated anthocyanins (derivatives of cyanidin [2] and peonidin [25]) were found (van Wyk 2003).

One of the physiological characteristics of phenolic compounds is their role as chemical barriers for ultraviolet radiation. Since the beginning of the twentieth century, reflective and absorbing properties of UV radiation have been known in many plant species, often imperceptible by human eyes, but distinguishable by bees and by other pollinating insects. Bees are able to discriminate the plants according to their UV reflection and absorbance patterns. In some Cesalpinieae, anthers contain flavonoids that absorb UVs, while filaments have flavonoids that reflect UVs. In some species, the absorbing pigments are chalcones, while in other are flavanones. In Cassieae, the UV absorbing molecules are mainly derivatives of quercetagetin (**21**), whereas in the petals of some Tephrosieae isoflavones, flavonoids and furanoflavones are found. Yellow flavonols are typical of Loteae or Genisteae petals (Harborne 1993).

Isoflavonoids are the characterizing flavonoids of Leguminosae and, in general, the unique occurrence of isoflavonoids in the subfamily Papilionoideae and perhaps the frequent absence of a hydroxyl group in position 5 are interesting features (van Wyk 2003). Isoflavonoids are similar to flavonoids in the structural formula, but the B aromatic ring is linked in position 3 rather than in position 2. Isoflavones are peculiar of Lotoideae and are uncommon in other subfamilies. In addition, they are also rare in other plant families, and for this reason they possess a remarkable chemotaxonomic value. Their structures range from simple molecules, such as genistein (**26**), to very complex molecules such as toxicarol isoflavone (**27**). There are many classes of isoflavones: the derivatives of genistein, of orobol (**28**), 6-hydroxylated isoflavones, and isoprenoid isoflavones (Veitch 2013). Two isoflavone glycosides, wistin (**29**) and ononin (**30**), were isolated as major constituents of *Glycyrrhiza* (the liquorice genus) species and found to be good chemotaxonomic markers (Kajiyama et al. 1993).

Interesting compounds related to isoflavonoids are rotenones, pterocarpans, coumestans, 2-arylbenzofurans, and coumaronochromones (Leuner et al. 2013; Marzouk et al. 2008; Shitamoto et al. 2010; Tang et al. 2002; Veitch 2013).

3.3.1.4 Other plant families

Flavonoids have been used for the chemotaxonomic characterization of many families. In some cases, as for the Anarthriaceae, results were perfectly in line with the phylogenetic assumptions gathered by biomolecular analyses (Williams et al. 1997). In Cuscutaceae some species accumulate derivatives of cinnamic acid, other flavonoids and others still have a cinnamates/flavonoids ratio of 1:1. More than 30 structures belonging to flavanones and chalcones are in the resins produced by the *Xanthorrhoea* genus (Xanthorroeaceae) and have been used to solve taxonomic problems.

Significant taxonomic markers of the Asphodelaceae family are the anthraquinones extracted from the roots. Due to the presence of peculiar anthraquinones such as aloesaponarin II (**31**) the *Lomatophyllum* genus was included in the *Aloe* genus (van Wyk et al. 1995). Over 90 flavonoid constituents have been discovered and characterized for their chemotaxonomic significance, including 38 new compounds from 15 species of the genus *Iris* (Iridaceae) and a checklist of the flavonoid compounds in *Iris* by species was published recently (Wang et al. 2010a). The pattern of phenolic compounds has been successfully used to solve controversial classification in several plant genera, including *Drosera* (Braunberger et al. 2015), *Anthurium* (Clark et al. 2014), *Aletris* (Li et al. 2014), *Crataegus* (Edwards et al. 2012), and *Bagassa* (Royer et al. 2010). Wood and bark from many species of both hardwood and softwood trees contain many types of flavonoid compounds. Most chemotaxonomic studies have been conducted on flavonoids in the extracts from softwoods such as *Podocarpus, Pinus, Pseudotsuga, Larix, Taxus, Libocedrus, Tsuja, Taxodium, Sequoia, Cedrus, Tsuga, Abies,* and *Picea,* as recently reviewed (Yazaki 2015). Figure 3.1 shows the chemical structure of some representative phenolic compounds.

3.3.2 Chemotaxonomy of terpenoids

Terpenoids are another important class of specialized metabolites, which has been widely used for chemotaxonomic studies. The number

Chapter three: Biodiversity and chemotaxonomic significance 33

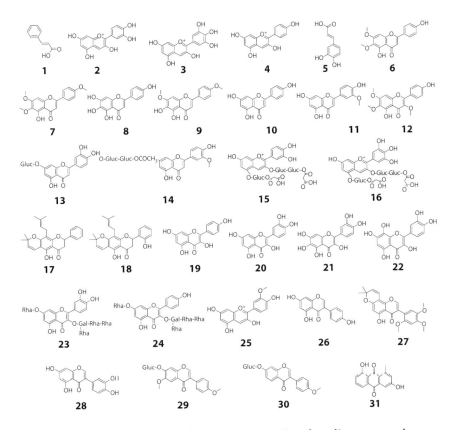

Figure 3.1 Chemical structure of some representative phenolic compounds.

of repetitions of a 5-carbon-atom isoprenoid building block defines several chemical groups within this class. For this reason, terpenoids are also known as isoprenoids. Instead of analyzing the chemotaxonomic significance of terpenoids in the different plant families, we will dissect these compounds into their different classes.

3.3.2.1 *Monoterpenes*

Monoterpenes are, along with sesquiterpenes (see below), the main constituents of essential oils extracted by steam distillation from many plant families, but especially in the Lamiaceae, Rutaceae, Apiaceae, Asteraceae, and in several gymnosperms. These low-molecular-weight compounds can be analyzed by gas chromatographic techniques, or following liquid nitrogen freezing of tissue and subsequent extraction with nonpolar solvents or with supercritical CO_2 (Bicchi and Maffei 2012; Capuzzo et al. 2013). The power of monoterpenes in determining and, sometimes, correcting the taxonomic status of plants is proven by several scientific reports. For instance, the family Pittosporaceae was classified in the Rosanae superorder, but a large number of species belonging to this family encompasses volatile compounds stored in resin channels and lysigenous pockets (Maffei 2010) (see also Chapter 2). However, this does not occur in other families ascribed to Rosanae, while it is common in those belonging to the superorder Aralianae, which includes the Apiaceae and the Araliaceae families. In gymnosperms, the most well-studied genera for chemotaxonomic purposes are *Pinus* (Nikolic et al. 2011; Sarac et al. 2013) and *Abies* (Niederbacher et al. 2015) (Pinaceae), *Juniperus* (Caramiello et al. 1995; Kim et al. 2015), *Thuja* (Tsiri et al. 2009) and *Cupressus* (Lohani et al. 2015) (Cupressaceae), and *Taxus* (van Rozendaal et al. 1999; Yasar 2013) (Taxaceae). In the gymnosperms, the most representative chemotaxonomic monoterpene markers are α-pinene (**32**), β-pinene (**33**), bornyl acetate (**34**), sabinene (**35**), α- thujone (**36**), β-thujone (**37**), fenchone (**38**), and *p*-cymene (**39**).

In angiosperms, the Lamiaceae is one of the most investigated families. This family, which includes more than 3500 species, can be easily divided into two categories: species that produce essential oils (Nepetoideae) and species where these compounds are less abundant (Lamioideae). However, the major distinction between the two families resides in the presence of iridoids, often glycosylated monoterpenes like the 5,9-epi-penstemoside (**40**), which are generally absent in the Nepetoideae (Delazar et al. 2004).

Among the Nepetoideae the most inspected genera for the chemotaxonomic significance of their essential oils are *Hyssopus* (Fathiazad and Hamedeyazdan 2011), *Lavandula* (Bajalan and Pirbalouti 2015; Bella et al. 2015), *Ocimum* (Grayer et al. 1996; Pino Benitez et al. 2009; Pirmoradi et al. 2013; Vieira and Simon 2006), *Origanum* (Gulsoy 2012; Lukas et al. 2015; Skoula et al. 1999), *Rosmarinus* (Elamrani et al. 2000), *Salvia* (Jassbi et al. 2012; Mathe et al. 2010), *Teucrium* (Perez et al. 2000; Radulovic et al. 2012), and *Thymus* (Jensen and Ehlers 2010; Pitarokili et al. 2014). In the genus *Mentha*, chemical varieties were identified virtually in every studied species (Baser et al. 2012). The chemotaxonomic survey based on monoterpenes has proven to be useful for understanding the intraspecific variation patterns, to define some species and to establish the degree of hybridization of some natural populations (Maffei 1990). Gas chromatographic techniques coupled with mass spectrometry and multivariate analyses of the data have been used to accomplish these results (Maffei et al. 1993c). The chemical distribution of the essential oils and the chemotaxonomy significance of these compounds have also been analyzed in the genera *Wiedemannia* (Kilic and Bagci 2014), *Nepeta* (Bisht et al. 2012), *Stachys* (Tundis et al. 2014), and *Micromeria* (Slavkovska et al. 2005). In terms of chemotaxonomic significance, the most representative monoterpenes in the Lamiaceae family are, among others, isopinocamphone (**41**), 1,8-cineole (**42**), borneol (**43**), camphor (**44**), linalool (**45**), linalyl acetate

(46), methyl chavicol (47), eugenol (48), methyl eugenol (49), menthol (50), pulegone (51), thymol (52), *p*-cymene (39), carvacrol (53), γ-terpinene (54), α-terpineol (55), isoiridomyrmecin (56), and menthofuran (57).

In the Asteraceae family monoterpenes are useful for species discrimination. In the genus *Achillea* the presence of oxygenated monoterpenes such as camphor (44), 1,8-cineole (42), artemisia alcohol (58), santolina alcohol (59), and terpinen-4-ol (60) was used for discerning different species (Maffei et al. 1989, 1993a; Muselli et al. 2007; Rahimmalek et al. 2009). In the same genus, β-pinene (33), camphor (44), and 1,8-cineole (42) have proved to separate species according to their geographical origin (Maffei et al. 1994). Monoterpenes are taxonomic markers in several other Asteraceae, such as in the genus *Chiliadenus*, in which camphor (44) is an excellent chemotaxonomic marker (Buhagiar et al. 2015; Sacco and Maffei 1987), or in the genus *Artemisia*, where the presence of β-thujone (37) has allowed to discriminate spontaneous populations, hardly distinguishable by morphological data (Blagojevic et al. 2015; Rubiolo et al. 2009). Moreover, monoterpenes are also chemotaxonomic markers in the genera *Chrysanthemum*, with camphor (44), chrysanthenone (61), safranal (62), and myrcene (63) (Sun et al. 2015); *Santolina*, containing borneol (43), santolina triene (64), and the irregular monoterpenes iso-lyratol (65) and lyratyl butyrate (66) (Gnavi et al. 2010b; Liu et al. 2007); *Tagetes*, containing (Z)-β-ocimene (67) (Singh et al. 2016); and *Eriocephalus*, with high contents of camphor (44) and γ-terpinene (54) (Viljoen et al. 2006).

Although monoterpenes are not found in Leguminosae leaves, they are quite frequent in flowers through the action of scent glands, or osmophores, which are predominantly floral secretory structures that secrete volatile substances during anthesis (see also Chapter 2) (Marinho et al. 2014). A larger quantity of monoterpenes is located in the *Acacia* flowers characterized by the presence of (Z)-β-ocimene (67) (Kotze et al. 2010), while a phenolic monoterpene, the bakuchiol (68), is exclusively present in *Psoralea corylifolia* (Park et al. 2005). Linalool (45) and its derivative *cis*-linalool pyran oxide (69) are characteristic floral scents of *Ceratonia siliqua* (Custodio et al. 2006) and *Cyathostegia mathewsii* (Lewis et al. 2003), whereas the chemical profile obtained from the floral scent of some *Parkia* species shows the presence of (Z)-β-ocimene (67) (Pettersson and Knudsen 2001; Piechowski et al. 2010). The latter compound is also typical of *Browneopsis disepala* floral scent, along with α-pinene (32) and β-pinene (33) (Knudsen and Klitgaard 1998). Iridoids are a group of monoterpenes, which show a particular importance for chemotaxonomic studies. These compounds are typical of major families including: Valerianaceae, Dipsacaceae, Rubiaceae, Gentianaceae, Apocynaceae, Oleaceae, Lamiaceae, and Ericaceae. However, they are only found in Rutanae, Rosanae, Cornanae, Loasanae, Gentiananae, Lamianae, and Ericanae superorders, indicating that the ability to synthesize these compounds occurred a few times in the

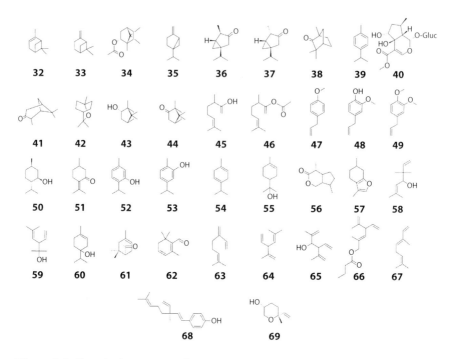

Figure 3.2 Chemical structure of some representative monoterpenes.

plants' phylogeny (Rosendal-Jensen 1991). Figure 3.2 shows the chemical structure of some representative monoterpenes.

3.3.2.2 Sesquiterpenes

As monoterpenes, sesquiterpenes are also components of essential oils and for this reason have been widely used as chemotaxonomic markers.

Several Lamiaceae store sesquiterpenes in their secretory tissues. Germacrene D (**70**), (E)-β-caryophyllene (**71**), δ-cadinene (**72**), (E)-β-farnesene (**73**), spathulenol (**74**), and γ-muurolene (**75**) are the major constituents in the genus *Stachys* (Aghaei et al. 2013; Bisht et al. 2008; Dimitrova-Dyulgerova et al. 2015), whereas *Pogostemon cablin* (patchouli) essential oil is unique because it consists of over 24 different sesquiterpenes, including patchouli alcohol (**76**) (Chen et al. 2014; Deguerry et al. 2006; Ding et al. 2011). The sesquiterpene 14-hydroxy-α-humulene (**77**) is the main constituent of the essential oil of *Salvia argentea* (Riccobono et al. 2016), whereas in the genus *Teucrium*, (E)-β-caryophyllene (**71**), germacrene D (**70**), and caryophyllene oxide (**78**) are the major chemotaxonomic markers (Djabou et al. 2012; Hammami et al. 2015; Kremer et al. 2015). α-Humulene (**79**) is characteristic of the genus *Hyptis* (Ashitani et al. 2015; Franco et al. 2011; McNeil et al. 2011), whereas E-E-α-farnesene

(80), and oplopanone (81) are frequently present in the genus *Origanum* (Lukas et al. 2015). α-Bisabolol oxide B (82) is a chemotaxonomic marker of *Anisomeles indica* (Batish et al. 2012), whereas aromadendrene (83), viridiflorene (84), β-selinene (85), and valencene (86) are typical of *Lepechinia paniculata* (Valarezo et al. 2012) and bicyclogermacrene (87) is a chemotaxonomic marker of the genus *Hypenia* (Silva et al. 2011). The genus *Salvia* is characterized by the presence of sesquiterpene lactones, which represent powerful chemotaxonomic markers. Germacrane sesquiterpenoids with an unusual Δ^3-15,6-lactone moiety, for example, scapiformolactones, are typical of *Salvia scapiformis* (Lai et al. 2013), whereas eudesmane-type sesquiterpenes, such as plebeiolide A (88) and plebeiafuran (89) are characteristic of *Salvia plebeia* (Dai et al. 2014).

In the Leguminosae, 4-α-copaenol (90), kaurene (91), cyclosativene (92), and *cis*-α-bergamotene (93) are characteristic compounds of the genus *Copaifera* (Leandro et al. 2012). Five typical sesquiterpenes are present in the heartwood of *Dalbergia odorifrea* (Wang et al. 2014), whereas in *Pterodon pubescens* the sesquiterpene oplopanone (81), which bears a modified cadinane skeleton, is being reported for the first time in this genus (Miranda et al. 2014). (*E*)-Nerolidol (94) is present in the flowering shoots of *Lupinus varius* (Al-Qudah 2013), whereas the guaiane sesquiterpene, (1β,6α,10α)-guai-4(15)-ene-6,7,10-triol (95), and a the lignan, (+)-lariciresinol 9′-stearate (96), are typical of the aerial parts of *Tephrosia vogelii* (Wei et al. 2009). Finally, (*E*)-β-farnesene (73) is present in the floral scent of *Cyathostegia mathewsii* (Lewis et al. 2003).

Sesquiterpene lactones are a particular feature of the Asteraceae, where they may have concentrations up to 2% of the dry weight of the plant (Maffei 2015). One of the advantages in using sesquiterpene lactones as chemotaxonomic markers instead of monoterpenes and other sesquiterpenes is their lower volatility, which allows a more efficient extraction. These compounds are particularly useful at low levels of the taxonomic hierarchy, as shown in the *Ambrosia psilostachya* (Miller et al. 1968). In the Asteraceae, more than 4000 structures of sesquiterpene lactones have been isolated from leaves and accumulated in the glandular trichomes (see Chapter 2). In the Heliantheae (Asteraceae) tribe, sesquiterpene lactones were used for the taxonomic discrimination of subtribes. In the Ambrosinae subtribe are found structures belonging to skeletons of germacranolides, melampolides, eudesmanolides, guaianolides, and ambrosanolides, while in *Bahiinae* there are only eliangolides and eudesmanolides. The Coreopsidinae and Milleriinae have only germacranolides, which in contrast are absent from Ecliptinae and Engelmaniinae. Melampolides and ambrosanolides are missing in Helianthinae, Neurolaeninae, Verbesinae, and Ziniinae. In the Astereae tribe and especially in the subtribe of Asterinae, guaianolides are widespread (Seaman 1982). In particular, in the Gonosperminae (Anthemideae, Asteraceae), the three genera *Gonospermum*, *Lugoa*, and *Inulanthera* are characterized by the presence of several eudesmane sesquiterpene lactones (Triana et al. 2010), whereas the germacranolide taraxinic acid β-glucopyranosyl ester (**97**), ainslioside (**98**), and crepidiaside B (**99**) are common constituents of *Taraxacum* species (Michalska and Kisiel 2008). A chemotaxonomic discussion based on the lactone profile of 18 species from genus *Anthemis* showed that the presence of various guaianolides, germacranolides, and antheindurolides, with the presence of hydruntinolides being components characteristic for sect. *Hiorthia* (Bruno et al. 2002; Staneva et al. 2008). The guaianolide leucodin (**100**) and the eudesmanolide tanacetin (**101**) are chemotaxonomic markers of *Cichorium spinosum* (Michalska and Kisiel 2007), whereas several guaianolides and germacranolides characterize the genus *Achillea* (Todorova et al. 2006). In *Viguiera radula* (Heliantheae), the co-occurrence of germacranolides and heliangolides, stored in glandular trichomes, shows the same general sesquiterpene lactone pattern of many other thus far investigated members of section *Paradosa* (Spring et al. 2003).

Several other plant families contain sesquiterpene lactones. For instance, costunolide (**102**), parthenolide (**103**), and lipiferolide (**104**) were isolated from the leaves of several *Magnolia* species, which belong to the Magnoliaceae family (Katekunlaphan et al. 2014), whereas the seco-prezizaane-type sesquiterpene, 3,4-dehydroneomajucin (**105**) is present in the fruits of *Illicium jiadifengpi* (Illiciaceae family) (Liu et al. 2016). Figure 3.3 shows the chemical structure of some representative sesquiterpenes.

3.3.2.3 Diterpenes

Several diterpene classes have been used for chemotaxonomic purposes using acyclic diterpenoids, and the bi-, tri- and tetracyclic diterpenoids including labdanes, halimanes, clerodanes, abietanes, pimaranes, kauranes, cembranes, tiglianes, salviatrienes, and many other compounds (Hanson 2015). In the Lamiaceae, the genus *Salvia* is one the most studied with respect to these compounds. About 800 compounds have been isolated from 130 *Salvia* species and more than 500 are diterpenes (Wu et al. 2012). *Salvia* plants are a rich source of tanshinones and royleanones, as well as their analogues and methods for the simultaneous qualitative and quantitative analysis of major and minor diterpenoids in different *Salvia* species have been developed (Zhou et al. 2009). For instance, the roots of *Salvia hypoleuca* contain the diterpenes manool (**106**) and 7α-acetoxyroyleanone (**107**) (Saeidnia et al. 2012), *Salvia gilliessi* is characterized by the presence of icetexone (**108**) and 12-hydroxy-11,14-diketo-6,8,12-abietatrien-19,20-olide (**109**) (Nieto et al. 2000), *Salvia divinorum* contains salvinorin A (**110**) (Hernandez-Bello et al. 2015), *Salvia cuspidata* contains the abietane 12-hydroxy-11,14-diketo-6,8,12-abietatrien-19,20-olide (HABTO, **111**) (Lozano et al. 2015), *Salvia pomifera* shows the presence of the diterpene carnosol (**112**), and salviol (**113**) (Trikka et al. 2015),

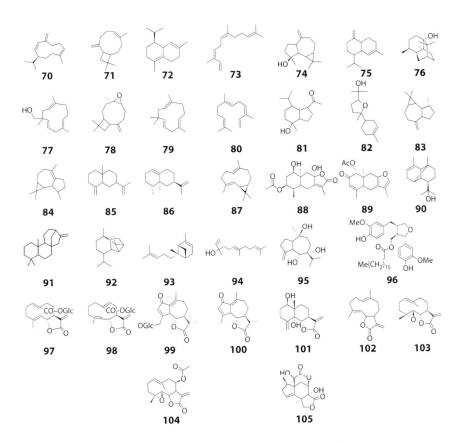

Figure 3.3 Chemical structure of some representative sesquiterpenes.

whereas the reddish root and rhizome of *Salvia miltiorrhiza* consist of abietane quinone diterpenoids including tanshinone I (**114**) and cryptotanshinone (**115**) (Hirata et al. 2015; Zhong et al. 2009). In the genus *Hyptis* (Lamiaceae) diterpenoids are extracted from a limited number of species and the main compounds are hyptol (**116**) and 15-β-methoxyfasciculatin (**117**) (Piozzi et al. 2009). The neo-clerodane diterpenes, teufruintin A (**118**), are characteristic of *Teucrium fruticans* (Lv et al. 2015), whereas the labdane-type diterpenoid 5-ethoxy-3-(2-((R)-4-hydroxy-2,5,5,8a-tetramethyl-3-oxo-3,5,6,7,8,8a-hexahydronaphthalen-1-yl)ethyl)furan-2(5H)-one (**119**) is present in *Leonurus japonicus* (Wu et al. 2015). Several diterpenes have been used as chemotaxonomic markers in the genus *Sideritis*, including 2-hydroxy-isophytol (**120**) (Fraga et al. 2009), whereas the labdane diterpenoid leoleorin (**121**) is typical of *Leonotis leonurus* (Wu et al. 2013).

Most diterpenes found in the Leguminosae have been isolated from the Cesalpinioideae subfamily. The majority of them show a labdanic skeleton. As for mono- and sesquiterpenes, quantitative changes in diterpenoids may have considerable chemotaxonomic significance, especially at the genus and species levels. In Detarieae (Cesalpinioideae) the distinguishing characteristic is the presence of bicyclic labdanes, while in Caesalpinieae are mainly the alcaloidic diterpenes (Langenheim 1981). For instance, *Caesalpinia sappan*, which is distributed in Southeast Asia, contains the diterpene phanginin (**122**) (Tran et al. 2015; Yodsaoue et al. 2008), *Caesalpinia echinata* produces cassane-type diterpenoids designated as echinalide H (**123**) (Mitsui et al. 2015), *Caesalpinia furfuracea* twigs contain the isopimarane diterpenes, caesalfurfuric acid A (**124**) (Siridechakorn et al. 2014), whereas *Caesalpinia pulcherrima* is characterized by the cassane-type diterpene isovouacapenol E (**125**) (Ragasa et al. 2003).

3.3.2.4 Triterpenes

Triterpenes include several classes of compounds, many of which are powerful deterrents toward herbivores. Phytosterols, pentacyclic triterpenes, saponins and sapogenins, cardiac glycosides, and limonoids are among the most represented categories. Some classes of triterpenoids (such as the sitosterols) are widespread while others (e.g., pentacyclic triterpenes) show taxonomic value only at a higher hierarchical level.

In the Lamiaceae, oleanolic acids and ursolic acids are common constituents. Derivatives of ursolic acid are characteristic of the Nepetoideae subfamily, principally in the genera *Teucrium*, *Lavandula*, and *Rosmarinus*, while in *Ajuga* spp. steroidal compounds such as ecdysone (**126**) and ajugasterone (**127**) are found (Cole 1992). Species with high oleanolic acid levels are mainly belonging to the *Salvia* genus and *Nepeta* x *faassenii*, whereas low oleanolic acid contents were found in *Marrubium alysson* and *Marrubium thessalum*. In general, the subfamily Lamioideae appears to be poorer in both oleanolic acids and ursolic acids than the subfamily Nepetoideae (Janicsak et al. 2006). The Macaronesian *Sideritis* normally contains pentacyclic triterpenes including rhoiptelenol (**128**), α-amyrin (**129**), and obtusifoliol (**130**) (Fraga et al. 2009).

In the Asteraceae, in the roots of the genus *Nannoglottis*, cycloartane-type, ursane-type, oleanane-type, and A-frideooleanane-type triterpenes are present (Meng et al. 2014). Triterpenes from *Saussurea* spp. comprise approximately 26 compounds of four types: lanostane, oleanane, ursane, and lupane (Wang et al. 2010b), whereas in *Tanacetum vulgare* monohydroxy triterpene alcohols and sterols are present (Wilkomirski and Kucharska 1992). Steroidal sapogenins with atypical structures such as pogosterol (**131**) are typical of some species of the *Vernonia* genus (Mungarulire et al. 1993), whereas triterpenoid sapogenin lactones are typical of *Grindelia* species (Kreutzer et al. 1990). These compounds demonstrated having deep relevance in chemotaxonomic surveys.

In the Leguminosae, most saponins are of the triterpene type, with steroidal saponins being rare. Only cardenolides are restricted to a few (but not all) members of the genera *Coronilla* and *Securigera* (Wink 2013).

Triterpenes are clearly the main chemotaxonomic markers for the major Ericales studied families. The triterpene skeletal types including oleanane, ursane, and lupane are found, especially in Marcgraviaceae, Lecythidaceae, Sapotaceae, and Theaceae. One of the most important classes of metabolites in the Ericales is represented by the saponins, especially of the triterpene type, and their occurrence is very common in many families of this order (Rocha et al. 2015).

Flower buds of some *Camellia sinensis* varieties produce typical oleanane-type triterpene oligoglycosides, floraassamsaponin I (**132**) (Ohta et al. 2015), whereas in the Rubiaceae family the pentacyclic triterpenes ursolic acids from the leaves and acyl lupeols from the stem bark have been reported from many species (Choze et al. 2010).

The chemosystematic significance of triterpenoid accumulation in the Meliaceae shows a wide distribution of seco-dammarane derivatives in *Aglaia* species. In addition, various biogenetic trends toward cycloartane, tirucallane, apotirucallane, and lupane basic skeletons have been described for the genus (Joycharat et al. 2008).

Triterpenes are also present in the epicuticular layers, mixed with waxes and alkanes. In Palmae there are several triterpenes with lupanic skeleton, which are characteristic in *Butia* and *Orbignya* genera (Garcia et al. 1995), whereas triterpenes were detected in waxes of *Sedum* species, the major triterpenes being β-amyrenyl acetate (**133**) (Stevens et al. 1994).

3.3.2.5 Tetraterpenes

The most common compounds in this class of terpenoids are carotenoids. In general, flower coloration in the range orange-red to red is determined by carotenoids; however, since their presence is virtually in all plant parts they are not usually involved in chemotaxonomic investigations. However, their distribution in flower petals is not universal and thus can be taken into account to solve some taxonomic problems. Excellent results were obtained by analyzing carotenoids in petals of the Rosaceae and the Asteraceae. The compounds used for these studies are the α- and the β-carotene, lutein, flavoxantine, and violaxanthine (Maffei 2015). In the Leguminosae, species belonging to the *Acacia* genus have only β-carotene (**134**) and lutein (**135**), while in the morphologically similar genus *Ulex*, taraxanthin (**136**) is also found. Genera such as *Genista*, *Cytisus*, and *Laburnum* are difficult to be discerned merely on morphological basis, but possess clear-cut patterns of variation in carotenoids which can be helpful for taxonomic investigations on their tribe, the Genisteae. Finally, studies on *Astragalus* designate the rubixanthin (**137**) as an outstanding chemotaxonomic marker because it is only found in this genus (Maffei 2015).

3.3.2.6 Polyterpenes

Two polymers belong to the polyterpenes class: rubber and gutta-percha. The latter looks like a brown, tough mass and is inelastic. The gutta-percha is typical of *Palaquium* trees (*P. obtusifolium*, *P. borneense*) native of Malaysia, while another kind of rubber, called gutta-balata rubber, is obtained by fermentation from *Mimusops balata*. Both *Mimusops* and Palaquium genus belong to *Sapotaceae* family.

Figure 3.4 shows the chemical structure of some representative diterpenes, triterpenes, and tetraterpenes.

3.3.3 Chemotaxonomy of specialized products containing nitrogen

Numerous specialized metabolites contain in their carbon skeletons one or more nitrogen atoms. Among the most important classes are alkaloids, glucosinolates, cyanogenic glycosides, and several nonprotein amino acids. The alkaloids are nitrogen heterocyclic compounds with a powerful pharmacological action, while the betalains are red or yellow alkaloids only found in the *Caryophyllales* and absent in plants producing anthocyanins (Brockington et al. 2011). Glucosinolates, mustard-oil glycosides, are compounds that possess a sulfur atom in addition to nitrogen. These compounds are only produced by Capparales (Zenk and Juenger 2007). The cyanogenic glycosides elicit a toxic compound (HCN) and are found in several food plants such as cassava. The nonprotein amino acids are widespread in the plant kingdom and play an ecologically crucial role in chemical defense.

3.3.3.1 Alkaloids

A taxonomic criterion applied to the study of the alkaloids is that of Waterman (1998), based on the origin of the amino acid precursor. The first category of alkaloids to be considered are the derivatives of tyrosine (and phenylalanine), formed because of decarboxylation of dihydroxyphenylalanine (DOPA). This group includes 1-benzylisoquinoline alkaloids, alkaloids of the Amaryllidaceae, alkaloids of *Erythrina* spp., and betalains. In the Rutaceae, benzylisoquinoline alkaloids were used at the genus level to outline a group of species within the family. Unfortunately, other attempts to use these compounds in chemotaxonomic analysis as distinctive traits of related groups of plant families failed. For instance, Rutaceae, Papaveraceae, and Berberidineae are related families containing benzylisoquinoline alkaloids, but these compounds are also found in the Euphorbiaceae, Buxaceae, and Rhamnaceae, which are phylogenetically separated from the previous families (Waterman 1998). However, some classes of benzylisoquinoline alkaloids are specific to some plant families only, such as the dibenzazonine alkaloids of the Menispermaceae and Leguminosae or spiroisoquinoline alkaloids of the Fumariaceae and

Chapter three: Biodiversity and chemotaxonomic significance

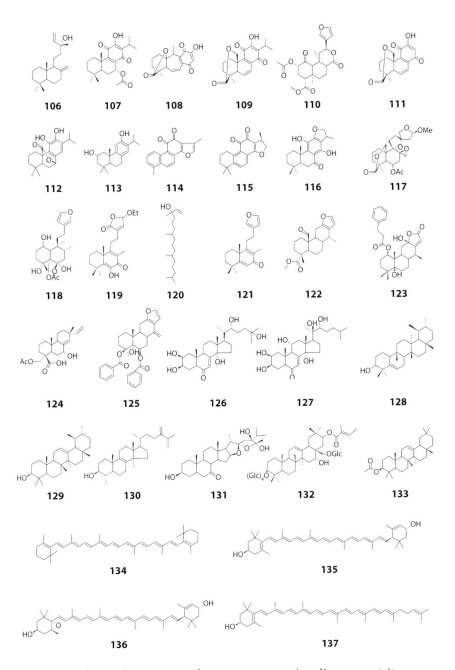

Figure 3.4 Chemical structures of some representative diterpenes, triterpenes, and tetraterpenes.

Papaverinae. In the Amaryllidaceae, a family of monocots angiosperms, there are alkaloids that share the same precursors of benzylisoquinoline alkaloids. Groups of plants producing these alkaloids, for instance norpluvine (**138**), separate from those that encompass benzylisoquinoline alkaloids, thus confirming the drift of monocotyledons from ancestors similar to Polycarpicae (Waterman 1998). Recently, several studies have dealt with alkaloids extracted from Leguminosae of the genus *Erythrina*, broadening the knowledge about quali-quantitative variations in species evolved from geographically distant areas. Erythroidine alkaloids, such as α-erythroidine (**139**), are present in the stem bark of *Erythrina poeppigiana* (Djiogue et al. 2014), whereas the dimeric *Erythrina* alkaloid, spirocyclic (6/5/6/6) erythrivarine A (**140**), has been isolated from *Erythrina variegata* (Zhang et al. 2014).

As already mentioned, the betalains are typical of Centrospermae (Caryophyllales) and are commonly found in Aizoaceae, Amaranthaceae, Basellaceae, Cactaceae, Chemopodiaceae, Didiereaceae, Nyctaginaceae, Phytolaccaceae, and Portulacaceae, but are absent in Caryophyllaceae and Molluginaceae. The chemotaxonomic value of betalains proved to be particularly useful in the classification of some Caryophyllales species of uncertain taxonomic position. However, there are still concerns in separating Caryophyllales into two groups according to the presence/absence of betalains because all other taxonomic criteria concur to treat them as one single and indivisible order (Brockington et al. 2011).

The second alkaloid category consists of indole-secologanin derivatives. The secologanin (**141**) is an iridoid monoterpene able to combine with an amine (usually the tryptamine) to form alkaloids. The indole-secologanin alkaloids are found in the Loganiaceae, Apocynaceae, and Rubiaceae families and in some taxa of the Cornales order. The basic structures of these alkaloids are of three types: I, II, and III whose most important representatives are ajmalicine (**142**), vincadifformine (**143**), and eglandine (**144**), respectively. Type III alkaloids are only found in Apocynaceae, making these molecules useful for chemotaxonomic classifications at the family or superior level. In many cases, the ability to accumulate these alkaloids is subordinate to possessing the biosynthetic pathway for the formation of seco- and iridoids precursors (Waterman 1998). The Asclepiadaceae are unfit to synthesize iridoids and indole-loganine alkaloids, instead they produce steroidal cardenolides, a class of compounds detectable only in a few genera belonging to the Apocinaceae (Maffei 2015).

Derivatives of anthranilic acid compose the third alkaloids category. These alkaloids are circumscribed to a large family, the Rutaceae, where furoquinoline are broadly represented. Unfortunately, this latter class of compounds also occurs occasionally in other families such as the Solanaceae, Asclepiadaceae, and Apocinaceae, while acridones are typical of Rutaceae only (Waterman 1998).

The alkaloids derived from the amino acids lysine and ornithine exemplify the fourth class of compounds. Tropane alkaloids belong to this group, and they are typical of the Solanaceae, Erythroxylaceae, Proteaceae, Euphorbiaceae, Rhizophoraceae, Convolvulaceae, and Cruciferae, but are also found, though rarely, in other families. Tropane alkaloids are characteristic of the genera *Datura*, *Brugmansia* (tree datura), and *Duboisia* of the Solanaceae (Griffin and Lin 2000). Chemotaxonomic studies demonstrated that the tropanic nucleus has evolved several times during the angiosperms phylogeny. However, in distinct plant families these alkaloids differ for substitutions and stereochemical arrangements on the structure of the tropanic core. Nevertheless, the use of these molecules is primarily at lower levels of the taxonomic hierarchy (Waterman 1998).

In addition, the pyrrolizidine alkaloids belong to the fourth category. They are a group of molecules formed by approximately 360 structures with a distribution restricted to certain taxa (Jenett-Siems et al. 1998; Langel et al. 2011; Trigo et al. 2003; Wink 2013). On the basis of taxonomical and biogenetic implications, we can distinguish five main classes of pyrrolizidine alkaloids: type I (or senecionine type) is typical of the tribe of Senecioneae (Asteraceae) and is formed by a group of more than 100 compounds in which necine base and necic acid form macrocyclic diesters; type II (or triangularine type) consists of more than 50 different structures and is typical of the Asteraceae and Boraginaceae families and is represented by open-chain diesters; type III (structurally similar to type I, moncrotaline type) comprises approximately 30 structures and is characteristic of the Fabaceae; type IV (or lycopsamine type) is made of monoesters scattered in more than 100 molecules of the Eupatorieae (Asteraceae) tribe; finally, type V (or phalaenopsine type) is typical of the Orchidaceae (Hartmann 1999).

According to many authors, the chemotaxonomic significance of quinolizidine alkaloids is high because they are primarily found in Sophoreae, Padalyrieae (*sensu lato*), and Genisteae (Papilionoideae, Leguminosae) tribes. In addition, chemotaxonomic studies allow to separate basal shrubs species that accumulate alkaloids such as racemic sparteine (**145**) and/or matrine (**146**), from more recent species that produce (–)-sparteine (Waterman 1998). The tribe Sophoreae contains the most diverse forms of quinolizidine alkaloids, and the variation in this compound production was also used to separate species of genus *Lupinus* (Wink et al. 1995). Other families able to produce these alkaloids are Berberidaceae, Ranunculaceae, Solanaceae, Chenopodiaceae, and Rubiaceae. For instance, the alkaloid pattern of *Leontice leontopetalum* (Berberidaceae) is characterized by quinolizidine alkaloids of the lupanine-type with lupanine (**147**) as the main compound, whereas *Leontice ewersmannii* accumulates quinolizidine alkaloids of the matrine-type and the α-pyridone-type as major compounds (Gresser et al. 1993). A chemical dichotomy was demonstrated for the species of the section Spartioides of the genus *Genista* (Fabaceae: Genisteae): one group of species contained the α-pyridone alkaloids cytisine (**148**), N-methylcytisine, and anagyrine (**149**) as major alkaloids, while the other group contained lupanine (**147**), 13-hydroxylupanine (**150**), and its esters as main compounds (Greinwald et al. 1995). Finally, the quinolizidine alkaloids retamine (**151**), 17-oxoretamine (**152**), and 12-α-hydroxylupanine (**153**) were detected in the aerial parts of *Genista ephedroides* (Pistelli et al. 2001), whereas jussiaeiine A (**154**) is recognized as a marker of the genus *Ulex* (Maximo et al. 2006).

3.3.3.2 Glucosinolates

As we already mentioned, this category of nitrogen compounds is restricted to Capparales order and show a fully dissimilar structure in the Brassicaceae family (Zenk and Juenger 2007). These compounds are extremely useful at low levels of the classification hierarchy and are identified primarily through gas or paper chromatography. In the genus *Aurinia* (Brassicaceae), glucoalyssin (**155**), glucobrassicanapin (**156**), and glucoberteroin (**157**) are the major compounds (Blazevic et al. 2013), whereas gluconasturtiin (**158**) and glucobrassicin (**159**) are typical of *Barbarea* (Brassicaceae) species (Agerbirk et al. 2003). Gluconapin (**160**) and progoitrin (**161**) are characteristic of seeds of oilseed turnip (*Brassica rapa* var *oleifera*) collected from several different geographic regions (Davik and Heneen 1993), whereas glucotropaeolin (**162**) is present in some *Lepidium* (Brassicaceae) species (Radulovic et al. 2008).

3.3.3.3 Cyanogenic glycosides

Cyanogenesis (i.e., the production of HCN from damaged plant tissue) requires the presence of two biochemical pathways, one that controls the synthesis of the cyanogen glycoside and the other that controls the production of a specific, hydrolyzing β-glucosidase. The cyanogenic glycosides of *Eucalyptus nobilis* is prunasin (**163**) (D-mandelonitril β-D-glucoside) (Gleadow et al. 2003), whereas a cyanogenic glucoside, 6'-O-galloyl sambunigrin (**164**), is present in the tropical tree *Elaeocarpus* sericopetalus (Elaeocarpaceae). This is the first formal characterization of a cyanogenic constituent in the Elaeocarpaceae family and the second for Malvales order. Moreover, *E. sericopetalus* contains the highest foliar concentrations of cyanogenic glycosides known for leaves of any tree (Miller et al. 2006b). Cyanogenic glycosides lucumin (**165**) and prunasin (**163**) have been reported for the first time in the Lamiaceae family, from foliage of the rare Australian endemic rainforest tree *Clerodendrum grayi* (Miller et al. 2006a), whereas passibiflorin (**166**), a bisglycoside containing the 6-deoxy-β-D-gulopyranosyl residue, was isolated from several *Passiflora* (Passifloraceae) species (Jaroszewski et al. 2002).

3.3.3.4 Nonprotein amino acids

While 20 amino acids are involved directly in protein structure, there are thousands of others, the nonprotein amino acids (NPAAs), which play no such role. However, nonprotein amino acids have various meanings in ecology and survival strategies of the producing plants and have been used for chemotaxonomic purposes as well (Vranova et al. 2011). In plants, NPAAs possess different roles including antiherbivory, antimicrobial and allelochemical activity, protection against stress, signaling, nitrogen storage, and as toxins against invertebrates and vertebrates (Bell 2003). There are hundreds of different structures in several families, some of which are quite common as intermediate metabolites.

Some structures are only found in a few plant families, while others are widespread among living organisms. For example, the γ-methyleneglutamic acid (**167**) is present in the Leguminosae, Liliaceae, and Cannabinaceae while the canavanine (**168**) has a very restricted distribution in the subfamily of Lotoideae (Leguminosae). For instance, the lathyrine (**169**) is only present in the genus *Lathyrus*. Among nonprotein amino acids, it is worthy to mention the pipecolic acid derivatives, many hydroxylated derivatives are exclusively found in Mimosoideae and are restricted to a few genera, with particular reference to the *Inga* genus. In the Leguminosae, there is a perfect match between the classification based on morphological cues and those using nonprotein amino acids as chemotaxonomic markers. This means that the mutations occurred in various biosynthetic pathways for the production of amino acids contributed, as mutations that have changed the morphology, to the evolution of Leguminosae (Bell 1971).

In seeds of several Australian *Acacia* species the major NPAAs are djenkolic acid (**170**), mimosine (**171**), and lanthionine (**172**) (Boughton et al. 2015), whereas azetidine-2-carboxylic acid (**173**) is a toxic and teratogenic nonprotein amino acid found in roots of sugar beets and garden beets (*Beta vulgaris*) (Rubenstein et al. 2006, 2009). The relationship between several *Bocoa* species with *Ateleia* and *Cyathostegia* is supported by the presence of the rare *Ateleia*-type nonprotein amino acid 2,4-methanoproline (**174**) (Kite and Ireland 2002), whereas O-oxalylhomoserine (**175**) was isolated from the aerial parts of *Lathyrus latifolius* (Bell et al. 1996).

3.3.4 Chemotaxonomic significance of fatty acids and surface alkanes

Among the lipophilic compounds the most used in chemotaxonomic studies are the fatty acids contained in seeds and leaves and the alkanes found in epicuticular layers. Seed fatty acids are in the form of oils, primarily as esters of glycerol and with carbon chains of variable length. The variability of fatty acids does not depend only on the number of carbon atoms, but also on the number of unsaturations, the presence of hydroxyl substituents or epoxy, allenic, and acetylenic groups.

The efficacy of seed fatty acids for taxonomical surveys has been proved at the genus and species level, while these compounds lose their chemotaxonomic meaning at the family or higher hierarchical level. The most abundant fatty acids are those with 16 and 18 carbon atoms, that is, palmitic acid (and its unsaturated derivatives, palmitoleic, and palmitolenic) and stearic acid (and its unsaturated derivatives oleate, linoleate, and linolenate).

In the Leguminosae, the linoleic acid (**176**) percentage varies between 40% and 67%, while fatty acids with chains of less than 14 carbon atoms are rare; in species belonging to the Taxaceae, Pinaceae, Taxodiaceae, and Cupressaceae, α-linolenic acid has been used as a chemotaxonomic marker to distinguish between two groups of families (Wolff and Bayard 1995; Wolff et al. 1996). Moreover, in conifer seeds, six Δ5-olefinic acids may occur in variable proportions depending on the family. These are the taxoleic (**177**), pinolenic (**178**), and sciadonic (**179**) acids (all double bonds in the *cis* configuration). In the Pinaceae and Cupressaceae, the seed fatty acids are rich in Δ5-olefinic acids, including acid of 20 carbon atoms with two, three, and four unsaturations. Chemotaxonomic grouping of the main Pinatae families (Pinaceae, Taxodiaceae, Taxaceae, and Cupressaceae) was based on the data obtained from these molecules, using multivariate statistical analysis (principal component analysis and discriminant analysis). With this procedure, it was possible to further discriminate many Pinaceae genera (*Pinus, Abies, Cedrus, Piceae,* and *Larix*) (Wolff et al. 1997). Unusual fatty acids of the seed are not always found in leaves and fruit of the same species. Thus, the study of "unusual" fatty acids becomes a powerful tool for chemotaxonomic investigations, as demonstrated in the Ranunculaceae, where some Δ-6 fatty acids were only found in one genus or a few closely related genera (Aitzetmuller and Tsevegsuren 1994).

Free fatty acids and fatty acids esterified with mono-, di-, and triglycerides are present in the nonpolar fraction of some Lamiaceae leaves (Maffei and Scannerini 1993). In the genus *Mentha,* fatty acids show several degrees of interspecific variability depending on the class of lipids analyzed (mono-, di-, or triglycerides) and the organ from which they are extracted (leaves, stems, or flowers). Hence, these compounds proved to be very useful in chemotaxonomic studies at low-level hierarchical classification (Maffei and Scannerini 1992a,b). The analysis carried out on *Lavandula* hybrids highlighted the existence of distinct chemotypes within a large population of wild plants scattered on alpine altitudinal gradients. In contrast, at higher hierarchical levels leaf and flower fatty acids show no taxonomic significance (Maffei and Peracino 1993).

Chapter three: Biodiversity and chemotaxonomic significance

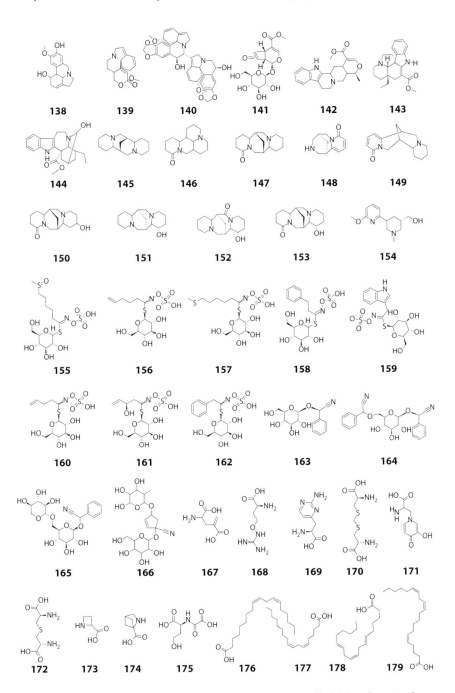

Figure 3.5 Chemical structures of some representative alkaloids, glucosinolates, cyanogenic glycosides, and fatty acids.

In the early 1960s, Eglinton published a series of works on the taxonomic significance of alkanes deposited as epicuticular layers on plant surfaces (Eglinton et al. 1962). The power of these substances is mainly due to their presence not only in plants, but also in bacteria, fungi, and animal kingdoms. Recently, a survey of more than 550 species belonging to several plant families supported the hypothesis of Eglinton, at family, subfamily, and tribe levels (Maffei et al. 2004). Epicuticular alkanes have a lower discriminating power at the species level because they vary considerably depending on the developmental stage of the plant, as well as on environmental factors (Maffei et al. 1993b).

A study on about nine main families of angiosperms reported that, in general, the chemical composition of epicuticular alkanes consists of linear and branched chains of 21–36 carbon atoms. In the Leguminosae and Apiaceae, 21–25 carbon atoms alkanes are the prevailing compounds (Maffei 1996c), whereas n-alkanes and iso-alkanes with a higher number of carbon atoms (34–36) are characteristic of the Gramineae (Maffei 1996b). The Boraginaceae and Solanaceae mostly produce alkanes with 29 carbon atoms, the Lamiaceae and Verbenaceae are characterized by alkanes of both 31 and 33 carbon atoms. Whereas in the Compositae and Scrophulariaceae alkanes, 31 carbon atoms prevail (Maffei 1994, 1996a).

Figure 3.5 shows the chemical structure of some representative alkaloids, glucosinolates, cyanogenic glycosides, and fatty acids.

References

Agerbirk, N., M. Orgaard, and J.K. Nielsen. 2003. Glucosinolates, flea beetle resistance, and leaf pubescence as taxonomic characters in the genus *Barbarea* (Brassicaceae). *Phytochemistry* 63: 69–80.

Aghaei, Y., M. Hossein Mirjalili, and V. Nazeri. 2013. Chemical diversity among the essential oils of wild populations of *Stachys lavandulifolia* VAHL (Lamiaceae) from Iran. *Chem. Biodivers.* 10: 262–273.

Aitzetmuller, K. and N. Tsevegsuren. 1994. Seed fatty-acids, front-end-desaturases and chemotaxonomy—A case study in the Ranunculaceae. *J. Plant Physiol.* 143: 538–543.

Alavez-Solano, D., R. Reyes-Chilpa, M. Jimenez-Estrada, F. Gomez-Garibay, I. Chavez-Uribe, and M. Sousa-Sanchez. 2000. Flavanones and 3-hydroxyflavanones from *Lonchocarpus oaxacensis*. *Phytochemistry* 55: 953–957.

Al-Qudah, M.A. 2013. Chemical composition of the essential oil from Jordanian *Lupinus varius* L. *Arab. J. Chem.* 6: 225–227.

Ashitani, T., S.S. Garboui, F. Schubert et al. 2015. Activity studies of sesquiterpene oxides and sulfides from the plant *Hyptis suaveolens* (Lamiaceae) and its repellency on *Ixodes ricinus* (Acari: Ixodidae). *Exp Appl. Acarol.* 67: 595–606.

Bajalan, I. and A.G. Pirbalouti. 2015. Variation in chemical composition of essential oil of populations of *Lavandula x intermedia* collected from Western Iran. *Ind. Crop Prod.* 69: 344–347.

Balandrin, M.F., S.M. Lee, and J.A. Klocke. 1988. Biologically active volatile organosulfur compounds from seeds of the neem tree, *Azadirachta indica* (Meliaceae). *J. Agr. Food Chem.* 36: 1048–1054.

Baser, K.H.C., M. Kurkcuoglu, B. Demirci, T. Ozek, and G. Tarimcilar. 2012. Essential oils of *Mentha* species from Marmara region of Turkey. *J. Essent. Oil Res.* 24: 265–272.

Bassi, P., A. Basile, A. Stefanini et al. 1995. Effects of lead on the nuclear repetitive DNA of the moss *Funaria hygrometrica* (Bryophyta). *Protoplasma* 188: 104–108.

Bates-Smith, E.C. 1962. The phenolic constituents of plants and their taxonomic significance. I. Dicotyledons. *Bot. J. Linn. Soc.* 58: 95–153.

Batish, D.R., H.P. Singh, M. Kaur, R.K. Kohli, and S. Singh. 2012. Chemical characterization and phytotoxicity of volatile essential oil from leaves of *Anisomeles indica* (Lamiaceae). *Biochem. Syst. Ecol.* 41: 104–109.

Bell, E.A. 1971. Comparative biochemistry of non-protein amino acids. In *Chemotaxonomy of the Leguminosae*, eds. J.B. Harborne, D. Boutler, and B.L. Turner, 179–206. London: Academic Press.

Bell, E.A. 2003. Nonprotein amino acids of plants: Significance in medicine, nutrition, and agriculture. *J. Agr. Food Chem.* 51: 2854–2865.

Bell, E.A., K.P.W.C. Perera, P.B. Nunn, M.S.J. Simmonds, and W.M. Blaney. 1996. Non-protein amino acids of *Lathyrus latifolius* as feeding deterrents and phagostimulants in *Spodoptera littoralis*. *Phytochemistry* 43: 1003–1007.

Bella, S., T. Tuttolomondo, G. Dugo et al. 2015. Composition and variability of the essential oil of the flowers of *Lavandula stoechas* from various geographical sources. *Nat. Prod. Commun.* 10: 2001–2004.

Bicchi, C. and M.E. Maffei. 2012. The plant volatilome: Methods of analysis. In *High Throughput Phenotyping in Plants. Methods and Protocols*, ed. J. Normanly, 289–310. Totowa, NJ: Humana Press.

Bisht, D.S., S.C. Joshi, R.C. Padalia, and C.S. Mathela. 2012. Isoiridomyrmecin rich essential oil from *Nepeta erecta* Benth. and its antioxidant activity. *Nat. Prod. Res.* 26: 29–35.

Bisht, D.S., R.C. Padalia, S.C. Joshi, K.K. Singh, and C.S. Mathela. 2008. Sesquiterpene hydrocarbons rich essential oil of *Stachys sericea* wall. *J. Essent. Oil Bear. Pl.* 11: 586–590.

Blagojevic, P.D., N.S. Radulovic, and D. Skropeta. 2015. (Chemotaxonomic) implications of postharvest/storage-induced changes in plant volatile profiles—The case of *Artemisia absinthium* L. essential oil. *Chem. Biodivers.* 12: 1237–1255.

Blazevic, I., G.R. De Nicola, S. Montaut, and P. Rollin. 2013. Glucosinolates in two endemic plants of the Aurinia genus and their chemotaxonomic significance. *Nat. Prod. Commun.* 8: 1463–1466.

Boughton, B.A., P. Reddy, M.P. Boland, U. Roessner, and P. Yates. 2015. Non-protein amino acids in Australian acacia seed: Implications for food security and recommended processing methods to reduce djenkolic acid. *Food Chem.* 179: 109–115.

Braunberger, C., M. Zehl, J. Conrad et al. 2015. Flavonoids as chemotaxonomic markers in the genus Drosera. *Phytochemistry* 118: 74–82.

Brockington, S.F., R.H. Walker, B.J. Glover, P.S. Soltis, and D.E. Soltis. 2011. Complex pigment evolution in the Caryophyllales. *New Phytol.* 190: 854–864.

Bruno, M., S. Rosselli, M.L. Bondi, T.E. Gedris, and W. Herz. 2002. Sesquiterpene lactones of *Anthemis alpestris*. *Biochem. Syst. Ecol.* 30: 891–895.

Buchs, W. 2003. Biodiversity and agri-environmental indicators—General scopes and skills with special reference to the habitat level. *Agr. Ecosyst. Environ.* 98: 35–78.

Buhagiar, J.A., M.T. Camilleri-Podesta, P. Cioni, G. Flamini, and L. Pistelli. 2015. Essential oil composition of summer and winter foliage of *Chiliadenus bocconei*. *Nat. Prod. Commun.* 10: 1323–1324.

Capuzzo, A., M.E. Maffei, and A. Occhipinti. 2013. Supercritical fluid extraction of plant flavors and fragrances. *Molecules* 18: 7194–7238.

Caramiello, R., A. Bocco, G. Buffa, and M. Maffei. 1995. Chemotaxonomy of *Juniperus communis*, *J. sibirica* and *J. intermedia*. *J. Essent. Oil Res.* 7: 133–145.

Chen, Y., Y.G. Wu, Y. Xu et al. 2014. Dynamic accumulation of sesquiterpenes in essential oil of *Pogostemon cablin*. *Rev. Bras. Farmacogn.* 24: 626–634.

Choze, R., P.G. Delprete, and L.M. Liao. 2010. Chemotaxonomic significance of flavonoids, coumarins and triterpenes of *Augusta longifolia* (Spreng.) Rehder, Rubiaceae-Ixoroideae, with new insights about its systematic position within the family. *Rev. Bras. Farmacogn.* 20: 295–299.

Clark, B.R., B.J. Bliss, J.Y. Suzuki, and R.P. Borris. 2014. Chemotaxonomy of Hawaiian *Anthurium* cultivars based on multivariate analysis of phenolic metabolites. *J. Agr. Food Chem.* 62: 11323–11334.

Cole, M.D. 1992. The significance of the terpenoids in the Labiatae. In *Advances in Labiatae Science*, eds. R.M. Harley and T. Reynolds, 315–324. Kew: Royal Botanic Gardens.

Crawford, D.J. 1978. Flavonoid chemistry and angiosperm evolution. *Bot. Rev.* 44: 421–456.

Crawford, D.J., M.E. Mort, and J.K. Archibald. 2005. Biosystematics, chromosomes and molecular data: Melding the old and the new. *Taxon* 54: 285–289.

Cropper, A. 1993. Convention on biological diversity. *Environ Conserv* 20: 364–364.

Custodio, L., H. Serra, J.M.F. Nogueira, S. Goncalves, and A. Romano. 2006. Analysis of the volatiles emitted by whole flowers and isolated flower organs of the carob tree using HS-SPME-GC/MS. *J. Chem. Ecol.* 32: 929–942.

Dai, Y.Q., L. Liu, G.Y. Xie et al. 2014. Four new eudesmane-type sesquiterpenes from the basal leaves of *Salvia plebeia* R. Br. *Fitoterapia* 94: 142–147.

Davik, J. and W.K. Heneen. 1993. Identification of oilseed turnip (*Brassica rapa* L var *oleifera*) cultivar groups by their fatty-acid and glucosinolate profiles. *J. Sci. Food Agr.* 63: 385–390.

Deguerry, F., L. Pastore, S.Q. Wu, A. Clark, J. Chappell, and M. Schalk. 2006. The diverse sesquiterpene profile of patchouli, *Pogostemon cablin*, is correlated with a limited number of sesquiterpene synthases. *Arch. Biochem. Biophys.* 454: 123–136.

Delazar, A., M. Byres, S. Gibbons et al. 2004. Iridoid glycosides from *Eremostachys glabra*. *J. Nat. Prod.* 67: 1584–1587.

Dimitrova-Dyulgerova, I., P. Merdzhanov, K. Todorov et al. 2015. Essential oils composition of *Betonica officinalis* L. and *Stachys sylvatica* L. (Lamiaceae) from Bulgaria. *Cr. Acad. Bulg. Sci.* 68: 991–998.

Ding, W.B., L.D. Lin, M.F. Liu, and X.Y. Wei. 2011. Two new sesquiterpene glycosides from *Pogostemon cablin*. *J. Asian Nat. Prod Res.* 13: 599–603.

Djabou, N., A. Muselli, H. Allali et al. 2012. Chemical and genetic diversity of two Mediterranean subspecies of *Teucrium polium* L. *Phytochemistry* 83: 51–62.

Djiogue, S., M. Halabalaki, D. Njamen et al. 2014. Erythroidine alkaloids: A novel class of phytoestrogens. *Planta med.* 80: 861–869.

Edwards, J.E., P.N. Brown, N. Talent, T.A. Dickinson, and P.R. Shipley. 2012. A review of the chemistry of the genus *Crataegus*. *Phytochemistry* 79: 5–26.
Eglinton, G., A.G. Gonzalez, R.J. Hamilton, and R.A. Raphael. 1962. Hydrocarbon constituents of the wax coatings of plant leaves: A taxonomic survey. *Phytochemistry* 1: 89–102.
Elamrani, A., S. Zrira, B. Benjilali, and M. Berrada. 2000. A study of Moroccan rosemary oils. *J. Essent. Oil Res.* 12: 487–495.
Emerenciano, V.R., K.O. Barbosa, M.T. Scotti, and M.J.R. Ferreira. 2007. Self-organizing maps in chemotaxonomic studies of Asteraceae: A classification of tribes using flavonoid data. *J. Braz. Chem. Soc.* 18: 891–899.
Fathiazad, F. and S. Hamedeyazdan. 2011. A review on *Hyssopus officinalis* L.: Composition and biological activities. *Afr. J. Pharm. Pharmacol.* 5: 1959–1966.
Fraga, B.M., M.G. Hernandez, C. Fernandez, and J.M.H. Santana. 2009. A chemotaxonomic study of nine Canarian *Sideritis* species. *Phytochemistry* 70: 1038–1048.
Franco, C.R.P., P.B. Alves, D.M. Andrade et al. 2011. Essential oil composition and variability in *Hyptis fruticosa*. *Rev. Bras. Farmacogn.* 21: 24–32.
Garcia, S., H. Heinzen, C. Hubbuch, R. Martinez, X. Devries, and P. Moyna. 1995. Triterpene methyl ethers from Palmae epicuticular waxes. *Phytochemistry* 39: 1381–1382.
Giannasi, D.E. 1978. Systematic aspects of flavonoid biosynthesis and evolution. *Bot. Rev.* 44: 339–429.
Gleadow, R.M., A.C. Veechies, and I.E. Woodrow. 2003. Cyanogenic *Eucalyptus nobilis* is polymorphic for both prunasin and specific beta-glucosidases. *Phytochemistry* 63: 699–704.
Gnavi, G., C.M. Bertea, and M.E. Maffei. 2010a. PCR, sequencing and PCR-RFLP of the 5S-rRNA-NTS region as a tool for the DNA fingerprinting of medicinal and aromatic plants. *Flav. Fragr. J.* 25: 132–137.
Gnavi, G., C.M. Bertea, M. Usai, and M.E. Maffei. 2010b. Comparative characterization of *Santolina insularis* chemotypes by essential oil composition, 5S-rRNA-NTS sequencing and EcoRV RFLP-PCR. *Phytochemistry* 71: 930–936.
Grayer, R.J., M.R. Eckert, A. Lever, N.C. Veitch, G.C. Kite, and A.J. Paton. 2010. Distribution of exudate flavonoids in the genus *Plectranthus*. *Biochem. Syst. Ecol.* 38: 335–341.
Grayer, R.J., G.C. Kite, F.J. Goldstone, S.E. Bryan, A. Paton, and E. Putievsky. 1996. Infraspecific taxonomy and essential oil chemotypes in sweet basil, *Ocimum basilicum*. *Phytochemistry* 43: 1033–1039.
Greinwald, R., I. Vanrensen, M. Veit, P. Canto, and L. Witte. 1995. A chemical dichotomy in quinolizidine alkaloid accumulation within the section *Spartioides* of the genus *Genista* (Fabaceae, Genisteae). *Biochem. Syst. Ecol.* 23: 89–97.
Gresser, G., P. Bachmann, L. Witte, and F.C. Czygan. 1993. Distribution and taxonomic significance of quinolizidine alkaloids in *Leontice leontopetalum* and *L. ewersmannii* (Berberidaceae). *Biochem. Syst. Ecol.* 21: 679–685.
Griffin, W.J. and G.D. Lin. 2000. Chemotaxonomy and geographical distribution of tropane alkaloids. *Phytochemistry* 53: 623–637.
Gulsoy, S. 2012. Evaluation of essential oils and phenolic compounds of some *Origanum* (Labiatae/Lamiaceae) taxonomy. *Asian J. Chem.* 24: 2479–2483.
Hadacek, F. 2002. Secondary metabolites as plant traits: Current assessment and future perspectives. *Crit. Rev. Plant Sci.* 21: 273–322.

Hammami, S., R. El Mokni, K. Faidi et al. 2015. Chemical composition and antioxidant activity of essential oil from aerial parts of *Teucrium flavum* L. subsp. *flavum* growing spontaneously in Tunisia. *Nat. Prod. Res.* 29: 2336–2340.
Hanson, J.R. 2015. Diterpenoids of terrestrial origin. *Nat. Prod. Res.* 32: 1654–1663.
Harborne, J.B. 1971. Distribution of flavonoids in the Leguminosae. In *Chemotaxonomy of the Leguminosae*, eds. J.B. Harborne, D. Boulter, and B.L. Turner, 31–71. London: Academic Press.
Harborne, J.B. 1993. *Introduction to Ecological Biochemistry*. London: Academic Press.
Harborne, J.B. 1994. Chemotaxonomy of anthocyanins and phytoalexins in the Compositae. In *Compositae: Systematics. Proceedings of the International Compositae Conference, Kew*, eds. D.J.N. Hind and H.J. Beentje, 207–218. Kew: Royal Botanic Gardens.
Hartmann, T. 1999. Chemical ecology of pyrrolizidine alkaloids. *Planta* 207: 483–495.
Hegnauer, R. and R.J. Grayerbarkmeijer. 1993. Relevance of seed polysaccharides and flavonoids for the classification of the leguminosae—A chemotaxonomic approach. *Phytochemistry* 34: 3–16.
Hegnauer, R. and M. Hegnauer. 2001. *Chemotaxonomie der Pflanzen*. Basel: Birkhäuser.
Hernandez-Bello, R., R.V. Garcia-Rodriguez, K. Garcia-Sosa et al. 2015. Salvinorin A content in legal high products of *Salvia divinorum* sold in Mexico. *Forensic. Sci. Int.* 249: 197–201.
Heywood, V.H. and R.T. Watson. 1995. *Global Biodiversity Assessment*. Cambridge: Cambridge University Press.
Hirata, A., S.Y. Kim, N. Kobayakawa, N. Tanaka, and Y. Kashiwada. 2015. Miltiorins A-D, diterpenes from *Radix Salviae miltiorrhizae*. *Fitoterapia* 102: 49–55.
Ilnicki, T. 2014. Plant biosystematics with the help of cytology and cytogenetics. *Caryologia* 67: 199–208.
Janicsak, G., K. Veres, A.Z. Kakasy, and I. Mathe. 2006. Study of the oleanolic and ursolic acid contents of some species of the Lamiaceae. *Biochem. Syst. Ecol.* 34: 392–396.
Jaroszewski, J.W., E.S. Olafsdottir, P. Wellendorph et al. 2002. Natural cyclopentanoid cyanohydrin glycosides, part 23. Cyanohydrin glycosides of *Passiflora*: Distribution pattern, a saturated cyclopentane derivative from *P. guatemalensis*, and formation of pseudocyanogenic alpha-hydroxyamides as isolation artefacts. *Phytochemistry* 59: 501–511.
Jassbi, A.R., M. Asadollahi, M. Masroor et al. 2012. Chemical classification of the essential oils of the Iranian *Salvia* species in comparison with their botanical taxonomy. *Chem. Biodivers.* 9: 1254–1271.
Jenett-Siems, K., T. Schimming, M. Kaloga et al. 1998. Phytochemistry and chemotaxonomy of the Convolvulaceae—Part 4—Pyrrolizidine alkaloids of *Ipomoea hederifolia* and related species. *Phytochemistry* 47: 1551–1560.
Jensen, C.G. and B.K. Ehlers. 2010. Genetic variation for sensitivity to a thyme monoterpene in associated plant species. *Oecologia* 162: 1017–1025.
Joycharat, N., H. Greger, O. Hofer, and E. Saifah. 2008. Flavaglines and triterpenes as chemical markers of *Aglaia oligophylla*. *Biochem. Syst. Ecol.* 36: 584–587.
Kajiyama, K., Y. Hiraga, K. Takahashi et al. 1993. Flavonoids and isoflavonoids of chemotaxonomic significance from *Glycyrrhiza pallidiflora* (Leguminosae). *Biochem. Syst. Ecol.* 21: 785–793.

Katekunlaphan, T., R. Chalermglin, T. Rukachaisirikul, and P. Chalermglin. 2014. Sesquiterpene lactones from the leaves of *Magnolia sirindhomiae*. *Biochem. Syst. Ecol.* 57: 152–154.

Kharazian, N. 2014. Chemotaxonomy and flavonoid diversity of *Salvia* L. (Lamiaceae) in Iran. *Acta Bot. Bras.* 28: 281–292.

Kilic, O. and E. Bagci. 2014. Essential oil composition of *Wiedemannia* Fisch. & C.A. Mey. genus from Turkey: A chemotaxonomic approach. *J. Essent. Oil Bear. Pl.* 17: 741–746.

Kim, M.G., N.H. Lee, J.M. Kim, S.G. Lee, and H.S. Lee. 2015. Chemical composition of essential oils extracted from five *Juniperus chinensis* varieties in Korea. *J. Essent. Oil Bear. Pl.* 18: 852–856.

Kite, G.C. and H. Ireland. 2002. Non-protein amino acids of *Bocoa* (Leguminosae; Papilionoideae). *Phytochemistry* 59: 163–168.

Knudsen, J.T. and B.B. Klitgaard. 1998. Floral scent and pollination in *Browneopsis disepala* (Leguminosae: Caesalpinioideae) in western Ecuador. *Brittonia* 50: 174–182.

Kotze, M.J., A. Jurgens, S.D. Johnson, and J.H. Hoffmann. 2010. Volatiles associated with different flower stages and leaves of *Acacia cyclops* and their potential role as host attractants for *Dasineura dielsi* (Diptera. Cecidomyiidae). *S. Afr. J. Bot.* 76: 701–709.

Kremer, D., S. Bolaric, D. Ballian et al. 2015. Morphological, genetic and phytochemical variation of the endemic *Teucrium arduini* L. (Lamiaceae). *Phytochemistry* 116: 111–119.

Kreutzer, S., O. Schimmer, and R. Waibel. 1990. Triterpenoid sapogenins in the genus *Grindelia*. *Planta Med.* 56: 392–394.

Lai, Y.J., Y.B. Xue, M.K. Zhang et al. 2013. Scapitormolactones A-I: Germacrane sesquiterpenoids with an unusual delta(3)-15,6-lactone moiety from *Salvia scapiformis*. *Phytochemistry* 96: 378–388.

Langel, D., D. Ober, and P.B. Pelser. 2011. The evolution of pyrrolizidine alkaloid biosynthesis and diversity in the Senecioneae. *Phytochem. Rev.* 10: 3–74.

Langenheim, J.H. 1981. Terpenoids of the Leguminosae. In *Advances in Legume Systematics*, eds. R.M. Polhill and P.H. Raven, 599–626. Kew: Royal Botanic Gardens.

Leandro, L.M., F.D. Vargas, P.C.S. Barbosa, J.K.O. Neves, J.A. da Silva, and V.F. da Veiga. 2012. Chemistry and biological activities of terpenoids from copaiba (*Copaifera* spp.) oleoresins. *Molecules* 17: 3866–3889.

Leuner, O., J. Havlik, J. Hummelova, E. Prokudina, P. Novy, and L. Kokoska. 2013. Distribution of isoflavones and coumestrol in neglected tropical and subtropical legumes. *J. Sci. Food Agric.* 93: 575–579.

Lewis, G.P., J.T. Knudsen, B.B. Klitgaard, and R.T. Pennington. 2003. The floral scent of *Cyathostegia mathewsii* (Leguminosae, Papilionoideae) and preliminary observations on reproductive biology. *Biochem. Syst. Ecol.* 31: 951–962.

Li, L.Z., M.H. Wang, J.B. Sun, and J.Y. Liang. 2014. Flavonoids and other constituents from *Aletris spicata* and their chemotaxonomic significance. *Nat. Prod. Res.* 28: 1214–1217.

Liu, J.F., F.Y. Liu, N.L. Zhang et al. 2016. Two new sesquiterpene lactones from the fruits of *Illicium jiadifengpi*. *Nat. Prod. Res.* 30: 322–326.

Liu, K., P.-G. Rossi, B. Ferrari, L. Berti, J. Casanova, and F. Tomi. 2007. Composition, irregular terpenoids, chemical variability and antibacterial activity of the essential oil from *Santolina corsica* Jordan et Fourr. *Phytochemistry* 68: 1698–1705.

Lohani, H., U. Bhandari, G. Gwari, S.Z. Haider, and N.K. Chauhan. 2015. Constituents of essential oils of *Cupressus arizonica* Greene from Uttarakhand Himalaya (India). *J. Essent. Oil Res.* 27: 459–463.

Lozano, E.S., R.M. Spina, C.E. Tonn, M.A. Sosa, and D.A. Cifuente. 2015. An abietane diterpene from *Salvia cuspidata* and some new derivatives are active against *Trypanosoma cruzi*. *Bioorg. Med. Chem. Lett.* 25: 5481–5484.

Lukas, B., C. Schmiderer, and J. Novak. 2015. Essential oil diversity of European *Origanum vulgare* L. (Lamiaceae). *Phytochemistry* 119: 32–40.

Lv, H.W., J.G. Luo, M.D. Zhu, H.J. Zhao, and L.Y. Kong. 2015. neo-Clerodane diterpenoids from the aerial parts of *Teucrium fruticans* cultivated in China. *Phytochemistry* 119: 26–31.

Maffei, M. 1988. Environmental factors affecting the oil composition of some *Mentha* species grown in North West Italy. *Flav. Fragr. J.* 3: 79–84.

Maffei, M. 1990. Plasticity and genotypic variation in some *Mentha* x *verticillata* hybrids. *Biochem. Syst. Ecol.* 18: 493–502.

Maffei, M. 1994. Discriminant-analysis of leaf wax alkanes in the Lamiaceae and 4 other plant families. *Biochem. Syst. Ecol.* 22: 711–728.

Maffei, M. 1996a. Chemotaxonomic significance of leaf wax alkanes in the Compositae. In *Compositae: Systematics. Proceedings of the International Compositae Conference*, eds. D.J.N. Hind and H.J. Beentje, 141–158. Kew: Royal Botanic Gardens.

Maffei, M. 1996b. Chemotaxonomic significance of leaf wax alkanes in the Gramineae. *Biochem. Syst. Ecol.* 24: 53–64.

Maffei, M. 1996c. Chemotaxonomic significance of leaf wax *n*-alkanes in the umbelliferae, Cruciferae and Leguminosae (Subf. Papilionoideae). *Biochem. Syst. Ecol.* 24: 531–545.

Maffei, M., S. Badino, and S. Bossi. 2004. Chemotaxonomic significance of leaf wax *n*-alkanes in the Pinales (Coniferales). *J. Biol. Res.* 1: 3–19.

Maffei, M., F. Chialva, and A. Codignola. 1989. Essential oils and chromosome numbers from Italian *Achillea* species. *J. Essent. Oil Res.* 2: 57–64.

Maffei, M., G. Doglia, F. Chialva, and F. Germano. 1993a. Essential oils, chromosome number and karyotypes from Italian *Achillea* species. Part II. *J. Essent. Oil Res.* 5: 61–70.

Maffei, M., M. Mucciarelli, and S. Scannerini. 1993b. Environmental factors affecting the lipid-metabolism in *Rosmarinus officinalis* L. *Biochem. Syst. Ecol.* 21: 765–784.

Maffei, M., M. Mucciarelli, and S. Scannerini. 1994. Essential oils from *Achillea* species of different geographic origin. *Biochem. Syst. Ecol.* 22: 679–687.

Maffei, M. and V. Peracino. 1993. Fatty acids from some *Lavandula* hybrids growing spontaneously in North-West Italy. *Phytochemistry* 33: 373–376.

Maffei, M., V. Peracino, and T. Sacco. 1993c. Multivariate methods for aromatic plants: An application to mint essential oils. *Acta Hortic.* 330: 159–169.

Maffei, M. and S. Scannerini. 1992a. Fatty-acid variability in some *Mentha* species. *Biochem. Syst. Ecol.* 20: 573–582.

Maffei, M. and S. Scannerini. 1992b. Seasonal-variations in fatty-acids from nonpolar lipids of developing peppermint leaves. *Phytochemistry* 31: 479–484.

Maffei, M. and S. Scannerini. 1993. Fatty-acid variability from nonpolar lipids in some Lamiaceae. *Biochem. Syst. Ecol.* 21: 475–486.

Maffei, M.E. 2010. Sites of synthesis, biochemistry and functional role of plant volatiles. *S. Afr. J. Bot.* 76: 612–631.

Maffei, M.E. 2015. *Molecole Bioattive delle Piante*. Rome: Gruppo Editoriale l'Espresso.
Maffei, M.E., J. Gertsch, and G. Appendino. 2011. Plant volatiles: Production, function and pharmacology. *Nat. Prod. Rep.* 28: 1359–1380.
Marinho, C.R., C.D. Souza, T.C. Barros, and S.P. Teixeira. 2014. Scent glands in legume flowers. *Plant Biol.* 16: 215–226.
Marzouk, M.S.A., M.T. Ibrahim, O.R. El-Gindi, and M.S. Abou Bakr. 2008. Isoflavonoid glycosides and rotenoids from *Pongamia pinnata* leaves. *Z. Naturforsch. C* 63: 1–7.
Mathe, I., A. Mathe, J. Hohmann, and G. Janicsak. 2010. Volatile and some nonvolatile chemical constituents of Mediterranean *Salvia* species beyond their native area. *Isr. J. Plant Sci.* 58: 273–277.
Maximo, P., A. Lourenco, A. Tei, and M. Wink. 2006. Chemotaxonomy of Portuguese *Ulex*: Quinolizidine alkaloids as taxonomical markers. *Phytochemistry* 67: 1943–1949.
McNeil, M., P. Facey, and R. Porter. 2011. Essential oils from the *Hyptis* genus—A review (1909–2009). *Nat. Prod. Commun.* 6: 1775–1796.
Meng, X.H., C.Z. Zou, X.J. Jin et al. 2014. New clerodane diterpenoid glycosides from the aerial parts of *Nannoglottis carpesioides*. *Fitoterapia* 93: 39–46.
Michalska, K. and W. Kisiel. 2007. Further sesquiterpene lactones and phenolics from *Cichorium spinosum*. *Biochem. Syst. Ecol.* 35: 714–716.
Michalska, K. and W. Kisiel. 2008. Sesquiterpene lactones from *Taraxacum erythrospermum*. *Biochem. Syst. Ecol.* 36: 444–446.
Miller, H.E., T.J. Mabry, B.L. Turner, and W.W. Payne. 1968. Infraspecific variation of sesquiterpene lactones in *Ambrosia psilostachya* (Compositae). *Am. J. Bot.* 55: 316–324.
Miller, R.E., M.J. McConville, and I.E. Woodrow. 2006a. Cyanogenic glycosides from the rare Australian endemic rainforest tree *Clerodendrum grayi* (Lamiaceae). *Phytochemistry* 67: 43–51.
Miller, R.E., M. Stewart, R.J. Capon, and I.E. Woodrow. 2006b. A galloylated cyanogenic glycoside from the Australian endemic rainforest tree *Elaeocarpus sericopetalus* (Elaeocarpaceae). *Phytochemistry* 67: 1365–1371.
Miranda, M.L.D., F.R. Garcez, A.R. Abot, and W.S. Garcez. 2014. Sesquiterpenes and other constituents from leaves of *Pterodon pubescens* Benth (Leguminosae). *Quimica Nova* 37: 473-+.
Mishio, T., K. Takeda, and T. Iwashina. 2015. Anthocyanins and other flavonoids as flower pigments from eleven *Centaurea* species. *Nat. Prod. Commun.* 10: 447–450.
Mitsui, T., R. Ishihara, K. Hayashi, N. Matsuura, H. Akashi, and H. Nozaki. 2015. Cassane-type diterpenoids from *Caesalpinia echinata* (Leguminosae) and their NF-kappa B signaling inhibition activities. *Phytochemistry* 116: 349–358.
Mungarulire, J., R.M. Munabu, C. Murasaki et al. 1993. A novel steroidal sapogenin, pogosterol from *Vernonia pogosperma*. *Chem. Pharm. Bull.* 41: 411–413.
Muselli, A., J.M. Desjobert, A.F. Bernardini, and J. Costa. 2007. Santolina alcohol as component of the essential oil of *Achillea ageratum* L. from Corsica island. *J. Essent. Oil Res.* 19: 319–322.
Niederbacher, B., J.B. Winkler, and J.P. Schnitzler. 2015. Volatile organic compounds as non-invasive markers for plant phenotyping. *J. Exp. Bot.* 66: 5403–5416.
Nieto, M., E.E. Garcia, O.S. Giordano, and C.E. Tonn. 2000. Icetexane and abietane diterpenoids from *Salvia gilliessi*. *Phytochemistry* 53: 911–915.

Nikolic, B., M. Ristic, V. Tesevic, P.D. Marin, and S. Bojovic. 2011. Terpene chemodiversity of relict conifers *Picea omorika*, *Pinus heldreichii*, and *Pinus peuce*, endemic to Balkan. *Chem. Biodivers.* 8: 2247–2260.

Nikolova, M.T., R.J. Grayer, E. Genova, and E.A. Porter. 2006. Exudate flavonoids from Bulgarian species of *Salvia*. *Biochem. Syst. Ecol.* 34: 360–364.

Ohta, T., S. Nakamura, S. Nakashima et al. 2015. Acylated oleanane-type triterpene oligoglycosides from the flower buds of *Camellia sinensis* var. *assamica*. *Tetrahedron* 71: 846–851.

Ohyama, M., T. Tanaka, J. Yokoyama, and M. Iinuma. 1995. Occurrence of prenylated flavonoids and oligostilbenes and its significance for chemotaxonomy of genus *Sophora* (Leguminosae). *Biochem. Syst. Ecol.* 23: 669–677.

Park, C.H., S.C. Chae, S.-Y. Park et al. 2015. Anthocyanin and carotenoid contents in different cultivars of *Chrysanthemum* (*Dendranthema grandiflorum* Ramat.) flower. *Molecules* 20: 11090–11102.

Park, E.J., Y.Z. Zhao, Y.C. Kim, and D.H. Sohn. 2005. Protective effect of (S)-bakuchiol from *Psoralea corylifolia* on rat liver injury *in vitro* and *in vivo*. *Planta Med.* 71: 508–513.

Perez, I., M.A. Blazquez, and H. Boira. 2000. Chemotaxonomic value of the essential oil compounds in species of *Teucrium pumilum* aggregate. *Phytochemistry* 55: 397–401.

Pettersson, S. and J.T. Knudsen. 2001. Floral scent and nectar production in *Parkia biglobosa* Jacq. (Leguminosae: Mimosoideae). *Bot. J. Linn. Soc.* 135: 97–106.

Pichersky, E. and D.R. Gang. 2000. Genetics and biochemistry of secondary metabolites in plants: An evolutionary perspective. *Trends Plant Sci.* 5: 439–445.

Piechowski, D., S. Doetterl, and G. Gottsberger. 2010. Pollination biology and floral scent chemistry of the Neotropical chiropterophilous *Parkia pendula*. *Plant Biol.* 12: 172–182.

Pino Benitez, N., E.M. Melendez Leon, and E.E. Stashenko. 2009. Eugenol and methyl eugenol chemotypes of essential oil of species *Ocimum gratissimum* L. and *Ocimum campechianum* Mill. from Colombia. *J. Chromatogr. Sci.* 47: 800–803.

Piozzi, F., M. Bruno, S. Rosselli, and A. Maggio. 2009. The diterpenoids from the genus *Hyptis* (Lamiaceae). *Heterocycles* 78: 1413–1426.

Pirmoradi, M.R., M. Moghaddam, and N. Farhadi. 2013. Chemotaxonomic analysis of the aroma compounds in essential oils of two different *Ocimum basilicum* L. varieties from Iran. *Chem. Biodivers.* 10: 1361–1371.

Pistelli, L., A. Bertoli, I. Giachi, I. Morelli, P. Rubiolo, and C. Bicchi. 2001. Quinolizidine alkaloids from *Genista ephedroides*. *Biochem. Syst. Ecol.* 29: 137–141.

Pitarokili, D., T. Constantinidis, C. Saitanis, and O. Tzakou. 2014. Volatile compounds in *Thymus* sect. *Teucrioides* (Lamiaceae): Intraspecific and interspecific diversity, chemotaxonomic significance and exploitation potential. *Chem. Biodivers.* 11: 593–618.

Radulovic, N., M. Dekic, M. Joksovic, and R. Vukicevic. 2012. Chemotaxonomy of Serbian *Teucrium* species inferred from essential oil chemical composition: The case of *Teucrium scordium* L. ssp. *scordioides*. *Chem. Biodivers.* 9: 106–122.

Radulovic, N., B. Zlatkovic, D. Skropeta, and R. Palic. 2008. Chemotaxonomy of the peppergrass *Lepidium coronopus* (L.) Al-Shehbaz (syn. Coronopus squamatus) based on its volatile glucosinolate autolysis products. *Biochem. Syst. Ecol.* 36: 807–811.

Ragasa, C.Y., J. Ganzon, J. Hofilena, B. Tamboong, and J.A. Rideout. 2003. A new furanoid diterpene from *Caesalpinia pulcherrima*. *Chem. Pharm. Bull.* 51: 1208–1210.

Rahimmalek, M., B.E.S. Tabatabaeb, N. Etemadi, S.A.H. Goli, A. Arzani, and H. Zeinali. 2009. Essential oil variation among and within six *Achillea* species transferred from different ecological regions in Iran to the field conditions. *Ind. Crop. Prod.* 29: 348–355.

Riccobono, L., A. Maggio, S. Rosselli, V. Ilardi, F. Senatore, and M. Bruno. 2016. Chemical composition of volatile and fixed oils from of *Salvia argentea* L. (Lamiaceae) growing wild in Sicily. *Nat. Prod. Res.* 30: 25–34.

Rocha, M.E.D., M.R. Figueiredo, M.A.C. Kaplan, T. Durst, and J.T. Arnason. 2015. Chemotaxonomy of the Ericales. *Biochem. Syst. Ecol.* 61: 441–449.

Rosendal-Jensen, S. 1991. Plant iridoids, their biosynthesis and distribution in angiosperms. In *Ecological Chemistry and Biochemistry of Plant Terpenoids*, eds. J.B. Harborne, F.A. Tomas-Barberan, and M.I. Gil, 133–158. Oxford: Oxford Science Publications.

Royer, M., G. Herbette, V. Eparvier, J. Beauchene, B. Thibaut, and D. Stien. 2010. Secondary metabolites of *Bagassa guianensis* Aubl. wood: A study of the chemotaxonomy of the Moraceae family. *Phytochemistry* 71: 1708–1713.

Rubenstein, E., T. McLaughlin, R.C. Winant et al. 2009. Azetidine-2-carboxylic acid in the food chain. *Phytochemistry* 70: 100–104.

Rubenstein, E., H.L. Zhou, K.M. Krasinska, A. Chien, and C.H. Becker. 2006. Azetidine-2-carboxylic acid in garden beets (*Beta vulgaris*). *Phytochemistry* 67: 898–903.

Rubiolo, P., M. Matteodo, C. Bicchi et al. 2009. Chemical and biomolecular characterization of *Artemisia umbelliformis* Lam., an important ingredient of the Alpine Liqueur "Genepi." *J. Agric. Food Chem.* 57: 3436–3443.

Sacco, T. and M. Maffei. 1987. Essential oil from *Chiliadenus lopadusanus* growing spontaneously in Lampedusa Island (Italy). *Planta Med.* 53: 582.

Saeidnia, S., M. Ghamarinia, A.R. Gohari, and A. Shakeri. 2012. Terpenes from the root of *Salvia hypoleuca* Benth. *J. Pharm. Sci.* 20: 6.

Saito, N. and J.B. Harborne. 1992. Correlations between anthocyanin type, pollinator and flower color in the Labiatae. *Phytochemistry* 31: 3009–3015.

Sarac, Z., S. Bojovic, B. Nikolic, V. Tesevic, I. Dordevic, and P.D. Marin. 2013. Chemotaxonomic significance of the terpene composition in natural populations of *Pinus nigra* J.F.Arnold from Serbia. *Chem. Biodivers.* 10: 1507–1520.

Seaman, F.C. 1982. Sesquiterpene lactones as taxonomic characters in the Asteraceae. *Bot. Rev.* 48: 121–595.

Seeligmabb, P. 1994. Flavonoids of the compositae as evolutionary parameters in the tribes which synthesize them: A critical approach. In *Compositae: Systematics. Proceedings of the International Compositae Conference, Kew*, eds. D.J.N. Hind and H.J. Beentje, 159–167. Kew: Royal Botanic Gardens.

Seitz, C., S. Ameres, K. Schlangen, G. Forkmann, and H. Halbwirth. 2015. Multiple evolution of flavonoid 3′,5′-hydroxylase. *Planta* 242: 561–573.

Shitamoto, J., K. Matsunami, H. Otsuka, T. Shinzato, and Y. Takeda. 2010. Crotalionosides A-C, three new megastigmane glucosides, two new pterocarpan glucosides and a chalcone C-glucoside from the whole plants of *Crotalaria zanzibarica*. *Chem. Pharm. Bull.* 58: 1026–1032.

Silva, J.G., M.T. Faria, E.R. Oliveira et al. 2011. Chemotaxonomic significance of volatile constituents in *Hypenia* (Mart. ex Benth.) R. Harley (Lamiaceae). *J. Braz. Chem. Soc.* 22: 955–U217.

Singh, P., A. Krishna, V. Kumar et al. 2016. Chemistry and biology of industrial crop *Tagetes* Species: A review. *J. Essent. Oil Res.* 28: 1–14.

Siridechakorn, I., S. Cheenpracha, T. Ritthiwigrom et al. 2014. Isopimarane diterpenes and flavan derivatives from the twigs of *Caesalpinia furfuracea*. *Phytochem. Lett.* 7: 186–189.

Skaltsa, H., P. Georgakopoulos, D. Lazari et al. 2007. Flavonoids as chemotaxonomic markers in the polymorphic *Stachys swainsonii* (Lamiaceae). *Biochem. Syst. Ecol.* 35: 317–320.

Skoula, M., P. Gotsiou, G. Naxakis, and C.B. Johnson. 1999. A chemosystematic investigation on the mono- and sesquiterpenoids in the genus *Origanum* (Labiatae). *Phytochemistry* 52: 649–657.

Skoula, M., R.J. Grayer, G.C. Kite, and N.C. Veitch. 2008. Exudate flavones and flavanones in *Origanum* species and their interspecific variation. *Biochem. Syst. Ecol.* 36: 646–654.

Slavkovska, V., M. Couladis, S. Bojovic et al. 2005. Essential oil and its systematic significance in species of *Micromeria* Bentham from Serbia & Montenegro. *Plant Syst. Evol.* 255: 1–15.

Spring, O., R. Zipper, J. Conrad, B. Vogler, I. Klaiber, and F.B. Da Costa. 2003. Sesquiterpene lactones from glandular trichomes of *Viguiera radula* (Heliantheae; Asteraceae). *Phytochemistry* 62: 1185–1189.

Staneva, J.D., M.N. Todorova, and L.N. Evstatieva. 2008. Sesquiterpene lactones as chemotaxonomic markers in genus *Anthemis*. *Phytochemistry* 69: 607–618.

Stevens, J.F., H. Thart, A. Bolck, J.H. Zwaving, and T.M. Malingre. 1994. Epicuticular wax composition of some European *Sedum* species. *Phytochemistry* 35: 389–399.

Sun, H.N., T. Zhang, Q.Q. Fan et al. 2015. Identification of floral scent in *Chrysanthemum* cultivars and wild relatives by gas chromatography-mass spectrometry. *Phytochemistry* 20: 5346–5359.

Takeda, K., J.B. Harborne, and R. Self. 1986. Identification of malonated anthocyanins in the liliaceae and Labiatae. *Phytochemistry* 25: 2191–2192.

Tanaka, Y. and F. Brugliera. 2013. Flower colour and cytochromes P450. *Philos. Trans. R. Soc. Lond. B Biol. Sci.* 368: 20120432.

Tang, Y.P., J. Hu, J.H. Wang, and F.C. Lou. 2002. A new coumaronochromone from *Sophora japonica*. *J. Asian Nat. Prod. Res.* 4: 1–5.

Todorova, M.N., B. Mikhova, A. Trendafilova, A. Vitkova, H. Duddeck, and M. Anchev. 2006. Sesquiterpene lactones from *Achillea asplenifolia*. *Biochem. Syst. Ecol.* 34: 136–143.

Toki, K., N. Saito, and T. Honda. 1998. Acylated anthocyanins from the bluepurple flowers of *Triteleia bridgesii*. *Phytochemistry* 48: 729–732.

Tomas-Barberan, F.A. and M.I. Gil. 1992. Chemistry and natural distribution of flavonoids in the Labiatae. In *Advances in Labiatae Science*, eds. R.M. Harley and T. Reynolds, 299–305. Kew: Royal Botanic Gardens.

Tomas-Barberan, F.A. and E. Wollenweber. 1990. Flavonoid aglycones from the leaf surfaces of some Labiatae species. *Plant Syst. Evol.* 173: 109–118.

Tran, M.H., M.T.T. Nguyen, H.D. Nguyen, T.D. Nguyen, and T.T. Phuong. 2015. Cytotoxic constituents from the seeds of Vietnamese *Caesalpinia sappan*. *Pharm. Biol.* 53: 1549–1554.

Triana, J., J.L. Eiroa, J.J. Ortega et al. 2010. Chemotaxonomy of *Gonospermum* and related genera. *Phytochemistry* 71: 627–634.

Trigo, J.R., I.R. Leal, N.I. Matzenbacher, and T.M. Lewinsohn. 2003. Chemotaxonomic value of pyrrolizidine alkaloids in southern Brazil Senecio (Senecioneae: Asteraceae). *Biochem. Syst. Ecol.* 31: 1011–1022.

Trikka, F.A., A. Nikolaidis, C. Ignea et al. 2015. Combined metabolome and transcriptome profiling provides new insights into diterpene biosynthesis in *S. pomifera* glandular trichomes. *BMC Genomics* 16: 19.

Tsiri, D., K. Graikou, L. Poblocka-Olech, M. Krauze-Baranowska, C. Spyropoulos, and I. Chinou. 2009. Chemosystematic value of the essential oil composition of *Thuja* species cultivated in Poland-antimicrobial activity. *Molecules* 14: 4707–4715.

Tundis, R., L. Peruzzi, and F. Menichini. 2014. Phytochemical and biological studies of *Stachys* species in relation to chemotaxonomy: A review. *Phytochemistry* 102: 7–39.

Valant-Vetschera, K.M., J.N. Roitman, and E. Wollenweber. 2003. Chemodiversity of exudate flavonoids in some members of the Lamiaceae. *Biochem. Syst. Ecol.* 31: 1279–1289.

Valarezo, E., A. Castillo, D. Guaya, V. Morocho, and O. Malagon. 2012. Chemical composition of essential oils of two species of the Lamiaceae family: *Scutellaria volubilis* and *Lepechinia paniculata* from Loja, Ecuador. *J. Essent. Oil Res.* 24: 31–37.

van Rozendaal, E.L.M., S.J.L. Kurstjens, T.A. van Beek, and R.G. van den Berg. 1999. Chemotaxonomy of *Taxus*. *Phytochemistry* 52: 427–433.

van Wyk, B.E. 2003. The value of chemosystematics in clarifying relationships in the genistoid tribes of papilionoid legumes. *Biochem. Syst. Ecol.* 31: 875–884.

van Wyk, B.E., A. Yenesew, and E. Dagne. 1995. Chemotaxonomic survey of anthraquinones and pre-anthraquinones in roots of *Aloe* species. *Biochem. Syst. Ecol.* 23: 267–275.

Veitch, N.C. 2013. Isoflavonoids of the Leguminosae. *Nat. Prod. Rep.* 30: 988–1027.

Veitch, N.C., P.C. Elliott, G.C. Kite, and G.P. Lewis. 2010. Flavonoid glycosides of the black locust tree, *Robinia pseudoacacia* (Leguminosae). *Phytochemistry* 71: 479–486.

Veitch, N.C., G.C. Kite, and G.P. Lewis. 2008. Flavonol pentaglycosides of *Cordyla* (Leguminosae: Papilionoideae: Swartzieae): Distribution and taxonomic implications. *Phytochemistry* 69: 2329–2335.

Vieira, R.F. and J.E. Simon. 2006. Chemical characterization of basil (*Ocimum* spp.) based on volatile oils. *Flavour. Frag. J.* 21: 214–221.

Viljoen, A.M., E.W. Njenga, S.F. van Vuuren, C. Bicchi, P. Rubiolo, and B. Sgorbini. 2006. Essential oil composition and in vitro biological activities of seven Namibian species of *Eriocephalus* L. (Asteraceae). *J. Essent. Oil Res.* 18: 124–128.

Vranova, V., K. Rejsek, K.R. Skene, and P. Formanek. 2011. Non-protein amino acids: Plant, soil and ecosystem interactions. *Plant Soil*. 342: 31–48.

Wang, H., Y.M. Cui, and C.Q. Zhao. 2010a. Flavonoids of the genus *Iris* (Iridaceae). *Mini. Rev. Med. Chem.* 10: 643–661.

Wang, H., W.H. Dong, W.J. Zuo et al. 2014. Five new sesquiterpenoids from *Dalbergia odorifera*. *Fitoterapia* 95: 16–21.

Wang, Y.F., Z.Y. Ni, M. Dong et al. 2010b. Secondary metabolites of plants from the genus *Saussurea*: Chemistry and biological activity. *Chem. Biodivers.* 7: 2623–2659.

Waterman, P.G. 1998. Chemical taxonomy of alkaloids. In *Alkaloids, Biochemistry, Ecology, and Medicinal Applications*, eds. M.F. Roberts and M. Wink, 87–107. New York: Plenum.
Wei, H.H., H.H. Xu, H.H. Xie, L.X. Xu, and X.Y. Wei. 2009. Sesquiterpenes and lignans from *Tephrosia vogelii*. *Helv. Chim. Acta* 92: 370–374.
Wilkomirski, B. and E. Kucharska. 1992. Triterpene chemotypes of some polish populations of *Tanacetum vulgare*. *Phytochemistry* 31: 3915–3916.
Williams, C.A., J.B. Harborne, J. Greenham, B.G. Briggs, and L.A.S. Johnson. 1997. Flavonoid evidence and the classification of the Anarthriaceae within the Poales. *Phytochemistry* 45: 1189–1196.
Wink, M. 2013. Evolution of secondary metabolites in legumes (Fabaceae). *S. Afr. J. Bot.* 89: 164–175.
Wink, M., C. Meissner, and L. Witte. 1995. Patterns of quinolizidine alkaloids in 56 species of the genus *Lupinus*. *Phytochemistry* 38: 139–153.
Wolff, R.L. and C.C. Bayard. 1995. Fatty-acid composition of some pine seed oils. *J. Am. Oil. Chem. Soc.* 72: 1043–1046.
Wolff, R.L., L.G. Deluc, and A.M. Marpeau. 1996. Conifer seeds: Oil content and fatty acid composition. *J. Am. Oil Chem. Soc.* 73: 765–771.
Wolff, R.L., L.G. Deluc, A.M. Marpeau, and B. Comps. 1997. Chemotaxonomic differentiation of conifer families and genera based on the seed oil fatty acid compositions: Multivariate analyses. *Trees Struct. Funct.* 12: 57–65.
Wollenweber, E., M. Dorr, H. Fritz, and K.M. ValantVetschera. 1997. Exudate flavonoids in miscellaneous Asteraceae. 2. Exudate flavonoids in several Asteroideae and Cichorioideae (Asteraceae). *Z. Naturforsch. C* 52: 137–143.
Wollenweber, T.E., M. Dorr, H. Fritz, and K.M. Valant-Vetschera. 1998. Exudate flavonoids in several Asteroideae and Cichorioideae (Asteraceae) (vol 52c, p. 137, 1997). *Z. Naturforsch. C* 53: 937–937.
Wollenweber, E., M. Dorr, A. Rustaiyan, J.N. Roitman, and E.H. Graven. 1992. Exudate flavonoids of some salvia and a *Trichostema* species. *Z. Naturforsch. C* 47: 782–784.
Wollenweber, E., S. Stern, J.N. Roitman, and G. Yatskievych. 1989. External leaf flavonoids of Polanisia-*Trachysperma*. *Phytochemistry* 28: 303–305.
Wu, H.K., J. Li, F.R. Fronczek et al. 2013. Labdane diterpenoids from *Leonotis leonurus*. *Phytochemistry* 91: 229–235.
Wu, Y.B., Z.Y. Ni, Q.W. Shi et al. 2012. Constituents from *Salvia* species and their biological activities. *Chem. Rev.* 112: 5967–6026.
Wu, H.K., S.S. Wang, H.J. Liu et al. 2015. Two new diterpenoids from *Leonurus japonicus*. *Rev. Bras. Farmacogn.* 25: 180–182.
Yasar, S. 2013. Volatile constituents of *Taxus baccata* L. leaves from western and southern Turkey. *Asian J. Chem.* 25: 9123–9125.
Yazaki, Y. 2015. Wood colors and their coloring matters: A review. *Nat. Prod. Commun.* 10: 505–512.
Yodsaoue, O., S. Cheenpracha, C. Karalai et al. 2008. Phanginin A-K, diterpenoids from the seeds of *Caesalpinia sappan* Linn. *Phytochemistry* 69: 1242–1249.
Zaidi, F., B. Voirin, M. Jay, and M.R. Viricel. 1998. Free flavonoid aglycones from leaves of *Mentha pulegium* and *Mentha suaveolens* (Labiatae). *Phytochemistry* 48: 991–994.
Zenk, M.H. and M. Juenger. 2007. Evolution and current status of the phytochemistry of nitrogenous compounds. *Phytochemistry* 68: 2757–2772.

Zhang, B.J., M.F. Bao, C.X. Zeng et al. 2014. Dimeric Erythrina alkaloids from the flower of *Erythrina variegata*. *Org. Lett.* 16: 6400–6403.

Zhong, G.-X., P. Li, L.-J. Zeng, J. Guan, D.-Q. Li, and S.-P. Li. 2009. Chemical characteristics of *Salvia miltiorrhiza* (Danshen) collected from different locations in China. *J. Agric. Food Chem.* 57: 6879–6887.

Zhou, Y., G. Xu, F.F.K. Choi et al. 2009. Qualitative and quantitative analysis of diterpenoids in *Salvia* species by liquid chromatography coupled with electrospray ionization quadrupole time-of-flight tandem mass spectrometry. *J. Chromatogr. A* 1216: 4847–4858.

chapter four

Biosynthesis and roles of Salicaceae salicylates

Riitta Julkunen-Tiitto and Virpi Virjamo

Contents

4.1 Introduction 65
4.2 Composition and variation of salicylates 67
4.3 Biosynthesis and turnover of salicylates 72
    4.3.1 Biosynthesis of salicylates 72
    4.3.2 Turnover of salicylates 74
4.4 Biological roles of salicylates 75
    4.4.1 Salicylates in human health 75
    4.4.2 Salicylates effective to herbivores 75
    4.4.3 Salicylates as deterrents to rusts 77
4.5 Conclusion 78
References 78

4.1 Introduction

The Salicaeae family consists of about 300 *Salix* and more than 30 *Populus* species largely located throughout the cooler areas of the Northern and Southern hemispheres. The growth forms of the *Salix* species are very variable from very slow- and short-growing to tree-like ones, while the *Populus* genus contains only tree-like growth forms. The *Salicaceae* species are a dioecious, insect-wind pollinated, rapid-growing species and generally, the components of the earliest succession phases of a vegetation having frequent and vigorous vegetative reproduction (e.g., Hämet-Ahti et al. 1998). Noteworthy is that *Salix* species are easily hybridized; that is the main source of the problems in species identification and classification (e.g., Argus 1973; Skvortsov 1999).

    Salicylates are typical compounds in the family of Salicaceae, genus *Salix* and *Populus*. These are regarded as low-molecular-weight phenolic glucosides that are variable with conjugated β-D-glucosides of salicyl alcohol. Free salicyl alcohol is rare in *in vivo* plants, but is readily produced under decomposition conditions due to, for example, improper preservation of plant tissues after harvesting for analyses (Julkunen-Tiitto and Sorsa 2001; Julkunen-Tiitto and Tahvanainen 1989). The number of salicylates detected thus far from the *Salix* and *Populus* species is more than 15 individual structures (Figure 4.1). Salicin, the simplest glucoside, is the most widespread, while usually found in low amounts in the species, it is the first phenolic glucoside ever found and extracted from plants. Salicin, β-D-glucoside of 2-hydroxybenzylalcohol, the sugar bound with the phenolic hydroxyl group of aglycone, was detected by Buchner in 1828, in the bark of *S. alba* and *S. incana* (Thieme 1963). The final structure of salicin was found by Irvine and Rose in 1906 (Thieme 1963). Reichardt et al. (1988) claimed that salicin is the most commonly found phenolic glucoside in plants; in addition to Salicaceae, it has been found in some members of the Rosaceae family. In the 1960s, during the most intensive research period of Salicaceae, a number of more abundant salicylates were discovered and defined (e.g., Pearl and Darling 1970, 1971; Thieme 1965a), such as salicortin and tremulacin, although these are only found thus far in the *Salicaceae* species. Later, other conjugated forms of salicylates, such as disalicortins, cinnamoyl-salicortins, and acetyl-tremulacins were identified (e.g., Abreu et al. 2011; Paajanen et al. 2011; Randriamanana et al. 2015a).

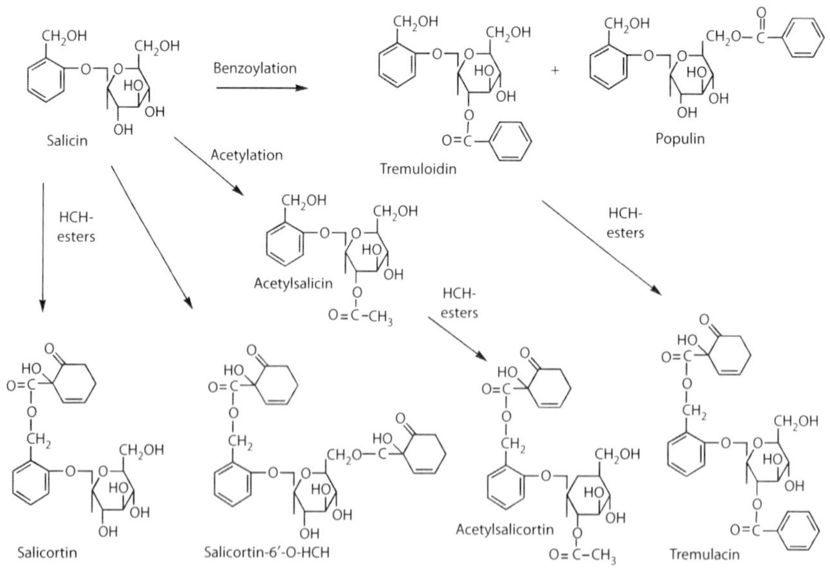

Figure 4.1 Main salicylates identified from *Salicaceae* species (arrows show conjugation with acetyl-, benzoyl-, or HCH-esters [6-hydroxy-2-cyclohexen-on-oyl group]).

4.2 Composition and variation of salicylates

Because salicylates, especially those with higher molecular mass compounds, are known to be very unstable, the interpretation of *in vivo* composition of salicylates has been difficult. Plant material destruction, inappropriate drying, certain organic solvents, and neutral and alkaline conditions during extraction are all possible factors inducing qualitative and quantitative changes in obtained salicylate profiles (e.g., Julkunen-Tiitto 1989a,b; Julkunen-Tiitto and Gebhardt 1992; Julkunen-Tiitto and Sorsa 2001; Lindroth and Koss 1996; Orians 1995). It seems that if it is not possible to make a chemical analysis from fresh material, then the next best choice would be to dry Salicaceae material in a freeze-dryer (Julkunen-Tiitto and Sorsa 2001; Orians 1995) or in a dry-air dryer at room temperature (Kosonen et al. 2012; Nybakken et al. 2012). The prober salicylate material handling is of utmost importance if the results are interpreted against abiotic and biotic factors (e.g., Hakulinen et al. 1995; Ikonen et al. 2002; Juntheikki et al. 1996; Nybakken et al. 2012), and also herbal drug use as well in chemotaxonomy (e.g., Julkunen-Tiitto and Meier 1992; Shao 1991).

Composition and the total amount of salicylates have been found to be rather species-specific and also genotype-specific (e.g., Boeckler et al. 2011; Förster et al. 2008; Hakulinen et al. 1995, 1998; Julkunen-Tiitto 1989b; Julkunen-Tiitto et al. 2015; Lindroth and Hwang 1996a; Meier 1988; Randriamanana et al. 2015a; Ruuhola et al. 2001; Shao 1991). There are only a few *Salix* species, for example, *S. triandra*, that are nearly free of salicylates in leaves and twigs and *S. myrtilloides* in leaves (Julkunen-Tiitto 1989a). Conjugated salicyl alcohol derivatives, such as salicortin, acetylsalicortin, and tremulacin, are the marker compounds for several *Salix* species. 2'-O-acetylsalicortin, a quite rare salicylate, is a marker for *S. pentandra* and *S. fragilis*, the former one is regarded to be the oldest *Salix* species. 2'-O-acetylsalicortin is also found in twigs of *S. myrsinifolia* and *S. aurita* (Förster et al. 2008; Julkunen-Tiitto 1989; Julkunen-Tiitto and Tahvanainen 1989). The high amount of tremulacin is typical for low-growing *Salix* species, such as *S. repens* and *S. myrsinites* (e.g., Meier et al. 1988) as well it is found together with salicortin as the most abundant salicylate in *Populus* species, such as *P. tremula* (Randriamanana et al. 2015a), *P. tremuloides* (Lindroth et al. 2002), their hybrid (Boeckler et al. 2014; Kosonen et al. 2012), and *P. trichocarpa* (Massad et al. 2014). These most abundant salicylates may be present in the species in more than 15%/DW in plant tissues (e.g., Julkunen-Tiitto 1989b, 1996; Lindroth et al. 1987; Meier 1988; Ruuhola and Julkunen-Tiitto 2003; Shao 1991). Salicin, acetylsalicin, and tremuloidin are found often in very small amounts giving doubts of their real existence in Salicaceae due to the weak stability of their conjugated forms (e.g., Egloff 1982) decomposing to salicin, 2'-O-acetylsalicortin, and tremuloidin (Julkunen-Tiitto and Tahvanainen 1989; Meier 1988; Orians 1995).

However, because in *Salicaceae* species, especially in genus *Salix*, easy hybridization may induce chemical variation, in pure species real interindividual variation appears to be only quantitative (e.g., Förster et al. 2008; Julkunen-Tiitto 1985, 1986; Julkunen-Tiitto and Tahvanainen 1988; Lindroth et al. 1987; Meier et al. 1988).

Salicylates are more often found in much higher quantities in twig tissues compared to the leaves of corresponding species and if leaves contain salicylates, then twigs also contain them (Förster et al. 2008; Julkunen-Tiitto 1989b). Salicylates are located fairly equally all over the leaf blade, while younger leaves are found to contain higher amounts compared to older leaves within the individual (Julkunen-Tiitto 1989b; Lindroth et al. 1987). Interestingly, leaf petioles contain salicylates while about 50% less compared to the leaf blade and composition of salicylate resembling corresponding twigs (Julkunen-Tiitto 1989b). Salicylates are also found in vegetative and more profound in flower buds, mostly resembling the corresponding species composition in leaves (Julkunen-Tiitto 1989a). For example, noteworthy are the species, such as *S. caprea*, *S. aurita*, and *S. cinerea*, of which the leaves or vegetative buds were almost free of salicylates, flower buds contained moderate or high levels of salicylates (Figure 4.2, Julkunen-Tiitto 1986, 1989a). Moreover, if leaves contained a high amount of salicylates (as in *S. myrsinifolia*, *S. purpurea*, and *S. repens*) buds also contained moderate amounts of salicylates (Julkunen-Tiitto 1986, 1989a;

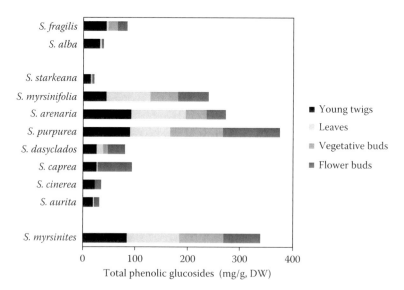

Figure 4.2 Salicylates found in different organs of *Salix* species. (According to Julkunen-Tiitto, R. 1989b. *Distribution of Certain Phenolics in Salix Species [Salicaceae]*. Joensuu: University of Joensuu.)

Julkunen-Tiitto and Tahvanainen 1989; Sivadasan et al. 2015). Salicylates are also found in the seeds of *S. myrsinifolia* (about 0.6%/DW) and *S. pentandra* (about 1.7%/DW) and the composition was found to resemble that of leaves (Julkunen-Tiitto 1989b; Randriamanana et al. 2015b). Similarly, the developing cotyledons of the germinating seeds contained salicylates similar in composition to leaves while the salicylate pattern of hypocotyls resembled those of twigs (Julkunen-Tiitto 1989b). The occurrence of salicylates in roots has not been studied thoroughly. However, there seems to be a difference in accumulation of salicylates in roots between *Salix* and *Populus* species. Roots of *Salix* species, such as *S. myrsinifolia*, *S. pentandra*, and *S. lapponum*, are totally devoid of salicylates (Julkunen-Tiitto 1989b) while at least roots of *P. tremula* contained a high amount and similar composition of salicylates compared to stems and leaves of the greenhouse grown seedlings (Randriamanana et al. 2014).

The production costs of salicylates based on glucose use in biosynthesis is claimed to be fairly high (Gershenzon 1994; Ruuhola and Julkunen-Tiitto 2003). It is consisting of the fact that the salicylate precursors are phenylalanine derived that is one of the most expensive amino acids produced by plants and that maintaining so high amount of salicylates found in Salicaceae should be costly. It has been shown that often plants are maintaining a steady-state level of secondary metabolites in tissues, meaning that synthesis and turnover are in balance (e.g., Barz and Koster 1981; Barz et al. 1985; Ruuhola and Julkunen-Tiitto 2003). However, Thieme (1965b) found marked diurnal fluctuation in leaf and twig salicylates. Controversially, there were no diurnal changes in salicylates found in current-growth twigs of northern *S. myrsinifolia* and only a small decrease in leaf salicylates was found to be studied over a three-day and three-night period (Julkunen-Tiitto 1989b). Moreover, salicylates are not reclaimed from the senescent leaves of *S. purpurea* (Julkunen-Tiitto 1989b) and only tended to decrease in four genotypes of *P. tremuloides* (Lindroth et al. 2002) before leaf abscission although all chlorophylls, nitrogen, and the main carbohydrates are mostly decreased, which indicates the low role of salicylate in overall metabolic costs.

Seasonal changes in many secondary plant compounds are found; small molecular weight compounds are often disappearing/decreasing and replaced by higher-molecular-weight compounds (e.g., Dement and Mooney 1974; Feeny 1970; Julkunen-Tiitto 1989b). *Salicaceae* species has expressed variable patterns in salicylate changes over the growing season. For instance, salicylates in the leaves of *S. dasyclados* (= *S.* "Acuatica") (Julkunen-Tiitto 1989b), *S. pentandra*, *S. purpurea*, and *P. candidans* (Thieme 1965b, 1971) and *P. grandidentata* (Lindroth et al. 1987), and in the bark of *S. daphnoides*, *S. pentandra*, and *S. purpurea* (Fröster et al. 2008) were highest at the beginning of the growing season while decreasing toward the end of the growing season. Instead, *P. deltoides* and *P. tremuloides* did not show any changes (Lindroth et al. 1987) while salicylates increased linearly over the season in the leaves of *S. sericea* and *S. eriocephala* (Fritz et al. 2001).

Salicylate content may change over plant ontogeny. Previously, Lindroth and Pajutee (1987) did not detect any salicylates in *S. pentandra* individuals, while Thieme (1965b), Meier (1988), Julkunen-Tiitto (1989a), and later, Ruuhola and Julkunen-Tiitto (2003), reported variable amounts of salicylates in the species. This discrepancy most probably has been induced by different ages of studied plants. *S. pentandra* and also *P. alba* are the species where phase change in salicylate chemistry will happen when plants are turning from seedlings to saplings or older individuals (Figure 4.3). When the individuals are young, seedling phase accumulation of salicylate in leaves is really high while it will decrease drastically when individuals reach sapling or adult phase: then salicylates are replaced by other phenolics, such as tannins, phenolic acids, and flavonoids.

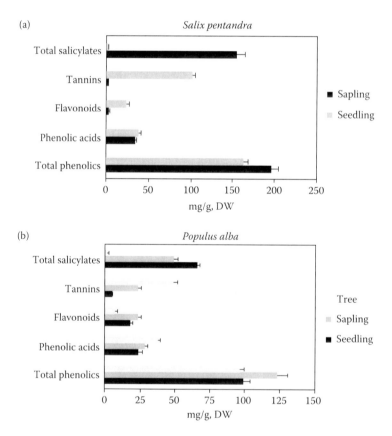

Figure 4.3 Ontogenetic changes of salicylates and other phenolics in the leaves of *Salix pentandra* (a) and *Populus alba* (b).

The effect of growing conditions, such as soil nutrients, light, temperature, and ultraviolet radiation (UVB) have been found to affect the salicylate content of Salicaceae plants to a certain extent. Nitrogen addition to soil decreased markedly the accumulation of salicylates/phenolics in leaves of *S. dasyclados* (= S. "Aquatica") (Julkunen-Tiitto 1989a), *S. dasyclados* (Larsson et al. 1986), and *P. tremuloides* (Bryant et al. 1987), and in the young twigs of *S. alaxensis* (Bryant 1987). Noteworthy was that relative amounts of salicylates were also changed; for instance, in *S. dasyclados* (= S. "Aquatica") the main salicylate, salicortin decreased while small molecular mass salicin slightly increased (Julkunen-Tiitto 1989a). Similarly, in *S. sericea* leaves, salicortin and 2'-O-cinnamoyl salicortin was found to decrease by nutrient addition, and showed interactions at low water availability (Lower and Orians 2003; Lower et al. 2003). Accordingly, *P. tremuloides* plants showed decline in salicortin and tremulacin at low nutrient supply toward the end of the growing season but this decrease was genotype dependent (Lindroth et al. 2002).

Some of the differences in salicylate content due to environmental factors may be induced by sexual dimorphism of Salicaceae. Randriamanana et al. (2014) studied sexual variation of six male and six female gentotypes of *P. tremula* in a greenhouse experiment under phosphorus and nitrogen addition. They did not detect any sexual difference in salicylate content while nitrogen addition to the soil increased salicylates in leaves. *Salicaceae* species have also been shown to respond differently to elevated temperature. For example, in young, greenhouse grown seedlings of *S. myrsinifolia* and of *P. tremula* x *P. tremuloides* leaf salicylate amount was markedly decreased at +2°C temperature enhancement (Kosonen et al. 2012; Veteli et al. 2002). In another greenhouse experiment, the corresponding decreasing effect of temperature in salicylates was found in twigs of both genders of *S. myrsinifolia* cutting plants (Nybakken and Julkunen-Tiitto 2013). On the contrary, *S. myrsinifolia* grown in the field site increased the content of most salicylates in the leaves due to temperature and UVB enhancement during the second (Nybakken et al. 2012) and third growing season (Randriamanana et al. 2015b) and due to UVB only after the third growing season (Randriamanana et al. 2015b). In the same study, where four male and four female genotypes were screened, no gender difference was found in salicylate content in leaves after the first two growing seasons or in seeds during the third growing season but after the third growing season, salicylates were induced in female leaves. This indicates that females could be more defended than males, and in future climate change this might affect the sex ratio and spatial distribution of *S. myrsinifolia* (Randriamanana et al. 2015b). Moreover, the effect of temperature and UVB has been shown to be different in *P. tremula*, where salicylates were increased by enhanced UVB and the effect was partly masked by enhanced temperature (Randriamanana et al. 2015a). There could be several explanations for these divergent responses, such as treatment duration, age of plants, different phase change and/or acclimation, that are clearly seen, for instance, in the same field experiment conducted over three growing seasons with *Betula pendula* plants (Tegelberg et al. 2002).

4.3 Biosynthesis and turnover of salicylates

4.3.1 Biosynthesis of salicylates

A biosynthetic pathway leading to salicin (glucoside of salicyl alcohol) has been studied to some extent while biosynthetic routes leading to higher-molecular-weight salicylates, such as acetylsalicortin or tremulacin, are almost lacking. It has been suggested that salicin is formed in leaf tissue from salicyl aldehyde via glucoside of salicyl aldehyde, helicin exluding salicyl alcohol as a direct precursor (Babst et al. 2010; Pridham and Saltmarsh 1963; Zenk 1967). Salicyl alcohol is derived from phenylpropanoid pathway either via benzoic acid or cinnamic acid (Figure 4.4) (Zenk 1967).

Some proof for cinnamic acid and benzoic acid in salicylate biosynthesis was obtained from studies with *S. pentandra* plantlets using 2-aminoindan-2-phosphonic acid that inhibit phenylalanine ammonialyase that converts phenylalanine to cinnamic acid. This inhibition strongly reduced the accumulation of salicylates in shoot tips of *S. pentandra* (Ruuhola and Julkunen-Tiitto 2003). Ruuhola and Julkunen-Tiitto also suggested that benzoic aldehyde and salicyl aldehyde could not be precursors *in vivo* for other salicylates due to their immediate toxicity and unstability. Salicyl alcohol was also converted in *S. purpurea* (Zenk 1967) and *P. nigra* leaves (Babst et al. 2010) into isosalicin (*o*-hydroxybenzyl-β-D-glucopyranoside). Accordingly, *S. daphnoides* shoots converted salicyl alcohol into isosalicin and only traces of salicin were found (Pridham and Saltmarsh 1963). In

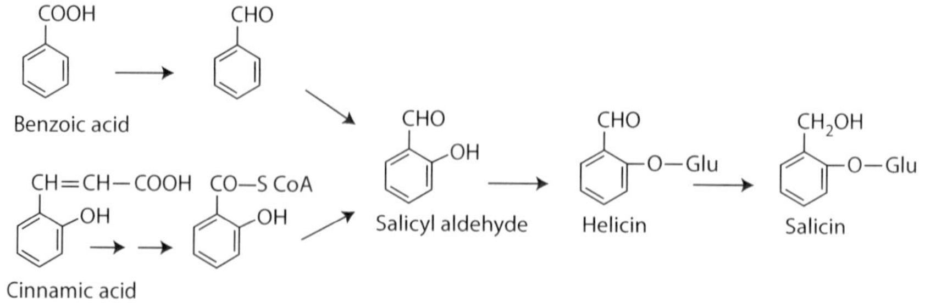

Figure 4.4 A schematic representation of salicin biosynthesis. (According to Zenk, M.H. 1967. *Phytochemistry* 6: 245–252.)

other species, such as in *Vicia faba* seeds, *Zea mays* seedlings (Pridham and Saltmarsh 1963) and *Geum urbanum* roots and old leaves (Psenák et al. 1969), salicyl alcohol is also converted to isosalicin. Moreover, *Geum urbanum* roots and old leaves infiltrated with salicin produced again isosalicin (Psenák et al. 1969).

Moreover, it has been suggested that cultured suspension cells are able to glucosylate phenolic substrates efficiently when the molecules are relatively small in size and have few substituents adjacent to hydroxy groups (Tabata et al. 1988). Cell suspension cultures of *S. matsudana* have been shown to be able to convert salicyl alcohol to both salicin and isosalicin, while salicyl aldehyde fed cells did not produce its β-D-glucoside, helicin, but instead aldehyde is reduced or oxidized to corresponding glucosides of salicyl alcohol and salicylic acid, respectively (Dombrowski 1993). In our study, salicylate-free cell cultures derived from several willow (*S. myrsinifolia, S. viminalis, S. pentandra, S. eleagnos,* and *S. matsudana*) species were all able to take up and convert feed salicyl alcohol, helicin, and salicin (Julkunen-Tiitto, Petersen, Alferman, unpublished results). Depending on the precursor and the species the major products were mono-β-D-glucosides (isosalicin and salicin). Moreover, the cells did not release the glucosylated products back into culture medium. There was free salicyl alcohol in cells fed with all precursors except in cells of *S. elaeagnos* and *S. matsudana* fed with salicyl alcohol. Both of these latter two species were capable of diglucosylation of salicyl alcohol. Interestingly, the cells of *S. viminalis* showed to be the most powerful for biotransformation of the precursors, although *in vivo* leaves of *S. viminalis* contain a very tiny amount of salicin or other salicylates (Julkunen-Tiitto 1989b; Julkunen-Tiitto et al. unpublished results). Instead, the cells of *S. myrsinifolia* which contain a moderate amount of salicin and a high amount of its derivative, salicortin *in vivo* (e.g., Julkunen-Tiitto 1989b; Nybakken et al. 2012; Paunonen et al. 2009) did take up and transform salicin and salicyl alcohol very slowly. The priority of the position of glucosylation in the substrate molecules showed to be willow species-specific and obviously determined by the phylogenetic development of the willow species. *S. pentandra*, the oldest and most primitive species that contains acetylsalicylates *in vivo* leaves (Egloff 1982; Julkunen-Tiitto 1989; Ruuhola and Julkunen-Tiitto 2003) diverged qualitatively from other species. Evolutively younger willow species produced mainly isosalicin while *S. pentandra* cells converted all precursors mainly to salicin. Helicin of the applied compounds was very quickly turned over mainly into isosalicin in *S. myrsinifolia, S. viminalis, S. elaeagnos* cells and salicin in *S. pentandra* cells (Julkunen-Tiitto et al. unpublished results). In willow screening studies done in our laboratory, helicin has not detected *in vivo* willow tissues in agreement with Babst et al. (2010) found in *P. nigra* leaf discs. Zenk (1967) has reported a small amount of helicin inside the cells of *S. purpurea* leaves and Babst et al. (2010) in salicylates fed *P. nigra* leaf discs. According to Babst et al. (2010) salicylaldehyde could be converted mostly to salicyl alcohol before conversion to salicin and helicin, found in trace amounts could be an artefact or salicyl aldehyde could be glycosylated to form helicin that could be immediately transformed to salicin. It is also curious that in our study, salicin-fed cells transform salicin to isosalicin, instead of keeping it as salicin. This may indicate that either the cells may have a limited pool size for salicin amount and thus, excess salicin should be transformed to isosalicin, or the production of salicin and isosalicin processes simultaneously so that production of isosalicin is favored due to more active alcohol group compared with the hydroxyl group in salicyl alcohol as suggested by Pridham and Saltmarsh (1963). Moreover, we did not find any higher molecular mass salicylates, such as salicortin or acetyl salicortin in cell suspensions fed with salicyl aldehyde, helicin or salicyl alcohol as was also found in *P. nigra* leaves by Babst et al. (2010).

Ruuhola and Julkunen-Tiitto (2003) showed that higher molecular mass salicylate, salicortin was strongly increased when benzoic acid was applied to the growth media of *S. pentandra* plantlets while salicin was not able to increase salicortin concentration. Babst et al. (2010) clarified for the first time the origin of 6-hydroxy-2-cyclohexen-on-oyl (HCH) moiety found in salicortins, acetylsalicortins, and tremulacins. HCH is an important part of those salicylates due to its suggested high defensive role against generalist herbivores (e.g., Lindroth and Hwang 1996b; Ruuhola et al. 2001). They fed *Populus nigra* leaves with isotope labeled cinnamic acid, benzoate and salicylates and measured label incorporation to salicin and salicortin. Babst et al. (2010) showed that cinnamic acid is not only a precursor of salicin but HCH was also cinnamic acid derived, with benzoic acid and benzaldehyde as intermediates in higher molecular mass salicylates, for example, salicortin. They also showed that salicylaldehyde or salicyl alcohol may not be the immediate precursors of salicortin while benzoic acid, benzaldehyde, and benzyl alcohol would be the most probable precursor of salicyl moiety of salicortin.

4.3.2 Turnover of salicylates

Salicylates with HCH moiety are regarded very labile compounds and start to decompose easily under improper prehandling of the plant material and tissue rupture (e.g., Clausen et al. 1989, 1991; Julkunen-Tiitto and Sorsa 2001; Julkunen-Tiitto and Tahvanainen 1988; Lindroth and Koss 1996; Pearl and Darling 1970; Ruuhola and Julkunen-Tiitto 2003). Degradation of salicortin will produce salicin, 6-HCH and catechol enzymaticly in alkaline conditions and also chemically (Julkunen-Tiitto and Meier 1992; Ruuhola and Julkunen-Tiitto 2003). Similarly, tremulacin and acetylsalicortin of leaf extract produce in alkaline conditions tremuloidin, acetylsalicin, salicin, 6-HCH, and catechol, while disalicortins already start to decompose at neutral conditions (Ruuhola and Julkunen-Tiitto, 2003). Noteworthy is that in neutral and alkaline conditions, the migration of acetyl and benzoyl groups from position 2' to position 6' will also happen, and for instance yields populin from tremuloidin (Pearl and Darling 1963; Ruuhola and Julkunen-Tiitto 2003).

4.4 Biological roles of salicylates

4.4.1 Salicylates in human health

The *Salix* species, especially, has a long use in herbal medicine, known to last for centuries as an analgesic, anti-inflammatory, antipyretic (used for, e.g., head, gut, and joints). They have been used to relieve pains though they do not heal illnesses (e.g., rheumatism). The earliest use of *Salix* in folk medicine originates about 3500 years ago, and is described in an Egyptian treatise (Ebens´Papyrus) (Pierpoint 1994; Setty and Sigal 2005). Edward Stone in 1763 introduced willow bark powder (called *Cortex salignus* or *Cortex salicis*) as an effective material in healing fever patients. In 1828, Bucher discovered and in 1838 Piria clarified an effective product from willows called "acide salicylic," while the final structure of salicin was discovered by Irvine and Rose (Thieme 1963). These works were the background for the development of synthetic aspirin (acetyl salicylic acid) by Fridrich Bayer & Co. in 1897. However, *Salix* drugs were regarded as better tolerable than aspirin, inducing no side effects (irritation of stomach) typical for aspirin (e.g., Förster et al. 2008). Later, especially, salicin, tremulacin, and salicortin were found to be the most effective compounds to relieve rheumatic disturbances, headache, and different infections: They are noninflammable, temperature-reducing, and pain-alleviating compounds (Chrubasik et al. 2000; Förster et al. 2008; Lardos et al. 2004). The positive mode of action of salicylates continues to increase the use of *Salix* drugs in herbal medicine (e.g., Förster et al. 2008; Setty and Sigal 2005); although other compounds in *Salicaceae* herbal drugs, such as tannins and flavonoids, may also contribute to the total response (Schmid et al. 2001).

4.4.2 Salicylates effective to herbivores

Salicaceae salicylates are known to have variable roles in immediate plant–herbivore interactions. Tissue destruction of *Salix* and the *Populus* plants will induce quick chemical changes in salicylate profiles and the release of HCH and other degradation products (e.g., Clausen et al. 1991; Julkunen-Tiitto and Meier 1992; Ruuhola and Julkunen-Tiitto 2003; Reichardt et al. 1988) (Figure 4.5): Higher molecular mass salicylates will degrade to smaller counterparts and even further more toxic products (e.g., catechol). HCH is known to be a powerful feeding deterrent to generalist herbivores

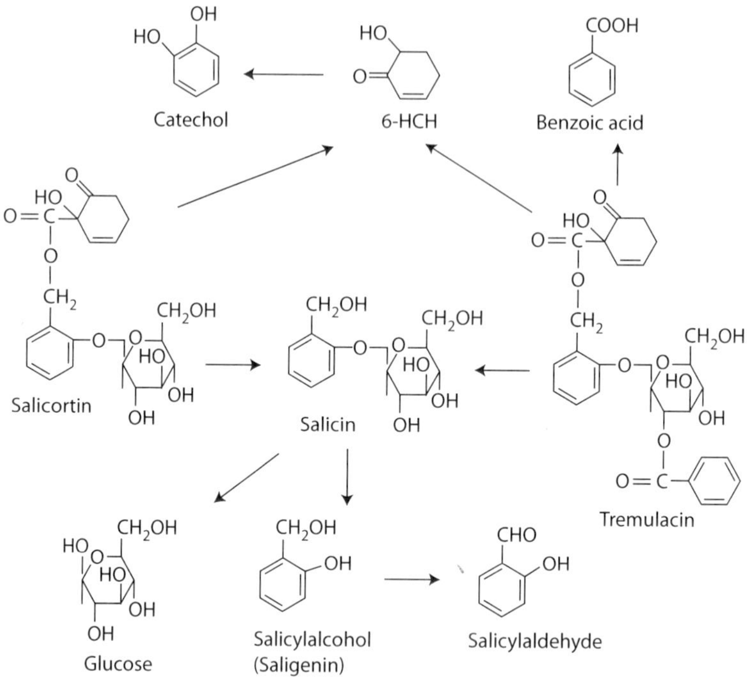

Figure 4.5 General decomposition routes of salicortin, tremulacin, and salicin.

(Clausen et al. 1989, 1991; Reichardt et al. 1990), and is an effective defensive mechanism for *Salicaceae* species against generalist herbivores. Ruuhola et al. (2001) tested decomposition of *S. myrsinifolia* and *S. pentandra* salicylates in generalist larvae, *Operophtera brumata*, and found that higher-molecular-weight salicylates were decomposed to salicin and catechol in the digestive tract, while salicin degradation was slow. The growth of larvae was also markedly reduced when fed with high salicylate leaves of *S. pentandra* and *S. myrsinifolia*. Similarly, it has been shown that salicylates affect food selection of *Salicaceae*-feeding insects and twigs browsing mountain hares and field voles, decreasing the growth, or acting as feeding deterrents of generalist herbivores and reducing damage of salicylate-producing plants (e.g., Boeckler et al. 2011, 2014; Bryant and Kuropat 1980; Clausen et al. 1989; Denno et al. 1990; Heiska et al. 2007; Lindroth et al. 1988; Reichardt et al. 1990; Tahvanainen et al. 1985a,b).

Although salicylates have a defensive role for salicylate-producing plants, there are also specialist herbivores that have overcome salicylates and their degradation products and are even able of using them as their own energy source (glucose) and transferring them to volatile defenses (salicylaldehyde) (Figure 4.5) (Kolehmainen et al. 1995; Pasteels et al. 1983;

Rank et al. 1998; Rowell-Rahier and Pasteels 1986). Contrasting results have been obtained recently in studies with salicylates of *P. tremuloides* and specialist leaf mining lepidopteran *Phyllocornis populiella* (Young et al. 2010). They showed that there is a threshold concentration of 27 mg/g DW of salicin and tremulacin and at higher concentrations herbivore damage was reduced. Generally, although more research is needed, especially for the induction of salicylates during damage (e.g., Clausen et al. 1989; Julkunen-Tiitto et al. 1996; Stevens and Lindroth 2005), for the role of salicylates, they could be regarded in plant–herbivore interactions as multifunctional plant compounds having a deterrent role and also a nutritional role when a defensive mode of action is lost against pests specialized to eat *Salicaceae* plant material.

4.4.3 Salicylates as deterrents to rusts

The rust fungi of the genus *Melampsora* Castagne (order Uredinales) are very common pathogenic fungi and cause serious rust disease and affect growth losses in *Salicaceae* species, especially in *Salix* species (Helfer 1992; Pei et al. 1996). Generally, it is known that there is a wide within and between species (Gullberg and Ryttman 1993; Julkunen-Tiitto et al. 1994; Pinon 1995) and temporal variation in rust frequency (Hakulinen et al. 1999; Heiska et al. 2007), while until now very little was known of rust-*Salicaceae* interactions. Julkunen-Tiitto et al. (1994) screened 10 *S. myrsinifolia* clones, growing in field conditions and showed that the salicortin concentration of leaves correlated negatively with rust intensity, while salicin did not indicate any correlation. It was concluded that the high constitutive level of salicortin in leaves could function as a nontoxic preexisting precursor as a possible postinfectional rust inhibitor. Catechol (Figure 4.5), a decomposition product of salicortin has been reported to have a high toxicity and to actively inhibit spore germination and hyphal growth (Vidhyasekaran 1988). This preformed rust-resistant role of salicortin could not be detected in another study, done for eight and older *S. myrsinifolia* clones over three successive years (Hakulinen and Julkunen-Tiitto 2000). Hakulinen (1998) has tested the rust-inoculation effect to the accumulation of phenolics, including salicylates in leaves of eight greenhouse growing *S. myrsinifolia* clones under low and optimum nitrogen levels. The rust infection induced significant accumulation of salicortin in leaves below the infected leaves in seven out of eight clones studied under low nitrogen fertilization, while there was no correlation between salicortin amounts and rust infection intensity (Hakulinen 1998). This indicates that local rust infection is able to systemic induction of salicortin and some other phenolics in the plant. However, rust infection was not increased although nitrogen addition reduced salicortin or other phenolics.

4.5 Conclusion

A wide variety of salicylates are signature compounds for *Salix* and *Populus* species, *Salicaceae*. They are also important due to the fact that they have markedly affected the health of humans for hundreds of years, and are obviously the most famous example of a still used medicine developed from an herbal source. Chemically, most of the salicylates are very labile compounds and care should be taken to preserve, extract, and analyze them. Salicylates show very wide variation by tissues, age, season, and environmental variables. Although they have been very important for humans, little is done to clarify their biosynthesis *in vivo* and their real role in plant physiology. Salicortin seems to respond to rust fungi, but salicylate role in the resistance of pathogens is still unsure. Because of their lability, higher-molecular-weight mass salicylates are decomposing easily to HCH (as one fragment), which has been shown to be high deterrents against generalist herbivores while specialist herbivores use salicylates for their own defense and energy. Finally, there are a lot of studies on salicylates in *Salicaceae* species although too little has been solved to get a clear picture of the overall role of this very abundant phenolic compound group in the species.

References

Abreu, I.N., M. Ahnlund, T. Moritz, and B.R. Albrectsen. 2011. UHPLC-ESI/TOFMS determination of salicylate-like phenolic glycosides in *Populus tremula* leaves. *J. Chem. Ecol.* 37: 857–870.

Argus, G.W. 1973. *The Genus* Salix *in Alaska and in Yukon*. Ottawa: National Museum of Canada.

Babst, B.A., S.A. Harding, and C.-J. Tsai. 2010. Biosynthesis of phenolic glycosides from phenylpropanoid and benzenoid precursors in *Populus*. *J. Chem. Ecol.* 36: 286–297.

Barz, W. and J. Köster. 1981. Turnover and degradation of secondary natural products. In *The Biochemistry of Plants, A Comprehensive Treatise*, Vol. 7, ed. E.E. Conn, 35–84. Orlando: Academic Press.

Barz, W., J. Köster, K.M. Weltring, and D. Strack. 1985. Recent advances in the metabolism and degradation of phenolic compounds in plants and animals. In *The Biochemistry of Plant Phenolics*, Vol. 25, ed. C.F. van Sumere, 35–84. Oxford: Clarenton Press.

Boeckler, G.A., J. Gershenzon, and S.B. Unsicker. 2011. Phenolic glycosides of the Salicaceae and their role as anti-herbivore defenses. *Phytochemistry* 72: 1497–1509.

Boeckler, G.A., M. Towns, S.B. Unsicker et al. 2014. Transgenic upregulation of condensed tannin pathway in poplar leads to a dramatic shift in leaf palatability for two tree-feeding Lepidoptera. *J. Chem. Ecol.* 40: 150–158.

Bryant, J.P. and P.J. Kuropat. 1980. Selection of winter forage by sub-arctic browsing vertebrates—Role of plant chemistry. *Ann. Rev. Ecol. Syst.* 11: 261–285.

Bryant, J.P. 1987. Feltleaf willow-snowhare hare interactions: Plant carbon/nutrient balance and foodplain succession. *Ecology* 68: 1319–1327.

Bryant, J.P., T.P. Clausen, P.B. Reichardt, M.C. McCarthy, and R.A. Werner. 1987. Effect of nitrogen fertilization upon secondary chemistry and nutritional value of quaking aspen (*Populus tremuloides* Michx.) leaves for the large aspen tortrix (*Choristoneura conflictana* [Walker]. *Oecologia* 73: 513–517.

Chrubasik, S., E. Eisenberg, E. Balan, T. Weinberger, R. Luzzati, and C. Conradt. 2000. Treatment of low back pain exacerbations with willow bark extract: A randomized double- blind study. *Am. J. Med.* 109: 9–14.

Clausen, T.P., P.B. Reichardt, J.P. Bryant, R.A. Werner, K. Post, and K. Frisby. 1989. Chemical model for short-term induction in quaking aspen (*Populus tremuloides*) foliage against herbivores. *J. Chem. Ecol.* 15: 2335–2346.

Clausen, T.P., P.B. Reichardt, J.P. Bryant, and R.A. Werner. 1991. Long-term and short-term induction in quaking aspen: Related phenomena? In *Phytochemical Induction by Herbivores*, eds. D.W. Tallamy and M.J. Raupp, 71–83. New York: John Wiley & Sons Incorporated.

Dement, W.A. and H.A. Mooney. 1974. Seasonal variation in the production of tannins and cyanogenic glucosides in chaparral shrub, *Heleromeles arbutifolia*. *Oecologia* (Berl.) 15: 6–76.

Denno, R.F., S. Larsson, and K.L. Olmstead. 1990. Role of enemy-free space and plant quality in host-plant selection by willow beetles. *Ecology* 71: 124–137.

Dombrowski, K. 1993. Phytochemische und enzymologische Untersuchungen zur Biotransformation von Salicylverbindungen durch Zellkulturen der Weidenart *Salix matsudana* f. *Tortuosa*. Ph.D. thesis, Düsseldorf University.

Egloff, C.P. 1982. Phenolglycoside einheimisscher *Salix*-Arten. Ph.D. thesis, Zürich, Nr. 7138.

Feeny, P. 1970. Seasonal change in oak leaf tannins and nutrients as a cause of spring feeding by winter moth caterpillars. *Ecology* 4: 565–581.

Fritz, R.S., C.G. Hochwender, D.A. Lewkiewicz, S. Bothwell, and C.M. Orians. 2001. Seedling herbivory by slugs in a willow hybrid system: Developmental changes in damage, chemical defense, and plant performance. *Oecologia* 129: 87–97.

Förster, N., C. Ulrichs, M. Zander, R. Kätze, and I. Mewis. 2008. Influence of the season on the salicylate and phenolic glycoside contents in the bark of *Salix daphnoides*, *Salix pentandra*, and *Salix purpurea*. *J. Appl. Bot. Food Qual.* 82: 99–102.

Gershenzon, J. 1994. The cost of plant chemical defense against herbivory: A biochemical perspective. In *Insect–Plant Interactions*, Vol. 5, ed. E.A. Bernays, 105–173. Boca Raton, FL: CRC Press.

Gullberg, U. and H. Ryttman. 1993. Genetics of field resistance to *Melampsora* in *Salix viminalis*. *Eur. J. For. Pathol.* 23: 75–84.

Hakulinen, J. and R. Julkunen-Tiitto. 2000. Variation in leaf phenolics of field cultivated willow (*Salix myrsinifolia* Salisb.) clones in relation to occurrence of *Melampsora* rust. *Eur. J. For. Path.* 30: 29–41.

Hakulinen, J., R. Julkunen-Tiitto, and J. Tahvanainen. 1995. Does nitrogen fertilization have an impact on the trade-off between willow growth and defensive secondary metabolism? *Trees* 9: 235–240.

Hakulinen, J., S. Sorjonen, and R. Riitta Julkunen-Tiitto. 1999. Leaf phenolic of the three willow clones differing in development of *Melampsora* rust infection. *Physiol. Plant.* 105: 662–669.

Hakulinen. J. 1998. Nitrogen induced reduction in leaf phenolic level is not accompanied by increased rust frequency in a compatible willow (*Salix myrsinifolia*)- *Melampsora* rust infection. *Physiol. Plant.* 102: 101–110.

Heiska, S., O.-P. Tikkanen, M. Rousi et al. 2007. The susceptibility of herbal willow to *Melampsora* rust and herbivores. *Eur. J. Plant. Pathol.* 118: 275–285.
Helfer, S. 1992. The rust disease of willows in Britain. *Proc. R. Soc. Edinburg* 98B:119–134.
Hämet-Ahti, L., J. Suominen, T. Ulvinen, and P. Uotila. 1998. *Retkeilykasvio.* Helsinki: Natural Central Museum, University Press.
Ikonen, A., J. Tahvanainen, and H. Roininen. 2002. Phenolic secondary compounds as determinants of the host plant preferences of the leaf beetle *Agelastica alni. Chemoecology* 12: 125–131.
Julkunen-Tiitto, R. 1985. Phenolic constituents in the leaves of Northern Willows: Methods for the analysis of certain phenolics. *J. Agr. Food Chem.* 33: 213–217.
Julkunen-Tiitto, R. 1986. A chemotaxonomic survey of phenolics in leaves of northern *Salicaceae* species. *Phytochemistry* 25: 663–667.
Julkunen-Tiitto, R. 1989a. Phenolic compounds of the genus *Salix*: A chemotaxonomical survey of further Finnish species. *Phytochemistry* 28: 2115–2125.
Julkunen-Tiitto, R. 1989b. *Distribution of Certain Phenolics in Salix Species (Salicaceae).* Joensuu: University of Joensuu.
Julkunen-Tiitto, R. 1996. Growth versus defense in photomixotrophic culture conditions: The effect of sucrose, nitrogen and pH on the phytomass and secondary phenolics in *S. myrsinifolia* plantlets. *Ecoscience* 3: 297–303.
Julkunen-Tiitto, R. and J. Tahvanainen. 1988. The effect of the sample preparation method on extractable phenolics of *Salicaceae* species. *Planta Med.* 55: 55–58.
Julkunen-Tiitto, R. and J. Tahvanainen. 1989. The effect of the sample preparation method on extractable phenolics of *Salicaceae* species. *Planta Med.* 55: 55–58.
Julkunen-Tiitto, R. and B. Meier. 1992. Enzymatic decomposition of salicin and its derivatives obtained from *Salicaceae* species. *J. Nat. Prod.* 55: 1204–1212.
Julkunen-Tiitto, R. and K. Gebhardt. 1992. Further studies on drying willow (*Salix*) twigs: The effect of low drying temperature on labile phenolics. *Planta Med.* 58: 385–386.
Julkunen-Tiitto, R., J. Hakulinen, and B. Meier. 1994. The response of growth and secondary metabolism to *Melampsora* rusts in field cultivated willow (*Salix*) clones. *Acta Horticulturae* 381: 679–682.
Julkunen-Tiitto, R., M. Rousi, J.P. Bryant, S. Sorsa, M. Keinänen, and H. Sikanen. 1996. Chemical diversity of several Betulaceae species: Comparison of phenolics and terpenoids in northern birch stems. *Trees* 11:16–22.
Julkunen-Tiitto, R. and S. Sorsa. 2001. Testing the drying methods for willow flavonoids, tannins and salicylates. *J. Chem Ecol.* 27: 779–789.
Julkunen-Tiitto, R., L. Nybakken, T. Randriamanana, and V. Virjamo. 2015. Boreal woody species resistance affected by climate change. In *Climate Change and Insect Best*, ed. C. Björkman and P. Niemelä, 54–73. Preston, United Kingdom: CAB International.
Juntheikki, M.-R., R. Julkunen-Tiitto, and A. Hagerman. 1996. Salivary tannin binding proteins in root vole (*Microtus oeconomus*). *Biochem. Syst. Ecol.* 24: 25–35.
Kolehmainen, J., R. Julkunen-Tiitto, H. Roininen, and J. Tahvanainen. 1995. Phenolic glycosides as feeding cues for willow-feeding beetles. *Entomol. Exp. Appl.* 74: 235–243.
Kosonen, M., S. Keski-Saari, T. Ruuhola, P. Constabel, and R. Julkunen-Tiitto. 2012. Effects of overproduction of condensed tannins and elevated temperature on chemical and ecological traits of genetically modified hybrid aspens (*Populus tremula* × *P. tremuloides*). *J. Chem. Ecol.* 38: 1235–1246.

Lardos, A., C.B. Schmidlin, M. Fischer et al. 2004. Wirksamkeit und Verträglichkeit eines wässrig ausgezogenen Weidenrindenextraktes bei Patienten mit Hüft- und Kniearthrose. *Zeitschrift für Phytotherapie* 25: 275–281.

Larsson, S., A. Wiren, L. Lundgren, and T. Ericsson. 1986. Effects of light and nutrient stress on phenolic chemistry in *Salix dasyclados* and susceptibility to *Galerucella lineola* (Coleoptera). *Oikos* 47: 205–210.

Lindroth, R.L., M.T.S. Hsia, and J.M. Scriber. 1987. Seasonal pattern in the phytochemistry of three *Populus* species. *Biochem. Syst. Ecol.* 15: 681–686.

Lindroth, R.L., J.M. Scriber, and M.T.S. Hsia. 1988. Chemical ecology of the tiger swallowtail: Mediation of host use by phenolic glycosides. *Ecology* 69: 814–822.

Lindroth, R.L. and P.A. Koss. 1996. Preservation of Salicaceae leaves for phytochemical analyses: Further assessment. *J. Chem. Ecol.* 22: 765–771.

Lindroth, R.L. and S.Y. Hwang. 1996a. Clonal variation of foliar chemistry of quaking aspen (*Populus tremuloides* Michx.). *Biochem. Syst. Ecol.* 24: 357–364.

Lindroth, R.L. and S.Y. Hwang. 1996b. Diversity, redundancy and multiplicity in chemical defense systems of aspen. *Recent Adv. Phytochem.* 30: 25–26.

Lindroth, R.L. and M.S. Pajutee. 1987. Chemical analysis of phenolic glycosides: Art, facts, and artifacts. *Oecologia* 74(1): 144–148.

Lindroth, R.L., T.L. Osier, H.R.H. Barnhill, and S.A.Wood. 2002. Effects of genotype and nutrient availability on phytochemistry of trembling aspen (*Populus tremuloides* Michx) during leaf senescence. *Biochem. System. Ecol.* 30: 297–307.

Lower, S.S. and C.M. Orians. 2003. Soil nutrients and water availability interact to influence willow growth and chemistry but not leaf beetle performance. *Entomol. Exp. Appl.* 107: 69–79.

Lower, S.S., S. Kirshenbaum, and C.M. Orians. 2003. Preference and performance of a willow feeding leaf beetle: Soil nutrient and flooding effects on host quality. *Oecologia* 136: 402–411.

Massad, T.J., S.E. Trumbore, G. Ganbat et al. 2014. An optimal defense strategy for phenolic glycoside production in *Populus trichocarpa*—Isotope labelling demonstrates secondary metabolite production in growing leaves. *New Phytol.* 203: 607–619.

Meier, B. 1988. *Analytik, chromatographische Verhalten und potentielle Wirksamkeit der inhaltstoffe salicylhaltiger Arzneipflanzen Mitteleuropas.* Habilitationschrift, Zürich: Eidgenöschishe Technische Hochschule.

Meier, B., R. Julkunen-Tiitto, J. Tahvanainen, and O. Sticher. 1988. Comparative HPLC and GLC determination of phenolic glucosides in *Salicaceae* species. *J. Chromatogr.* 442: 175–186.

Nybakken, L., R. Hörkkä, and R. Julkunen-Tiitto. 2012. Combined enhancements of temperature and UVB influence growth and phenolics in clones of the sexually dimorphic *Salix myrsinifolia*. *Physiologica Plantarum* 145: 551–564.

Nybakken, L. and R. Julkunen-Tiitto. 2013. Gender differences in *Salix myrsinifolia* at the pre-reproductive stage are little affected by simulated climatic change. *Physiol. Plant.* 147: 465–476.

Orians, C.M. 1995. Preserving leaves for tannin and phenolic glucoside analyses: A comparison of methods using three willow taxa. *J. Chem Ecol.* 21: 1235–1343.

Paajanen, R., R. Julkunen-Tiitto, L. Nybakken et al. 2011. Dark-leaved willow (*Salix myrsinifolia*) is resistant to three-factor climate change. *New Phytol.* 190: 161–168.

Pasteels, J.M., M. Rowell-Rahier, J.C. Braekman, and A. Dupont. 1983. Salicin from host plants as precursors of salicylaldehyde in defensive secretion of Chrysomelinea larvae. *Physiol. Entomol.* 8: 307–314.

Paunonen, R., R. Julkunen-Tiitto, M. Rousi, R. Tegelberg, and S. Heiska. 2009. Salicylate and biomass yield, and leaf phenolics of dark-leaved willow (*Salix myrsinifolia* Salisb.) clones under different cultivation methods after the second cultivation cycle. *Industrial Crops Products* 29: 261–268.

Pearl, I.A. and S.F. Darling. 1963. Studies on leaf of the family Salicaceae. III. Migration of acyl groups during isolation of glycoside from *Populus grandidentata* leaves. *Arch. Biochem. Biophys.* 102: 33.38.

Pearl, I.A. and S.F. Darling. 1970. Phenolic extractives of *Salix purpurea* bark. *Phytochemistry* 9: 1277–1281.

Pearl, I.A. and S.F. Darling. 1971. The structures of salicortin and tremulacin. *Phytochemistry* 10: 3161–3166.

Pei, M.H., D.J. Royle, and T. Hunter. 1996. Pathogenic specialization in *Melampsora epitea* var. *epitea* on *Salix*. *Plant Pathol.* 45: 679–690.

Pierpoint, W. 1994. Salicyl acid and its derivatives in plants: Medicines, metabolites and messenger molecules. *Adv. Bot. Res.* 20: 163–235.

Pinon, J. 1995. Variability in the genus *Populus* in sensitivity to *Melampsora* rusts. *Silv. Genet.* 41: 25–34.

Pridham, J.B. and M.J. Saltmarsh. 1963. The biosynthesis of phenolic glucosides in plants. *Biochem. J.* 87: 218–224.

Psenák, M., P. Kovács, and A. Jindra. 1969. Glucose transferase in roots of *Geum urbanum*. *Phytochemistry* 8: 1655–1670.

Randriamanana, T.R., L. Nybakken, A. Lavola, P.J. Aphalo, K. Nissinen, and R. Julkunen-Tiitto. 2014. Sex-related differences in growth and carbon allocation to defence in *Populus tremula* as explained by current plant defence theories. *Tree Physiol.* 34: 471–487.

Randriamanana, T.R., A. Lavola, and R. Julkunen-Tiitto. 2015a. Interactive effects of supplemental UV-B and temperature in European aspen seedlings: Implications for growth, leaf traits, phenolic defense and associated organisms. *Plant Physiol. Biochem.* 93: 84–93.

Randriamanana, T.R., K. Nissinen, J. Moilanen, L. Nybakken, and R. Julkunen-Tiitto. 2015b. Long-term UV-B and temperature enhancements suggest that females of *Salix myrsinifolia* plants are more tolerant to UV-B than males. *Environ. Exp. Bot.* 109: 296–305.

Rank, N.E., A. Köpf, R. Julkunen-Tiitto, and J. Tahvanainen. 1998. Host preference and larval performance of the salicylate-using leaf beetle *Phratora vitellinae*. *Ecology* 79: 618–631.

Reichardt, P.B., T.P. Clausen, and J.P. Bryant. 1988. A critical look at the role of phenol glycosides in plant defense. In *Biologically Active Natural Products*, ed. H.G. Cutler, 130–142. Washington D.C.: ACS Symposium Series. Natural Products for Potential Use in Agriculture.

Reichardt, P.B., J.P. Bryant, B.R. Mattes, T.P. Clausen, F.S.III Chapin, and M. Meyer. 1990. Winter chemical defense of Alaskan balsam poplar against snowshoe hares. *J. Chem. Ecol.* 16: 1941–1959.

Rowell-Rahier, M. and J.M. Pasteels. 1986. Economics of chemical defense in Chrysomelinae. *J. Chem. Ecol.* 12: 1189–1203.

Ruuhola, T.M., O.-P. Tikkanen, and J. Tahvanainen. 2001. Differences in host use efficiency of larvae of a generalist moth, *Operothera brumata* (Lepidoptera: Geometridae) on three chemically divergent *Salix* species. *J. Chem. Ecol.* 27: 1595–1615.

Ruuhola, T. and R. Julkunen-Tiitto. 2003. Trade-off between the synthesis of salicylates and the growth of micropropagated *Salix pentandra* plants. *J. Chem. Ecol.* 29: 1565–1588.
Schmid, B., R. Lüdtke, H.-K. Selbmann et al. 2001. Efficacy and tolerability of a standardized willow bark extract in patients with Osteoarthritis: Randomized placebo-controlled, double blind clinical trial. *Phytother. Res.* 15: 344–350.
Setty, A. and L. Sigal. 2005. Herbal medications commonly used in the practice of rheumatology: Mechanisms of action, efficacy and side effects. *Seminar in Arthritis and Rheumatism* 43: 773–784.
Shao, Y. 1991. Phytochemischer Atlas der Schweizer Weiden. Dissertation ETH, Nr. 9532, Zürich.
Sivadasan, U., T. Randriamanana, L. Nybakken, and R. Julkunen-Tiitto. 2015. Responses of vegetative buds of *Salix myrsinifolia* to temperature and UVB elevation. *Plant Physiol. Biochem.* 93: 66–73.
Skvortsov, A.K. 1999. *Willows in Russia and Adjacent Countries. Taxonomical and Geographical Revision*. Joensuu: University of Joensuu, Faculty of Mathematics and Natural Sciences, Report Series 39.
Stevens, M.T. and R.L. Lindroth. 2005. Induced resistance in the intermediate growth of aspen (*Populus tremuloides*). *Oecologia* 145: 298–306.
Tabata, M., Y. Umetani, M. Ooya, and S. Tanaka. 1988. Glucosylation of phenolic compounds by plant cell cultures. *Phytochemistry* 27: 809–813.
Tahvanainen, J., E. Helle, R. Julkunen-Tiitto, and A. Lavola. 1985a. Phenolic compounds of willow bark as deterrents against feeding by mountain hare. *Oecologia* 65: 319–323.
Tahvanainen, J., R. Julkunen-Tiitto, and J. Kettunen. 1985b. Phenolic glycosides govern the food selection pattern of willow feeding leaf beetles. *Oecologia* 67: 52–56.
Tegelberg, R., R. Julkunen-Tiitto, and P.J. Aphalo. 2002. The effects of long-term elevated ultraviolet-B radiation on phytochemicals in the bark of silver birch (*Betula pendula*). *Tree Physiol.* 22: 1257–1263.
Thieme, H. 1963. Die Phenolglycoside der Salicaceen. 1. Mitteilung: Allgemeine Übersicht. *Pharmazie* 19: 770–774.
Thieme, H. 1965a. Die Phenolglycoside der Salicaceen. 5. Mitteilung: Untersuchungen über neu isolierte Glycoside und neuere Arbeiten zur Structuraufklärerung: Nachweis und Bestimmung der neuen Glycoside. *Pharmazie* 20: 570–574.
Thieme, H. 1965b. Die phenolglycoside der Salicaceaen. 6. Mitteilung: Untersuchungen uber die jahreszeitlich bedingten Veränderungen de Glycosidkonzentrationen, uber die Abbhängigkeit de glycosidgehalts von der Tageszeit und vom Alte der Planzenorgane. *Pharmazie* 20: 688–691.
Veteli, T.O., K. Kuokkanen, R. Julkunen-Tiitto, H. Roininen, and J. Tahvanainen. 2002. Effects of elevated CO_2 and temperature on plant growth and herbivore defensive chemistry. *Glob. Change Biol.* 8: 1240–1252.
Vidhyasekaran, P. 1988. *Physiology of Disease Resistance in Plants*. Boca Raton, FL: CRC Press.
Young, B., D. Wagner, P. Doak, and T. Clausen. 2010. Within-plant distribution of phenolic glycosides and extrafloral nectaries in trembling aspen (*Populus tremuloides*, Salicaceae). *Am. J. Bot.* 97: 601–610.
Zenk, M.H. 1967. Pathways of salicyl alcohol and salicin formation in *Salix purpurea* L. *Phytochemistry* 6: 245–252.

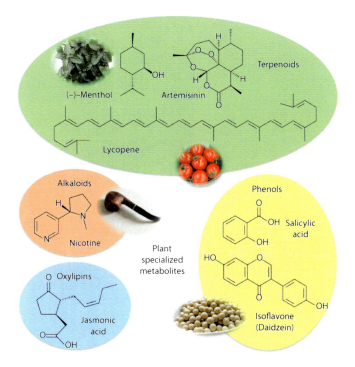

Figure 1.1 Representative plant specialized metabolites.

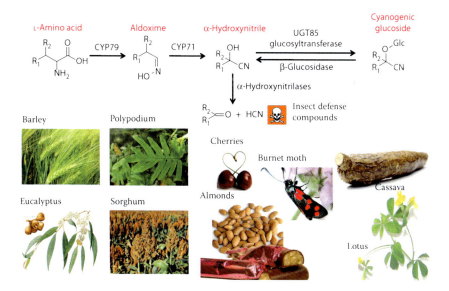

Figure 7.1 Schematic illustration of biosynthesis and bioactivation of cyanogenic glucosides in crop plants.

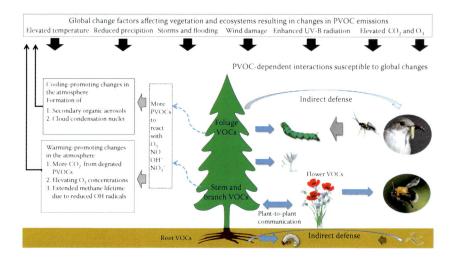

Figure 9.5 Effect of global change factors on PVOC emissions from vegetation, PVOC-based feedback response to the atmosphere, and PVOC-dependent interactions susceptible to global changes in the ecosystem.

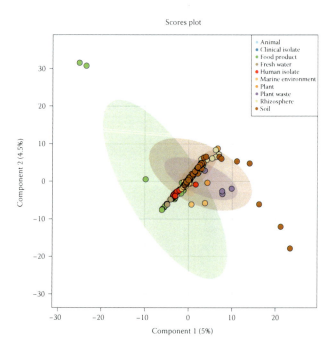

Figure 10.1 Principal component analysis of mVOC-emitting microorganisms of different habitats. (From Xia, J. et al. 2015. *Nucl. Acid Res.* 43: W251–W257.)

Figure 11.2 Photographs of dietary crops, which produce volatile glycosides. Gray, blue, and green shading indicates glucose, xylose, and apiose, respectively. Yellow circle shading indicates volatile aglycone.

Figure 12.2 Herbivory-induced volatile organic compound (VOC) emission mediating direct and indirect resistance. (Photos: Danny Kessler (a) André Kessler (b).)

Figure 12.3 The wild tomato, *Solanum peruvianum* (Solanaceae) with one of its pollinators, a Halictidae bee. (From Kessler, A., R. Halitschke, and K. Poveda. et al. 2011. *Ecology* 92: 1769–1780. Photo: André Kessler.)

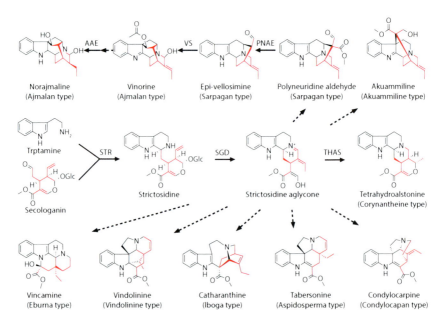

Figure 13.7 Formation of various MIAs from central precursor strictosidine. The C10 monoterpene moiety is labeled in red. Solid arrows show the identified catalytic steps, and the dotted arrows show the unidentified steps.

chapter five

Alkaloid biosynthesis and regulation in plants

Tsubasa Shoji

Contents

5.1 Introduction 85
5.2 Benzylisoquinoline alkaloids 86
    5.2.1 Pathways and enzymes 86
    5.2.2 Trafficking in opium poppies and *C. japonica* 91
    5.2.3 Transcriptional regulation 92
5.3 Monoterpenoid indole alkaloids 92
    5.3.1 Pathways and enzymes 93
    5.3.2 Trafficking in *C. roseus* 97
    5.3.3 Transcriptional regulation 99
5.4 Nicotine and tropane alkaloids (TAs) 100
    5.4.1 Pathways and enzymes 100
    5.4.2 Tobacco nicotine transporters 103
    5.4.3 Regulation of nicotine biosynthesis 103
5.5 Purine alkaloids 104
5.6 Perspectives 106
Acknowledgment 108
References 108

5.1 Introduction

Alkaloids are a diverse group of specialized metabolites with nitrogen-containing heterocyclic rings (Roberts and Wink 1998). Approximately 12,000 different alkaloids have been identified in nature, not just majorly in plants but also in microorganisms and animals. Over 20% of flowering plant species are estimated to produce alkaloids. Owing to their biological activities, many alkaloids have been used as medicines, poisons, and narcotics; even today natural alkaloids are considered important lead compounds for designing novel pharmaceuticals (Mishra and Tiwari 2011). Alkaloids accumulated in plants typically act as defense toxins against herbivores and pathogens, conferring the plants' adaptive advantages and allowing for their explosive expansion and diversification. Alkaloids are generally synthesized from amino acids or nucleotides and can be classified into a number of structurally related groups based on the precursors and pathways from which they originated. Molecular and genomic approaches have been increasingly applied to studies regarding alkaloid biosynthesis in plants, facilitating the discovery of the genes involved and providing a large array of molecular tools for more flexible and effective metabolic engineering in plants and microbial hosts (Facchini and De Luca 2008; Glenn et al. 2013; Ziegler and Facchini 2008).

5.2 Benzylisoquinoline alkaloids

Benzylisoquinoline alkaloids (BIAs) are a diverse group of alkaloids with approximately 2500 known structures, including the analgesics morphine and codeine, antibacterial sanguinarine and berberine, muscle relaxant papaverine, cough suppressant and anticancer agent noscapine, and other pharmacologically valuable substances. BIAs mainly occur in the Papaveraceae, Ranunculaceae, Berberidaceae, and Menispermaceae families. As such, elucidation of BIA biosynthesis has been intensively studied using opium poppies (*Papaver somniferum*, Papaveraceae), California poppies (*Eschscholzia californica*, Papaveraceae), *Coptis japonica* (Ranunculaceae), and *Thalictrum* species (Ranunculaceae) (Figure 5.1; Beaudoin and Facchini 2014; Hagel and Facchini 2013).

5.2.1 Pathways and enzymes

Two tyrosine derivatives, dopamine and 4-hydroxyphenylacetaldehyde (4-HPAA), are formed from tyrosine via tyramine and 4-hydroxyphenylpyruvate (4-HPP), respectively. Decarboxylation of tyrosine to tyramine and transamination of tyrosine to 4-HPP are catalyzed by tyrosine decarboxylase (TYDC) (Facchini and De Luca 1994) and tyrosine aminotransferase (TyrAT) (Lee and Facchini 2011), respectively. (S)-Norcoclaurine is generated through the Pictet–Spengler condensation of 4-HPAA and dopamine catalyzed by norcoclaurine synthase (NCS). NCS is a protein of the pathogenesis-related 10 (PR10) and Bet v 1 allergen protein family in opium poppies and *T. flavum* (Lee and Facchini 2010; Liscombe et al. 2005; Samanani et al. 2004). *Coptis japonica* has two types of NCS, PR10-type CjPR10A and CjCNS, a protein with a similarity to 2-oxoglutarate-dependent dioxygenase (ODD) but lacking a conserved 2-oxoglutarate-binding motif (Lee and Facchini 2010; Minami et al. 2007). (S)-Norcoclaurine is converted to (S)-reticuline through a series of modifications catalyzed by norcoclaurine 6-O-methyltransferase (6OMT) (Facchini and Park 2003;

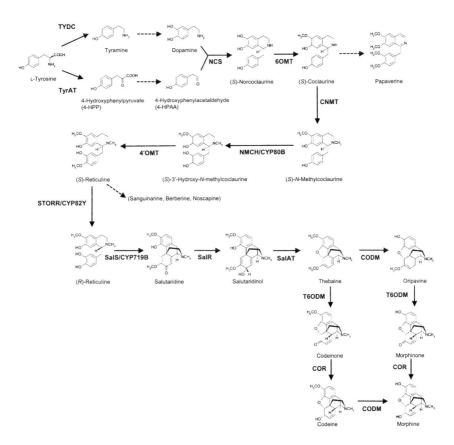

Figure 5.1 Biosynthetic pathways for benzylisoquinoline alkaloids. Enzymes for which cDNAs have been cloned are shown in bold. (Abbreviations: TYDC, tyrosine decarboxylase; TyrAT, tyrosine aminotransferase; 4-HPP, 4-hydroxyphenylpyruvate; 4-HPAA, 4-hydroxyphenylacetaldehyde; NCS, norcoclaurine synthase; 6OMT, norcoclaurine 6-*O*-methyltransferase; CNMT, coclaurine *N*-methyltransferase; NMCH, *N*-methylcoclaurine 3′-hydroxylase; 4′OMT, 3′-hydroxy-*N*-methylcoclaurine 4′-*O*-methyltransferase; STORR, (*S*)- to (*R*)-reticuline; SalS, salutaridine synthase; SalR, salutaridine reductase; SalAT, salutaridine 7-*O*-acetyltransferase; CODM, codeine *O*-demethylase; T6ODM, thebaine 6-*O*-demethylase; COR, codeinone reductase; BBE, berberine bridge enzyme; CFS, cheilanthifoline synthase; SPS, stylopine synthase; TNMT, tetrahydroprotoberberine *cis*-*N*-methyltransferase; MSH, (*S*)-*cis*-*N*-methylstylopine 14-hydroxylase; P6H, protopine 6-hydroxylase; DBOX, dihydrobenzophenanthridine oxidase; SanR, sanguinarine reductase; SOMT, scoulerine 9-*O*-methyltranfease; CAS, canadine synthase; STOX, (*S*)-tetrahydroberberine oxidase; AT1, acetyltransferase1; CXE1, carboxyesterase1; NOS, noscapine synthase.)

(*Continued*)

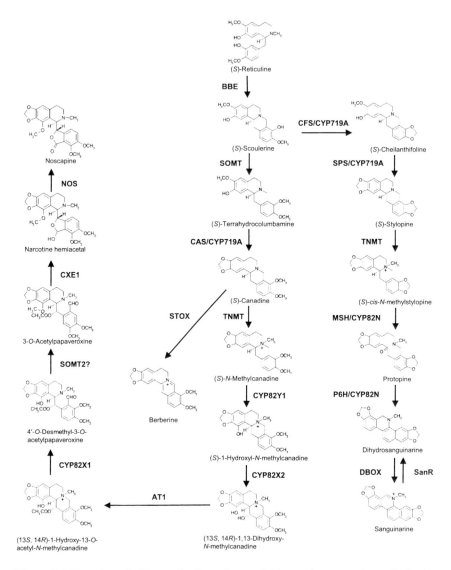

Figure 5.1 (Continued) Biosynthetic pathways for benzylisoquinoline alkaloids. Enzymes for which cDNAs have been cloned are shown in bold. (Abbreviations: TYDC, tyrosine decarboxylase; TyrAT, tyrosine aminotransferase; 4-HPP, 4-hydroxyphenylpyruvate; 4-HPAA, 4-hydroxyphenylacetaldehyde; NCS, norcoclaurine synthase; 6OMT, norcoclaurine 6-O-methyltransferase; CNMT, coclaurine N-methyltransferase; NMCH, N-methylcoclaurine 3'-hydroxylase; 4'OMT, 3'-hydroxy-N-methylcoclaurine 4'-O-methyltransferase; STORR, (S)- to (R)-reticuline; SalS, salutaridine synthase; SalR, salutaridine reductase; SalAT, salutaridine 7-O-acetyltransferase; CODM, codeine O-demethylase; T6ODM, thebaine 6-O-demethylase; COR, codeinone reductase; BBE, berberine bridge

Morishige et al. 2000; Ounaroon et al. 2003), coclaurine *N*-methyltransferase (CNMT) (Choi et al. 2002; Facchini and Park 2003), CYP80B subfamily *N*-methylcoclaurine 3′-hydroxylase (NMCH) (Haung and Kutchan 2000; Ikezawa et al. 2003; Pauli and Kutchan 1998), and 3′-hydroxy-*N*-methylcoclaurine 4′-*O*-methyltransferase (4′OMT) (Facchini and Park 2003; Morishige et al. 2000; Ziegler et al. 2005). (*S*)-Reticuline serves as a central intermediate among pathways leading to different structural subgroups of BIAs, as some BIAs such as papaverine are synthesized via upstream intermediates of reticuline (Desgagné-Penix and Facchini 2012; Pathak et al. 2013).

The pathway leading to codeine, morphine, and other morphinan alkaloids begins with a two-step epimerization of (*S*)-reticuline to its *R* isomer via 1,2-dehydroreticuline. The genetic locus responsible for this epimerization, known as (*S*)- *to* (*R*)-*reticuline* (STORR), encodes a fusion protein comprising CYP82Y2 and aldo–keto reductase (AKR) modules (Winzer et al. 2015). The N-terminal CYP82Y2 mediates the oxidation of (*S*)-reticuline to 1,2-dehydroreticuline, while subsequent reduction to (*R*)-reticuline depends on the C-terminal AKR-encoding portion (Farrow et al. 2015; Winzer et al. 2015). The modular assembly of STORR implies the occurrence of substrate channeling for sequential reactions. (*R*)-Reticuline is converted to salutaridine through C–C phenol coupling catalyzed by salutaridine synthase (SalS) of the CYP719B subfamily (Gesell et al. 2009). Salutaridine is reduced by a short-chain dehydrogenase/reductase (SDR), salutaridine reductase (SalR), to salutaridinol (Higashi et al. 2011; Ziegler et al. 2006), which then undergoes *O*-acetylation via salutaridinol 7-*O*-acetyltransferase (SalAT) (Grothe et al. 2001). Spontaneous cyclization of the resultant salutaridinol-7-*O*-acetate affords the pentacyclic BIA, thebaine. Thebaine is subsequently converted to morphine through two pathways. In the major pathway, thebaine is initially demethylated by an ODD-family enzyme, thebaine 6-*O*-demethylase (T6ODM), to neopione, which is spontaneously rearranged to codeinone (Hegal and Facchini 2010). Codeinone is then reduced to codeine via an NADPH-dependent AKD-family codeinone reductase (COR) bearing sequence similarity to the C-terminal module of STORR (Unterlinner et al. 1999). Finally, codeine is demethylated by an ODD similar to T6ODM, codeine *O*-demethylase (CODM), to afford morphine (Hagel and Facchini 2010). In a minor

---

enzyme; CFS, cheilanthifoline synthase; SPS, stylopine synthase; TNMT, tetrahydroprotoberberine *cis*-*N*-methyltransferase; MSH, (*S*)-*cis*-*N*-methylstylopine 14-hydroxylase; P6H, protopine 6-hydroxylase; DBOX, dihydrobenzophenanthridine oxidase; SanR, sanguinarine reductase; SOMT, scoulerine 9-*O*-methyltranfease; CAS, canadine synthase; STOX, (*S*)-tetrahydroberberine oxidase; AT1, acetyltransferase1; CXE1, carboxyesterase1; NOS, noscapine synthase.)

alternative route, CODM catalyzes the initial demethylation, converting thebaine to oripavine, which is followed by T6ODM- and COR-catalyzed steps, converting oripavine to morphine via morphinone (Brochmann-Hanssen 1984). In low-morphine varieties of opium poppy, including *top1*mutations (Millgate et al. 2004), the two ODD-family demethylase genes, *T6ODM* and *CODM*, were found to be down-regulated at the transcriptional level (Hagel and Facchini 2010), supporting the roles of encoded enzymes in the pathway.

(S)-Scoulerine is a branch point intermediate that is metabolized into multiple types of BIAs, including sanguinarine, berberine, and noscapine. Conversion of (S)-reticuline into (S)-scoulerine occurs through linking of an isoquinoline ring with a benzyl ring, thereby forming a berberine bridge. This conversion is catalyzed by berberine bridge enzyme (BBE), a unique type of flavoprotein containing a bi-covalently attached flavin adenine dinucleotide (FAD) (Dittrich and Kutchan 1991; Facchini et al. 1996; Winkler et al. 2008). In sanguinarine biosynthesis, (S)-scoulerine is converted to (S)-stylopine via (S)-cheilanthifoline through sequential formation of two methylenedioxy bridges catalyzed by CYP719A subfamily enzymes, cheilanthifoline synthase (CFS) and stylopine synthase (SPS) (Ikezawa et al. 2007, 2009). (S)-Stylopine is N-methylated by tetrahydroprotoberberine *cis*-N-methyltransferase (TNMT), for which cDNA was isolated based on its homology to CNMT, yielding a quaternary ammonium compound, (S)-*cis*-N-methylstylopine (Liscombe and Facchini 2007). Hydroxylation of (S)-*cis*-N-methylstylopine catalyzed by (S)-*cis*-N-methylstylopine 14-hydroxylase (MSH), which belongs to the CYP82N subfamily (Beaudoin and Facchini 2013), is followed by a ring tautomerization, yielding protopine. Protopine is further hydroxylated by another CYP82N subfamily enzyme, protopine 6-hydroxylase (P6H), to 6-hydroxyprotopine, which is nonenzymatically rearranged to form dihydrosanguinarine (Takemura et al. 2013). Finally, dihydrosanguinarine is oxidized to sanguinarine via FAD-linked dihydrobenzophenanthridine oxidase (DBOX) (Hagel et al. 2012). The reverse conversion of sanguinarine to dihydrosanguinarine, which might prevent the cytotoxic effects of sanguinarine, is mediated by sanguinarine reductase (SanR) (Vogel et al. 2010; Weiss et al. 2006).

(S)-Scoulerine is methylated by scoulerine 9-O-methyltransferase (SOMT) (Takeshita et al. 1995) to (S)-tetrahydrocolumbamine, which is subsequently converted to canadine by a methylenedioxy bridge-forming CYP719A, canadine synthase (CAS) (Ikezawa et al. 2003). (S)-Canadine is then oxidized to berberine by FAD-dependent (S)-tetrahydroberberine oxidase (STOX) (Gesell et al. 2011).

Canadine is also used in the formation of noscapine, one of the major BIAs in opium poppy latex, with morphine (Chen et al. 2015). Molecular cloning of SOMT (Dang and Facchini 2012; Winzer et al.

2012) and CAS (Dang and Facchini 2014a) from opium poppies confirmed that the part of the pathway is shared between noscapine and berberine biosynthesis. In the first step of the noscapine branch of the pathway, canadine is *N*-methylated by TNMT (Liscombe and Facchini 2007) to (*S*)-*N*-methylcanadine. A cluster of 10 genes in the opium poppy genome potentially includes all biosynthesis genes involved in the formation of noscapine from scoulerine, with the exception of TNMT (Winzer et al. 2012). Functional characterization of enzymes encoded by these clustered genes helps to understand the rest of the pathway. (*S*)-*N*-methylcanadine is hydroxylated at C-1 by CYP82Y1 (Dang and Facchini 2014b) to form (*S*)-1-hydroxy-*N*-methylcanadine, followed by another hydroxylation at C-14 by CYP82X2 (Dang et al. 2015) to form (13*S*,14*R*)-1,13-dihydroxy-*N*-methylcanadine. Following the transfer of an acetyl group to the C-13 oxygen of the CYP82X2 reaction product by acetyltransferase AT1, C-8 hydroxylation of the acetylated product (13*S*,14*R*)-1-hydroxy-13-*O*-acetyl-*N*-methylcanadine occurs via CYP82X1, leading to a ring opening reaction that yields 4'-*O*-desmethyl-3-*O*-acetylpapaveroxine, which contains an aldehyde group (Dang et al. 2015). When the protecting acetyl group introduced by AT1 is removed by carboxyesterase CXE1, hemiacetal formation occurs, converting 4'-*O*-desmethyl-3-*O*-acetylpapaveroxine, or 3-*O*-acetylpapaveroxine if *O*-methylation occurred via SOMT2 (Dang and Facchini 2012), to narcotine hemiacetal (Dang et al. 2015). The last enzyme of the SDR family, noscapine synthase (NOS), catalyzes the formation of noscapine from narcotine hemiacetal (Chen and Facchini 2014).

5.2.2 *Trafficking in opium poppies and* C. japonica

In opium poppies, BIA biosynthesis and accumulation occur in three distinct cell types, companion cells, sieve elements, and lactifers. Lactifers are highly specialized cells found adjacent or proximal to sieve elements of the phloem. As has been demonstrated for most BIA biosynthesis genes, gene transcripts and their corresponding enzymes localize in companion cells and sieve elements, respectively (Bird et al. 2003; Samanani et al. 2005, 2006; Weid et al. 2004). Such localization patterns suggest gene expression in companion cells and subsequent translocation of the enzymes to adjunct sieve elements, which is the site of BIA biosynthesis. The BIAs are then transported to and stored in the cytoplasm of lactifers, or latex. Recent applications of proteomics revealed that enzymes involved in the final three steps of morphine biosynthesis, T6ODM, COR, and CODM, and the last enzyme in the noscapine pathway, NOS, primarily accumulate in lactifers rather than in sieve elements, indicating separation of the later steps from upstream enzymes and possible intercellular translocation of relevant pathway intermediates (Chen and Facchini 2014;

Onoyovwe et al. 2013). Both symplasmic and apoplasmic routes have been considered for metabolite translocation from sieve elements to lactifiers because of the frequent presence of plasmodesmata connecting the two cell types. Membrane transporters involved in BIA transport have not been reported in opium poppies.

In *C. japonica*, berberine is synthesized in the roots and translocated through xylems to the rhizomes, where it is taken up into the cells and stored as an antibacterial compound in central vacuoles. ATP-binding cassette (ABC) proteins (Shoji 2014) of the B or multidrug resistance (MDR) subfamilies, CjABCB1/CjMDR1 and related CjABCB2, mediate the cellular uptake of berberine through plasma membranes in the rhizomes (Shitan et al. 2003, 2013). Vacuolar sequestration of berberine by proton/berberine antiporters was biochemically characterized in *C. japonica* cultured cells (Otani et al. 2005).

5.2.3 Transcriptional regulation

Transcript profiling has led to the identification of transcription factors involved in regulation of BIA pathways. In *C. japonica* cultured cells, a set of transcription factor genes including the WRKY-family *CjWRKY1* and bHLH-family *CjbHLH1* were found to be coordinately expressed with BIA biosynthesis genes and thus elicited by JA (Kato et al. 2007; Yamada et al. 2011). Transient RNAi-mediated suppression and overexpression experiments using *C. japonica* protoplasts indicated regulation of BIA biosynthesis genes by *CjWRKY1* and *CjbHLH1* (Kato et al. 2007; Yamada et al. 2011). In stable, transgenic lines of California poppy cultured cells, suppression of *CjbHLH1* homologs, *CpbHLH1* and *CpbHLH2*, resulted in corresponding decreases in BIA level and associated gene expression (Yamada et al. 2015).

5.3 Monoterpenoid indole alkaloids

Monoterpenoid indole alkaloids (MIAs) are a large group of complex alkaloids that includes over 3000 compounds mainly from the Apocynaceae, Loganiaceae, and Rubiaceae families. The anticancer drugs vinblastine and vincristine and the antiarrhythmic drug ajmaline are prominent examples of clinically important MIAs. *Catharanthus roseus* (Apocynaceae), also known as Madagascar periwinkle, produces a series of nearly 200 different MIAs, including the dominant catharanthine and vindoline, whose heterodimers are vinblastine and vincristine, which are less abundant and thus highly valuable. Molecular and cellular studies on MIA biosynthesis have centered on this commercial source of dimeric MIAs, which are useful in chemotherapy (De Luca et al. 2014; Dugé de Bernonville et al. 2015).

5.3.1 Pathways and enzymes

MIAs comprise indole and monoterpenoid moieties derived from tryptamine and secologanin, respectively (Figure 5.2). Tryptamine is derived from tryptophan through a single decarboxylation step catalyzed by pyridoxal-dependent tryptophan decarboxylase (TDC) (De Luca et al.

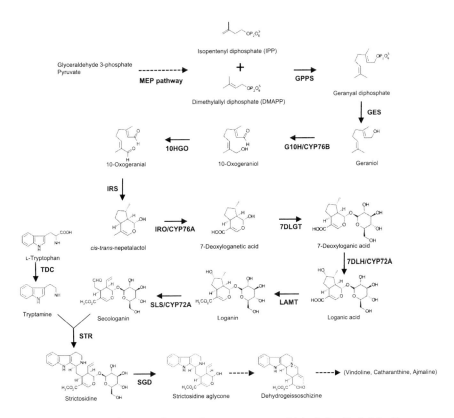

Figure 5.2 Biosynthetic pathways for monoterpenoid indole alkaloids. Enzymes for which cDNAs have been cloned are shown in bold. (Abbreviations: MEP, methylerythritol 4-phosphate; IPP, isopentenyl diphosphate; DMAPP, dimethylallyl diphosphate; GPPS, geranyl diphosphate synthase; GES, geraniol synthase; G10H, geraniol 10-hydroxylase; 10HGO, 10-hydroxygeraniol oxidoreductase; IRS, iridoid synthase; IRO, iridoid oxidase; 7DLGT, 7-deoxyloganetic acid glucosyltransferase; 7DLH, 7-deoxyloganic acid 7-hydroxylase; LAMT, loganic acid methyltransferase; SLS, secologanin synthase; TDC, tryptophan decarboxylase; STR, strictosidine synthase; SGD, strictosidine β-glucosidase; T16H, tabersonine 16-hydroxylase; 16OMT, 16-hydroxytabersonine *O*-methyltransferase; NMT, *N*-methyltransferase; D4H, desacetoxyvindoline 4-hydroxylase; DAT, deacetylvindoline 4-*O*-acetyltransferase; PRX, peroxidase; PNAE, polyneuridine aldehyde esterase; VS, vinorine synthase; AAE, acetylajmalan esterase.) *(Continued)*

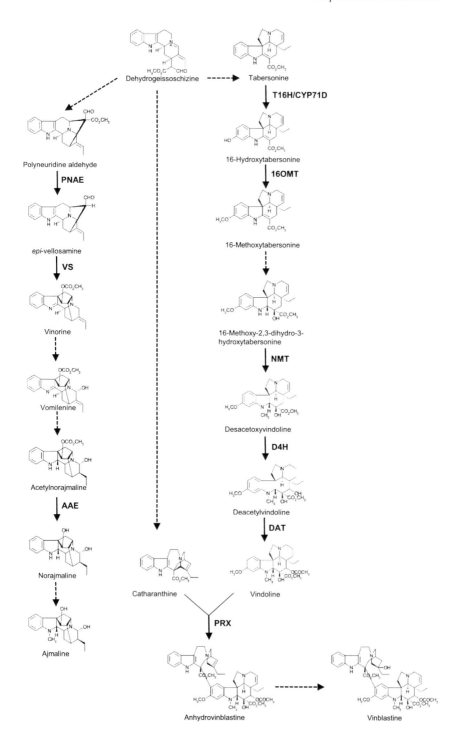

1989). Secologanin is part of a large group of unusual monoterpenoids known as iridoids. While secologanin serves as an alkaloid precursor, most iridoids act as defense compounds in plant–herbivore interactions in a broad range of plants (Dinda and Debnath 2013; Oudin et al. 2007a). Via the plastid-localized methylerythritol 4-phosphate (MEP) pathway, glyceraldehyde 3-phosphate and pyruvate are converted into isoprenoid units isopentenyl diphosphate (IPP) and dimethylallyl diphosphate (DMAPP) (Chahed et al. 2000; Oudin et al. 2007b; Veau et al. 2000). A monoterpenoid known as geraniol is generated via head-to-tail condensation of MEP pathway-derived IPP and DMAPP catalyzed by geranyl diphosphate (GPP) synthase (GPPS) (Rai et al. 2013) and a subsequent step catalyzed by geraniol synthase (GES) (Simkin et al. 2013). Geraniol is hydroxylated to a diol, 10-hydroxygeraniol, and then oxidized to 10-oxogeraniol by a bifunctional CYP76B enzyme, geraniol 10-hydroxylase (G10H) (Collu et al. 2001; Höfer et al. 2013). An alcohol dehydrogenase, known as 10-hydroxygeraniol oxidoreductase (10HGO), oxidizes 10-oxogeraniol to dialdehyde 10-oxogeranial (Miettinen et al. 2014). In contrast to typical monoterpenoids formed though cyclization of GPP, 10-oxogeranial is cyclized by a unique cyclase recruited from the SDR family, iridoid synthase (IRS), to form the iridoid, *cis-trans*-nepetalactol (Geu-Flores et al. 2012). The cyclization product is then converted to 7-deoxyloganetic acid through a three-step oxidation catalyzed by the CYP76A subfamily iridoid oxidase (IRO) (Miettinen et al. 2014; Salim et al. 2014) A UDP-glucose glycosyltransferase (UGT), 7-deoxyloganetic acid glucosyltransferase (7DLGT), transfers a glucosyl group to 7-deoxyloganetic acid to generate 7-deoxyloganic acid (Asada et al. 2013), which is then hydroxylated by CYP72A subfamily 7-deoxyloganic acid 7-hydroxylase (7DLH) (Miettinen et al. 2014; Salim et al. 2013) to loganic acid. Loganic acid is methylated by SAM-dependent loganic acid methyltransferase (LAMT) (Murata et al. 2008) to loganin, which is subsequently converted to secologanin through a ring opening reaction catalyzed by secologanin synthase (SLS) of the CYP72A subfamily (Irmler et al. 2000).

Figure 5.2 (Continued) Biosynthetic pathways for monoterpenoid indole alkaloids. Enzymes for which cDNAs have been cloned are shown in bold. (Abbreviations: MEP, methylerythritol 4-phosphate; IPP, isopentenyl diphosphate; DMAPP, dimethylallyl diphosphate; GPPS, geranyl diphosphate synthase; GES, geraniol synthase; G10H, geraniol 10-hydroxylase; 10HGO, 10-hydroxygeraniol oxidoreductase; IRS, iridoid synthase; IRO, iridoid oxidase; 7DLGT, 7-deoxyloganetic acid glucosyltransferase; 7DLH, 7-deoxyloganic acid 7-hydroxylase; LAMT, loganic acid methyltransferase; SLS, secologanin synthase; TDC, tryptophan decarboxylase; STR, strictosidine synthase; SGD, strictosidine β-glucosidase; T16H, tabersonine 16-hydroxylase; 16OMT, 16-hydroxytabersonine *O*-methyltransferase; NMT, *N*-methyltransferase; D4H, desacetoxyvindoline 4-hydroxylase; DAT, deacetylvindoline 4-*O*-acetyltransferase; PRX, peroxidase; PNAE, polyneuridine aldehyde esterase; VS, vinorine synthase; AAE, acetylajmalan esterase.)

Strictosidine, a precursor of all other MIAs, is formed through the Pietet–Spengler condensation of tryptamine and secologanin catalyzed by strictosidine synthase (STR) (Kutchan et al. 1988; McNight et al. 1990). STR is not homologous to NCS, which catalyzes the same type of condensation in BIA biosynthesis. Strictosidine β-glucosidase (SGD) deglucosylates the monoterpenoid glucoside strictosidine to its aglycone, a reactive hemiacetal intermediate (Geerlings et al. 2000; Gerasimenko et al. 2002). This reactive aglycone is then converted via unstable intermediates, perhaps spontaneously, to dehydrogeissoschizine, which serves as a branch intermediate leading to diverse MIA pathways. Most of the downstream pathways originating from the branch intermediate, such as those leading to tabersonine and catharanthine, have remained uncharacterized.

Vindoline is one of the main MIAs that accumulates in *C. roseus* leaves and is formed through a six-step reaction from tabersonine derived from dehydrogeissoschizine. All but one of the six steps involved in vindoline formation has been defined at the molecular level. A methoxy group is initially introduced into tabersonine through a hydroxylation performed by tabersonine 16-hydroxylase (T16H) of the CYP71D subfamily (Besseau et al. 2013; Schröder et al. 1999) and subsequent *O*-methylation by SAM-dependent 16-hydroxytabersonine *O*-methyltransferase (16OMT) (Lavac et al. 2008). 16-Methoxytabersonine is metabolized by an uncharacterized hydratase and *N*-methylated at the indole nitrogen by an enzyme similar to tocopherol C *N*-methyltransferase (NMT) (Liscombe et al. 2010) to yield desacetoxyvindoline. Finally, vindoline is formed from desacetoxyvindoline through a hydroxylation catalyzed by the ODD-family desacetoxyvindoline 4-hydroxylase (D4H) (Vazquez-Flota et al. 1997) and subsequent *O*-acetylation by an acetyl-CoA-dependent acetyltransferase of the BAHD family, deacetylvindoline 4-*O*-acetyltransferase (DAT) (St-Pierre et al. 1998). Vindoline and catharanthine, two abundant MIAs in *C. roseus* leaves, are coupled by a vacuolar class III peroxidase (PRX) to form anhydrovinblastine, a common precursor of all dimeric MIAs, which is further converted to vinblastine and vincristine (Costa et al. 2008).

Rauwolfia serpentina (Apocynaceae), a medicinal plant native to India, produces various MIAs, including ajmaline, and has been studied for the elucidation of ajmaline biosynthesis (Ruppert et al. 2005a). In the first step specific to ajmaline biosynthesis, the formation of a sarpagan bridge in dehydrogeissoschizine generates polyneuridine aldehyde. Polyneuridine aldehyde is then hydrolyzed by polyneuridine aldehyde esterase (PNAE) of the α/β hydrolase family to an acid intermediate, which spontaneously transforms to *epi*-vellosamine (Dogru et al. 2000). Vellosamine is acetylated by an acetyl-CoA-dependent acetyltransferase of the BAHD family, vinorine synthase (VS), to give the ajmalan-type alkaloid, vinorine (Bayer et al. 2004; Ma et al. 2005), which is then hydroxylated to vomilenine. Vomilenine is then subjected to two reduction steps leading to acetylnorajmaline. Acetylnorajmaline is deacetylated by acetylajmalan esterase (AAE) of the GDSL lipase family to norajmaline (Ruppert et al. 2005b), which is finally *N*-methylated to form ajmaline. Genes encoding PNAE, VS, and AAE have been cloned.

5.3.2 Trafficking in C. roseus

The complex spatial organization of the MIA pathway in *C. roseus* has become clear through *in situ* hybridization and immunolocalization analyses, complemented with omics-based approaches and isolation of specific cell types by laser capture microdissection and carborundum abrasion (Courdavault et al. 2014; De Luca and St-Pierre 2000; Dugé de Bernonville et al. 2015). In *C. roseus* leaves, MIA biosynthesis has been found to rely on the intricate communication between at least four different cell types, internal phloem associated parenchyma (IPAP), epidermis, lactifier, and idioblast (Figure 5.3). The early portion of the pathway, including the entire plastidial MEP pathway and the first eight steps of the iridoid pathway up to the formation of loganic acid, is localized in IPAP cells (Burlat et al. 2004; Geu-Flores et al. 2012; Miettinen et al. 2014; Oudin et al. 2007a,b). The close proximity of IPAP cells to the phloem and their plasmodesmata connection is believed to enable efficient allocation of photosynthetic assimilates to alkaloid biosynthesis. Loganic acid moves from IPAP to leaf epidermis cells, where subsequent steps are carried out, including the formation of the first common pathway intermediate, strictosidine (Irmler et al. 2000; Lavac et al. 2008; Murata et al. 2008; St-Pierre et al. 1999). Of particular interest is the subcellular separation of nuclear SGD and its substrate, strictosidine, formed by vacuolar STR, which is considered a prerequisite for herbivory-induced, explosive production of a highly toxic strictosidine aglycone with protein cross-linking activity (Guirimand et al. 2010). Such system dependence on enzyme-substrate compartmentalization and its sudden disruption by cell damage, known as a "nuclear time bomb," is considered to be an example of preformed defense mechanisms.

Catharanthine is an MIA possibly formed in the epidermis, though molecular details of its formation remain unclear, and it accumulates in wax exudates on the leaf surface (Roepke et al. 2010). An ABC protein transporter from *C. roseus*, TPT2, mediates the ATP-driven active export of catharanthine from epidermal cell layers to its surface (Yu and De Luca 2013). In accordance with its exporter function, *TPT2* is expressed predominantly in the epidermis and up-regulated by JA and catharanthine.

Vindoline, one of the most abundant MIAs in *C. roseus* leaves, does not form in the leaf epidermis. While T16H and 16OMT are involved in the formation of 16-methoxytabersonine from tabersonine, they reside in the epidermal cells (Besseau et al. 2013; Lavac et al. 2008). D4H and DAT,

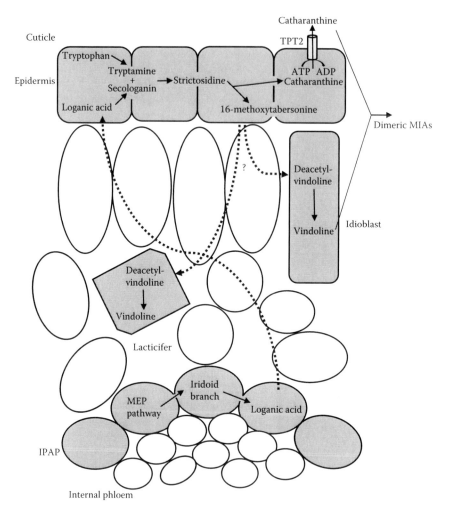

Figure 5.3 Trafficking of monoterpenoid indole alkaloids in *Catharanthus roseus* leaves. (Abbreviations: MEP, methylerythritol 4-phosphate; IPAP, internal phloem associated parenchyma.)

the last two enzymes of vindoline formation, are restricted in lactifier and idioblast cells (St-Pierre et al. 1999), which are distributed in the internal parts of leaves and thus generally separate from the epidermal layer, suggesting intercellular transport of relevant intermediates. The separate accumulation of two monomeric MIAs, catharanthine secreted to the leaf surface and vindoline produced in the inner part of the leaf, is a likely reason for such limited accumulation of dimeric MIAs, such as clinically important vinblastine and vincristine. As these dimeric MIAs depend on coupling of monomeric MIAs, their formation seems to occur only when the two tissues are mixed because of leaf damage triggered by wounding or herbivory (Roepke et al. 2010).

5.3.3 Transcriptional regulation

JA and fungal elicitor coordinately up-regulate *STR* and *TDC* genes that encode key enzymes in the MIA pathway. Studies on transcriptional regulation of *STR* and *TDC* have revealed functionally important *cis*-elements in their promoter regions, including a JA- and elicitor-responsive element (JERE) (Menke et al. 1999). A one-hybrid yeast assay using a JERE found that the *STR* promoter as bait led to the isolation of an AP2/ERF family transcription factor, octadecanoid-responsive Catharanthus AP2-domain 2 (ORCA2), from *C. roseus* (Menke et al. 1999). *ORCA2* is induced in response to JA and elicitor treatment, with transient overexpression of *ORCA2* in *C. roseus* protoplasts activating the *STR* promoter. A handful of transcription factors of various families, including Zn-finger transcriptional repressors ZCT1, ZCT2, and ZCT3 (Pauw et al. 2004), have been reported based on their ability to interact with regulatory elements other than JERE in *STR* and *TDC* promoters (Sibéril et al. 2001; van der Fits et al. 2000), but *in planta* functionalities of the DNA-protein interactions remain obscure.

ORCA3, a transcription factor closely related to ORCA2, has been identified through a T-DNA activation tagging approach using *C. roseus* cultured cells and aimed at isolating regulators of TDC activity (van der Fits and Memelink 2000). Overexpression of *ORCA3* in *C. roseus* cultured cells up-regulated multiple, but not all, MIA pathway genes, including *TDC, STR, D4H*, and genes involved in primary metabolic steps leading to the MIA pathway. This implies that ORCA3 is a transcriptional activator of some, but not all, steps in the entire MIA-supplying metabolism. In *ORCA3*-overexpressing cells, most likely because the iridoid-supplying branch of the pathway remains limited, MIAs increase only after the cells are fed with loganin.

A bHLH transcription factor, *C. rosues* MYC2 (CrMYC2), of which Arabidopsis ortholog MYC2 is one of the key components in JA signaling (Kazan and Manners 2013), directly regulates *ORCA3* by recognizing a G-box element in its promoter (Zhang et al. 2011), which is required for JA-responsive expression of the gene (Vom Endt et al. 2007). In a transient assay, *CrMYC2* activates the *ORCA3* promoter, while transgenic suppression of *CrMYC2* decreases the JA-induced expression of *ORCA3* and *ORCA2* in *C. roseus* cultured cells (Zhang et al. 2011). In addition to CrMYC2, AT-hook DNA-binding proteins were found to interact with an A/T-rich portion of the *ORCA3* promoter, which is important for high-level expression of *ORCA3* (Vom Endt et al. 2007).

A bHLH transcription factor from *C. roseus*, bHLH iridoid synthesis 1 (BIS1), was found to activate all genes encoding enzymes involved in the conversion of geranyl diphosphate to loganic acid, indicating that BIS1 is a transcriptional activator for the IPAP-localized iridoid branch of the MIA pathway (Van Moerkercke et al. 2015). This iridoid branch-specific regulatory role of BIS1 is considered complementary to that of ORCA3, which seemingly acts much less on this branch (see above). Overexpression of *BIS1* alone, in contrast to ORCA3, increases the iridoid and MIA accumulation in *C. roseus* cultured cells, indicating the potential of *BIS1* as a molecular tool for metabolic engineering of the MIA pathway.

A WRKY family transcription factor, CrWRKY1, is involved in MIA regulation in *C. roseus* roots (Suttipanta et al. 2011). *CrWRKY1* is preferentially expressed in the roots and induced by JA, ethylene, and gibberellin. It has the ability to bind to W-box elements found in the *TDC* promoter. Overexpression of *CrWRKY1* in *C. roseus* hairy roots up-regulates *TDC* and transcription repressors *ZCT1*, *ZCT2*, and *ZCT3*, while transcription activators *ORCA2*, *ORCA3*, and *CrMYC2* are confined in the hairy root lines. Increased accumulation of serpentina in the transgenic hairy roots implies the role of CrWRKY1 in directing root-specific accumulation of this MIA.

5.4   Nicotine and tropane alkaloids (TAs)

Nicotine and TAs are derived from the unnatural amino acid ornithine via shared intermediates, diamine putrescine and *N*-methylpyrrolinium cation, and are particularly rich in a number of Solanaceae species (Figure 5.4; Shoji and Hashimoto 2015a). Nicotine is a toxic alkaloid found in tobacco (*Nicotiana tabacum*, Solanaceae), playing a defensive role against insect herbivory (Steppuhn et al. 2004), with its induced production mainly mediated by JA signaling (Shoji and Hashimoto 2013; Shoji et al. 2000). Medicinal plants of *Atropa*, *Datura*, *Duboisia*, and *Hyoscyamus*, all belonging to Solanaceae, are often sources of TAs, including clinically important anticholinergic hyoscyamine and scopolamine.

5.4.1   Pathways and enzymes

Ornithine is decarboxylated to putrescine by pyridoxal 5′-phosphate-dependent ornithine decarboxylase (ODC) (Imanishi et al. 1998).

Figure 5.4 Biosynthetic pathways for nicotine and tropane alkaloids. Enzymes for which cDNAs have been cloned are shown in bold. (Abbreviations: ODC, ornithine decarboxylase; PMT, putrescine *N*-methyltransferase; MPO, *N*-methylputrescine oxidase; BBL, berberine bridge enzyme-like; TR, tropinone reductase; H6H, hyoscyamine 6β-hydroxylase.)

Chapter five: Alkaloid biosynthesis and regulation in plants

Transgenic studies supported that the alkaloid pathway depends on the ODC-mediated formation of putrescine from ornithine (Chintapakorn and Hamill 2007; De Boer et al. 2011a). Putrescine is N-methylated by an S-adenosylmethionine-dependent putrescine N-methyltransferase (PMT), which is assumed to originate from spermidine synthase that catalyzes propyl group transfer to putrescine, a substrate shared with PMT, from decarboxylated SAM (Hibi et al. 1994; Suzuki et al. 1999; Teuber et al. 2007). N-methylputrescine is oxidatively deaminated to 4-methylaminobutanal by a copper and topaquinone-containing N-methylputrescine oxidase (MPO) of the diamine oxidase family (Heim et al. 2007; Katoh et al. 2007; Maliwan et al. 2014). 4-Methylaminobutanal is believed to spontaneously cyclize to N-methylpyrrolinium cation, which is incorporated into nicotine as a pyrrolidine ring and tropane alkaloids as a tropane ring. The pathways for the two types of alkaloids branch off following formation of this five-membered ring cation.

Nicotine is composed of two heterocyclic rings, a pyrrolidine ring derived from the N-methylpyrrolinium cation and a pyridine ring from nicotinic acid. Nicotinic acid is an intermediate of a primary metabolic pathway supplying nicotinamide adenine dinucleotide (NAD) originating from aspartate in plants (Katoh and Hashimoto 2004). The later part of the nicotine pathway, after nicotinic acid and N-methylpyrrolinium cation, including a ring coupling, was postulated to involve steps mediated by two oxidoreductases, a NADH-dependent PIP-family enzyme, A622 (De Boer et al. 2009; Hibi et al. 1994; Kajikawa et al. 2009), and a FAD-dependent berberine bridge enzyme-like protein (BBL) (Kajikawa et al. 2011). Biochemical details of the A622 and BBL-catalyzed reactions and others involved in the later steps remain undefined.

In TA-producing plants, the reactive intermediate, N-methylpyrrolinium cation, is converted to tropinone, which contains a bicyclic tropane ring, through less defined steps. A ketone group in tropinone is stereospecifically reduced to tropine and oppositely configured pseudotropine by two distinct, but structurally related, SDR family tropinone reductases (TRs), TR-I and TR-II (Nakajima et al. 1993, 1998). Pseudotropine is purported to be metabolized to calystegines, a group of polyhydroxylated and nonesterified TAs with a nortropane ring (Biastoff and Dräger 2007), while tropine is esterified with phenylalanine-derived phenyllactate to generate littorine. Littorine is intramolecularly rearranged to hyoscyamine aldehyde by the CYP80F family littorine mutase (CYP80F) (Li et al. 2006). Hyoscyamine aldehyde is oxidized to hyoscyamine by an undefined alcohol dehydrogenase. A two-step reaction, including hydroxylation and subsequent epoxidation, is catalyzed by an ODD, hyoscyamine 6β-hydroxylase (H6H), generating scopolamine from hyoscyamine via 6β-hydroxyhyoscyamine (Matsuda et al. 1991; Yun et al. 1992).

5.4.2 Tobacco nicotine transporters

In tobacco, nicotine is synthesized in the roots and transported through the xylems to the leaves, where it is stored in central vacuoles to avoid cytotoxicity. Two distinct types of multidrug and toxic compound extrusion (MATE) family transporters (Shoji 2014), JA-induced alkaloid transporter (JAT1) (Morita et al. 2009) and highly homologous MATE1 and MATE2 (Shoji et al. 2009), were identified as tonoplast-localized transporters involved in vacuolar sequestration of nicotine in tobacco. JAT1 plays a transport role in leaves (Morita et al. 2009), whereas MATE1 and MATE2 are expressed along with nicotine metabolic genes in the roots (Shoji et al. 2009). Tobacco nicotine uptake permease 1 (NUP1) is a membrane transporter of the purine permease (PUP) family that mediates inward transport of nicotine at the plasma membrane mainly in root epidermis cells (Hildreth et al. 2011; Kato et al. 2014). Similar to its Arabidopsis homolog, AtPUP1, tobacco NUP1 also has the ability to transport vitamin B6 (Kato et al. 2015). In addition to, and possibly independent from, its function as a membrane transporter, NUP1 was found to be involved in the regulation of nicotine biosynthesis by controlling the expression of the ERF189 transcription factor (Hildreth et al. 2011; Kato et al. 2014), suggesting an unexplored aspect of this transporter.

5.4.3 Regulation of nicotine biosynthesis

JA-induced nicotine formation in tobacco depends on coordinated transcriptional induction of a series of nicotine metabolic and transport genes. Two types of JA-responsive transcription factors, represented by tobacco ERR189 and MYC2 (NtMYC2), were shown to directly up-regulate genes encoding key enzymes in the pathway by being bound to their promoter regions (De Boer et al. 2011b; De Sutter et al. 2005; Shoji and Hashimoto 2011a,b; Shoji et al. 2010; Todd et al. 2010; Zhang et al. 2012).

A transcription factor of the AP2/ERF family, ERF189, is closely related with ORCAs from *C. roseus* involved in MIA regulation (see above). DNA-binding specificities of ERF189, ORCA3, and related ERFs, which are similar but distinct, have been studied (Shoji and Hashimoto 2012; Shoji et al. 2013) and found to be largely defined by a few critical amino acid residues in the DNA-binding domains (Shoji et al. 2013). In tobacco, the *ERF189* gene was found to be clustered with genes of its close homologs on the genetic locus *NICOTINE2* (*NIC2*), which is responsible for nicotine content in tobacco, and deleted in the *nic2* mutant that has been used to breed a low-nicotine variety of tobacco, supporting the role of ERF189 as a pathway-specific regulator (Shoji et al. 2010). A number of clustered *NIC2*-locus *ERF* genes other than *ERF189* are inducible not only with JA but also by salt stress, implying their involvement in an abiotic stress response rather than nicotine regulation (Shoji and Hashimoto 2015a,b).

A tobacco bHLH family transcription factor, NtMYC2, plays more general regulatory roles in JA signaling, not restricted to regulation of nicotine formation, binding defined E-box elements in promoter regions of downstream genes (De Boer et al. 2011b; Shoji and Hashimoto 2011a; Zhang et al. 2012). In addition to direct regulation of biosynthetic genes, NtMYC2 also regulates transcription of *ERF189*, acting upstream of the pathway-regulating ERF factor (Shoji and Hashimoto 2011a). Such a regulatory relationship between these two factors in tobacco is similar to that found between their counterparts CrMYC2 and ORCA3 in *C. roseus*, which form a cascade where CrMYC2 directly regulates *ORCA3* transcription but not MIA biosynthesis genes (see above).

Conserved components in JA signaling, COI1 and JAZ proteins, are required for JA-induced nicotine formation in the *Nicotiana* species (Paschold et al. 2007; Shoji et al. 2008), acting upstream of the two key transcription factors ERF189 and NtMYC2; a direct interaction between at least tobacco JAZ and NtMYC2 proteins was demonstrated (Shoji and Hashimoto 2011a). A mitogen-activated protein kinase kinase, JA-factor stimulating MAPKK1 (JAM1), was found to stimulate the activities of tobacco ERF189-type and NtMYC2 factors in a transient expression assay using tobacco protoplasts, indicating possible involvement of a JA-elicited phosphorylation cascade in nicotine regulation (De Boer et al. 2011).

5.5 Purine alkaloids

Caffeine and related purine alkaloids (PAs) are derived from purine nucleotides and serve as stimulatory ingredients in coffee, tea, and other nonalcoholic beverages (Ashihara 2008). PAs are found in a broad range of plant species, though significant accumulation is restricted to some species such as tea (*Camellia sinensis*) and coffee (*Coffea arabica*). *In planta* roles of PAs in chemical defense and allelopathy have been proposed.

Xanthosine is considered to be a key metabolite connecting primary purine metabolism and PA biosynthesis and is readily supplied via multiple routes from purine nucleotides. It is converted to caffeine through three *N*-methylation steps and one ribose removal step (Figure 5.5). Xanthosine is initially *N*-methylated at the N7 position by 7-methylxanthosine synthase (XMT) (Mizuno et al. 2003a, Uefuji et al. 2003). Removal of ribose from 7-methylxanthosine to give 7-methylxanthine has yet to be fully elucidated; however, catalysis by XMT, coupled with the initial methylation, has been suggested (McCarthy and McCarthy 2007). The two consecutive *N*-methylation steps at N3 and N1 convert 7-methylxanthine to caffeine

Figure 5.5 Biosynthetic pathways for purine alkaloids. Enzymes for which cDNAs have been cloned are shown in bold. (Abbreviations: XMT, xanthosine methyltransferase; TS, theobromine synthase; CS, caffeine synthase.)

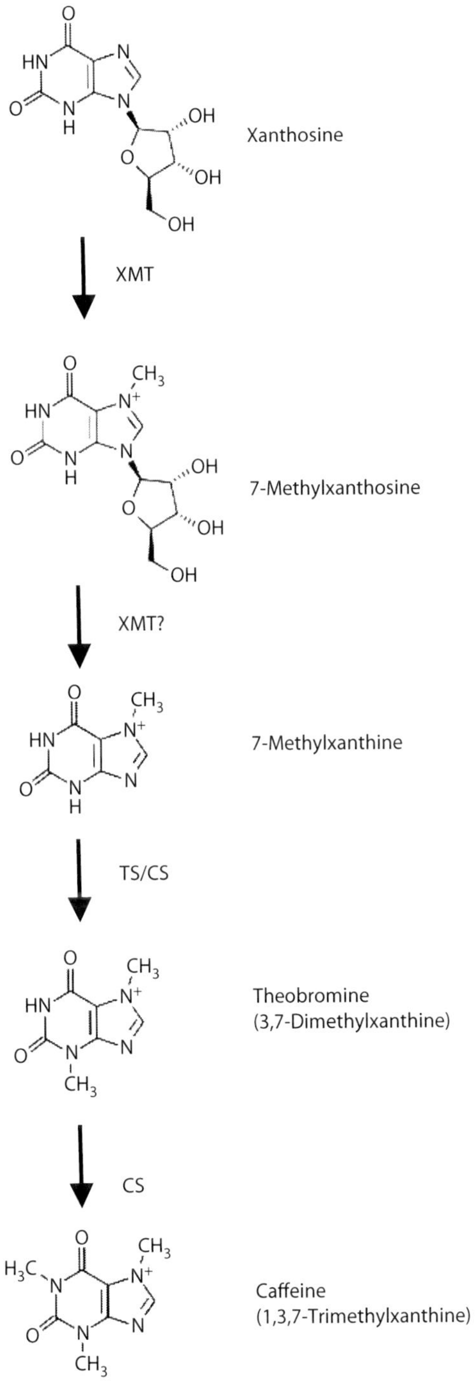

Chapter five: Alkaloid biosynthesis and regulation in plants

Xanthosine

↓ XMT

7-Methylxanthosine

↓ XMT?

7-Methylxanthine

↓ TS/CS

Theobromine
(3,7-Dimethylxanthine)

↓ CS

Caffeine
(1,3,7-Trimethylxanthine)

via theobromine. The N3 methylation is catalyzed by theobromine synthase (TS) (Ogawa et al. 2001; Uefuji et al. 2003; Yoneyama 2006), whereas a bifunctional caffeine synthase (CS) mediates both N-methylations (Kato et al. 2000; Mizuno et al. 2003b; Uefuji et al. 2003). The caffeine biosynthetic methyltransferases, XMT, TS, and CS, are structurally related to SAM-dependent N-methyltransferases with distinct substrate specificities, which belong to a unique subfamily with C-methyltransferases for salicylic, jasmonic, and benzoic acids, rather than other N-methyltransferases (Kato and Mizuno 2004). The identification of N-methyltransferase genes allows us to genetically manipulate caffeine biosynthesis (Ogita et al. 2003) and implement this pathway into heterogeneous hosts (Uefuji et al. 2005).

5.6 Perspectives

Recent applications of genomic technologies, combined with molecular and biochemical approaches, have contributed to the identification of an increasingly large number of genes involved in alkaloid production, as clearly demonstrated by the nearly complete elucidation of pathways for important BIAs at the molecular level (Beaudoin and Facchini 2014; Chen et al. 2015). The invention of next-generation sequencing has enabled the high-throughput sequencing of a comprehensive set of transcripts, or transcriptome, without prior genomic information, providing unprecedented opportunities to readily mine candidate genes from nonmodel, alkaloid-producing plants. Transcriptomics-based gene mining efforts have been largely successful, especially when applied to secondary metabolic genes. This is in part due to the tight regulation of genes at the transcript level and the concerted expression of sets of genes involved in given pathways. Comparative analyses with the help of metabolite profiling and the use of appropriate materials, such as specific tissues devoted to alkaloid production (Lavac et al. 2008; Murata et al. 2008) and genetic variants with altered alkaloid contents (Hegel and Facchini 2010; Hibi et al. 1994; Kajikawa et al. 2011; Katoh et al. 2007; Shoji et al. 2009), have proved promising in this endeavor. Accumulating knowledge of these pathways has revealed that a handful of biosynthetic enzymes belonging to particular subgroups within larger enzyme families act as catalysts in the same pathways by accepting structurally related substrates (e.g., CNMT and TNMT, CYP82N, and CYP72A) (see above). Genomic clustering of genes encoding nonhomologous enzymes involved in secondary pathways have become widely recognized in plants (Nützmann and Osbourn 2014) and alkaloid pathways are no exception, as exemplified in the case of a gene cluster for the noscapine pathway in opium poppies (Winzer et al. 2012). Genome sequencing of alkaloid-producing plants, which has been completed for coffee (Denoeud et al. 2014) and tobacco (Sierro et al. 2014) and will become common for other species (Kellner et al. 2015), provides many cases of such clusters and thereby introduces the possibility of discovering novel genes based on their clustering with known metabolic genes.

The chemical synthesis of complex alkaloids bearing chiral centers is unrealistic on an industrial scale and alkaloid extraction from plant materials is a common practice even today. Biotechnological applications are feasible solutions to overcome the scarcity of high-value phytochemicals. Molecular cloning of pathway genes allows us to engineer these pathways in plants and microorganisms (Facchini et al. 2012). As such, multistep metabolic pathways for plant alkaloids have been reconstructed in yeast (Brown et al. 2015; Fossati et al. 2014) and *Escherichia coli* (Nakagawa et al. 2011). It is important to use strains engineered to produce sufficient metabolic precursors that sustain the implemented alkaloid pathways. Once the platform strains producing the central intermediates in the pathway are established, stepwise introduction of genes to extend the pathway becomes possible. Introducing novel catalytic activities created through engineering native plant enzymes or deployed from heterogynous systems is a promising combinatorial strategy to expand the metabolic diversities of host organisms. Such synthetic approaches are also applicable to metabolic engineering in plants (Glenn et al. 2013; Runguphan and O'Connor 2009; Runguphan et al. 2010).

Gene regulation and metabolite transport, which are both essential for proper accumulation of alkaloids in plants, have begun to be elucidated and will continue to be a major research focus in coming years. As a result of the cell-type-specific, and often multisite, and elicitor-inducible properties of alkaloid formation, an understanding of regulatory and transport mechanisms is required for effective metabolic engineering in multicellular plant systems. As reflected in successful transcriptomics approaches, transcriptional regulation of alkaloid biosynthesis genes encoding metabolic enzymes, as well as membrane transporters for metabolites, generally depends on master transcription factors that control either the entire or portions of the pathway (Shoji et al. 2010; van der Fits and Memelink 2000; Van Moerkercke et al. 2015). We must address how the activities of pathway-controlling factors are regulated by external and internal cues to better understand the metabolic processes in both physiological and developmental contexts. Molecular frameworks of hormonal and other signaling in plants, evolutionarily conserved and shared among species, have been intensively studied in model species and have begun to become clear. How are the conserved signaling cascades anchored to a diverse series of downstream metabolisms, each of which have typically evolved in a restricted plant lineage? The involvement of the structurally related transcription factors ERF189 from tobacco (Shoji et al. 2010) and ORCAs from *C. roseus* (van der Fits and Memelink 2000) in JA-dependent regulation of unrelated alkaloid pathways is a feasible example of molecular components integrating downstream defense metabolisms into the basic framework of JA signaling.

Acknowledgment

I thank Dr. Takashi Hashimoto of the Nara Institute of Science and Technology for critically reading the manuscript.

References

Asada, K., V. Salim, S. Masada-Atsumi et al. 2013. A 7-deoxyloganetic acid glucosyltransferase contributes a key step in secologanin biosynthesis in Madagascar periwinkle. *Plant Cell*. 25: 4123–4134.

Ashihara, H., H. Sano, and A. Crozier. 2008. Caffeine and related purine alkaloids: Biosynthesis, catabolism, function and genetic engineering. *Phytochemistry* 69: 841–856.

Bayer, A., X. Ma, and J. Stöckigt. 2004. Acetyltransfer in natural product biosynthesis- functional cloning and molecular analysis of vinorine synthase. *Bioorg. Med. Chem.* 12: 2787–2795.

Beaudoin, G.A.W. and P.J. Facchini. 2013. Isolation and characterization of a cDNA encoding (S)-cis-N-methylstylopine 14-hydroxylase from opium poppy. *Biochem. Biophys. Res. Commun.* 431: 597–603.

Beaudoin, G.A.W. and P.J. Facchini. 2014. Benzylisoquinoline alkaloid biosynthesis in opium poppy. *Planta* 240: 19–32.

Besseau, S., F. Kellner, A. Lanoue et al. 2013. A pair of tabersonine 16-hydroxylase initiates the synthesis of vindoline in an organ-dependent manner in *Catharanthus roseus*. *Plant Physiol.* 163: 1792–1803.

Biastoff, S. and B. Dräger. 2007. Calystegines. *Alkaloids Chem. Biol.* 64: 49–102.

Bird, D.A., V.R. Franceschi, and P.J. Facchini. 2003. A tale of three cell types: Alkaloid biosynthesis is localized to sieve elements in opium poppy. *Plant Cell* 15: 2626–2635.

Brochmann-Hanssen, E. 1984. A second pathway for the terminal steps in the biosynthesis of morphine. *Planta Med.* 50: 343–345.

Brown, S., M. Clastre, V. Courdavault, and S.E. O'Conner. 2015. De novo production of the plant-derived alkaloid strictosidine in yeast. *Proc. Natl. Acad. Sci. USA* 112: 3205–3210.

Burlat, V., A. Oudin, M. Courtois, M. Rideau, and B. St-Pierre. 2004. Co-expression of three MEP pathway genes and geraniol 10-hydroxylase in internal phloem parenchyma of *Catharanthus roseus* implicates multicellular translocation of intermediates during the biosynthesis of monoterpene indole alkaloids and isoprenoid-derived primary metabolites. *Plant J.* 38: 131–141.

Chahed, K., A. Oudin, N. Guivarc'h et al. 2000. 1-Deoxy-D-xylulose 5-phosphate synthase from periwinkle: cDNA identification and induced gene expression in terpenoid indole alkaloid-producing cells. *Plant Physiol. Biochem.* 38: 559–566.

Chen, X., T.T. Dang, and P.J. Facchini. 2015. Noscapine comes of age. *Phytochemistry* 111: 7–13.

Chen, X. and P.J. Facchini. 2014. Short-chain dehydrogenase/reductase catalysing the final step of noscapine biosynthesis is localized to lactifers in opium poppy. *Plant J.* 77: 173–184.

Chintapakorn, Y. and J.D. Hamill. 2007. Antisense-mediated reduction in ADC activity casues minor alterations in the alkaloid profile of cultured hairy roots and regenerated transgenic plants of *Nicotiana tabacum*. *Phytochemistry* 68: 2465–2479.

Choi, K.-B., T. Morishige, N. Shitan, K. Yazaki, and F. Sato. 2002. Molecular cloning and characterization of coclaurine N-methyltransferase from cultured cells of *Coptis japonica*. *J. Biol. Chem.* 277: 830–835.

Collu, G., N. Unver, A.M. Peltenburg-Looman, R. van der Heijden, R. Verpoorte, and J. Memelink. 2001. Geraniol 10-hydroxylase, a cytochrome P450 enzyme involved in terpenoid indole alkaloid biosynthesis. *FEBS Lett.* 508: 215–220.

Costa, M.M., F. Hilliou, P. Duarte et al. 2008. Molecular cloning of characterization of a vacuolar class III peroxidase involved in the metabolism of anticancer alkaloids in *Catharanthus roseus*. *Plant Physiol.* 146: 403–417.

Courdavault, V., N. Papon, M. Clastre, N. Giglioli-Guivarc'h, B. St-Pierre, and V. Burlat. 2014. A look inside an alkaloid multisite plant: *Catharanthus* logistics. *Curr. Opin. Plant Biol.* 19: 43–50.

Dang, T.T.T., X. Chen, and P.J. Facchini. 2015. Acetylation serves as a protective group in noscapine biosynthesis in opium poppy. *Nat. Chem. Biol.* 11: 104–106.

Dang, T.T.T. and P.J. Facchini. 2012. Characterization of three O-methyltransferases involved in noscapine biosynthesis in opium poppy. *Plant Physiol.* 159: 618–631.

Dang, T.T.T. and P.J. Facchini. 2014a. Cloning and characterization of canadine synthase involved in noscapine biosynthesis in opium poppy. *FEBS Lett.* 588: 198–204.

Dang, T.T.T. and P.J. Facchini. 2014b. Characterization of N-methylcanadine 1-hydroxylase, a cytochrome P450 catalyzing the first commited step of noscapine biosynthesis in opium poppy. *J. Biol. Chem.* 289: 2013–2028.

De Boer, K.D., H.L. Dalton, F.J. Edward, and J.D. Hamill. 2011a. RNAi-mediated down-regulation of ornithine decarboxylase (ODC) leads to reduced nicotine and increased anatabine levels in transgenic *Nicotiana tabacum* L. *Phytochemistry* 72: 344–355.

De Boer, K.D., J.C. Lye, C.D. Aitken, A.K. Su, and J.D. Hamill. 2009. The *A622* gene in *Nicotiana glauca* (tree tobacco): Evidence for a functional role in pyridine alkaloid synthesis. *Plant Mol. Biol.* 69: 299–312.

De Boer, K.D., S. Tilleman, L. Pauwels et al. 2011b. APETALA2/ETHYLENE RESPONSE FACTOR and basic helix-loop-helix tobacco transcription factors cooperatively mediate jasmonate-elicited nicotine biosynthesis. *Plant J.* 66: 1053–1065.

De Luca, V., C. Marineau, and N. Brisson. 1989. Molecular cloning and analysis of cDNA encoding a plant tryptophan decarboxylase: Comparison with animal dopa decarboxylase. *Proc. Natl. Acad. Sci. USA* 86: 2582–2586.

De Luca, V., V. Salim, A. Thamm, S.A. Matsuda, and F. Yu. 2014. Making iridoids/secoiridoids and Monoterpenoid indole alkaloids: Progress on pathway elucidation. *Curr. Opin. Plant Biol.* 19: 35–42.

De Luca, V. and B. St-Pierre. 2000. The cell and developmental biology of alkaloid biosynthesis. *Trends Plant Sci.* 5: 168–173.

Denoeud, F., L. Carretero-Paulet, A. Dereeper et al. 2014. The coffee genome provides insight into the convergent evolution of caffeine biosynthesis. *Science* 345: 1181–1184.
Desgagné-Penix, I. and P.J. Facchini. 2012. Systematic silencing of benzylisoquinoline alkaloid biosynthesis genes reveals the major route to papaverine in opium poppy. *Plant J.* 72: 331–344.
De Sutter, V., R. Vanderhaeghen, S. Tilleman et al. 2005. Exploration of jasmonate signalling via automated and standardized transient expression assays in tobacco cells. *Plant J.* 44: 1065–1076.
Dinda, B. and S. Debnath. 2013. Monoterpenes: Iridoids. In *Natural Products*, eds. K.G. Ramawat and J.N. Merillon, 3009–3067. Heidelberg: Springer.
Dittrich, H. and T.M. Kutchan. 1991. Molecular cloning, expression, and induction of berberine bridge enzyme, an enzyme essential to the formation of benzophenanthridine alkaloids in the response of plants to pathogen attack. *Proc. Natl. Acad. Sci. USA* 88: 9969–9973.
Dogru, E., H. Warzecha, F. Seibel, S. Haebel, F. Lottspeich, and J. Stöckigt. 2000. The gene encoding polyneuridine aldehyde esterase of monoterpenoid indole alkaloid biosynthesis in plants is an ortholog of the α/β hydrolase super family. *Eur. J. Biochem.* 267: 1397–1406.
Dugé de Bernonville, T., M. Clastre, S. Besseau et al. 2015. Phytochemical genetics of the Madagascar periwinkle: Unravelling the last twists of the alkaloid engine. *Phytochemistry* 113: 9–23.
Facchini, P.J., J. Bohlmann, P.S. Covello et al. 2012. Synthetic biosynthesis for the production of high-value plant metabolites. *Trends Plant Sci.* 30: 127–131.
Facchini, P.J. and V. De Luca. 1994. Differential and tissue-specific expression of a gene family for tyrosine/dopa decarboxylase in opium poppy. *J. Biol. Chem.* 269: 26684–26690.
Facchini, P.J. and V. De Luca. 2008. Opium poppy and Madagascar periwinkle: Model non-model systems to investigate alkaloid biosynthesis in plants. *Plant J.* 54: 763–784.
Facchini, P.J. and S.U. Park. 2003. Developmental and inducible accumulation of gene transcripts involved in alkaloid biosynthesis in opium poppy. *Phytochemistry* 64: 177–186.
Facchini, P.J., C. Penzes, G. Johnson, and D. Bull. 1996. Molecular characterization of berberine bridge enzyme genes from opium poppy. *Plant Physiol.* 112: 1669–1677.
Farrow, S.C., J.M. Hagel, G.A.W. Beaudoin, D.C. Burns, and P.J. Facchini. 2015. Stereochemical inversion of (S)-reticuline by a cytochrome P450 fusion in opium poppy. *Nat. Chem. Biol.* 11: 728–734.
Fossati, E., A. Ekins, L. Narcross et al. 2014. Reconstitution of a 10-gene pathway for synthesis of the plant alkaloid dihydrosanguinarine in *Saccharomyces cerevisiae*. *Nat. Commun.* 5: 3283.
Geerlings, A., M.M. Ibañez, J. Memelink, R. van Der Heijden, and R. Verpoorte. 2000. Molecular cloning and analysis of strictosidine β-D-glucosidase, an enzyme in terpenoid indole alkaloid biosynthesis in *Catharanthus roseus*. *J. Biol. Chem.* 275: 3051–3056.
Gerasimenko, I., Y. Sheludko, X. Ma, and J. Stöckigt. 2002. Heterologous expression of a *Rauvolfia* cDNA encoding strictosidine glucosidase, a biosynthetic key to over 2000 monoterpenoid indole alkaloids. *Eur. J. Biochem.* 269: 2204–2213.

Gesell, A., M.L. Díaz Chávez, R. Kramell, M. Piotrowski, P. Macheroux, and T.M. Kutchan. 2011. Heterologus expression of two FAD-dependent oxidases with (s)-tetrahydroprotoberberine oxidase activity from *Argemone mexicana* and *Berberis wilsoniae* in insect cells. *Planta* 233: 1185–1197.

Gesell, A., M. Rolf, J. Ziegler, M.L. Díaz Chávez, F.C. Huang, and T.M. Kutchan. 2009. CYP719B1 is salutaridine synthase, the C-C phenol-coupling enzyme of morphine biosynthesis in opium poppy. *J. Biol. Chem.* 284: 24432–24442.

Geu-Flores, F., N.H. Sherden, V. Courdavault et al. 2012. An alternative route to cyclic terpenes by reductive cyclization in iridoid biosynthesis. *Nature* 492: 138–142.

Glenn, W.S., W. Runguphan, and S.E. O'Connor. 2013. Recent progress in the metabolic engineering of alkaloids in plant systems. *Curr. Opin. Biotechnol.* 24: 354–365.

Grothe, T., R. Lenz, and T.M. Kutchan. 2001. Molecular characterization of the salutaridinol 7-O-acetyltransferase involved in morphine biosynthesis in opium poppy *Papaver somniferum*. *J. Biol. Chem.* 276: 30717–30723.

Guirimand, G., V. Courdavault, A. Lanoue et al. 2010. Strictosidine activation in Apocynaceae: Towards a "nuclear time bomb"? *BMC Plant Biol.* 10: 182.

Hagel, J.M., G. Beaudoin, E. Fossati, A. Ekins, V.J.J. Martin, and P.J. Facchini. 2012. Characterization of a flavoprotein oxidase from opium poppy catalysing the final steps in sanguinarine and papaverine biosynthesis. *J. Biol. Chem.* 287: 42972–42983.

Hagel, J.M. and P.J. Facchini. 2010. Dioxygenases catalyze the O-demethylatuion steps in morphine biosynthesis in opium poppy. *Nat. Chem. Biol.* 6: 273–275.

Hagel, J.M. and P.J. Facchini. 2013. Benzylisoquinoline alkaloid metabolism: A century of discovery and a brave new world. *Plant Cell Physiol.* 54: 647–672.

Haung, F.C. and T.M. Kutchan. 2000. Distribution of morphinan and benzo[c] phenanthridine alkaloid gene transcript accumulation in *Papaver somniferum*. *Phytochemistry* 53: 555–564.

Heim, W.G., K.A. Sykes, S.B. Hildreth, J. Sun, R.H. Lu, and J.G. Jelesko. 2007. Cloning and characterization of a *Nicotiana tabacum* methylputrescine oxidase transcript. *Phyochemistry* 68: 454–463.

Hibi, N., S. Higashiguchi, T. Hashimoto, and Y. Yamada. 1994. Gene expression in tobacco low-nicotine mutants. *Plant Cell* 6: 723–735.

Higashi, Y., T.M. Kutchan, and T.J. Smith. 2011. Atomic structure of salutaridine reductase from the opium poppy (*Papaver somniferrum*). *J. Biol. Chem.* 286: 6532–6541.

Hildreth, S.B., E.A. Gehman, H. Yang et al. 2011. Tobacco nicotine uptake permease (NUP1) affects alkaloid metabolism. *Proc. Natl. Acad. Sci. USA* 108: 18179–18184.

Höfer, R., L. Dong, F. André et al. 2013. Geraniol hydroxylase and hydroxygeraniol oxidase activities of the CYP76 family of cytochrome P450 enzymes and potential for engineering the early steps of the (seco)iridoid pathway. *Metab. Eng.* 20: 221–232.

Ikezawa, N., K. Iwase, and F. Sato. 2007. Molecular cloning and characterization of methylene bridge-forming enzymes involved in stylopine biosynthesis in *Eshscholzia californica*. *FEBS J.* 274: 1019–1035.

Ikezawa, N., K. Iwase, and F. Sato. 2009. CYP719A subfamily of cytochrome P450 oxygenases and isoquinoline alkaloid biosynthesis in *Eschscholzia californica*. *Plant Cell Rep.* 28: 123–133.

Ikezawa, N., M. Tanaka, M. Nagayoshi et al. 2003. Molecular cloning and characterization of CYP719, a methylenedioxy bridge-forming enzyme that belongs to a novel P450 family, from cultured *Coptis japonica* cells. *J. Biol. Chem.* 278: 38557–38565.

Imanishi, S., K. Hashizume, M. Nakakita et al. 1998. Differential induction of methyl jasmonate of genes encoding ornithine decarboxylase and other enzymes involved in nicotine biosynthesis in tobacco cultured cells. *Plant Mol. Biol.* 38: 1101–1111.

Irmler, S., G. Schröder, B. St-Pierre et al. 2000. Indole alkaloid biosynthesis in *Catharanthus roseus*: New enzyme activities and identification of cytochrome P450 CYP72A1 as secologanin synthase. *Plant J.* 24: 797–804.

Kajikawa, M., N. Hirai, and T. Hashimoto. 2009. A PIP-family protein is required for biosynthesis of tobacco alkaloids. *Plant Mol. Biol.* 69: 287–298.

Kajikawa, M., T. Shoji, A. Katoh, and T. Hashimoto. 2011. Vacuole-localized berberine bridge enzyme-like proteins are required for a late step of nicotine biosynthesis in tobacco. *Plant Physiol.* 155: 2010–2022.

Kato, K., N. Shitan, T. Shoji, and T. Hashimoto. 2015. NUP1 transports both tobacco alkaloids and vitamin B6. *Phytochemistry* 112: 33–40.

Kato, K., T. Shoji, and T. Hashimoto. 2014. Tobacco nicotine uptake permease regulates expression of the key transcription factor gene for nicotine biosynthesis pathway. *Plant Physiol.* 166: 2195–2204.

Kato, M. and K. Mizuno. 2004. Caffeine synthase and related methyltransferases in plants. *Front. Biosci.* 9: 1833–1842.

Kato, M., K. Mizuno, A. Crozier, T. Fujimura, and H. Ashihara. 2000. Caffeine synthase gene from tea leaves. *Nature* 406: 956–957.

Kato, N., E. Dubouzet, Y. Kokabu et al. 2007. Identification of a WRKY protein as a transcriptional regulator of benzylisoquinoline alkaloid biosynthesis in *Coptis japonica*. *Plant Cell Physiol.* 48: 8–18.

Katoh, A. and T. Hashimoto. 2004. Molecular biology of pyridine nucleotide and nicotine biosynthesis. *Front Biosci.* 9: 1577–1586.

Katoh, A., T. Shoji, and T. Hashimoto. 2007. Molecular cloning of *N*-methylputrescine oxidase from tobacco. *Plant Cell Physiol.* 48: 550–554.

Kazan, K. and J.M. Manners. 2013. MYC2: The master in action. *Mol. Plant.* 6: 686–703.

Kellner, F., J. Kim, B.J. Clavijo et al. 2015. Genome-wide investigation of plant natural product biosynthesis. *Plant J.* 82: 680–692.

Kutchan, T.M., N. Hampp, F. Lottspeich, K. Beyreuther, and M.H. Zenk. 1988. The cDNA clone for strictosidine synthase from *Rauvolfia serpentina*. DNA sequence determination and expression in *Escherichia coli*. *FEBS Lett.* 237: 40–44.

Lavac, D., J. Murata, W.S. Kim, and V. De Luca. 2008. Application of carborundum abrasion for investigating the leaf epidermis: Molecular cloning of *Catharanthus roseus* 16-hydroxytabersonine-16-*O*-methyltransferase. *Plant J.* 53: 225–236.

Lee, E.J. and P.J. Facchini. 2010. Norcoclaurine synthase is a member of the pathogen-related 10/Bet v1 protein family. *Plant Cell* 22: 3489–3503.

Lee, E.J. and P.J. Facchini. 2011. Tyrosine aminotransferase contributes to benzylisoquinoline alkaloid biosynthesis in opium poppy. *Plant Physiol.* 157: 1067–1078.

Li, R., D.W. Reed, E. Liu et al. 2006. Functional genomics analysis of alkaloid biosynthesis in *Hyoscyamus niger* reveals a cytochrome P450 involved in littorine rearrangement. *Chem. Biol.* 13: 513–530.

Liscombe, D.K. and P.J. Facchini. 2007. Molecular cloning and characterization of tetrahydroprotoberberine cis-N-methyltransferase, an enzyme involved in alkaloid biosynthesis in opium poppy. *J. Biol. Chem.* 282: 14741–14751.

Liscombe, D.K., B.P. MacLeod, N. Loukanina, O.I. Nandi, and P.J. Facchini. 2005. Evidence for the monophyletic evolution of benzylisoquinoline alkaloid biosynthesis in angiosperms. *Phytochemistry* 66: 2501–2520.

Liscombe, D.K., A.R. Usera, and S.E. O'Conner. 2010. Homolog of tocopherol C methyltransferases catalyzes N methylation in anticancer alklaloid biosynthesis. *Proc. Natl. Acad. Sci. USA.* 107: 18793–18798.

Ma, X., J. Koepke, S. Panjikar, G. Fritzsch, and J. Stöckigt. 2005. Crystal structure of vinorine synthase, the first representative of the BAHD superfamily. *J. Biol. Chem.* 280: 13576–13583.

Maliwan, N., K. Kato, T. Shoji, and T. Hashimoto. 2014. Molecular evolution of N-methylputrescine oxidase in tobacco. *Plant Cell Physiol.* 55: 436–444.

Matsuda, J., J. Okada, T. Hashimoto, and Y. Yamada. 1991. Molecular cloning of hyoscyamine 6β-hydroxylase, a 2-oxoglutarate-dependent dioxygenase, from cultured roots of *Hyoscyamus niger*. *J. Biol. Chem.* 25: 9460–9464.

McCarthy, A.A. and J.G. McCarthy. 2007. The structure of two N-methyltransferases from the caffeine biosynthetic pathway. *Plant Physiol.* 144: 879–889.

McNight, T.D., C.A. Roessner, R. Devagupta, A.I. Scott, and C.L. Nessler. 1990. Nucleotide sequence of a cDNA encoding the vacuolar protein strictosidine synthase from *Catharanthus roseus*. *Nucleic Acids Res.* 18: 4939.

Menke, F.L.H., A. Champion, J.W. Kijne, and J. Memelink. 1999. A novel jasmonate- and elicitor-responsive element in the periwinkle secondary metabolite biosynthetic gene *Str* interacts with a jasmonate- and elicitor-inducible AP2-domain transcription factor, ORCA2. *EMBO J.* 18. 4455–4463.

Miettinen, K., L. Dong, N. Navrot et al. 2014. The seco-iridoid pathway from *Catharanthus roseus*. *Nat. Commun.* 5: 3065.

Millgaqte, A.G., B.J. Pogson, I.W. Wilson et al. 2004. Analgesia: Morphine-pathway block in *top1* poppies. *Nature* 431:413–414.

Minami, H., E. Dubouzet, K.B. Choi, K. Yazaki, and F. Sato. 2007. Functional analysis of norcoclaurine synthase in *Coptis japonica*. *J. Biol.Chem.* 282: 6274–6282.

Mishra, B.B. and V.K. Tiwari. 2011. Natural products: An evolving role in future drug discovery. *Eur. J. Med. Chem.* 46: 4769–4807.

Mizuno, K., M. Kato, F. Irino, N. Yoneyama, T. Fujimura, and H. Ashihara. 2003a. The first committed step reaction of caffeine biosynthesis: 7-Methylxanthosine synthase is closely homologous to caffeine synthase in coffee (*Coffea Arabica* L.). *FEBS Lett.* 547: 56–60.

Mizuno, K., A. Okuda, M. Kato et al. 2003b. Isolation of a new dual-functional caffeine synthase gene encoding an enzyme for the conversion of 7-methylxanthine to caffeine from coffee (*Coffea Arabica* L.). *FEBS Lett.* 534: 75–81.

Morishige, T., T. Tsujita, Y. Yamada, and F. Sato. 2000. Molecular characterization of the S-adenosyl-L-methionine:3′-hydroxy-N-methylcoclaurine-4′-O-methyltransferase of isoquinoline alkaloid biosynthesis in *Coptis japonica*. *J. Biol. Chem.* 275: 23398–23405.

Morita, M., N. Shitan, K. Sawada et al. 2009. Vacuolar transport of nicotine is mediated by a multidrug and toxic compound extrusion (MATE) transporter in *Nicotiana tabacum*. *Proc. Natl. Acad. Sci. USA* 106: 2447–2452.

Murata, J., J. Roepke, H. Gordon, and V. De Luca. 2008. The leaf epidermome of *Catharanthus roseus* reveals its biochemical specialization. *Plant Cell* 20: 524–542.

Nakagawa, A., H. Minami, J.S. Kim et al. 2011. A bacterial platform for fermentive production of plant alkaloids. *Nat. Commun.* 2: 326.

Nakajima, K., T. Hashimoto, and Y. Yamada. 1993. Two tropinone reductases with different stereospecificities are short-chain dehydrogenases evolved from a common ancestor. *Proc. Natl. Acad. Sci. USA* 95: 9591–9595.

Nakajima, K., A. Yamashita, H. Akama et al. 1998. Crystal stuctures of two tropinone reductases: Different reaction sterospecificities in the same protein fold. *Proc. Natl. Acad. Sci. USA* 95: 4876–4881.

Nützmann, H.W. and A. Osbourn. 2014. Gene clustering in plant specialized metabolism. *Curr. Opin. Biotechnol.* 26: 91–99.

Ogawa, M., Y. Herai, N. Koizumi, T. Kusano, and H. Sano. 2001. 7-Methylxanthine methyltransferase of coffee plants. *J. Biol. Chem.* 276: 8213–8218.

Ogita, S., H. Uefuji, Y. Yamaguchi, N. Koizumi, and H. Sano. 2003. Producing decaffeinated coffee plants. *Nature* 423: 823.

Onoyovwe, A., J.M. Hagel, X. Chen, M.F. Khan, D.C. Schriemer, and P.J. Facchini. 2013. Morphine biosynthesis in opium poppy involves two cell types: Sieve elements and laticifers. *Plant Cell* 25: 4110–4122.

Otani, M., N. Shitan, K. Sakai, E. Martinola, F. Sato, and K. Yazaki. 2005. Characterization of vacuolar transport of the endogenous alkaloid berberine in *Coptis japonica*. *Plant Physiol.* 138: 1939–1946.

Oudin, A., M. Courtois, M. Rideau, and M. Clastre. 2007a. The iridoid pathway in *Catharanthus roseus* alkaloid biosynthesis. *Phytochem. Rev.* 6: 259–276.

Oudin, A., S. Mahroug, V. Courdavault, N. Hervouet, C. Zelwer, M. Rodríguez-Concepción, B. St-Pierre, and V. Burlat. 2007b. Spatial distribution and hormonal regulation of gene products from methyl erythritol phosphate and monoterpene-secoiridoid pathways in *Catharanthus roseus*. *Plant Mol. Biol.* 65: 13–30.

Ounaroon, A., G. Decker, J. Schmidt, F. Lottspeich, and T.M. Kutchan. 2003. (R, S)-Reticuline 7-O-methyltransferase and (R, S)-norcoclaurine 6-O-methyltransferase of *Papaver somniferum*-cDNA cloning and characterization of methyl transfer enzymes of alkaloid biosynthesis in opium poppy. *Plant J.* 36: 808–819.

Paschols, A., R. Halitschke, and I.T. Baldwin. 2007. Co(i)-ordinating defences: NaCOI1 mediates herbivore-induced resistance in *Nicotiana attenuata* and reveals the role of herbivore movement in avoiding defences. *Plant J.* 51: 79–91.

Pathak, S., D. Lakhwani, P. Gupta et al. 2013. Comparative transcriptome analysis using high papaverine mutant of *Papaver somniferum* reveals pathway and uncharacterized steps of Papaverine biosynthesis. *PLoS One* 8: e65622.

Pauli, H.H. and T.M. Kutchan. 1998. Molecular cloning and functional heterologous expression of two alleles encoding (S)-N-methylcoclaurine hydroxylase (CYP80B1), a new methyl jasmonate-inducible cytochrome P-450-dependent mono-oxygenase of benzylisoquinoline alkaloid biosynthesis. *Plant J.* 13: 793–801.

Pauw, B., F.A. Hiliou, V.S. Martin et al. 2004. Zinc finger proteins act as transcriptional repressors of alkaloid biosynthesis genes in *Catharanthus roseus*. *J. Biol. Chem.* 279: 52940–52948.

Rai, A., S.S. Smita, A.K. Singh, K. Shanker, and D.A. Nagegowda. 2013. Heteromeric and homomeric geranyl diphosphate synthases from *Catharanthus roseus* and their role in monoterpene indole alkaloid biosynthesis. *Mol. Plant* 6: 1531–1549.

Roberts, M.F. and R. Wink. 1998. *Alkaloids: Biochemistry, Ecology, and Medicinal Applications*. London: Plenum Press.

Roepke, J., V. Salim, M. Wu et al. 2010. Vinca drug components accumulate exclusively in leaf exudates of Madagascar periwinkle. *Proc. Natl. Acad. Sci. USA* 107: 15287–15292.

Runguphan, W. and S.E. O'Conner. 2009. Metabolic reprogramming of periwinkle plant culture. *Nat. Chem. Biol.* 5: 151–153.

Runguphan, W., X. Qu, and S.E. O'Conner. 2010. Integrating carbon-halogen bond formation into medicinal plant metabolism. *Nature* 468: 461–464.

Ruppert, M., X. Ma, and J. Stöckigt. 2005a. Alkaloid biosynthesis in *Rauvolfia*-cDNA cloning of major enzymes of the ajmaline pathway. *Curr. Org. Chem.* 9: 1431–1444.

Ruppert, M., J. Woll, A. Giritch, E. Genady, X. Ma, and J. Stöckigt. 2005b. Functional expression of an ajmaline pathway-specific esterase from *Rauvolfia* in a novel plant-virus expression system. *Planta* 222: 888–898.

Salim, V., B. Wiens, S. Masada-Atsumi, F. Yu, and V. De Luca. 2014. 7-Deoxyloganetic acid synthase catalyzes a key 3 step oxidation to form 7-deoxyloganetic acid in *Catharanthus roseus* iridoid biosynthesis. *Phytochemistry* 101: 23–31.

Salim, V., F. Yu, J. Altarejos-Caballero, and V. De Luca. 2013. Virus induced gene silencing identifies *Catharanthus roseus* 7-deoxyloganic acid 7-hydroxylase, a step in iridoid and monoterpenoid indole alkaloid biosynthesis. *Plant J.* 76: 754–765.

Samanani, N., J. Alcantara, R. Bourgault, K.G. Zulak, and P.J. Facchini. 2006. The role of phloem sieve elements and laticifers in the biosynthesis and accumulation of alkaloids in opium poppy. *Plant J.* 47: 547–563.

Samanani, N., S.U. Park, and P.J. Facchini. 2005. Cell type-specific localization of transcripts encoding nine consecutive enzymes involved in protoberberine alkaloid biosynthesis. *Plant Cell* 17: 915–926.

Samanani, N., D.K. Liscombe, and P.J. Facchini. 2004. Molecular cloning and characterization of norcoclaurine synthase, an enzyme catalysing the first committed step in benzylisoquinoline alkaloid biosynthesis. *Plant J.* 40: 302–313.

Schröder, G., E. Unterbusch, M. Kaltenbach et al. 1999. Light-induced cytochrome P450-dependent enzyme in indole alkaloid biosynthesis: Tabersonine 16-hydroxylase. *FEBS Lett.* 458: 97–102.

Shitan, N., I. Bazin, K. Dan, K. et al. 2003. Involvement of CjMDR1, a plant multidrug-resistance-type ATP-binding cassette protein, in alkaloid transport in *Coptis japonica*. *Proc. Natl. Acad. Sci. USA* 100: 751–756.

Shitan, N., F. Dalmas, K. Dan et al. 2013. Characterization of *Coptis japonica* CjABCB2, an ATP-binding cassette protein involved in alkaloid transport. *Phytochemistry* 91: 109–116.

Shoji, T. 2014. ATP-binding cassette and multidrug and toxic compound extrusion transporters in plants: A common theme among diverse detoxification mechanisms. *Int. Rev. Cell Mol. Biol.* 309: 303–346.

Shoji, T. and T. Hashimoto. 2011a. Tobacco MYC2 regulates jasmonate-inducible nicotine biosynthesis genes directly and by way of the *NIC2*-locus *ERF* genes. *Plant Cell Physiol.* 52: 1117–1130.

Shoji, T. and T. Hashimoto. 2011b. Recruitment of a duplicated primary metabolism gene into the nicotine biosynthesis regulon in tobacco *Plant J.* 67: 949–959.

Shoji, T. and T. Hashimoto. 2012. DNA-binding and transcriptional activation properties of tobacco *NIC2*-locus ERF189 and related proteins. *Plant Biotechnol.* 29: 35–42.

Shoji, T. and T. Hashimoto. 2013. Smoking out the masters: Transcriptional regulators for nicotine biosynthesis in tobacco. *Plant Biotechnol.* 30: 217–224.

Shoji, T. and T. Hashimoto. 2015a. Polyamine-derived alkaloids in plants: Molecular elucidation of biosynthesis. In *Polyamines, a Universal Molecular Nexus for Growth, Survival and Specialized Metabolism*, eds. T. Kusano and H. Suzuki, 189–200. Heidelberg: Springer.

Shoji, T. and T. Hashimoto. 2015b. Stress-induced expression of *NICOTINE2*-locus genes and their homologs encoding Ethylene Response Factor transcription factors in tobacco. *Phytochemistry* 113: 41–49.

Shoji, T., K. Inai, K. Yazaki et al. 2009. Multidrug and toxic compound extrusion-type transporters implicated in vacuolar sequestration of nicotine in tobacco roots. *Plant Physiol.* 149: 708–718.

Shoji, T., M. Kajikawa, and T. Hashimoto. 2010. Clustered transcription factors regulate nicotine biosynthesis in tobacco. *Plant Cell* 22: 3390–3409.

Shoji, T., M. Mishima, and T. Hashimoto. 2013. Divergent DNA-binding specificities of a group of ETHYLENE RESPONSE FACTOR transcription factors involved in plant defense. *Plant Physiol.* 162: 977–990.

Shoji, T., T. Ogawa, and T. Hashimoto. 2008. Jasmonate-induced nicotine formation in tobacco is mediated by tobacco *COI1* and *JAZ1* genes. *Plant Cell Physiol.* 49: 1003–1012.

Shoji, T., Y. Yamada, and T. Hashimoto. 2000. Jasmonate induction of putrescine *N*-methyltransferase genes in the root of *Nicotiana sylvestris*. *Plant Cell Physiol.* 41: 1072–1076.

Sibéril, Y., S. Benhamron, J. Memelink et al. 2001. *Catharanthus roseus* G-box binding factor 1 and 2 act as repressors of strictosidine synthase gene expression in cell cultures. *Plant Mol. Biol.* 45: 477–488.

Sierro, N., J.D.N. Battey, S. Ouadi et al. 2014. The tobacco genome sequence and its comparison with those of tomato and potato. *Nat. Commun.* 5: 3833.

Simkin, A.J., K. Miettinen, P. Claudel et al. 2013. Characterization of the plastidial geraniol synthase from Madagascar periwinkle which initiates the monoterpenoid branch of the alkaloid pathway in internal phloem associated parenchyma. *Phytochemistry* 85: 36–43.

Steppuhn, A., K. Gase, B. Krock, R. Halitschke, and I.T. Baldwin. 2004. Nicotine's defensive function in nature. *PLoS Biol.* 2: 1074–1080.

St-Pierre, B., P. Laflamme, A.M. Alarco, and V. De Luca. 1998. The terminal *O*-acetyltransferase involved in vindoline biosynthesis defines a new class of proteins responsible for coenzyme A-dependent acyl transfer. *Plant J.* 14: 703–713.

St-Pierre, B., F.A. Vazquez-Flota, and V. De Luca. 1999. Multicellular compartmentation of *Catharanthus roseus* alkaloid biosynthesis predicts intercellular translocation of a pathway intermediate. *Plant Cell* 11: 887–900.

Suttipanta, N., S. Pattanaik, M. Kulshrestha, B. Patra, S.K. Singh, and L. Yuan. 2011. The transcription factor CrWRKY1 positively regulates the terpenoid indole alkaloid biosynthesis in *Catharanthus roseus*. *Plant Physiol.* 157: 2081–2093.

Suzuki, K., Y. Yamada, and T. Hashimoto. 1999. Expression of *Atropa belladonna* putrescine *N*-methyltransferase genes in root pericycle. *Plant Cell Physiol.* 40: 289–297.
Takemura, T., N. Ikezawa, K. Iwase, and F. Sato. 2013. Molecular cloning and characterization of a cytochrome P450 in sanguinarine biosynthesis from *Eschscholzia californica* cells. *Phytochemistry* 91: 100–108.
Takeshita, N., H. Fujiwara, H. Mimura, J.H. Fitchen, Y. Yamada, and F. Sato. 1995. Molecular cloning and characterization of *S*-adenosyl-L-methionine:scoulerine-9-*O*-methyltransferase from cultured cells of *Coptis japonica*. *Plant Cell Physiol.* 36: 29–36.
Teuber, M., M.E. Azemi, F. Namjoyan et al. 2007. Putrescine *N*-methyltransferase; a structure-function analysis. *Plant Mol. Biol.* 63: 787–801.
Todd, A.T., E. Liu, S.L. Polvi, R.T. Pammett, and J.E. Page. 2010. A functional genomics screen identifies diverse transcription factors that regulate alkaloid biosynthesis in *Nicotiana benthamiana*. *Plant J.* 62: 589–600.
Uefuji, H., S. Ogita, Y. Yamaguchi, N. Koizumi, and H. Sano. 2003. Molecular cloning and functional characterization of three distinct *N*-methyltransferases involved in the caffeine biosynthetic pathway in coffee plants. *Plant Physiol.* 132: 372–380.
Uefuji, H., Y. Tatsumi, M. Morimoto, P. Kaothien-Nakayama, S. Ogita, and H. Sano. 2005. Caffeine production in tobacco plants by simultaneous expression of three coffee *N*-methyltransferases and its potential as a pest repellant. *Plant Mol. Biol.* 59: 221–227.
Unterlinner, B., R. Lenz, and T.M. Kutchen. 1999. Molecular cloning and functional expression of codeinone reductase: The penultimate enzyme in morphine biosynthesis in the opium poppy *Papaver somniferum*. *Plant J.* 18. 465–475.
van der Fits, L. and J. Memelink. 2000. ORCA3, a jasmonate-responsive transcriptional regulator of plant primary and secondary metabolism. *Science* 289: 295–297.
van der Fits, L., H. Zhang, F.L.H. Menke, M. Deneka, and J. Memelink. 2000. A *Catharanthus roseus* BPF-1 homologue interacts with an elicitor-responsive region of the secondary metabolite biosynthetic gene *Str* and is induced by elicitor via a JA-independent signal transduction pathway. *Plant Mol. Biol.* 44: 675–685.
Van Moerkercke, A., P. Steensma, F. Schweizer et al. 2015. The bHLH transcription factor BIS1 controls the iridoid branch of the monoterpenoid indole alkaloid pathway in *Catharanthus roseus*. *Proc. Natl. Acad. Sci. USA* 112: 8130–8135.
Vazquez-Flota, F., E. De Carolis, A.M. Alarco, and V. De Luca. 1997. Molecular cloning and characterization of desacetoxyvindoline-4-hydroxylase, a 2-oxoglutarate dependent-dioxygenase involved in the biosynthesis of vindoline in *Catharanthus roseus* (L.) G. Don. *Plant Mol. Biol.* 34: 935–948.
Veau, B., M. Courtois, A. Oudin, J.C. Chénieux, M. Rideau, and M. Clastre. 2000. Cloning and expression of cDNAs encoding two enzymes of the MEP pathway in *Catharanthus roseus*. *Biochem. Biophys. Acta.* 1517: 159–163.
Vogel, M., M. Lawson, W. Sippl, U. Conrad, and W. Roos. 2010. Structure and mechanism of sanguinarine reductase, an enzyme of alkaloid detoxification. *J. Biol. Chem.* 285: 18397–18406.

Vom Endt, D., M. Soares e Silva, J.W. Kijne, G. Pasquali, and J. Memelink. 2007. Identification of a bipartite jasmonate-responsive promoter element in the *Catharanthus roseus* ORCA3 transcription factor gene that interacts specifically with AT-hook DNA-binding proteins. *Plant Physiol.* 144: 1680–1689.

Weid, M., J. Ziegler, and T.M. Kutchan. 2004. The roles of latex and the vascular bundle in morphine biosynthersis in the opium poppy, *Papaver somniferum*. *Proc. Natl. Acad. Sci. USA* 101: 13957–13962.

Weiss, D., A. Baumert, M. Vogel, and W. Roos. 2006. Sanguinarine reductase, a key enzyme of benzophenanthridine detoxification. *Plant Cell Environ.* 29: 291–302.

Winkler, A., A. Lyskowski, S. Riedl et al. 2008. A concerted mechanism for berberine bridge enzymes. *Nat. Chem. Biol.* 4: 739–741.

Winzer, T., V. Gazda, Z. He et al. 2012. A *Papaver somniferum* 10-gene cluster for synthesis of the anticancer alkaloid noscapine. *Science* 336: 1704–1708.

Winzer, T., M. Kern, A.J. King et al. 2015. Morphinan biosynthesis in opium poppy requires a P450-oxidoreductase fusion protein. *Science* 349: 309–312.

Yamada, Y., Y. Kokabu, K. Chaki et al. 2011. Isoquinoline alkaloid biosynthesis is regulated by a unique bHLH-type transcription factor in *Coptis japonica*. *Plant Cell Physiol.* 51: 1131–1141.

Yamada, Y., Y. Motomura, and F. Sato. 2015. CjbHLH1 homologs regulate sanguinarine biosynthesis in *Eschscholzia californica* cells. *Plant Cell Physiol.* 56: 1019–1030.

Yoneyama, N., H. Morimoto, C.X. Ye, H. Ashihara, K. Mizuno, and M. Kato. 2006. Substrate specificity of N-methyltransferase involved in purine alkaloids synthesis is dependent upon one amino acid residue of the enzyme. *Mol. Gen. Genomics.* 275: 125–135.

Yu, F. and V. De Luca. 2013. ATP-binding cassette transporter controls leaf surface secretion of anticancer drug components in *Catharanthus roseus*. *Proc. Natl. Acad. Sci. USA* 110: 15830–15850.

Yun, D.J., T. Hashimoto, and Y. Yamada. 1992. Metabolic engineering of medicinal plants; transgenic *Atropa belladonna* with an improved alkaloid composition. *Proc. Natl. Acad. Sci. USA* 89: 11799–11803.

Zhang, H., S. Hedhili, G. Montiel et al. 2011. The basic helix-loop-helix transcription factor CrMYC2 controls the jasmonate-responsive expression of the *ORCA* genes that regulate alkaloid biosynthesis in *Catharanthus roseus*. *Plant J.* 67: 61–71.

Zhang, H.B., M.T. Bokowiec, P.J. Rushton, S.C. Han, and M.P. Timko. 2012. Tobacco transcription factors NtMYC2a and NtMYC2b form nuclear complexes with the NtJAZ1 repressor and regulate multiple jasmonate-inducible steps in nicotine biosynthesis. *Mol. Plant* 5: 73–84.

Ziegler, J., M.L. Diaz Chavez, R. Kramell, C. Ammer, and T.M. Kutchan. 2005. Comparative microarray analysis of morphine containing *Papaver somniferum* and eight morphine free *Papaver* species identifies an O-methyltransferase involved in benzylisoquinoline biosynthesis. *Planta* 222: 458–471.

Ziegler, J. and P.J. Facchini. 2008. Alkaloid biosynthesis: Metabolism and trafficking. *Annu. Rev. Plant Biol.* 59: 735–769.

Ziegler, J., S. Voigtlander, J. Schmidt et al. 2006. Comparative transcript and alkaloid profiling in *Papaver* species identifies a short chain dehydrogenase/reductase involved in morphine biosynthesis. *Plant J.* 48: 177–192.

chapter six

Pinaceae alkaloids

Virpi Virjamo and Riitta Julkunen-Tiitto

Contents

6.1 Introduction 119
6.2 Biosynthesis of Pinaceae piperidine alkaloids 121
    6.2.1 Biosynthetic route 121
    6.2.2 Control of biosynthesis 123
6.3 Appearance of Pinaceae piperidine alkaloids 124
    6.3.1 Genera and species-specific differences in the alkaloid chemistry of conifers 124
    6.3.2 Amounts and localization of Pinaceae piperidine alkaloids 126
6.4 Biological roles of Pinaceae piperidine alkaloids 127
6.5 Conclusion 128
References 128

6.1 Introduction

Secondary chemistry of evergreen coniferous trees is separated from deciduous trees in several ways, and one notable difference is that several Pinaceae family genera, including pine (*Pinus*), spruce (*Picea*), and fir (*Abies*), accumulate alkaloids in addition to phenolics and terpenoids (e.g., Stermitz et al. 1994, 2000). These little studied Pinaceae alkaloids are often considered as minor compounds because of their relatively small total yields if compared to phenolics and terpenoids. However, Pinaceae piperidines are highly interesting because alkaloids are typically effective defensive compounds, even in minor concentrations. Moreover, nitrogen-based defense is somewhat unexpected in the Northern boreal forests, which is occupied with alkaloid-accumulating coniferous species, as nitrogen is typically considered a limiting factor.

    Pinaceae piperidine alkaloid compounds are water soluble and volatile (Tawara et al. 1993). Still, they can be considered antifeedants rather than volatile deterrents as no evidence has been found that coniferous trees will emit alkaloids. Structurally, Pinaceae alkaloids are 2,6-disubstituted piperidines, nitrogen located in a heterocyclic ring (Stermitz et al. 1990; Tallent and Horning 1956). Based on the biosynthetic route, Pinaceae

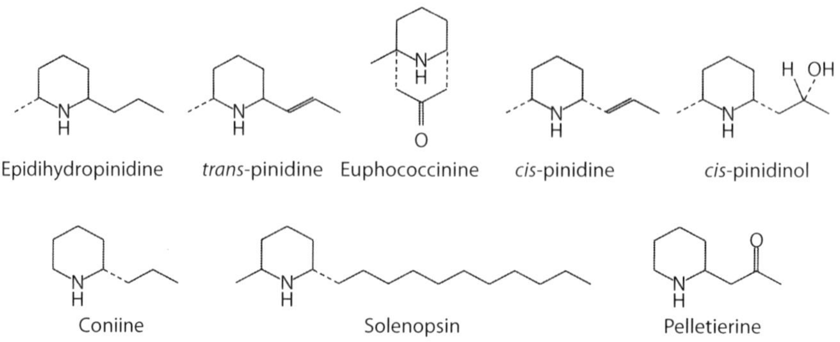

Figure 6.1 Typical piperidine alkaloid end products of coniferous species (top row) and examples of some structurally similar compounds detected from poison hemlocks (*Conium maculatum*), fire ants (*Solenopsis* sp.), and pomegranate (*Punica granatum*), respectively (bottom row).

alkaloids can also be defined as pseudo-alkaloids, as their carbon backbone is not originated from lysine as in true piperidine alkaloids (Leete and Juneau 1969; Leete et al. 1975). Pinaceae alkaloid compounds have small molecular weights from 137 to 171 g/moL depending on possible double bond, oxygen and hydroxyl group typically located on 6–carbon attached side chain (Gerson and Kelsey 2004; Stermitz et al. 1994; Tawara et al. 1993, 1999) (Figure 6.1). However, double bond, oxygen or hydroxyl group can also be attached to the heterocyclic ring structure (Tawara et al. 1993, 1999).

Some structurally similar and highly toxic alkaloid compounds are monosubstituted piperidines described from poison hemlock (*Conium maculatum*), and 2,6-disubstituted piperidines described from fire ants (*Solenopsis* sp.) (Green et al. 2012; Jones et al. 1990) (Figure 6.1). Moreover, 2,6-disubstituted piperidines are produced by Coccinellidae family beetles (e.g., *Epilachna varivestis* and *Cryptolaemus montrouzieri*) and by endemic Polynesian costal plant species (*Euphorbia atoto*) (Brown and Moore 1982; Eisner et al. 1986; Hart et al. 1967). The occurrence of homogeneous or closely related compounds in the range of evolutionarily distinct taxon suggest that biosynthesis of these compounds has evolved separately. However, within the Pinaceae family, biosynthesis of 2,6-piperidine alkaloids is considered as an evolutionary old feature (Stermitz et al. 2000).

To clarify, trivial names are commonly used for Pinaceae alkaloids. The first compound identified from coniferous species was (-)-*cis*-2-methyl-6-(2-propenyl)-piperidine from grey pine (*Pinus sabianana*) (Tallent and Horning 1956; Tallent et al. 1955). This compound was named *cis*-pinidine in honor of its pine origin (Tallent et al. 1955), and other compounds found more recently have trivial names referring to this first coniferous alkaloid compound. Euphococcinine, however, is an exception. It was originally found and named after the Australian coastal plant, *Euphorbia atoto* (Hart et al. 1967).

6.2   Biosynthesis of Pinaceae piperidine alkaloids

6.2.1   Biosynthetic route

Pinaceae species accumulate a wide range of alkaloid compounds, and to date more than 20 different compounds have been identified or tentatively identified (e.g., Tawara et al. 1993, 1999; Todd et al. 1995). In the future, a number of known coniferous piperidine alkaloids can be expected to increase while trace compounds can be identified with more accurate methodological setups.

A major part of the compounds is intermediate products of biosynthetic route or simple modifications of the end products (Tawara et al. 1993, 1995). The major end products of the *cis*-piperidine biosynthesis route are *cis*-pinidine and euphococcinine, while the end products of the *trans*-piperidine biosynthesis route are *trans*-pinidine and epidihydropinidine (Tawara et al. 1993; Todd et al. 1995) (Figure 6.2). However, the most abundant compounds are not always the end products of the biosynthetic route, and in some cases, species capacity to produce all types of piperidine alkaloids is lacking (Tawara et al. 1993; Virjamo and Julkunen-Tiitto 2014).

Biosynthesis of Pinaceae alkaloids starts with a polyketide precursor (Leete and Juneau 1969; Leete et al. 1975) (Figure 6.2). This is consistent with biosynthesis of poisonous hemlock alkaloids (Leete and Olson 1972). The difference between the biosynthetic route to monosubstituted piperiridines such as coniine and disubstituted piperidines such as *cis*-pinidine is a number of acetate units: there are five linear acetate units in the case of *cis*-pinidine and four acetate units in the case of coniine (Leete and Juneau 1969). However, it should be noted that not all pinidine-resembling structures are formed from acetate units. For example, monosubstituted pelletierine isolated from pomegranate (Figure 6.1) is originated from lysine precursor (Gupta and Spenser 1969).

Nitrogen is attached to a polyketidine structure after loss of one carboxyl group to form a heterocyclic ring structure (Leete and Juneau 1969). The first actual piperidine alkaloid intermediate in the biosynthetic route to *cis*-piperidines is thought to be 1,2-dehydropinidinone (Tawara et al. 1993, 1995) (Figure 6.2). Unlike data for actual precursors and some more late steps of synthesis, this first intermediate is not confirmed by labeling, but is supported by structural features and strong experimental data (Tawara et al. 1995; Todd et al. 1995; Virjamo and Julkunen-Tiitto 2014). From 1,2-dehydropinidine the two major end products, euphococcinine and *cis*-pinidine, are formed, *cis*-pinidine through several intermediate

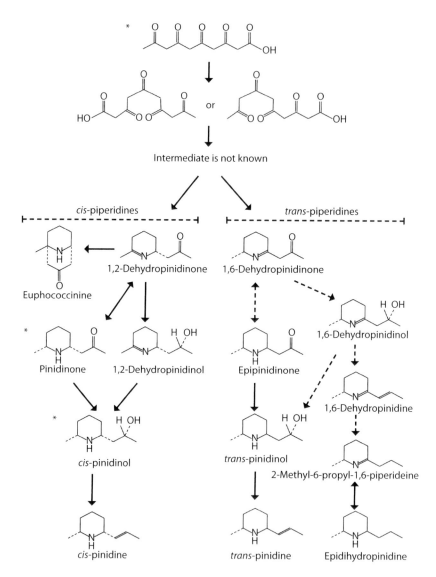

Figure 6.2 Biosynthetic route to Pinaceae piperidine alkaloid end products. (According to Leete, E. and K.N. Juneau. 1969. *J. Am. Chem. Soc.* 24: 5614–5618; Tawara, J.N. et al. 1993. *J. Org. Chem.* 58: 4813–4818; Tawara, J.N., F.R. Stermitz, and A.V. Blokhin. 1995. *Phytochemistry* 39: 705–708; Todd, F.G., F.R. Stermitz, and A.V. Blokhin. 1995. *Phytochemistry* 40: 401–406; Gerson, E.A. and R.G. Kelsey. 2004. *Biochem. Syst. Ecol.* 32: 62–74; Virjamo, V. and R. Julkunen-Tiitto. 2014. *Trees* 28: 427–437.) Dotted lines refer to tentatively identified modifications. Compounds marked with an asterisk are proven as Pinaceae piperidine biosynthesis intermediates by labeling tests. (According to Leete, E. and K.N. Juneau. 1969. *J. Am. Chem. Soc.* 24: 5614–5618; Tawara, J.N., F.R. Stermitz, and A.V. Blokhin. 1995. *Phytochemistry* 39: 705–708.)

compounds (Tawara et al. 1993, 1995) (Figure 6.2). The biosynthetic route to *cis*-pinidines is well established, while the biosynthesis route to *trans*-piperidines is still tentative. Experimental data, however, suggest that *trans*-piperidines are biosynthesized separately from *cis*-piperidines (Virjamo and Julkunen-Tiitto 2014). Separated routes are also supported by coniine biosynthesis studies, where dehydrogeneration of coniine to γ-coniceine is reported to be highly stereospecific (Leete and Olson 1972). Based on structural features and accumulation order of *trans*-piperidines, it is likely that ring structure and 6-carbon attached side chains are going through similar modifications than described for the *cis*-pinidine route, at least for *trans*-pinidine (Todd et al. 1995, Virjamo and Julkunen-Tiitto 2014) (Figure 6.2). Thus, the first intermediate at the *trans*-piperidine route would be 1,6-dehydropinidinone (Virjamo and Julkunen-Tiitto 2014). Formation of another end product, and the main compound in many spruce species, epidihydropinidine, can also be tentatively presented based on structure and accumulation of possible intermediate compounds. We have earlier suggested that epidihydropinidine would be formulated from the same precursor 1,6-dehydropinidinone than *trans*-pinidine through two 1,6-imine intermediates (Virjamo and Julkunen-Tiitto 2014). However, based on structural features and accumulation data from young needles (Virjamo, unpublished data) it is likely that 1,6-dehydropinidine tentatively reported by Gerson and Kelsey (2004) is also the intermediate of epidihydropinidine biosynthesis (Figure 6.2).

6.2.2 Control of biosynthesis

As expected, biosynthesis of Pinaceae piperidine alkaloids seems to be strongly dependent on availability of nitrogen (Gerson and Kelsey 1999). Otherwise, regulation of biosynthesis Pinaceae piperidines is poorly known. Thus far it has been shown that accumulation of Norway spruce (*Picea abies*) alkaloids is controlled by temperature, but not UV-B light (Virjamo et al. 2014). Moreover, common garden studies have revealed that biosynthesis of coniferous alkaloids is dependent on genetic factors (Gerson et al. 2009; Virjamo and Julkunen-Tiitto 2016). Interestingly, unlike biosynthetically and structurally related coniine alkaloids, no evidence has been found that accumulation of coniferous piperidine alkaloids would be induced by herbivory (Castells et al. 2005; Gerson and Kelsey 1998; Schiebe et al. 2012).

Biosynthesis of piperidine alkaloids starts at an early phase. For example, in ponderosa pine (*Pinus ponderosa*), seedlings accumulate alkaloids already at 6-day old plantlets, at the stage where differentiation of stems or needles is not completed (Tawara et al. 1995). In blue spruce (*Picea pungens*), alkaloids have been reported from 9-day old seedlings (Todd et al. 1995). However, a major part of alkaloids in the youngest coniferous seedlings is intermediates of biosynthesis (Tawara et al. 1995). Similar results have been reported from young vegetative parts of the adult trees. Piperidine alkaloids accumulate in developing needles and stems at an early stage, and at that time, alkaloid yield is first dominated by intermediate compounds (Todd et al. 1995; Virjamo and Julkunen-Tiitto 2014).

6.3 Appearance of Pinaceae piperidine alkaloids

6.3.1 Genera and species-specific differences in the alkaloid chemistry of conifers

While most of the studied spruce (*Picea*) and pine (*Pinus*) species and several fir (*Abies*) species have been shown to accumulate Pinaceae alkaloids, there are several remarkable differences between genera (Table 6.1). Most importantly, there is a major difference in how these species biosynthesize *cis*- and *trans*-piperidines. With some minor exceptions, it can be generalized that pine (*Pinus*) and fir (*Abies*) species accumulate *cis*-2,6-disubstituted piperidines while spruce (*Picea*) species accumulate both *cis*- and *trans*-2,6-disubstituted piperidines (Stermitz et al. 2000; Tawara et al. 1993). Another major difference is that alkaloids are far less widespread in firs than in pines or spruces, where nearly all studied species have been shown to accumulate alkaloids (e.g., Gerson and Kelsey 2004; Stermitz et al. 1994; Stermitz et al. 2000) (Table 6.1).

From 33 coniferous species investigated thus far for alkaloid chemistry (Table 6.1), very few atypical species-specific alkaloid compositions have been recognized. Ponderosa pine (*Pinus ponderosa*) has been shown to produce *trans*-oriented epipinidinone in trace amounts, atypical for pine species (Gerson et al. 2009). Also, some species, including Scots pine (*Pinus sylvestris*), which has major economical and ecological importance in northern latitudes, seems to lack the capacity to produce *cis*-pinidine, one of the two typical main products in the pine species (Tawara et al. 1993). However, the most remarkable difference in alkaloid chemistry has been recorded from two endemic spruce species: North American Brewer's weeping spruce (*Picea breweriana*) and rare European species Serbian spruce (*Picea omorica*). *Picea breweriana* is able to accumulate partly typical Pinaceae alkaloids (euphococcinine), but the species also accumulates pyrrolidine alkaloids and lysine-originated true piperidine alkaloids, identical to pomegranate (*Punica granatum*) alkaloids (Schneider et al. 1995) (Figure 6.3). On the other hand, the main component for *Picea omorica* is a *cis*-pinidinone type of compound with oxygen attached to the ring structure at the 4-carbon position (Stermitz et al. 2000) (Figure 6.3). Biosynthesis of 4-oxy-2,6-piperidines in conifers is not studied, but the compound has earlier recognized a trace component for *Picea pungens* (Tawara et al. 1999).

Chapter six: Pinaceae alkaloids

Table 6.1 Main alkaloid end products from needles of Pinaceae species

Species	Main end products	Refs.
Abies balsamea	Euphococcinine	Stermitz et al. (2000)
Abies concolor	Euphococcinine	Stermitz et al. (2000)
Abies grandis	ND[a]	Stermitz et al. (2000)
Abies lowiana	ND	Stermitz et al. (2000)
Abies procera	ND	Stermitz et al. (2000)
Pinus cembroides	*Cis*-pinidine, euphococcinine	Gerson and Kelsey (2004)
Pinus contorta	Euphococcinine	Gerson and Kelsey (1998)
Pinus discolor	ND	Gerson and Kelsey (2004)
Pinus durangensis	*Cis*-pinidine	Gerson and Kelsey (2004)
Pinus edulis	*Cis*-pinidine, euphococcinine	Stermitz et al. (1994)
Pinus engelmannii	*Cis*-pinidine	Gerson and Kelsey (2004)
Pinus flexilis	Euphococcinine	Stermitz et al. (1994)
Pinus jeffreyi	*Cis*-pinidine, euphococcinine	Tallent et al. (1955), Stermitz et al. (1994)
Pinus leiophylla	*Cis*-pinidine, euphococcinine	Gerson and Kelsey (2004)
Pinus lumholtzii	Euphococcinine	Gerson and Kelsey (2004)
Pinus monophylla	*Cis*-pinidine, euphococcinine	Gerson and Kelsey (2004)
Pinus nigra	Euphococcinine	Stermitz et al. (1994)
Pinus pinea	*Cis*-pinidinol	Stermitz et al. (1994)
Pinus ponderosa	*Cis*-pinidine	Stermitz et al. (1994), Tawara et al. (1995), Gerson and Kelsey (1998, 2004), Gerson et al. (2009)
Pinus sabiniana	*Cis*-pinidine	Tallent et al. (1955)
Pinus sylvestris	Euphococcinine	Stermitz et al. (1994)
Pinus torreyana	*Cis*-pinidine	Tallent et al. (1955)
Picea abies	Epidihydropinidine, *cis*-pinidinol	Stermitz et al. (1994), Virjamo et al. (2013, 2014), Virjamo and Julkunen-Tiitto (2014)
Picea brachytyla	Epidihydropinidine, *cis*-pinidinol	Schneider et al. (1991)
Picea breweriana	Euphococcinine, hygroline, N-methylsedridine	Schneider et al. (1995)
Picea chihuahuana	Epidihydropinidine, *cis*-pinidinol	Schneider et al. (1991)
Picea engelmannii	Epidihydropinidine, *cis*-pinidinol	Schneider et al. (1991), Chantel et al. (1998)
Picea glauca	Epidihydropinidine, *cis*-pinidinol	Schneider et al. (1991), Stermitz et al. (1994)

(*Continued*)

Table 6.1 (Continued) Main alkaloid end products from needles of Pinaceae species

Species	Main end products	Refs.
Picea likiangensis	Epidihydropinidine, *cis*-pinidinol	Schneider et al. (1991)
Picea mariana	Epidihydropinidine, *cis*-pinidinol	Schneider et al. (1991)
Picea omorica	4-oxy-2,6-*cis*-pinidinol	Stermitz et al. (2000)
Picea pungens	Epidihydropinidine, *cis*-pinidinol, euphococcinine	Schneider et al. (1991), Stermitz et al. (1994), Todd et al. (1995)
Picea sitchensis	Epidihydropinidine, *cis*-pinidinol	Stermitz et al. (1994, 2000), Gerson and Kelsey (2002)

[a] ND stands for no alkaloids detected.

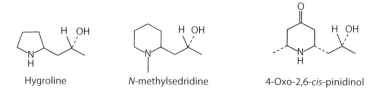

Hygroline N-methylsedridine 4-Oxo-2,6-*cis*-pinidinol

Figure 6.3 Examples of atypical structures detected from *Picea omorica* and *Picea breweriana*. (According to Schneider, M.J., S. Brendze, and J.A. Montali. 1995. *Phytochemistry* 39: 1387–1390; Stermitz, F.R., C.D. Kamm, and J.N. Tawara. 2000. *Biochem. Syst. Ecol.* 28: 177–181.)

6.3.2 Amounts and localization of Pinaceae piperidine alkaloids

Cell level localization of Pinaceae piperidine alkaloids is lacking. However, a reasonable amount of tissue level knowledge has accumulated. Piperidine alkaloids appear in both needles and twigs of coniferous trees, but also in roots, pistils, cones, and dormant buds (Stermitz et al. 1994; Todd et al. 1995; Virjamo and Julkunen-Tiitto 2014). In addition, alkaloids have been reported from resin (Stermitz et al. 1994). Concentrations of piperidine alkaloids are often lower than typically reported from phenolics and terpenoids. However, concentration of individual compounds may be as high as 0.4 mg/g (dry weight) (e.g., Gerson et al. 2009; Virjamo and Julkunen-Tiitto 2014), and generally, total alkaloids may vary from 0.03 to 0.08% of fresh weight of needles (Tawara et al. 1993). Most abundant components are often, but not always, the same as the end products of biosynthesis, for pine (*Pinus*) species they are typically are *cis*-pinidine, *cis*-pinidinol, and euphococcinine; and for spruce (*Picea*) species they are epidihydropinidine, *cis*-pinidinol, and euphococcinine (Stermitz et al. 1994; Tawara et al. 1993). In developing structures, intermediate compounds can occupy the alkaloid yield, but in mature plant parts it is typical that the main component or components are covering more than half of the total alkaloids (e.g., Gerson et al. 2009; Virjamo and Julkunen-Tiitto 2014).

6.4 Biological roles of Pinaceae piperidine alkaloids

Based on structure, Pinaceae alkaloids can be expected to be defensive compounds. However, activity can be compound specific. For example, activity of structurally related coniine is strongly dependent on its stereochemistry (Lee et al. 2008). As pines and spruces have different main components, Pinaceae alkaloid compounds may also present partly different functions for these genera. This underlines the importance to report compound-specific information in addition to total alkaloid yields in defensive studies. Accordingly, attempts to explain herbivory with total alkaloid amounts have been unsuccessful (Gerson and Kelsey 2002; Schiebe et al. 2012).

Biological activities of several individual Pinaceae piperidine compounds have been reported. Toxicity predicted from the structure is supported by results obtained with pine alkaloids as one major component; *cis*-pinidine has proven to be highly toxic in a frog embryo test (Stermitz et al. 1994; Tawara et al. 1993). It is also suggested that *cis*-pinidine is teratogenic (Tawara et al. 1993). However, most studies have focused on antifeedant effect toward herbivorous insects instead of toxicity. A mixture of the main components of spruces, *cis*-pinidinol and epidihydropinidine, as well as nonvolatile form of dihydropinidine has shown high antifeedant activity and growth reduction of herbivorous insects (*Choristoneura fumiferena*, *Hylobious abietus*, and *Peridroma saucia*, respectively) (Schneider et al. 1991; Shtykova et al. 2008; Stermitz et al. 1994). Euphococcinine is also a compound that is present in all three genera of the Pinaceae family where alkaloids have been reported thus far, is likely participating in defense against insects. Euphococcinine is also produced by some Coccinellidae family beetles where it has been reported to function as an active deterrent against spiders (Eisner et al. 1986). Interestingly, rasemic (±)-epidihydropinidine shows a high and wide spectrum of antibacterial activity (Fyhrquist et al. unpublished results), whereas *cis*-pinidinol has reported no antimicrobial activity and euphococcinine only weak activity against gram-negative bacterium (Tawara et al. 1993). Thus far, the only feeding experiment with mammals has been conducted with field voles, where voles prefer seedlings that simultaneously did have the highest nitrogen and the highest epidihydropinidine concentrations suggesting no antifeedant activity (Virjamo et al. 2013).

6.5 Conclusion

Despite small concentrations, Pinaceae alkaloid compounds clearly participate on tree defense together with other defensive factors. This emphasizes the importance of what is not yet known about Pinaceae piperidines, such as cell level localization, turnover rate of the main components, genes, and enzymes involved in Pinaceae alkaloid biosynthesis, and regulation mechanisms of accumulation levels. Another important missing knowledge is antifeedant activity and toxicity of Pinaceae alkaloids against mammalian herbivores. For example, antifeedant activity against moose, which cause significant species-selective damage especially to young pine plantations in Northern Europe where coniferous trees have a significant economical role, have not been investigated thus far.

References

Brown, W.V. and B.P. Moore. 1982. Defensive alkaloids of *Cryptolaemus montrouzieri* (Coleoptera: Coccinellidae). *Aust. J. Chem.* 35: 1255–1261.

Castells, E., M.A. Berhow, S.F. Vaughn, and M.R. Berenbaum. 2005. Geographic variation in alkaloid production in *Conium maculatum* populations experiencing differential herbivory by *Agonopterix alstroemeriana*. *J. Chem. Ecol.* 31: 1693–709.

Eisner, T., M. Goetz, D. Aneshansley, G. Ferstandig-Arnold, and J. Meinwald. 1986. Defensive alkaloid in blood of Mexican bean beetle (*Epilachna varivestis*). *Experentia* 42: 204–207.

Gerson, E.A. and R.G. Kelsey. 1998. Variation of piperidine alkaloids in ponderosa (*Pinus ponderosa*) and lodgepole pine (*P. contorta*) foliage from central Oregon. *J. Chem. Ecol.* 24: 815–827.

Gerson, E.A. and R.G. Kelsey. 1999. Piperidine alkaloids in nitrogen fertilized *Pinus ponderosa*. *J. Chem. Ecol.* 25: 2027–2039.

Gerson, E.A. and R.G. Kelsey. 2002. Piperidine alkaloids in sitka spruce with varying levels of resistance to white pine weevil (Coleoptera: Curculionidae). *J. Econ. Entomol.* 95: 608–613.

Gerson, E.A. and R.G. Kelsey. 2004. Piperidine alkaloids in North American Pinus taxa: Implications for chemosystematics. *Biochem. Syst. Ecol.* 32: 62–74.

Gerson, E.A., R.G. Kelsey, and J.B. St Clair. 2009. Genetic variation of piperidine alkaloids in Pinus ponderosa: A common garden study. *Ann. Bot.* 103: 447–457.

Green, B.T., S.T. Lee, K.E. Panter, and D.R. Brown. 2012. Piperidine alkaloids: Human and food animal teratogens. *Food Chem. Toxicol.* 50: 2049–2055.

Gupta, R.N. and I.D. Spenser. 1969. Specifically ring-labelled (±)-pelletierine. *Can. J. Chem.* 47: 445–447.

Hart, N.K., S.R. Johns, and J.A. Lamberton. 1967. (+)-9-Aza-1-methylbicyclo[3,3,1] nonan-3-one, a new alkaloid from *Euphorbia atoto* Forst. *Aust. J. Chem.* 20: 561–563.

Jones, T.H., M.S. Blum, and G.H. Robertson. 1990. Novel dialkylpiperidines in the venom of the ant *Monomorium Delagoense*. *J. Nat. Prod.* 53: 429–435.

Lee, S.T., B.T. Green, K.D. Welch, J.A. Pfister, and K.E. Panter. 2008. Stereoselective potencies and relative toxicities of coniine enantiomers. *Chem. Res. Toxicol.* 21: 2061–2064.
Leete, E. and K.N. Juneau. 1969. Biosynthesis of pinidine. *J. Am. Chem. Soc.* 24: 5614–5618.
Leete, E. and J.O. Olson. 1972. Biosynthesis and metabolism of Hemlock alkaloids. *J. Am. Chem. Soc.* 94: 5427–5477.
Leete, E., J.C. Lechleiter, and R.A. Carver. 1975. Determination of the "starter" acetate unit in the biosynthesis of pinidine. *Tetrahedron let.* 44: 3779–3782.
Schiebe, C., A. Hammerbacher, G. Birgersson et al. 2012. Inducibility of chemical defenses in Norway spruce bark is correlated with unsuccessful mass attacks by the spruce bark beetle. *Oecologia* 170: 183–198.
Schneider, M.J., S. Brendze, and J.A. Montali. 1995. Alkaloids of *Picea breweriana*. *Phytochemistry* 39: 1387–1390.
Schneider, M.J., J.A. Montali, D. Hazen, and C.E. Stanton. 1991. Alkaloids of *Picea*. *J. Nat. Prod.* 54: 905–909.
Shtykova, L., M. Masuda, C. Eriksson et al. 2008. Latex coatings containing antifeedants: Formulation, characterization, and application for protection of conifer seedlings against pine weevil feeding. *Prog. Org. Coat.* 63: 160–166.
Stermitz, F.R., Miller, M.M., and M.J. Schneider. 1990. The Stereochemistry of (-)-Pinidinol, a Piperidine Alkaloid from *Picea engelmannii*. *J. Nat. Prod.* 53: 1019–1020.
Stermitz, F.R., J.N. Tawara, M. Boeckl, M. Pomeroy, T.A. Foderaro, and F.G. Todd. 1994. Piperidine alkaloid content of *Picea* (spruce) and *Pinus* (pine). *Phytochemistry* 35: 951–953.
Stermitz, F.R., C.D. Kamm, and J.N. Tawara. 2000. Piperidine alkaloids of spruce (*Picea*) and fir (*Abies*) species. *Biochem. Syst. Ecol.* 28: 177–181.
Tallent, W.H. and E.C. Horning. 1956. The structure of pinidine. *J. Am. Chem. Soc.* 78: 4467–4469.
Tallent, W.H., V.L. Stromberg, and E.C. Horning. 1955. Pinus alkaloids. The alkaloids of *P. sabiniana* Dougl. and related species. *J. Am. Chem. Soc.* 77: 6361–6364.
Tawara, J.N., A. Blokhin, T.A. Foderaro, F.R. Stermitz, and H. Hope. 1993. Toxic piperidine alkaloids from pine (*Pinus*) and spruce (*Picea*) trees. New structures and a biosynthetic hypothesis. *J. Org. Chem.* 58: 4813–4818.
Tawara, J.N., P. Lorenz, and F.R. Stermitz. 1999. 4-hydroxylated piperidines and N-methyleuphococcinine (1-methyl-3-granatanone) from *Picea* (spruce) species. Identification and synthesis. *J. Nat. Prod.* 62: 321–323.
Tawara, J.N., F.R. Stermitz, and A.V. Blokhin. 1995. Alkaloids of young ponderosa pine seedlings and the late steps in the biosynthesis of pinidine. *Phytochemistry* 39: 705–708.
Todd, F.G., F.R. Stermitz, and A.V. Blokhin. 1995. Piperidine alkaloid content of *Picea pungens* (Colorado blue spruce). *Phytochemistry* 40: 401–406.
Virjamo, V. and R. Julkunen-Tiitto. 2014. Shoot development of Norway spruce (*Picea abies*) involves changes in volatile alkaloids and condensed tannins. *Trees* 28: 427–437.
Virjamo, V. and R. Julkunen-Tiitto. 2016. Variation in piperidine alkaloid chemistry of Norway spruce (*Picea abies*) foliage in diverse geographic origins grown in the same area. *Can J. For. Res.* 46: 456–460.

Virjamo, V., R. Julkunen-Tiitto, H. Henttonen et al. 2013. Differences in vole preference, secondary chemistry and nutrient levels between naturally regenerated and planted Norway spruce seedlings. *J. Chem. Ecol.* 39: 1322–1334.

Virjamo, V., S. Sutinen, and R. Julkunen-Tiitto. 2014. Combined effect of elevated UVB, elevated temperature and fertilization on growth, needle structure and phytochemistry of young Norway spruce (*Picea abies*) seedlings. *Glob. Chang. Biol.* 20: 2252–2260.

chapter seven

Biosynthesis, regulation, and significance of cyanogenic glucosides

Lasse Janniche Nielsen, Nanna Bjarnholt, Cecilia Blomstedt, Roslyn M. Gleadow, and Birger Lindberg Møller

Contents

7.1 Introduction	131
7.2 Structure and metabolism of cyanogenic glucosides	137
7.2.1 Structural diversity of cyanogenic glycosides	137
7.2.2 Biosynthesis of cyanogenic glycosides	139
7.3 Multifunctional cyanogenic glucosides	140
7.3.1 Dynamic defense systems	140
7.3.2 Turnover of cyanogenic glucosides and the possible metabolic pathways involved	143
7.3.3 Resource allocation	145
Acknowledgments	148
References	148

7.1 Introduction

Embryophytes (green land plants) are a highly successful group of multicellular eukaryotes belonging to the kingdom Plantae. They form a clade that includes flowering plants, conifers and other gymnosperms, ferns, clubmosses, hornworts, liverworts, and mosses. The largest clade, the angiosperms (flowering plants), have an estimated 450,000 species and are found on all continents of the world (Pimm and Joppa 2015). All these different plant species originate from early plants that have evolved and coexisted with insects and fungi for more than 400 million years (Furstenberg-Hagg et al. 2013). It has been estimated that at least one herbivorous species of insects exists for each plant. This predator–prey relationship has given rise to the coevolutionary theory, which proposes that insect herbivory is one of the main drivers for both plant and insect diversification (Ehrlich and Raven 1964). Plants are sessile organisms incapable of fleeing the potential dangers posed by herbivores and pathogenic fungi threatening their existence and have evolved a range of different defense mechanisms. These include both indirect and direct defenses (Furstenberg-Hagg et al. 2013). The indirect defense mechanisms involve the attraction of predatory organisms by volatiles and nourishing predators in food bodies and housing in specially adapted plant parts (Heil et al. 1997; Odowd 1980). Examples of attraction by volatiles include the attraction of predatory mites to aliphatic and aromatic oximes released by the cucumber (*Cucumis sativus*), golden chain (*Laburnum anagyroides*), *Robinia pseudoacacia,* and eggplant (*Solanum melongena*) as a response to invading spider mites (Agrawal et al. 2002; Takabayashi et al. 1994; Van Den Boom et al. 2004). In maize (*Zea mays*), lima beans (*Phaseolus lunatus*), and black poplar (*Populus nigra*) tissue damage caused by feeding caterpillars also elicits the release of volatile oximes as attractants of parasitic insects (Irmisch et al. 2013; Takabayashi et al. 1995; Wei et al. 2006).

The direct defenses involve physical barriers that prevent herbivores from feeding on the plant, that is, epicuticular wax films, trichomes, and accumulation of abrasive salt crystals, latex, and resins (Dussourd and Hoyle 2000; Eigenbrode and Espelie 1995; Phillips and Croteau 1999). An important part of the direct defense strategy used by plants is based on the production of bioactive natural products that are unpalatable, have antinutritional or toxic effects. The bioactive natural products include alkaloids, benzoxazinoids, glucosinolates, nonprotein amino acids, phenolics, terpenoids, and cyanogenic glycosides (CNglcs). Many of these bioactive compounds may be part of entangled dynamic trade-off systems where the action of the different compounds partly compensates for each other. This is seen in several cyanogenic species where CNglcs and condensed tannins are found at different time points during ontogeny (Briggs 1990; Dement and Mooney 1974; Goodger et al. 2006). The chemical class of CNglcs are of particular interest due to their presence in several different crop plants and fruit trees, that is, barley (*Hordeum vulgare*) (Nielsen et al. 2002), wheat (*Triticum aestivum*) (Erb et al. 1981; Jones 1998), sorghum (*Sorghum bicolor*) (Kojima et al. 1979), cassava (*Manihot esculenta*) (Jørgensen et al. 2011), almonds (*Prunus dulcis*) (Sánchez-Pérez et al. 2008), cherries (*Prunus* spp.) (Nahrstedt 1970), apples *(Malus pumilia* hybrids) (Dziewanowska et al. 1979), macadamia nuts (*Macadamia ternifolia*) (Dahler et al. 1995), and passionfruit (*Passiflora* spp.) (Jaroszewski et al. 2002; Spencer and Seigler 1983) (Figure 7.1).

In addition to being found in angiosperms, the presence of CNglcs in ferns and gymnosperms indicates that the ability to produce CNglcs is at least 300 million years old (Zagrobelny et al. 2008). The evolutionary

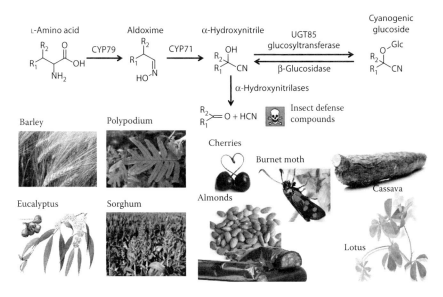

Figure 7.1 (**See color insert.**) Schematic illustration of biosynthesis and bioactivation of cyanogenic glucosides in crop plants. Some insects are also able to *de novo* biosynthesize cyanogenic glucosides or to sequester cyanogenic glucosides from their host plants.

edge provided by CNglcs has resulted in at least 3000 different cyanogenic plant species (Gleadow and Møller 2014). This is an estimated 0.7% of all the 450,000 plant species found in the whole plant kingdom (Pimm and Joppa 2015), although there have been few comprehensive surveys. In natural communities, where multiple individuals of each species have been tested, about 5% of plant species are cyanogenic (Adsersen and Adsersen 1993; Miller et al. 2006a,b; Thomsen and Brimer 1997). Survey of species within large genera estimate that about 4%–5% contain cyanogenic glucosides as observed in, for example, *Acacia* (Conn 1980) and *Eucalyptus* (Gleadow et al. 2008) (Figure 7.2). Surveys that included an overrepresentation of cultivated plants showed that up to 11% of the analyzed plant species were cyanogenic (Jones 1998), leading to the suggestion that there has been deliberate selection for cyanogenic plants by humans, perhaps because they are more resistant to pests (Jones 1998; McKey et al. 2010)

The toxic effects of CNglcs are caused by their ability to release hydrogen cyanide (HCN) when the cellular integrity of the plant tissues is compromised (Figure 7.1). The toxicity of HCN is caused by its inhibition of metalloenzymes and in particular cytochrome *c* oxidase, the key enzyme in the mitochondrial respiratory electron transport chain

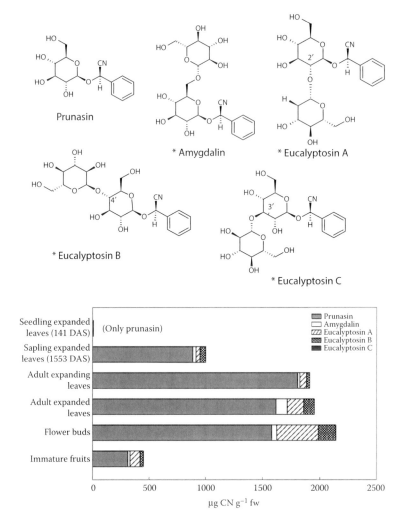

Figure 7.2 Phenylalanine-derived cyanogenic mono- and diglucosides present in selected Eucalyptus species in the course of plant ontogeny and their differential distribution in the tissues.

(Leavesley et al. 2008). This process, often referred to as cyanogenesis, the "cyanide bomb" or the bioactivation pathway (Morant et al. 2008), occurs when HCN is released together with *p*-hydroxybenzaldehyde via the sequential actions of a β-glucosidase (BGD) and an α-hydroxynitrile lyase (HNL) (Gleadow and Møller 2014) (Figure 7.1). The HCN released from this pathway, in principle, can be detoxified and reincorporated into the primary metabolism by two different pathways through the action

Chapter seven: Biosynthesis, regulation, and significance 135

of rhodanese or β-cyanoalanine synthase (CAS) (Siegien and Bogatek 2006). Rhodaneses are not found ubiquitously in plants and when present only display low activity (Selmar et al. 1988; Siegien and Bogatek 2006). The second and more general detoxification pathway incorporates HCN into β-cyanoalanine with the release of hydrogen sulfide, by the action of a β-cyanoalanine synthase (CAS) (Figure 7.3). β-Cyanoalanine is converted by nitrilases (NITs) belonging to the NIT4 family to asparagine and

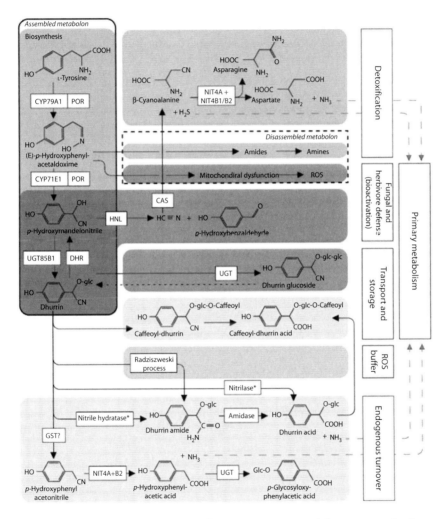

Figure 7.3 The biosynthesis and endogenous turnover of the cyanogenic glucoside dhurrin and the possible metabolic pathways involved and resulting physiological functions.

aspartate with the concomitant release of ammonia (Jenrich et al. 2007; Piotrowski et al. 2001). The released ammonia may subsequently be incorporated into the primary metabolism (Figure 7.3).

In addition to its possible release from the bioactivation of CNglcs, HCN is also formed in stoichiometric amounts with the plant hormone ethylene, which is involved in several physiological processes, including growth, development, and senescence (Iqbal et al. 2013; Matilla 2000). Ethylene is produced from methionine via the two intermediates S-adenosyl L-methionine (SAM) and 1-aminocyclopropane-1-carboxylic acid (ACC) and the sequential actions of SAM synthase, the rate limiting step catalyzed by ACC synthase and finally by ACC oxidase (Iqbal et al. 2013). The ethylene production is thus critically dependent on the activity of CAS to detoxify the released HCN. With ACC oxidase located in the cytoplasm (Siegien and Bogatek 2006) and the primary activity of CAS enzymes located in the mitochondria, small local increases in HCN concentrations are bound to occur as it diffuses to the different cellular compartments (Grossmann 1996; Siegien and Bogatek 2006). In these cellular compartments, HCN may act as a signaling molecule (Siegien and Bogatek 2006).

The control of cyanogenesis is dependent on the compartmentalization and separation of the different components of the system to avoid autotoxicity. In the cell, CNglcs are stored in vacuoles (Saunders and Conn 1978) or maybe in de-differentiated chloroplasts (Brillouet et al. 2014) separate from the degradative enzymes. The BGD may be present in the apoplastic space, the cytoplasm, in small vesicles, in the chloroplast, or bound to the cell wall dependent on the plant species (Gleadow and Møller 2014; Siegien and Bogatek 2006). The location of HNL is less well studied, but has thus far primarily been found in the cytoplasm (Hickel et al. 1996; Kojima et al. 1979). Separation of the different components of the bioactivation pathway is not only observed intracellularly, but also on a tissue-specific level. In the sorghum leaf blade, dhurrin is primarily found in the cells of the epidermal layer, whereas dhurrinase (DHR), HNL, and CAS are primarily located in the mesophyll cells (Kojima et al. 1979; Saunders and Conn 1978; Thayer and Conn 1981; Wurtele et al. 1984). In barley, CNglc epiheterodendrin and other β- and γ-hydroxynitrile glucosides are likewise located in the epidermal cells of the leaf, but contrary to sorghum, the highest CAS activity is also found here (Wurtele et al. 1985). In barley, the BGD activity hydrolyzing the hydroxynitrile glucosides is restricted to the endosperm of the germinating grain (Nielsen et al. 2002). In cassava, the CNglc linamarin is accumulated in 20-fold higher concentrations in the leaves compared to the tubers (White et al. 1998). The highest activity of BGD and HNL is also found in the leaves, while the highest activity of CAS is found in the tubers (Elias et al. 1997). These examples suggest that the separation of the different components both intracellularly and on the tissue level ensures that the CNglcs, BGD and HNL are only brought into contact during tissue disruption caused by herbivore feeding.

Despite the predominance of cyanogenic plants and the toxicity of the HCN released, several studies have failed to prove the effectiveness of CNglcs in herbivore defense (Hruska 1988). This variation in the effectiveness can be explained by the five factors described by Gleadow and Woodrow (2002a) and Møller (2010b). First, concentrations of CNglcs that accumulate in plants vary between individual plants and may be too low to assert a toxic effect. Second, the herbivore may have evolved to become a specialized feeder capable of coping with high amounts of HCN. Third, herbivores may avoid cyanogenic plants, when alternatives are available. Fourth, the importance of feeding style to avoid the release of HCN by a mixing of the different compartments containing CNglcs and hydrolytic enzymes. Fifth, some plant tissues may be acyanogenic due to the lack of BGD activity (Pentzold et al. 2015). Recent research has shown that the rate of release of hydrogen cyanide from CNglcs may determine their efficacy against herbivores, rather than the concentration of CNglcs per se (Ballhorn et al. 2010). Despite these important factors, high concentrations of CNglcs still seem to be an effective inhibitor of feeding generalist herbivores (Gleadow and Woodrow 2002a,b).

7.2 Structure and metabolism of cyanogenic glucosides

7.2.1 Structural diversity of cyanogenic glycosides

CNglcs are amino acid derived β-glycosides of α-hydroxynitriles (Figures 7.1 and 7.4). The aromatic amino acids phenylalanine and tyrosine and the aliphatic amino acids isoleucine, leucine, and valine are the most common precursors of CNglcs, but the nonproteinogenic amino acid 2-cyclopentenyl glycine serves as a precursor for CNglc formation in *Passiflora* spp. (Jaroszewski et al. 2002).

Some CNglcs are quite species-specific, while others are found to be present within several different plant species. The most common interspecific CNglcs are the aromatic monoglucosides, prunasin found in almond (Sánchez-Pérez et al. 2008) and eucalyptus (*Eucalyptus cladocalyx*) (Gleadow and Woodrow 2000), dhurrin found in sorghum (Busk and Møller 2002) and sugarcane (*Saccharum officinarum*) (Rosa Junior et al. 2007), and the aliphatic monoglucosides lotaustralin and linamarin found in *Lotus japonicus* (Forslund et al. 2004), cassava (Lykkesfeldt and Møller 1994), lima bean (Ballhorn et al. 2009), rubber tree (*Hevea brasiliensis*) (Selmar et al. 1988), and white clover (*Trifolium repens*) (Foulds and Grime 1972) (Figure 7.1). These CNglcs are derived from phenylalanine,

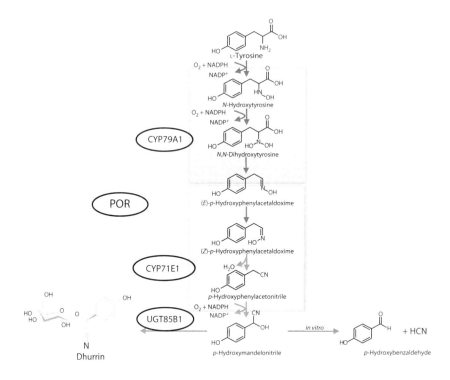

Figure 7.4 Biosynthetic pathway for the tyrosine-derived cyanogenic glucoside dhurrin.

tyrosine, isoleucine, and valine, respectively, and are characteristic for CNglcs in general with the core cyanohydrin always being stabilized by a β-glucosidic linkage to D-glucose. A cyanogenic monoglucoside (CNmglc) may be converted to the corresponding diglycoside (CNdglc) by putative glycosyltransferases (GTs), which add the second sugar moiety. In linustatin (linamarin-6′-glucoside) in rubber tree and amygdalin (prunasin-6′-glucoside) in almond, the secondary sugar is also a glucose attached by a β-1.6 linkage (Sánchez-Pérez et al. 2008; Selmar et al. 1988). Other linkage types to the secondary glucose also exist, such as the eucalyptosin A, B, and C from *Eucalyptus camphora*, which harbors a β-1.2, β-1.4, and β-1.3 linkage, respectively (Neilson et al. 2011) (Figure 7.2). In some CNdglcs, the second sugar moiety is not necessarily a glucose unit but could be, that is, arabinose as in vicianin (Ahn et al. 2007), xylose in lucumin (Miller et al. 2006b), apiose as in oxyanthin (Rockenbach et al. 1992), linamarin apioside and lotaustralin apioside and pentose in prunasin pentoside (Pičmanová et al. 2015). Beside cyanogenic diglycosides, cyanogenic triglycosides also exist, for example, xeranthin in the fruits of *Xeranthemum cylmdraceum* (Schwind et al. 1990) and amygdalin-6′-glucoside, amygdalin-4′-glucoside and amygdalin-2′-glucoside found in almond (Pičmanová et al. 2015).

7.2.2 Biosynthesis of cyanogenic glycosides

The biosynthetic pathway for the formation of CNglcs (Figure 7.1) was first elucidated in *S. bicolor* (Møller and Conn 1980) (Figure 7.4). The first committed step in the pathway involves a cytochrome P450 (CYP79A1), which via two *N*-hydroxylations convert tyrosine to *N*-hydroxytyrosine and *N,N*-dihydroxytyrosine (Møller and Conn 1979). The *N,N*-dihydroxy intermediate is spontaneously dehydrated to form 2-nitroso-3-(*p*-hydroxyphenyl) propionate, which is decarboxylated to form (*E*)-*p*-hydroxyphenylacetaldoxime (pOHPOx) (Sibbesen et al. 1994, 1995). The first enzyme (CYP79A1) is the signature enzyme of the pathway and appears to constitute the rate limiting step in the biosynthesis and controls the flow to the second P450 and the UDP-glucosyltransferase (UGT) (Busk and Møller 2002). After CYP79A1 has formed the E-isomer of pOHPOx, CYP71E1 rearranges it into the Z-isomer and catalyzes an NADPH-dependent dehydration reaction of (*Z*)-pOHPOx to *p*-hydroxyphenylacetonitrile (pOHPCN) (Bak et al. 1998; Clausen et al. 2015). A C-hydroxylation reaction then converts pOHPCN to *p*-hydroxymandelonitrile (pOHMN) (Kahn et al. 1997). The pOHMN is labile and will quickly dissociate into HCN and *p*-hydroxybenzaldehyde if not stabilized by a glucosylation of the hydroxyl group by UGT85B1 to form the CNglc, dhurrin (Jones et al. 1999) (Figure 7.4). Three isoforms of NADPH-dependent cytochrome P450 oxidoreductase (PORa, PORb, and PORc) may serve as electron donors to the two P450s described above (Halkier and Møller 1990; Jensen and Møller 2010). The expression of the three POR isoforms has not previously been investigated in sorghum, but as they serve as electron donors for multiple P450s in sorghum, they are not likely to specifically correlate with the expression of CYP79A1 or CYP71E1.

Evidence from several studies shows that the biosynthetic enzymes form a dynamic protein complex called a *metabolon* (Jensen et al. 2011; Kristensen et al. 2005; Laursen et al. 2015; Møller and Conn 1980). To form the metabolon, it is thought that the three endoplasmic reticulum (ER) membrane-bound enzymes CYP79A1, CYP71E1, and POR combine and then recruit the soluble UGT85B1, which stabilizes the complex (Jensen and Møller 2010; Laursen et al. 2015). The metabolon channels intermediates in the biosynthetic pathway and ensures that the toxic compounds pOHPOx and pOHMN are not released to the cytoplasm (Halkier et al. 1989; Møller and Conn 1980). Besides the metabolon formation, the lipid milieu may also play a role in relation to the catalytic activities. Reconstitution of sorghum CYP79A1 and POR in liposomes of varying lipid compositions gave up to fourfold difference in enzyme activity between the different lipid mixtures (Sibbesen et al. 1995).

Similar classes of intermediates like those involved in the dhurrin biosynthesis are also found in other cyanogenic species, such as white clover (Hughes 1991), flax (*Linum usitassimum*) (Cutler and Conn 1981), *L. japonicus* (Forslund et al. 2004), almond (Sánchez-Pérez et al. 2008), and cassava (Andersen et al. 2000) (Figure 7.1). Thus, the chemical classes of intermediates involved in CNglc biosynthesis appear to be similar across the different plant species investigated, including monocots and dicots (Gleadow and Møller 2014). The three biosynthetic genes involved in the biosynthesis of CNglcs have, since the characterization in sorghum, also been characterized in cassava (Andersen et al. 2000; Jørgensen et al. 2011; Kannangara et al. 2011) and *L. japonicus* (Forslund et al. 2004; Takos et al. 2011). In other species, individual genes involved in CNglc biosynthesis have been characterized, that is, a CYP79 in white clover (Olsen et al. 2008) and a UGT in almond (Franks et al. 2008). In cyanogenic plant species where all the biosynthetic genes have been characterized, a CYP79, CYP71, and UGT85 are found in all species except for *L. japonicus*, where a CYP736 is found instead of the CYP71 (Takos et al. 2011). Although these P450 enzymes all belong to the 71 clade and are sequence related, they may not be orthologous. This recruitment of homologous genes to perform a similar function constitutes an example of "repeated evolution" (Takos et al. 2011). In all the species described above, the biosynthetic genes form genomic clusters, which may help the plant preserve the defensive trait as a single entity, avoid self-toxicity due to segregation of the genes during reproduction, and allow coordinated expression of all genes (Takos et al. 2011) (Figure 7.5).

7.3 Multifunctional cyanogenic glucosides

7.3.1 Dynamic defense systems

Cyanogenic plants are not only exposed to the activity of herbivores but are also attacked by several specialized microorganisms. Although HCN should have a similar toxic effect on these organisms, specialized fungal pathogens have coevolved to counteract and exploit the HCN release (Fry and Evans 1977). The HCN released on fungal attack is metabolized into formamide and hydrolyzed to release ammonia which the fungi can use as a source of reduced nitrogen (N) (Knowles and Bunch 1986). In the rubber tree and lima bean, infection with the two specialized fungal pathogens *Microcyclus ulei* and *Colletotrichum gloeosporioides*, respectively, is more prominent in individuals with higher cyanide potential (HCNp) (Ballhorn et al. 2010; Kadow et al. 2012). The HCN released is both utilized by the fungi as described above but may also inhibit the synthesis in

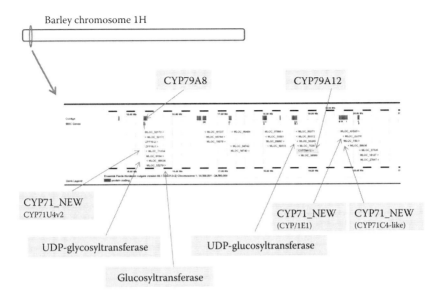

Figure 7.5 Genomic clusters on barley chromosome 1H containing genes possibly involved in the biosynthesis of hydroxynitrile glucosides in barley.

the host plant of antifungal compounds such as scopoletin and scopolin (Garcia et al. 1995; Lieberei et al. 1989). In barley, the barley powdery mildew (*Blumeria graminae* f. sp. *hordei*) is specialized to infect the epidermal cell layer of the leaf where the CNglc epiheterodendrin is stored (Nielsen et al. 2002). The leaves of barley are acyanogenic due to lack of BGD activity and the fungi is likely able to utilize the CNglc as an N and glucose source without the liberation of HCN (Møller 2010b). Reconstitution of cyanogenesis in barley by expression of sorghum BGD activity resulted in a 35%–60% reduction in the colonization rate (Nielsen et al. 2006). This indicates that the fungi do not possess the capacity to detoxify the high local concentrations of HCN released from epiheterodendrin and that the absence of a hydrolyzing enzyme probably enables powdery mildew to infect barley.

The paradox is therefore that accumulation of CNglcs in cyanogenic plants may confer resistance to generalist herbivores while at the same time render them vulnerable to specialized pathogenic fungi. However, the pathway for the biosynthesis of CNglcs may have evolved to combat these fungi as suggested by Møller (2010a). As described earlier, the biosynthesis of CNglcs involves the formation of metabolons to effectively channel the toxic intermediates into the formation of the CNglcs. Following disassembly of the metabolon at the physical site of fungal infection, the E-aldoxime produced by the CYP79 enzyme would then not be channeled to the CYP71 and instead accumulate at the site of infection and could in this way help to combat the infection (Møller 2010a) (Figures 7.1 and 7.4). Oximes and derivatives thereof are generally toxic to fungi and are used as fungicides to treat fungal diseases (Choudhary et al. 2007). Oxime accumulation can also inhibit mitochondrial oxidases and consequently promote lipid peroxidation and the formation of toxic reactive oxygen species (ROS) in the fungus (Rusch et al. 2009; Sakurada et al. 2009). Alternatively, production of carbonyl nitriles derived from the cyanohydrin may serve as protectants (Rajniak et al. 2015) (Figure 7.3).

In *Arabidopsis*, the biosynthesis of glucosinolates also involves an oxime intermediate produced by a CYP79 contained in a metabolon (Jensen et al. 2014). Small amounts of amines have been detected in these plants and it is possible that these compounds are the remnants of detoxification of oximes in cells adjacent to infected cells (Bednarek et al. 2009). The oximes could be converted to amides through Beckmann rearrangement and then reduced to amines (Møller 2010a). It is not clear what could trigger the disassembly of the metabolon with the concomitant release of oximes, but the oxidative burst with rapid production of large amounts of ROS during pathogen attack (Wojtaszek 1997) is likely to inactivate CYP71E1, which is less stable compared to CYP79A1 (Kahn et al. 1999).

The hydrogen peroxide produced in the oxidative burst during fungal attack may also be envisioned to react with CNglcs to form primary amide glucosides in the Radziszweski process (Figure 7.3) (Møller 2010b). Similarly, ROS released during other forms of oxidative stress, for example, air-drying, excessive light exposure, or senescence, could react with CNglcs, which would thus function as ROS scavengers (Møller 2010b; Sendker and Nahrstedt 2009). Studies in rubber trees show that the concentrations of CNglcs exhibit a diurnal pattern, with the highest concentration occurring before sunrise, followed by a rapid decrease on exposure to sunlight (Kongsawadworakul et al. 2009). Similarly in sorghum seedlings, biosynthesis and accumulation have been shown to occur at a higher rate in the dark (Adewusi 1990). In this way, the CNglcs could benefit the plant during excessive light or other forms of oxidative stress, but could also quench the ROS produced as a defensive measure during fungal attack (Møller 2010b) (Figure 7.3).

Several acyanogenic plant species have been reported to emit volatile oximes from vegetative tissue as a defense against herbivores or to attract pollinators. In black poplar, the formation of two aliphatic oximes and one aromatic from the amino acids isoleucine, leucine, and phenylalanine have been shown to originate from the actions of two P450s, CYP79D6, and CYP79D7 in the presence of POR and NADPH as the electron donor system (Irmisch et al. 2013). Volatile nitriles are also released from poplar and have been shown to originate from the conversion of the oximes by the actions of two other P450s named CYP71B40v3 and CYP71B41v2, without the need for NADPH and POR (Irmisch et al. 2014). The P450 enzyme families involved in the formation of oximes and nitriles are thus similar to the ones found in sorghum and cassava (Andersen et al. 2000; Bak et al. 1998; Jørgensen et al. 2011; Sibbesen et al. 1994). *In vivo* aldoxime release resulting directly from the operation of a cyanogenic glucoside biosynthetic pathway has not been reported. However, in mutated plants where the UGT protein is not functional and metabolon formation has been abolished, aldoxime was released and converted into numerous other metabolites in detoxification reactions (Blomstedt et al. 2016; Kristensen et al. 2005).

7.3.2 Turnover of cyanogenic glucosides and the possible metabolic pathways involved

The biosynthesis and accumulation of CNglcs often occur at the highest rate in young and developing tissues, presumably because these softer tissues with higher nutrient content are otherwise more attractive and easily digested by herbivores (Kursar and Coley 2003; Read and Stokes 2006). In this way, accumulation of CNglcs in high concentrations could be considered a transient necessity, which may occupy large percentages of the total organic matter available in N and sugar (Neilson et al. 2013). In the young leaves of *E. cladocalyx*, up to 20% of the total leaf N may be allocated to the CNglc prunasin (Gleadow et al. 1998). As the leaves mature, the HCNp in the older leaves decreases (Gleadow et al. 1998; Gleadow and Woodrow 2000). In etiolated sorghum seedlings, as much as 30% of the dry weight is stored as dhurrin (Halkier and Møller 1989). In sorghum, the content of CNglcs also decreases as the plant matures (Adewusi 1990; Busk and Møller 2002). In these projects, whole plants of sorghum and eucalyptus were analyzed, which exclude the possibility of transport or dilution by growth as an explanation for CNglc decrease. The CNglcs must have been metabolized via bioactivation or endogenous turnover (Figure 7.3). This does not rule out the possibility of internal transport between the site of biosynthesis and other tissues. In the rubber tree, linamarin accumulates in the mature seed. On germination, the CNglc is converted into the cyanogenic diglucoside (CNdglc) linustatin and transported to the seedling where it is turned over (Selmar et al. 1988). A similar situation occurs in bitter almonds where the CNmglc prunasin is biosynthesized in the tegument of the almond fruit throughout development and at the end of maturation is transported to the developing cotyledons of the kernel where it is converted and stored as the CNdglc amygdalin (Sánchez-Pérez et al. 2012). On germination of the kernel, the amygdalin is probably catabolized and used by the seedling in the primary metabolism (Sánchez-Pérez et al. 2008).

The endogenous turnover of CNglcs has been expected to occur via the bioactivation pathway (Figure 7.1) (Poulton 1990). As described previously, the enzymes involved in the bioactivation pathway are not co-located with CAS in mitochondria. If the bioactivation pathway is the primary route for turnover of CNglcs, the intracellular separation would make the turnover ineffective as it would rely on the diffusion of HCN to the mitochondria. As the diffusion is also not restricted to the mitochondria, the turnover would not be very precisely controlled and could exert toxic effects on other cellular compartments and potentially be deadly to the plant (Hay 2008; Siegien and Bogatek 2006). Although there may be evolutionary advantages based on reduced enzyme allocation costs in a shared biochemical pathway (Neilson et al. 2013), the associated disadvantages described point to an alternative endogenous turnover pathway which altogether avoids the release of HCN.

The occurrence of two potential alternative turnover pathways that bypasses the release of the toxic intermediates HCN and β-cyanoalanine has been proposed (Busk and Møller 2002; Jenrich et al. 2007; Pičmanová et al. 2015) (Figure 7.3). In the alternative pathway described by Jenrich et al. (2007) in sorghum, dhurrin is converted to pOHPCN via a proposed novel BGD/protein cofactor complex. No evidence has been reported since confirming the presence of such a BGD/protein cofactor complex. Instead, results from our lab point to the formation of pOHPCN through the action of glutathione-*S*-transferases (GSTs). The presence of a putative GST (Sobic. 001G012500) detected close to a single nucleotide polymorphism (SNP) associated with dhurrin content and in the vicinity of the sorghum biosynthetic gene cluster supports the potential involvement of a GST (Hayes et al. 2015). Following the conversion of dhurrin, pOHPCN is hydrolyzed to *p*-hydroxyphenylacetic acid (pOHPAAc) and ammonia by the action of a heteromeric enzyme complex composed of the enzymes NIT4A and NIT42B (Jenrich et al. 2007) (Figure 7.3). These enzymes are the same as those involved in the detoxification of HCN from the bioactivation pathway, but only the NIT4A/B2 constellation and not NIT4A/B1 can convert pOHPCN to pOHPAAc. Finally, the pOHPAAc can be further glucosylated to give *p*-glucosyloxyphenylacetic acid (pGlcPAAc) which is found in sorghum plants that have passed the seedling stage (Blomstedt et al. 2016; Jenrich et al. 2007; Pičmanová et al. 2015).

The second possible turnover pathway described by Pičmanová et al. (2015) is based on the findings of the same set of "unifying" metabolites of CNglcs in sorghum, cassava, and almond. The pathway for sorghum is depicted in Figure 7.3 and probably involves the conversion of the CNglc to a CNglc amide by bifunctional NITs or a nitrile hydratase (Pičmanová et al. 2015). The CNglc amide formed is likely hydrolyzed by amidases to form CNglc acid and ammonia, although such enzymes remain to be identified. Another possible way of producing the carboxylic acids is through the direct action of specific NITs, although such enzymes also remain to be discovered (Pičmanová et al. 2015). The carboxylic acid may be further decarboxylated to release carbon dioxide and to form a compound corresponding to the CNglc devoid of the nitrile group. Such a compound has been detected in cassava and almond but not in sorghum (Pičmanová et al. 2015). Thus far, the only enzymes from the two alternative turnover pathways which have been characterized are the heteromeric enzyme complex composed of the enzymes NIT4A and NIT4B2 (Jenrich et al. 2007). This complex has furthermore only been studied *in vitro* and it is therefore not known how the expression of these enzymes correlates with the concentration of dhurrin. A whole transcriptome time-series analysis could thus provide valuable new knowledge to the understanding of how CNglcs are metabolized in cyanogenic plants.

7.3.3 Resource allocation

If cyanogenic glucosides are an effective defense against herbivores, and also play other roles in plant metabolism, why are they only found in about 5% of all plant species (Gleadow et al. 2008; Miller et al. 2006a)? The most likely explanation is that there are costs involved with their synthesis, maintenance, and turnover, or even related to the cellular space they occupy (Coley et al. 1985; Gleadow and Møller 2014, King et al. 2006). The resources needed to maintain and synthesize defensive compounds, including cyanogenic glucosides, have been investigated in several different studies (Neilson et al. 2013). However, the actual costs have been hard to measure in both cyanogenic and acyanogenic plants (Gleadow and Møller 2014; Neilson et al. 2013). There are also major uncertainties about the costs in terms of reproductive fitness (i.e., evolutionary pressure) and plasticity (Ballhorn et al. 2011; Coley et al. 1985; Zangerl and Bazzaz 1992). Studies that have compared the relative costs in plants with high and low background concentrations of, for example, cyanogenic glucosides based on genotype differences have concluded that the costs may be simply too low to detect (Gershenzon 1984; Simon et al. 2010; Woodrow et al. 2002). The considerable phenotypic plasticity in the deployment of CNglc strongly supports the argument that under some circumstances, production and storage of these compounds is relatively cheap. For example, when plant growth is limited by lack of phosphorus, water, or due to osmotic stress, the concentration of cyanogenic glucosides increases both on a per mass and per nitrogen basis (Ballhorn et al. 2011; Ballhorn and Elias 2014; Ballhorn et al. 2014; Burns et al. 2012; Gleadow et al. 2016; Gleadow and Woodrow 2002b; Miller et al. 2014; Neilson et al. 2015; O'Donnell et al. 2013; Vandegeer et al. 2013). When growth is unconstrained, however, costs should be low. For example, C3 plants grown at an elevated CO_2 divert nitrogen resources away from the photosynthetic machinery (accounting for 25% of foliar nitrogen) to cyanogenic glucosides because of the increases in efficiency of the carbon fixation enzyme Ribulose bisphosphate carboxylase (Cavagnaro et al. 2011; Gleadow et al. 1998; Gleadow et al. 2009a,b; Rosenthal et al. 2012). Interestingly, no change in allocation has been observed in C4 plants, which do not have improved photosynthetic efficiency at elevated CO_2 (Gleadow et al. 2016).

One of the issues in calculating trade-offs is how to accurately define the nature of the costs involved, and design experiments which incorporate all the biotic and abiotic environmental factors that a plant encounters under natural conditions (Neilson et al. 2013). This is exemplified in white clover, where both cyanogenic and acyanogenic morphs occur, with the *Ac/ac* and *Li/li* genes controlling the production of CNglcs and presence of BGD, respectively (Olsen et al. 2008). During experimental conditions of drought stress, the fatalities of the *Acac* morphs increase by a factor of three compared to the *acac* morph, while under moist conditions the *acac* morph has an increased growth compared to the *Acac* morph (Foulds and Grime 1972). Similarly, it was found that *acac* morphs produced more seeds and flowers than the *Acac* morphs in the absence of herbivory (Kakes 1997). Nonetheless, when a mixed population of plants was subjected to the slug *Arion hortensis*, the majority of *acac* morphs suffered lethal damage while only a few *Acac* morphs shared a similar fate (Horrill and Richards 1986). Thus, the *Acac* morphs may show a decreased growth potential and reduced reproduction but have better defensive properties compared to the *acac* morph.

CNglcs may serve additional physiological roles *in planta* (Møller 2010b; Selmar and Kleinwächter 2013). Initially, the synthesis and storage of CNglcs is resource demanding. When produced, their presence may help to mitigate stress by quenching ROS and to reduce herbivory and disease attacks because of their toxicity (Figures 7.1 and 7.3). In addition, CNglc accumulation may also decrease the nutritional value of the plant tissues by sequestering free amino acids, sugars, and other nutrients in an unpalatable form (Gleadow and Møller 2014; Neilson et al. 2013). As described in the previous section, the CNglcs are not dead-end products, but can on demand be remobilized and allocated to growth and development of the plant. In the rubber tree, linamarin accumulates in the mature seed (Selmar et al. 1988). On germination, the CNglc is converted into the CNd-glc linustatin and transported to the cotyledons where it is catabolized and the N and sugar released incorporated into primary metabolism (Selmar 1993; Selmar et al. 1988). As a diglucoside, linustatin is protected from hydrolytic cleavage by those endogenous BGDs capable of cleaving linamarin. By sequential hydrolysis of the gentiobiose moiety of linustatin, linamarin can be reformed and stored in the tissues of the cotyledons for either defensive purposes or as a storage form of N and sugar

(Selmar 1993). In older rubber trees, linustatin is also transported from the leaves to the inner bark of the trunk, where it is catabolized to release N and glucose to be used for establishing the biosynthetic platform for latex and rubber production (Kongsawadworakul et al. 2009). Thus, CNglcs in rubber trees may serve a dual purpose in both herbivore defense and as N and glucose buffers in primary metabolism. The aliphatic CNglc linamarin is also found in cassava, where it is synthesized in the shoot apex of the plant and transported to the roots (Jørgensen et al. 2005). Small amounts of linustatin have also been found in this plant and a similar transport system to the one found in the rubber tree have been hypothesized to function in cassava (Pičmanová et al. 2015). In almond, a similar diglucoside based transport system of CNglcs is also found (Sánchez-Pérez et al. 2008, 2009). In the almond fruit, prunasin is biosynthesized in the tegument throughout development. In the bitter variants, the CNglc is transported into the developing cotyledons of the kernel where it is converted and stored as the CNdglc amygdalin (Møller 2010b). In parallel with amygdalin accumulation and turnover in seeds and seedlings of black cherry (*Prunus serotina* Ehrh.), it may be speculated that the amygdalin stored in the cotyledon of the bitter almond variants is similarly catabolized and used in the primary metabolism for growth and development (Swain and Poulton 1994).

In some plant species, the dynamic roles of CNglcs may have evolved to become an intricate and necessary part of the normal development at specific stages of ontogeny. In cassava, transgenic lines with a CNglc reduction of more than 25% in the tubers, displayed stunted growth. Acyanogenic seedlings (approximately 1% of wild-type level) were not viable (Jørgensen et al. 2005). Totally cyanide deficient 1 (*tcd1*) sorghum lines with a single amino change in CYP79A1 inactivating the enzyme and thus preventing the biosynthesis of dhurrin showed reduced growth at the seedling stage compared to wild type plants (Blomstedt et al. 2012). The same is true of cyanogenic insects. The *Zygaena filipendulae* moths that obtain their CNglcs by sequestering them from the plants that they feed on have a higher growth rate than larvae fed on acyanogenic plants (Zagrobelny et al. 2007a,b)

The results obtained by studying different cyanogenic plant species show that CNglcs are not mere defense compounds but dynamic secondary metabolites, which may play an active role in balancing the primary metabolism in the course of ontogeny. To understand these highly dynamic processes, knowledge of the enzymes involved is a prerequisite. While the enzymes involved in CNglc biosynthesis are well-known and extensively studied, only a single enzyme complex involved in the putative turnover pathways has been characterized (Jenrich et al. 2007). Within the next few years, the endogenous metabolism of CNglcs is envisioned to be resolved.

Acknowledgments

This work was supported by a grant from the VILLUM Foundation to the Research Center of Excellence "Plant Plasticity" (to BLM), by the Center for Synthetic Biology "bioSYNergy" supported by the UCPH Excellence Program for Interdisciplinary Research (to BLM), by an ERC Advanced Grant (to BLM) (ERC-2012-ADG_20120314, Project No: 323034), by an Australian Research Council (ARC) Linkage Grant LP0774941 (to RMG and BLM) with Pacific Seeds Pty. Ltd. and by ARC Discovery Grant DP130101049 (to RMG).

References

Adsersen, A. and H. Adsersen. 1993. Cyanogenic plants in the Galápagos Islands: Ecological and evolutionary aspects. *Oikos* 66: 511–520.

Adewusi, S.R.A. 1990. Turnover of dhurrin in green sorghum seedlings. *Plant Physiol.* 94: 1219–1224.

Agrawal, A.A., A. Janssen, J. Bruin, M.A. Posthumus, and M.W. Sabelis. 2002. An ecological cost of plant defence: Attractiveness of bitter cucumber plants to natural enemies of herbivores. *Ecol. Lett.* 5: 377–385.

Ahn, Y.O., H. Saino, M. Mizutani, B.-i. Shimizu, and K. Sakata. 2007. Vicianin hydrolase is a novel cyanogenic β-glycosidase specific to β-vicianoside (6-O-α-L-arabinopyranosyl-β-D-glucopyranoside) in seeds of *Vicia angustifolia*. *Plant Cell Physiol.* 48: 938–947.

Andersen, M.D., P.K. Busk, I. Svendsen, and B.L. Møller. 2000. Cytochromes P-450 from cassava (*Manihot esculenta* Crantz) catalyzing the first steps in the biosynthesis of the cyanogenic glucosides linamarin and lotaustralin—Cloning, functional expression in *Pichia pastoris*, and substrate specificity of the isolated recombinant enzymes. *J. Biol. Chem.* 275: 1966–1975.

Bak, S., R.A. Kahn, H.L. Nielsen, B.L. Møller, and B.A. Halkier. 1998. Cloning of three A-type cytochromes p450, CYP71E1, CYP98, and CYP99 from *Sorghum bicolor* (L.) Moench by a PCR approach and identification by expression in *Escherichia coli* of CYP71E1 as a multifunctional cytochrome p450 in the biosynthesis of the cyanogenic glucoside dhurrin. *Plant Mol. Biol.* 36: 393–405.

Ballhorn, D.J., A. Pietrowski, and R. Lieberei. 2010. Direct trade-off between cyanogenesis and resistance to a fungal pathogen in lima bean (*Phaseolus lunatus* L.). *J. Ecol.* 98: 226–236.

Ballhorn, D.J., A.L. Godschalx, S.M. Smart, S. Kautz, and M. Schädler. 2014. Chemical defense lowers plant competitiveness. *Oecologia* 176: 811–824.

Ballhorn, D.J. and J.D. Elias. 2014. Salinity-mediated cyanogenesis in white clover (*Trifolium repens*) affects trophic interactions. *Ann Bot.* 114: 357–366.

Ballhorn, D.J., S. Kautz, M. Heil, and A.D. Hegeman. 2009. Cyanogenesis of wild lima bean (*Phaseolus lunatus* L.) is an efficient direct defence in nature. *PLoS ONE* 4(5): e5450.

Ballhorn, D.J., S. Kautz, M. Jensen, I. Schmitt, M. Heil, and A.D. Hegeman. 2011. Genetic and environmental interactions determine plant defences against herbivores. *J. Ecol.* 99: 313–326.

Bednarek, P., M. Pislewska-Bednarek, A. Svatos et al. 2009. A glucosinolate metabolism pathway in living plant cells mediates broad-spectrum antifungal defense. *Science* 323: 101–106.

Blomstedt, C.K., N.H. O'Donnell, N. Bjarnholt, et al. 2016. Metabolic consequences of knocking out UGT85B1, the gene encoding the glucosyltransferase required for synthesis of dhurrin in *Sorghum bicolor* (L. Moench). *Plant Cell Physiol.* 57(2): 373–386.

Blomstedt, C.K., R.M. Gleadow, N. O'Donnell et al. 2012. A combined biochemical screen and TILLING approach identifies mutations in *Sorghum bicolor* L. Moench resulting in acyanogenic forage production. *Plant Biotechnol. J.* 10: 54–66.

Briggs, M.A. 1990. Chemical defense production in *Lotus corniculatus* L. I. The effects of nitrogen source on growth, reproduction and defense. *Oecologia* 83: 27–31.

Brillouet, J.-M., J.-L. Verdeil, E. Odoux, M., Lartaud, M. Grisoni, and G. Conéjéro. 2014. Phenol homeostasis is ensured in vanilla fruit by storage under solid form in a new chloroplast-derived organelle, the phenyloplast. *J. Exp. Bot.* 65: 2427–2435.

Burns, A.E., R.M. Gleadow, A. Zacarias, C.E. Cuambe, R.E. Miller, and T.R. Cavagnaro. 2012. Variations in the chemical composition of cassava (*Manihot esculenta* Crantz) leaves and roots as affected by genotypic and environmental variation. *J. Agric. Food Chem.* 60: 4946–4956.

Busk, P.K. and B.L. Møller. 2002. Dhurrin synthesis in sorghum is regulated at the transcriptional level and induced by nitrogen fertilization in older plants. *Plant Physiol.* 129: 1222–1231.

Cavagnaro, T.R., R.M. Gleadow, and R.M. Miller. 2011. Viewpoint: Plant nutrition in a high CO_2 world. *Funct. Plant Biol.* 38: 87–96.

Choudhary, G., S. Walla, J. Kumar, B. Kumar, and B.S. Parmar. 2007. Synthesis and antifungal activity of citral oxime esters against two phytopathogenic fungi *Rhizoctonia solani* and *Sclerotium rolfsii*. *Pestic. Res. J.* 19: 15–19.

Clausen, M., R.M. Kannangara, C.E. Olsen et al. 2015. The bifurcation of the cyanogenic glucoside and glucosinolate biosynthetic pathways. *Plant J.* 84: 558–573.

Coley, P.D., J.P. Bryant, and F. S. Chapin III. 1985. Resource availability and plant antiherbivore defense. *Science* 230: 895–899.

Conn, E.E. 1980. Cyanogenic compounds. *Annu. Rev. Plant Physiol. Plant Molec. Biol.* 31: 433–451.

Cutler, A.J. and E.E. Conn. 1981. The biosynthesis of cyanogenic glucosides in *Linum usitatissimum* (linen flax) *in vitro*. *Arch. Biochem. Biophys.* 212: 468–474.

Dahler, J.M., C.A. McConchie, and C.G.N. Turnbull. 1995. Quantification of cyanogenic glycosides in seedlings of three Macadamia (Proteaceae) species. *Aust. J. Bot.* 43: 619–628.

Dement, W.A. and H.A. Mooney. 1974. Seasonal variation in the production of tannins and cyanogenic glucosides in the chaparral shrub, *Heteromeles arbutifolia*. *Oecologia* 15: 65–76.

Dussourd, D.E. and A.M. Hoyle. 2000. Poisoned plusiines: Toxicity of milkweed latex and cardenolides to some generalist caterpillars. *Chemoecology* 10: 11–16.

Dziewanowska, K., I. Niedzwiedz, and S. Lewak. 1979. Hydrogen cyanide and cyanogenic compounds in seeds. III. Degradation of cyanogenic glucosides during apple seed stratification. *Physiologie Végétale* 17: 687–695.

Ehrlich, P.R. and P.H. Raven. 1964. Butterflies and plants: A study in coevolution. *Evolution* 18: 586–608.
Eigenbrode, S.D. and K.E. Espelie. 1995. Effects of plant epicuticular lipids on insect herbivores. *Annu. Rev. Entomol.* 40: 171–194.
Elias, M., B. Nambisan, and P.R. Sudhakaran. 1997. Catabolism of linamarin in cassava (*Manihot esculenta crantz*). *Plant Sci.* 126: 155–162.
Erb, N., H.D. Zinsmeister, and A. Nahrstedt. 1981. The cyanogenic glycosides of Triticum, Secale and Sorghum. *Planta Med.* 41: 84–89.
Forslund, K., M. Morant, B. Jørgensen et al. 2004. Biosynthesis of the nitrile glucosides rhodiocyanoside A and D and the cyanogenic glucosides lotaustralin and linamarin in *Lotus japonicus*. *Plant Physiol.* 135: 71–84.
Foulds, W. and J.P. Grime. 1972. The response of cyanogenic and acyanogenic phenotypes of *Trifolium repens* to soil moisture supply. *Heredity* 28: 181–187.
Franks, T.K., A. Yadollahi, M.G. Wirthensohn et al. 2008. A seed coat cyanohydrin glucosyltransferase is associated with bitterness in almond (*Prunus dulcis*) kernels. *Funct. Plant Biol.* 35: 236–246.
Fry, W.E. and P.H. Evans. 1977. Association of formamide hydro-lyase with fungal pathogenicity to cyanogenic plants. *Phytopathology* 67: 1001–1006.
Furstenberg-Hagg, J., M. Zagrobelny, and S. Bak. 2013. Plant defense against insect herbivores. *Int. J. Mol. Sci.* 14: 10242–10297.
Garcia, D., C. Sanier, J.J. Macheix, and J. Dauzac. 1995. Accumulation of scopoletin in *Hevea brasiliensis* infected by *Microcyclus ulei* (P. Henn.) V. ARX and evaluation of its toxicity for three leaf pathogens of rubber tree. *Physiol. Mol. Plant Pathol.* 47: 213–223.
Gershenzon, J. 1984. Changes in the levels of plant secondary metabolites under water and nutrient stress. In *Recent Advances in Phytochemistry*, eds. C. Steelink, B.N. Timmermann, and F.A. Loewus, vol 18, 273–320. New York: Plenum Press.
Gleadow, R.M. and B.L. Møller. 2014. Cyanogenic glycosides: Synthesis, physiology, and phenotypic plasticity. *Annu. Rev. Plant Biol.* 65: 155–185.
Gleadow, R.M. and I.E. Woodrow. 2000. Temporal and spatial variation in cyanogenic glycosides in *Eucalyptus cladocalyx*. *Tree Physiol.* 20: 591–598.
Gleadow, R.M. and I.E. Woodrow. 2002a. Constraints on effectiveness of cyanogenic glycosides in herbivore defense. *J. Chem. Ecol.* 28: 1301–1313.
Gleadow, R.M. and I.E. Woodrow. 2002b. Defence chemistry of Eucalyptus cladocalyx seedlings is affected by water supply. *Tree Physiol.* 22: 939–945.
Gleadow, R.M., E. Edwards, and J.R. Evans. 2009a. Changes in nutritional value of cyanogenic *Trifolium repens* at elevated CO_2. *J. Chem. Ecol.* 35: 476–47.
Gleadow, R.M., J. Haburjak, J.E. Dunn, M.E. Conn, and E.E. Conn. 2008. Frequency and distribution of cyanogenic glycosides in *Eucalyptus* L'Hérit. *Phytochemistry* 69: 1870–1874
Gleadow, R.M., J.R. Evans, S. McCaffrey, and T.R. Cavagnaro. 2009b. Growth and nutritive value of cassava (*Manihot esculenta* Cranz.) are reduced when grown at elevated CO_2. *Plant Biol.* 11 Suppl. 1: 76–82.
Gleadow, R.M., M.J. Ottman, B.A. Kimball et al. 2016. Drought–induced changes in nitrogen partitioning in sorghum are not moderated by elevated CO_2 in FACE studies. *Field Crop Res.* 185: 97–102.
Gleadow, R.M., W.J. Foley, and I.E. Woodrow. 1998. Enhanced CO_2 alters the relationship between photosynthesis and defence in cyanogenic *Eucalyptus cladocalyx* F. Muell. *Plant Cell Environ.* 21: 12–22.

Goodger, J.Q.D., R.M. Gleadow, and I.E. Woodrow. 2006. Growth cost and ontogenetic expression patterns of defence in cyanogenic *Eucalyptus* spp. *Trees Struct. Funct.* 20: 757–765.
Grossmann, K. 1996. A role for cyanide, derived from ethylene biosynthesis, in the development of stress symptoms. *Physiol. Plant* 97: 772–775.
Halkier, B.A. and B.L. Møller. 1989. Biosynthesis of the cyanogenic glucoside dhurrin in seedlings of *Sorghum bicolor* (L.) Moench and partial-purification of the enzyme-system involved. *Plant Physiol.* 90: 1552–1559.
Halkier, B.A. and B.L. Møller. 1990. The biosynthesis of cyanogenic glucosides in higher plants. Identification of three hydroxylation steps in the biosynthesis of dhurrin in *Sorghum bicolor* (L.) Moench and the involvement of 1-*ACI*-nitro-2-(*p*-hydroxyphenyl)ethane as an intermediate. *J. Biol. Chem.* 265: 21114–21121.
Halkier, B.A., C.E. Olsen, and B.L. Møller. 1989. The biosynthesis of cyanogenic glucosides in higher plants. The (*E*)- and (*Z*)-isomers of *p*-hydroxyphenylacetaldehyde oxime as intermediates in the biosynthesis of dhurrin in *Sorghum bicolor* (L.) Moench. *J. Biol. Chem.* 264: 19487–19494.
Hay, R.W. 2008. Plant metalloenzymes. In *Plants and the Chemical Elements*, ed. M.E. Farago, 107–148. Weinheim: Wiley-VCH Verlag GmbH.
Hayes, C.M., G.B. Burow, P.J. Brown, C. Thurber, Z. Xin, and J.J. Burke. 2015. Natural variation in synthesis and catabolism genes influences dhurrin content in sorghum. *Plant Genome.* 8(2): 1–9. doi: 10.3835/plantgenome2014.09.0048.
Heil, M., B. Fiala, K.E. Linsenmair, G. Zotz, P. Menke, and U. Maschwitz. 1997. Food body production in *Macaranga triloba* (Euphorbiaceae): A plant investment in anti-herbivore defence via symbiotic ant partners. *J. Ecol.* 85: 847–861.
Hickel, A., M. Hasslacher, and H. Griengl. 1996. Hydroxynitrile lyases: Functions and properties. *Physiol. Plant* 98: 891–898.
Horrill, J.C. and A.J. Richards. 1986. Differential grazing by the mollusc *Arion hortensis* Fér. on cyanogenic and acyanogenic seedlings of the white clover, *Trifolium repens* L. *Heredity* 56: 277–281.
Hruska, A.J. 1988. Cyanogenic glucosides as defense compounds: A review of the evidence. *J. Chem. Ecol.* 14: 2213–2217.
Hughes, M.A. 1991. The cyanogenic polymorphism in *Trifolium repens* L. (white clover). *Heredity* 66: 105–115.
Iqbal, N., A. Trivellini, A. Masood, A. Ferrante, and N.A. Khan. 2013. Current understanding on ethylene signaling in plants: The influence of nutrient availability. *Plant Physiol. Biochem.* 73: 128–138.
Irmisch, S., A.C. McCormick, G.A. Boeckler et al. 2013. Two herbivore-induced cytochrome p450 enzymes CYP79D6 and CYP79D7 catalyze the formation of volatile aldoximes involved in poplar defense. *Plant Cell* 25: 4737–4754.
Irmisch, S., A.C. McCormick, J. Gunther et al. 2014. Herbivore-induced poplar cytochrome P450 enzymes of the CYP71 family convert aldoximes to nitriles which repel a generalist caterpillar. *Plant J.* 80: 1095–1107.
Jaroszewski, J.W., E.S. Olafsdottir, P. Wellendorph et al. 2002. Cyanohydrin glycosides of *Passiflora*: Distribution pattern, a saturated cyclopentane derivative from *P. guatemalensis*, and formation of pseudocyanogenic α-hydroxyamides as isolation artefacts. *Phytochemistry* 59: 501–511.
Jenrich, R., I. Trompetter, S. Bak, C.E. Olsen, B.L. Møller, and M. Piotrowski. 2007. Evolution of heteromeric nitrilase complexes in Poaceae with new functions in nitrile metabolism. *Proc. Natl. Acad. Sci. USA* 104: 18848–18853.

Jensen, K. and B.L. Møller. 2010. Plant NADPH-cytochrome P450 oxidoreductases. *Phytochemistry* 71: 132–141.

Jensen, K., S.A. Osmani, T. Hamann, P. Naur, and B.L. Møller. 2011. Homology modeling of the three membrane proteins of the dhurrin metabolon: Catalytic sites, membrane surface association and protein-protein interactions. *Phytochemistry* 72: 2113–2123.

Jensen, L.M., B.A. Halkier, and M. Burow. 2014. How to discover a metabolic pathway? An update on gene identification in aliphatic glucosinolate biosynthesis, regulation and transport. *Biol. Chem.* 395: 529–543.

Jones, D.A. 1998. Why are so many food plants cyanogenic? *Phytochemistry* 47: 155–162.

Jones, P.R., B.L. Møller, and P.B. Høj. 1999. The UDP-glucose:*p*-hydroxymandelonitrile-*O*-glucosyltransferase that catalyzes the last step in synthesis of the cyanogenic glucoside dhurrin in *Sorghum bicolor*—Isolation, cloning, heterologous expression, and substrate specificity. *J. Biol. Chem.* 274: 35483–35491.

Jørgensen, K., A.V. Morant, M. Morant et al. 2011. Biosynthesis of the cyanogenic glucosides linamarin and lotaustralin in cassava: Isolation, biochemical characterization, and expression pattern of CYP71E7, the oxime-metabolizing cytochrome P450 enzyme. *Plant physiol.* 155: 282–292.

Jørgensen, K., S. Bak, P.K. Busk et al. 2005. Cassava plants with a depleted cyanogenic glucoside content in leaves and tubers. Distribution of cyanogenic glucosides, their site of synthesis and transport, and blockage of the biosynthesis by RNA interference technology. *Plant Physiol.* 139: 363–374.

Kadow, D., K. Voss, D. Selmar, and R. Lieberei. 2012. The cyanogenic syndrome in rubber tree *Hevea brasiliensis*: Tissue-damage-dependent activation of linamarase and hydroxynitrile lyase accelerates hydrogen cyanide release. *Ann. Bot.* 109: 1253–1262.

Kahn, R.A., S. Bak, I. Svendsen, B.A. Halkier, and B.L. Møller. 1997. Isolation and reconstitution of cytochrome P450ox and *in vitro* reconstitution of the entire biosynthetic pathway of the cyanogenic glucoside dhurrin from sorghum. *Plant Physiol.* 115: 1661–1670.

Kahn, R.A., T. Fahrendorf, B.A. Halkier, and B.L. Møller. 1999. Substrate specificity of the cytochrome P450 enzymes CYP79A1 and CYP71E1 involved in the biosynthesis of the cyanogenic glucoside dhurrin in *Sorghum bicolor* (L.) Moench. *Arch. Biochem. Biophys.* 363: 9–18.

Kakes, P. 1997. Difference between the male and female components of fitness associated with the gene Ac in *Trifolium repens*. *Acta Bot. Neerl.* 46: 219–223.

Kannangara, R., M.S. Motawia, N.K.K. Hansen et al. 2011. Characterization and expression profile of two UDP-glucosyltransferases, UGT85K4 and UGT85K5, catalyzing the last step in cyanogenic glucoside biosynthesis in cassava. *Plant J.* 68: 287–301.

King, D.K., R.M. Gleadow, and I.E. Woodrow. 2006. The accumulation of terpenoid oils does not incur a growth cost in *Eucalyptus polybractea* seedlings. *Funct. Plant Biol.* 33: 497–505.

Knowles, C.J. and A.W. Bunch. 1986. Microbial cyanide metabolism. *Ad. Microb. Physiol.* 27: 73–111.

Kojima, M., J.E. Poulton, S.S. Thayer, and E.E. Conn. 1979. Tissue distributions of dhurrin and of enzymes involved in its metabolism in leaves of *Sorghum bicolor*. *Plant Physiol.* 63: 1022–1028.

Kongsawadworakul, P., U. Viboonjun, P. Romruensukharom, P. Chantuma, S. Ruderman, and H. Chrestin. 2009. The leaf, inner bark and latex cyanide potential of *Hevea brasiliensis*: Evidence for involvement of cyanogenic glucosides in rubber yield. *Phytochemistry* 70: 730–739.

Kristensen, C., M. Morant, C.E. Olsen et al. 2005. Metabolic engineering of dhurrin in transgenic *Arabidopsis* plants with marginal inadvertent effects on the metabolome and transcriptome. *Proc. Natl. Acad. Sci. USA* 102: 1779–1784.

Kursar, T.A. and P.D. Coley. 2003. Convergence in defense syndromes of young leaves in tropical rainforests. *Biochem. Syst. Ecol.* 31: 929–949.

Laursen, T., B.L. Møller, and J.-E. Bassard. 2015. Plasticity of specialized metabolism as mediated by dynamic metabolons. *Trends Plant Sci.* 20: 20–32.

Leavesley, H.B., L. Li, K. Prabhakaran, J.L. Borowitz, and G.E. Isom. 2008. Interaction of cyanide and nitric oxide with cytochrome c oxidase: Implications for acute cyanide toxicity. *Toxicol. Sci.* 101: 101–111.

Lieberei, R., B. Biehl, A. Giesemann, and N.T.V. Junqueira. 1989. Cyanogenesis inhibits active defense reactions in plants. *Plant Physiol.* 90: 33–36.

Lykkesfeldt, J. and B.L. Møller. 1994. Cyanogenic glycosides in cassava, *Manihot esculenta* Crantz. *Acta Chem. Scand.* 48: 178–180.

Matilla, A.J. 2000. Ethylene in seed formation and germination. *Seed Sci. Res.* 10: 111–126.

McKey, D., T.R. Cavagnaro, T. Cliff, and R.M. Gleadow. 2010. Chemical ecology in coupled human and natural systems: People, manioc, multitrophic interactions and global change. *Chemoecology* 20: 109–133.

Miller, R.E., M.J. McConville, and I.E. Woodrow. 2006b. Cyanogenic glycosides from the rare Australian endemic rainforest tree *Clerodendrum grayi* (Lamiaceae). *Phytochemistry* 67: 43–51.

Miller, R.E., R. Jensen, I.E. Woodrow. 2006a. Frequency of cyanogenesis in tropical rainforests of Far North Queensland, Australia. *Ann. Bot.* 97: 1017–44.

Miller, R.E., R.M. Gleadow, and T.R. Cavagnaro. 2014. Age versus stage: Does ontogeny modify the effect of phosphorus and arbuscular mycorrhizas on above- and below-ground defence in forage sorghum? *Plant Cell Environ.* 37: 929–942.

Møller, B.L. 2010a. Dynamic metabolons. *Science* 330: 1328–1329.

Møller, B.L. 2010b. Functional diversifications of cyanogenic glucosides. *Curr. Opin. Plant Biol.* 13: 338–347.

Møller, B.L. and E.E. Conn. 1979. The biosynthesis of cyanogenic glucosides in higher plants. N-Hydroxytyrosine as an intermediate in the biosynthesis of dhurrin by *Sorghum bicolor* (Linn) Moench. *J. Biol. Chem.* 254: 8575–8583.

Møller, B.L. and E.E. Conn. 1980. The biosynthesis of cyanogenic glucosides in higher plants. N-Hydroxytyrosine as an intermediate in the biosynthesis of dhurrin by *Sorghum bicolor* (Linn) Moench. *J. Biol. Chem.* 255: 3049–3056.

Morant, A.V., K. Jørgensen, C. Jørgensen et al. 2008. β-Glucosidases as detonators of plant chemical defense. *Phytochemistry* 69: 1795–1813.

Nahrstedt, A. 1970. Cyanogenesis in *Prunus avium*. *Phytochemistry* 9: 2085–2089.

Neilson, E.H., C.K. Blomstedt, A.E. Edwards, B. Berger, B.L. Møller, and R.M. Gleadow. 2015. Utilization of high-throughput shoot imaging to examine the phenotypic response of a C_4 cereal to nutrient and water deficiency. *J. Exp. Bot.* 66: 1817–1832.

Neilson, E.H., J.Q.D. Goodger, I.E. Woodrow, and B.L. Møller. 2013. Plant chemical defense: At what cost? *Trends Plant Sci.* 18: 250–258.

Neilson, E.H., J.Q.D. Goodger, M.S. Motawia et al. 2011. Phenylalanine derived cyanogenic diglucosides from *Eucalyptus camphora* and their abundances in relation to ontogeny and tissue type. *Phytochemistry* 72: 2325–2334.

Nielsen, K.A., C.E. Olsen, K. Pontoppidan, and B.L. Møller. 2002. Leucine-derived cyano glucosides in barley. *Plant Physiol.* 129: 1066–1075.

Nielsen, K.A., M. Hrmova, J.N. Nielsen et al. 2006. Reconstitution of cyanogenesis in barley (*Hordeum vulgare* L.) and its implications for resistance against the barley powdery mildew fungus. *Planta* 223: 1010–1023.

O'Donnell, N.H., C.K. Blomstedt, A.D. Neale, D.bHamill, B.L. Møller, and R.M. Gleadow. 2013. PEG-induced osmotic stress affects growth and toxicity of forage sorghum. *Plant Physiol.* 73: 83–92.

Odowd, D.J. 1980. Pearl bodies of a neotropical tree, *Ochroma pyramidale*: Ecological implications. *Am. J. Bot.* 67: 543–549.

Olsen, K.M., S.-C. Hsu, and L.L. Small. 2008. Evidence on the molecular basis of the Ac/ac adaptive cyanogenesis polymorphism in white clover (*Trifolium repens* L.). *Genetics* 179: 517–526.

Pentzold, S., M. Zagrobelny, N. Bjarnholt, J. Kroymann, H. Vogel, C.E. Olsen, B.L. Møller, and S. Bak. 2015. Metabolism, excretion and avoidance of cyanogenic glucosides in insects with different feeding specialisations. *Insect Biochem. Mol. Biol.* 66: 119–128.

Phillips, M.A. and R.B. Croteau. 1999. Resin-based defenses in conifers. *Trends Plant Sci.* 4: 184–190.

Pičmanová, M., E.H. Neilson, M.S. Motawia et al. 2015. A recycling pathway for cyanogenic glycosides evidenced by the comparative metabolic profiling in three cyanogenic plant species. *Biochem J.* 469: 375–89.

Pimm, S.L. and L.N. Joppa. 2015. How many plant species are there, where are they, and at what rate are they going extinct? *Ann. Mo. Bot. Gard.* 100: 170–176.

Piotrowski, M., S. Schonfelder, and E.W. Weiler. 2001. The *Arabidopsis thaliana* isogene *NIT4* and its orthologs in tobacco encode β-cyano-L-alanine hydratase/nitrilase. *J. Biol. Chem.* 276: 2616–2621.

Poulton, J.E. 1990. Cyanogenesis in plants. *Plant Physiol.* 94: 401–405.

Rajniak, J., B. Barco, N.K. Clay, and E.S. Sattely. 2015. A new cyanogenic metabolite in *Arabidopsis* required for inducible pathogen defence. *Nature* 525: 376–379.

Read, J. and A. Stokes. 2006. Plant biomechanics in an ecological context. *Am. J. Bot.* 93: 1546–1565.

Rockenbach, J., A. Nahrstedt, and V. Wray. 1992. Cyanogenic glycosides from *Psydrax* and *Oxyanthus* species. *Phytochemistry* 31: 567–570.

Rosa Jr., V.E.d., F.T.S. Nogueira, P. Mazzafera, M.G.A. Landell, and P. Arruda. 2007. Sugarcane *dhurrin*: Biosynthetic pathway regulation and evolution. *XXVI Congress, International Society of Sugar Cane Technologists*, 958–962, ICC, Durban, South Africa.

Rosenthal, D.M., R.A. Slattery, R.E. Miller et al. 2012. Cassava about-FACE: Greater than expected yield stimulation of cassava (*Manihot esculenta*) by future CO_2 levels. *Glob. Chang. Biol.* 18: 2661–2675.

Rusch, G.M., A. Tveit, I.D.H. Waalkens-Berendsen, A.P.M. Wolterbeek, and G. Armour. 2009. Comparative reprotoxicity of three oximes. *Drug Chem. Toxicol.* 32: 381–394.

Sakurada, K., H. Ikegaya, H. Ohta, H. Fukushima, T. Akutsu, and K. Watanabe. 2009. Effects of oximes on mitochondrial oxidase activity. *Toxicol. Lett.* 189: 110–114.

Sánchez-Pérez, R., F. Sáez, J. Borch, F. Dicenta, B.L. Møller, and K. Jørgensen. 2012. Prunasin hydrolases during fruit development in sweet and bitter almonds. *Plant Physiol.* 158: 1916–1932.

Sánchez-Pérez, R., K. Jørgensen, C.E. Olsen, F. Dicenta, and B.L. Møller. 2008. Bitterness in almonds. *Plant Physiol.* 146: 1040–1052.

Sánchez-Pérez, R., K. Jørgensen, M.S. Motawia, F. Dicenta, and B.L. Møller. 2009. Tissue and cellular localization of individual β-glycosidases using a substrate-specific sugar reducing assay. *Plant J.* 60: 894–906.

Saunders, J.A. and E.E. Conn. 1978. Presence of cyanogenic glucoside dhurrin in isolated vacuoles from Sorghum. *Plant Physiol.* 61: 154–157.

Schwind, P., V. Wray, and A. Nahrstedt. 1990. Structure elucidation of an acylated cyanogenic triglycoside, and further cyanogenic constituents from *Xeranthemum cylindraceum*. *Phytochemistry* 29: 1903–1911.

Selmar, D. 1993. Transport of cyanogenic glucosides: Linustatin uptake by *Hevea cotyledons*. *Planta* 191: 191–199.

Selmar, D. and M. Kleinwächter. 2013. Stress enhances the synthesis of secondary plant products: The impact of stress-related over-reduction on the accumulation of natural products. *Plant Cell Physiol.* 54: 817–826.

Selmar, D., R. Lieberei, and B. Biehl. 1988. Mobilization and utilization of cyanogenic glycosides—the linustatin pathway. *Plant Physiol.* 86: 711–716.

Sendker, J. and A. Nahrstedt. 2009. Generation of primary amide glucosides from cyanogenic glucosides. *Phytochemistry* 70: 388–393.

Sibbesen, O., B. Koch, B.A. Halkier, and B.L. Møller. 1994. Isolation of the heme-thiolate enzyme cytochrome P-450TYR, which catalyzes the committed step in the biosynthesis of the cyanogenic glucoside dhurrin in *Sorghum bicolor* (L.) Moench. *Proc. Natl. Acad. Sci. USA* 91: 9740–9744

Sibbesen, O., B. Koch, B.A. Halkier, and B.L. Møller. 1995. Cytochrome P-450TYR is a multifunctional heme-thiolate enzyme catalyzing the conversion of L-tyrosine to *p*-hydroxyphenylacetaldehyde oxime in the biosynthesis of the cyanogenic glucoside dhurrin in *Sorghum-bicolor* (L.) Moench. *J. Biol. Chem.* 270: 3506–3511.

Siegien, I. and R. Bogatek. 2006. Cyanide action in plants—From toxic to regulatory. *Acta Physiol. Plant* 28: 483–497.

Simon, J., R.M. Gleadow, and I.E. Woodrow. 2010. Allocation of resources to chemical defence and plant functional traits is constrained by soil N. *Tree Physiol.* 30: 1111–1117.

Spencer, K.C. and D.S. Seigler. 1983. Cyanogenesis of *Passiflora edulis*. *J. Agric. Food Chem.* 31: 794–796.

Swain, E. and J.E. Poulton. 1994. Utilization of amygdalin during seedling development of *Prunus serotina*. *Plant Physiol.* 106: 437–445.

Takabayashi, J., M. Dicke, S. Takahashi, M.A. Posthumus, and T.A. Vanbeek. 1994. Leaf age affects composition of herbivore-induced synomones and attraction of predatory mites. *J. Chem. Ecol.* 20: 373–386.

Takabayashi, J., S. Takahashi, M. Dicke, and M.A. Posthumus. 1995. Developmental stage of herbivore *Pseudaletia separata* affects production of herbivore-induced synomone by corn plants. *J. Chem. Ecol.* 21: 273–287.

Takos, A.M., C. Knudsen, D. Lai et al. 2011. Genomic clustering of cyanogenic glucoside biosynthetic genes aids their identification in *Lotus japonicus* and suggests the repeated evolution of this chemical defence pathway. *Plant J.* 68: 273–286.

Thayer, S.S. and E.E. Conn. 1981. Subcellular localization of dhurrin β-glucosidase and hydroxynitrile lyase in the mesophyll cells of sorghum leaf blades. *Plant Physiol.* 67: 617–622.

Thomsen, K. and L. Brimer. 1997. Cyanogenic constituents in woody plants in natural lowland rain forest in Costa Rica. *Bot. J. Linn. Soc.* 124: 273–294.

Van Den Boom, C.E.M., T.A. Van Beek, M.A. Posthumus, A. De Groot, and M. Dicke. 2004. Qualitative and quantitative variation among volatile profiles induced by *Tetranychus urticae* feeding on plants from various families. *J. Chem. Ecol.* 30: 69–89.

Vandegeer, R., R.M. Miller, M. Bain, R.M. Gleadow, and T.R. Cavagnaro. 2013. Drought adversely affects tuber development and nutritional quality of the staple crop cassava (*Manihot esculenta* Crantz). *Funct. Plant Biol.* 40: 195–200.

Wei, J.N., J.W. Zhu, and L. Kang. 2006. Volatiles released from bean plants in response to agromyzid flies. *Planta* 224: 279–287.

White, W.L.B., D.I. Arias-Garzon, J.M. McMahon, and R.T. Sayre. 1998. Cyanogenesis in cassava—The role of hydroxynitrile lyase in root cyanide production. *Plant Physiol.* 116: 1219–1225.

Wojtaszek, P. 1997. Oxidative burst: An early plant response to pathogen infection. *Biochem. J.* 322: 681–692.

Woodrow, I.E., J.D. Slocum, R.M. Gleadow. 2002. Influence of water stress on cyanogenic capacity in Eucalyptus cladocalyx. *Funct. Plant Biol.* 29: 103–110.

Wurtele, E.S., B.J. Nikolau, and E.E. Conn. 1984. Tissue distribution of β-cyanoalanine synthase in leaves. *Plant Physiol.* 75: 979–982.

Wurtele, E.S., B.J. Nikolau, and E.E. Conn. 1985. Subcellular and developmental distribution of β-cyanoalanine synthase in barley leaves. *Plant Physiol.* 78: 285–290.

Zagrobelny, M., S. Bak, and B.L. Møller. 2008. Cyanogenesis in plants and arthropods. *Phytochemistry* 69: 1457–1468.

Zagrobelny, M., S. Bak, C.E. Olsen, and B.L. Møller. 2007a. Intimate roles for cyanogenic glucosides in the life cycle of *Zygaena filipendulae* (Lepidoptera, Zygaenidae). *Insect Biochem. Mol. Biol.* 37: 1189–1197.

Zagrobelny, M., S. Bak, C.T. Ekstrøm, C.E. Olsen, and B.L. Møller. 2007b. The cyanogenic glucoside composition of *Zygaena filipendulae* (Lepidoptera: Zygaenidae) as affected by feeding wild-type and transgenic Lotus populations with variable cyanogenic glucoside profiles. *Insect Biochem. Mol. Biol.* 37: 10–18.

Zangerl, A.R. and F.A. Bazzaz. 1992. Theory and pattern in plant defense allocation. In *Plant Resistance to Herbivores and Pathogens*, eds. R.S. Fritz and E.L. Simms. 363–391. Chicago: The University of Chicago Press.

chapter eight

Glucosinolate biosynthesis and functional roles

Tomohiro Kakizaki

Contents

8.1 Introduction 157
8.2 Chemical structure and degradation products of glucosinolates 158
8.3 Biosynthesis pathways 160
    8.3.1 First step: Side-chain elongation 160
    8.3.2 Second step: Core structure formation 162
    8.3.3 Third step: Secondary modification of the side chain 165
    8.3.4 Conservation of the biosynthesis pathway in Brassicaceae vegetables 166
8.4 Gene regulation 167
    8.4.1 MYBs 167
    8.4.2 MYCs 168
8.5 Transport 168
8.6 Variation in glucosinolate composition between and within plants ... 169
8.7 For the human diet 172
    8.7.1 Anticarcinogenic activity 172
    8.7.2 Induction of phase 2 detoxification enzymes 172
    8.7.3 Histone modification 173
8.8 For insects 173
8.9 Breeding for a glucosinolate profile 174
    8.9.1 Rapeseed 174
    8.9.2 Broccoli 175
    8.9.3 Radish 175
8.10 Conclusion 176
References 176

8.1 Introduction

Glucosinolates are sulfur-rich secondary metabolites abundantly contained in Brassicales that have important biological and economic roles in plant defense and human health. They are largely found in the Brassicaceae, which include important crops and vegetables such as rapeseed (*Brassica napus*), cabbage (*B. oleracea*), and the model plant Arabidopsis (*A. thaliana*). Extensive studies of *Arabidopsis* as a plant model of glucosinolate biosynthesis have increased our understanding of these important metabolites. The availability of the *Arabidopsis* genome sequence has permitted functional genomics approaches for identification of genes involved in glucosinolate biosynthesis. The commonality and specificity of glucosinolate synthesis pathways among Brassicaceae plants are becoming clearer. In addition, the functions of isothiocyanate, a degradation product of glucosinolate, have become clear at the molecular level, giving promise for the promotion of human health.

This chapter presents a summary of topics related to glucosinolate biosynthesis and functional roles in the Brassicaceae, along with a summary of chemical structures, biosynthesis pathway, transport, functions against other organisms, and importance for plant breeding.

8.2 Chemical structure and degradation products of glucosinolates

Plants produce approximately 200 different glucosinolates, which are structurally classified into three classes: aliphatic, indolic, and aromatic glucosinolates, based on precursor amino acids (Clarke 2010; Fahey et al. 2001; Halkier and Gershenzon 2006). Methionine (or alanine, isoleucine, leucine, and valine), tryptophan, and phenylalanine are precursors of aliphatic, indolic, and aromatic glucosinolates, respectively (Figure 8.1). The common structure of glucosinolates consists of a β-D-glucopyranose

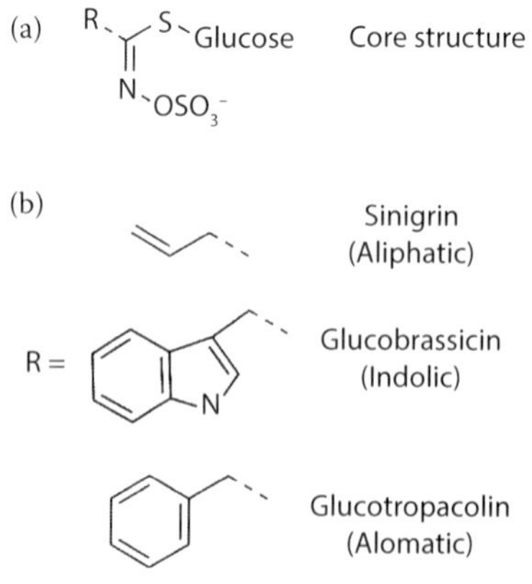

Figure 8.1 Core structure (a) and typical variation in the structure of the side chain (b) of glucosinolate. (R, variable side chain.)

residue linked by a sulfur atom to a (Z)-N-hydroximinosulfate ester, plus a variable side chain derived from precursor amino acids (Figure 8.1).

When plant cells are attacked by herbivores, the "glucosinolate–myrosinase system" is activated. Myrosinase is conformed as a dimer with molecular weight in the range of 62–75 kDa per subunit. Most myrosinases catalyze the hydrolysis of multiple glucosinolates as substrates (Bernardi et al. 2003; Chen and Halkier 1999). Glucosinolates are hydrolyzed by myrosinases to isothiocyanates, thiocyanates, nitriles, or epithionitriles, depending on pH and the presence of epithiospecifier protein (Bones and Rossiter 1996) (Figure 8.2). Isothiocyanates are

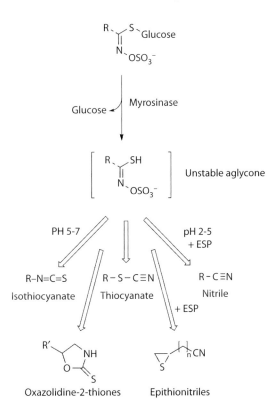

Figure 8.2 Scheme of glucosinolate hydrolysis and structure of possible degradation products. Glucosinolates are hydrolyzed by myrosinase when tissue is damaged. At neutral pH (pH 5–7), the unstable aglycones change to isothiocyanates. If the glucosinolate side chain is hydroxylated at the third carbon, spontaneous cyclization of the isothiocyanate results in the formation of oxazolidine-2-thione. Some glucosinolates can be hydrolyzed to thiocyanates. The interaction of a protein called epithiospecifier protein with myrosinase diverts the reaction toward the production of epithionitriles or nitriles depending on the glucosinolate structure and pH. (R, variable side chain.)

a main product of myrosinase hydrolysis and have many functions (described in Sections 8.7 and 8.8). If the side chain contains a double bond, in the presence of epithiospecifier protein and ferrous ions, the thiohydroximates rearranges to produce an epithionitrile and a nitrile (Bones and Rossiter 1996).

Myrosinase is mainly localized in special idioblasts, called myrosin cells, specifically along leaf veins in *Arabidopsis* (Andreasson et al. 2001; Husebye et al. 2002). In contrast, major glucosinolate is accumulated in glucosinolate-rich cell type (S-cells) which are ideally adjacent to phloem cells (Koroleva et al. 2010). The separate physical locations of myrosinase and glucosinolates prevent the generation of isothiocyanate prior to the occurrence of feeding damage.

8.3 Biosynthesis pathways

The biosynthesis pathway of glucosinolate is divided into three independent steps: side-chain elongation, formation of the core structure, and secondary modification of the side chain (Figure 8.3).

Given that more than 20 enzymes are involved in the biosynthesis pathway in *Arabidopsis*, polymorphisms in individual genes and their combinations determine the profile in different accessions of *Arabidopsis* (Kliebenstein et al. 2001a). In this section, the three biosynthesis steps revealed by studies using *A. thaliana*, which has approximately 40 types of glucosinolates mainly derived from methionine and tryptophan, are described (Table 8.1). A list of glucosinolate biosynthesis-associated genes appears in Table 8.2.

8.3.1 First step: Side-chain elongation

Aliphatic glucosinolates have various lengths of side chain, determined by the side-chain elongation step (Figure 8.3a). This process starts with deamination by a branched-chain amino acid aminotransferase (BCAT), giving rise to a 2-keto acid (Schuster et al. 2006). The 2-keto acid is imported into chloroplasts by bile acid transporter 5 (BAT5), also called bile acid:sodium symporter family protein 5 (BASS5), which is a plastidic transporter of 2-keto acids (Gigolashvili et al. 2009; Sawada et al. 2009b). Next, the 2-keto acid enters a side-chain elongation cycle. This cycle consists of condensation with acetyl-CoA by a methylthioalkylmalate synthase (MAM1-3), isomerization by an isopropylmalate isomerase (IPMI), and oxidative decarboxylation by an isopropylmalate dehydrogenase (IPMDH1) (Field et al. 2004; Knill et al. 2009; Sawada et al. 2009a; Wentzell et al. 2007). The product of these three reactions is a 2-keto acid that has been elongated by a single methylene. BCAT3 most likely catalyzes the terminal steps in the chain elongation process leading to short-chain glucosinolates (Knill

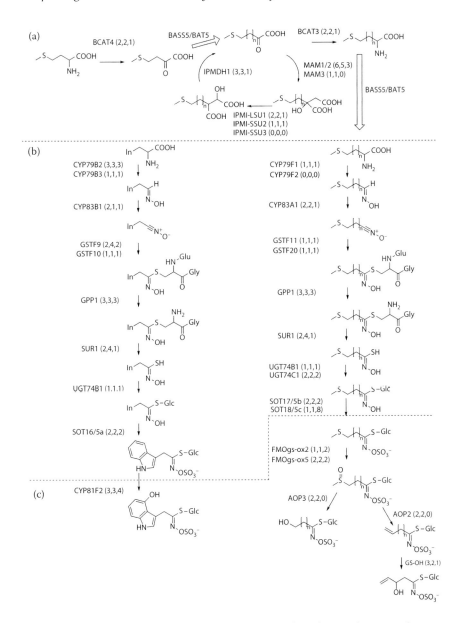

Figure 8.3 The aliphatic and indolic glucosinolate biosynthetic pathways. (a) Chain elongation steps. (b) Biosynthesis of core glucosinolate structure. (c) Secondary modifications. (The copy numbers of glucosinolate biosynthetic genes in *Brassica rapa*, *B. oleracea*, and *Raphanus sativus* are shown in parentheses.)

Table 8.1 Chemical name and abbreviation of typical glucosinolates

Glucosinolate name	Common name	Abbreviation
Aliphatic		
2-Propenyl	Sinigrin	2-Propenyl
3-Butenyl	Gluconapin	3-Butenyl
3-Hydroxypropyl		3OHP
4-Hydroxybutyl		4OHB
3-Methylsulfinylpropyl	Glucoiberin	3MSOP
4-Methylsulfinylbutyl	Glucoraphanin	4MSOB
5-Methylsulfinylpentyl	Glucoalyssin	5MSOP
6-Methylsulfinylhexyl	Glucohesperin	6MSOH
7-Methylsulfinylheptyl	Glucoibarin	7MSOH
8-Methylsulfinyloctyl	Glucohirsutin	8MSOO
3-Methylthiopropyl	Glucoibervirin	3MTP
4-Methylthiobutyl	Glucoerucin	4MTB
6-Methylthiohexyl	Glucosquerellin	6MTH
7-Methylthioheptyl		7MTH
8-Methylthiooctyl		8MTO
Indolic		
Indol-3-ylmethyl	Glucobrassicin	I3M
4-Methoxyindol-3-ylmethyl	4-Methoxyglucobrassicin	4MO-I3M
1-Methoxyindol-3-ylmethyl	Neoglucobrassicin	1MO-I3M
4-Hydroxyindol-3-ylmethyl	4-Hydroxyglucobrassicin	4OH-I3M
Aromatic		
Benzyl	Glucotropaeolin	Benzyl
2-Phenylethyl	Gluconastrutiin	2-Phenylethyl

et al. 2008). Elongated 2-keto acids are transported to the cytosol by BAT5/BASS5 (Gigolashvili et al. 2009; Sawada et al. 2009b).

8.3.2 Second step: Core structure formation

Elongated 2-keto acids are converted to aldoximes by P450 cytochromes of the CYP79 monooxygenase family (Figure 8.3b). Depending on precursor amino acid and length of the side chain, specific CYP79s perform catalysis (Chen et al. 2003; Hansen et al. 2001b; Reintanz et al. 2001). Next, aldoximes are oxidized to activated forms by P450 cytochromes of the CYP83 family. CYP83B1 metabolizes indolic acetaldoximes, and CYP83A1 converts aliphatic aldoximes (Bak and Feyereisen 2001; Hansen et al. 2001a; Hemm et al. 2003). The activated forms are transformed to thiohydroximates via

Chapter eight: Glucosinolate biosynthesis and functional roles

Table 8.2 List of glucosinolate biosynthesis related genes

Pathway	Name		AGI code
Aliphatic Pathway			
Transcription factors			
	MYB28	MYB DOMAIN PROTEIN 28	At5g61420
	MYB29	MYB DOMAIN PROTEIN 29	At5g07690
	MYB76	MYB DOMAIN PROTEIN 76	At5g07700
	MYC2		At1g32640
	MYC3		At1g32640
	MYC4		At4g17880
Biosynthesis genes			
	BCAT3	BRANCHED-CHAIN AMINOTRANSFERASE 3	At3g49680
	BCAT4	BRANCHED-CHAIN AMINOTRANSFERASE 4	At3g19710
		ARABIDOPSIS ISOPROPYLMALATE DEHYDROGENASE 3	At1g31180
	BAT5	BILE ACID:SODIUM SYMPORTER FAMILY PROTEIN 5	At4g12030
	MAM1	METHYLTHIOALKYLMALATE SYNTHASE 1	At5g23010
	MAM2	METHYLTHIOALKYLMALATE SYNTHASE 2	Not found in Col-0
	MAM3	METHYLTHIOALKYLMALATE SYNTHASE 3	At5g23020
	IPMI LSU1	ISOPROPYL MALATE ISOMERASE LARGE SUBUNIT 1	At4g13430
	IPMI SSU2	ISOPROPYL MALATE ISOMERASE 2	At2g43100
	IPMI SSU3	ISOPROPYL MALATE ISOMERASE 3	At3g58990
	IPMDH1	ISOPROPYL MALATE DEHYDROGENASE 1	At5g14200
	IPMDH3	ISOPROPYL MALATE DEHYDROGENASE 3	At1g31180
	CYP79F1	CYTOCHROME P450 79F1	At1g16410
	CYP79F2	CYTOCHROME P450 79F2	At1g16400
	CYP83A1	CYTOCHROME P450 83A1	At4g13770

(*Continued*)

Table 8.2 (Continued) List of glucosinolate biosynthesis related genes

Pathway	Name		AGI code
	GSTF11	GLUTATHIONE S-TRANSFERASE F11	At3g03190
	GSTU20	GLUTATHIONE S-TRANSFERASE TAU20	At1g78370
	GGP1	GAMMA-GLUTAMYL PEPTIDASE 1	At4g30530
	SUR1	SUPERROOT1	At2g20610
	UGT74C1	UGT74C1	At2g31790
	SOT18	DESULFO-GLUCOSINOLATE SULFOTRANSFERASE 18	At1g74090
	SOT17	DESULFO-GLUCOSINOLATE SULFOTRANSFERASE 17	At1g18590
	FMO-GSOX1	FLAVIN-MONOOXYGENASE GLUCOSINOLATE S-OXYGENASE 1	At1g65860
	FMO-GSOX2	FLAVIN-MONOOXYGENASE GLUCOSINOLATE S-OXYGENASE 2	At1g62540
	FMO-GSOX3	FLAVIN-MONOOXYGENASE GLUCOSINOLATE S-OXYGENASE 3	At1g62560
	FMO-GSOX4	FLAVIN-MONOOXYGENASE GLUCOSINOLATE S-OXYGENASE 4	At1g62570
	FMO-GSOX5	FLAVIN-MONOOXYGENASE GLUCOSINOLATE S-OXYGENASE 5	At1g12140
	AOP3	2-Oxoglutarate–Dependent Dioxygenase	At4g03050
	AOP2	2-Oxoglutarate–Dependent Dioxygenase	At4g03060
	GS-OH	2-Oxoglutarate–Dependent Dioxygenase	At2g25450
Indolic and Aromatic Pathways			
Transcription factors			
	MYB34	MYB DOMAIN PROTEIN 34	At5g60890
	MYB51	MYB DOMAIN PROTEIN 51	At1g18570
	MYB122	MYB DOMAIN PROTEIN 122	At1g74080

(*Continued*)

Chapter eight: Glucosinolate biosynthesis and functional roles 165

Table 8.2 (Continued) List of glucosinolate biosynthesis related genes

Pathway	Name		AGI code
Biosynthesis genes			
	CYP79A2	CYTOCHROME P450 79A2	At5g05260
	CYP79B2	CYTOCHROME P450 79B2	At4g39950
	CYP79B3	CYTOCHROME P450 79B3	At2g22330
	CYP83B1	CYTOCHROME P450 83B1	At4g31500
	GSTF9	GLUTATHIONE S-TRANSFERASE PHI 9	At2g30860
	GSTF10	GLUTATHIONE S-TRANSFERASE PHI 10	At2g30870
	GGP1	GAMMA-GLUTAMYL PEPTIDASE 1	At4g30530
	SUR1	SUPERROOT1	At2g20610
	UGT74B1	UGT74B1	At1g24100
	SOT16	SULFOTRANSFERASE 16	At1g74100
	CYP81F2	CYTOCHROME P450 81F2	At5g57220

glutathione conjugation and the C-S lyase (SUR1) reaction (Mikkelsen et al. 2004). The thiohydroximates are converted to the glucosinolate structure by *S*-glucosyltransferases of the UGT74 family and sulfotransferases (SOTs). Thiohydroximates are in turn *S*-glucosylated by glucosyltransferases of the UGT74 family to form desulfoglucosinolates. UGT74B1 metabolizes Phe-derived thiohydroximates (Grubb et al. 2004), and UGT74C1 has been suggested to glucosylate Met-derived substrates, based on its coexpression with genes in the biosynthesis of aliphatic glucosinolates (Gachon et al. 2005). The SOT (also called AtST5) family catalyzes the final step in the biosynthesis of the glucosinolate core structure through the sulfation of desulfoglucosinolates (Piotrowski et al. 2004). SOT16 metabolizes Trp- and Phe-derived desulfoglucosinolates, whereas SOT18 and SOT17 prefer long-chained aliphatic substrates (Piotrowski et al. 2004).

8.3.3 *Third step: Secondary modification of the side chain*

After glucosinolate structure formation, the side chains are modified by oxygenation, hydroxylation, alkenylation, benzoylation, and methoxylation (Figure 8.3c). In aliphatic glucosinolates, oxygenation is first step of secondary modification. The S-oxygenation of aliphatic glucosinolates is a common modification performed by flavin monooxygenases, FMO_{GS-OXs}. Five homologous FMOs ($FMO_{GS-OX1-5}$) are encoded in the *Arabidopsis* genome, and their substrate specificities have been characterized (Hansen et al. 2007; Li et al. 2008). FMO_{GS-OX1} catalyzes the S-oxygenation of methylthioalkyl to methylsulfinylalkyl glucosinolates (Hansen et al. 2007). Likewise, $FMO_{GS-OX2-4}$ have broad substrate specificity, without specificity for chain length, and catalyze the conversion from methylthioalkyl to the corresponding methylsulfinylalkyl glucosinolate. FMO_{GS-OX5} shows accurate specificity for long-chain 8-methylthiooctyl glucosinolate (Li et al. 2008). Alkenylation is the next step of modification. Alkenyl glucosinolates such as sinigrin are produced by 2-oxoglutarate-dependent dioxygenases (AOPs) from S-oxygenated glucosinolates. As shown by *in vitro* enzyme assay, AOP2 is capable of catalyzing both the 3-methylsulfinylalkyl and 4-methylsulfinylalkyl glucosinolates. In contrast to AOP2, AOP3 catalyzes the formation of hydroxyalkyl glucosinolates (Kliebenstein et al. 2001b). GS-OH also encodes 2-oxoglutarate-dependent dioxygenase, which is responsible for the biosynthesis of hydroxylated alkenyl glucosinolates such as 2-hydroxybut-3-enyl glucosinolate (Hansen et al. 2008). Kliebenstein et al. (2001a) reported that natural variation in aliphatic glucosinolate profiles among 39 accessions is controlled by a combination of genotypes of four loci: GS-Elong, AOP2, AOP3, and GS-OH.

8.3.4 Conservation of the biosynthesis pathway in Brassicaceae vegetables

With the spread of next-generation sequencers, genomic and transcriptome data have accumulated for Brassicaceae vegetables including cabbage, radish, Chinese cabbage, and rapeseed (Kitashiba et al. 2014; Liu et al. 2014; Mitsui et al. 2015; Wang et al. 2011). Transcriptome analysis of genes with inferred homology to those characterized in *Arabidopsis* has revealed that the majority of glucosinolate biosynthesis genes are highly conserved in Brassicaceae vegetables (Carpio et al. 2014; Gao et al. 2014; Mitsui et al. 2015; Pino Del Wang et al. 2013). Interestingly, most of the glucosinolate genes in Brassicaceae vegetables have been identified as multiple genes (Figure 8.3). One of the reasons is considered to be that the whole-genome triplication of ancestral species of these vegetables occurred after the divergence from the *Arabidopsis* ancestor. Divergence of the *Arabidopsis* lineage and the *Brassica–Raphanus* ancestor from a common ancestor is reported to have occurred 38.8 million years ago. After the divergence, whole-genome triplication in the *Brassica–Raphanus* ancestor may have occurred between 15.6 and 28.3 million years ago (Mitsui et al. 2015). Similarly, after the whole-genome triplication, *B. oleracea* and *B. rapa* diverged from a common ancestor ~4 million years ago (Liu et al. 2014). It is considered that the glucosinolate-associated genes of *Brassica* and *Raphanus* species have higher copy number and it results in complex glucosinolate profiles.

Chapter eight: Glucosinolate biosynthesis and functional roles   167

8.4 Gene regulation

Glucosinolate contains sulfur in its core structure, and its production and degradation pathways are affected by sulfur nutrition. The glucosinolate pathway and sulfur assimilation are the two important metabolic pathways affected by sulfur nutrition. Genes involved in glucosinolate biosynthesis are identified and characterized by genetic mapping and omics-based approaches using natural variation in glucosinolate profiles in *Arabidopsis* accessions (Sonderby et al. 2010b). Glucosinolates are accumulated constitutively (Wittstock and Halkier 2002) but are also synthesized in response to wounding or herbivory (Mewis et al. 2006; Mikkelsen et al. 2003). Under sulfur starvation, glucosinolates are degraded by myrosinase. Induction of myrosinase under low-sulfur conditions is suggested to be a metabolic response that serves for reuse of the sulfur pool in glucosinolates. To adapt to these environmental changes, transcription factors fine-tune the expression of the methionine and glucosinolate synthesis pathways. Several transcriptome analyses have suggested that activation of sulfate acquisition and repression of glucosinolate production are responses to sulfur limitation (Hirai et al. 2003; Maruyama-Nakashita et al. 2003, 2005; Nikiforova et al. 2003). The genes responsible for glucosinolate biosynthesis are strictly regulated by transcription factors, including R2R3-MYBs and helix-loop-helix MYCs (Gigolashvili et al. 2008; Hirai et al. 2007; Schweizer et al. 2013; Sonderby et al. 2010a). By contrast, SLIM1, an EIL-family transcription factor, negatively regulates both aliphatic and indolic biosynthesis pathways under sulfur-limiting conditions (Maruyama-Nakashita et al. 2006).

8.4.1 MYBs

The expression of glucosinolate-associated genes is positively regulated by six R2R3-MYB transcription factors. Biosynthesis of aliphatic glucosinolates derived from methionine is coordinated by MYB28, MYB29, and MYB76 complex (Gigolashvili et al. 2007b; Hirai et al. 2007; Sonderby et al. 2010a). Based on studies of knockout and ectopically overexpressing lines of these genes, these *MYB* genes have been suggested to regulate one another (Gigolashvili et al. 2007b, 2008). For example, all aliphatic glucosinolates were completely absent in the *myb28 myb29* double mutant (Beekwilder et al. 2008; Li et al. 2013). Interestingly, accumulation of long-chain aliphatic glucosinolates was almost completely abolished in the *myb28* single mutant, while short-chain aliphatic glucosinolates were reduced by approximately 50% in both the *myb28* and the *myb29* single mutants (Beekwilder et al. 2008; Hirai et al. 2007). The expression level of *MYB76* is low compared with that of *MYB28* and *MYB29* in *Arabidopsis* rosette leaves, but an *MYB76*-overexpressing plant accumulated aliphatic glucosinolates (Gigolashvili et al. 2008). Indolic glucosinolates, derived from tryptophan, are regulated by MYB34, MYB51, and MYB122 (Gigolashvili et al. 2007a; Malitsky et al. 2008; Skirycz et al. 2006).

8.4.2 MYCs

Schweizer et al. (2013) reported that glucosinolate biosynthesis genes were downregulated in a triple mutant lacking *MYC2*, *MYC3*, and *MYC4* (*myc234*), both short- and long-chained glucosinolates (Schweizer et al. 2013). Protein-binding microarray and promoter analysis have shown that MYC2, MYC3, and MYC4 bind to the core G-box (CACGTG), and similar motifs (Dombrecht et al. 2007; Fernandez-Calvo et al. 2011; Godoy et al. 2011). Nine genes involved in the aliphatic glucosinolate pathway are downregulated in the *myc234* triple mutant (Schweizer et al. 2013).

In response to exogenous and endogenous cues, plants synthesize various signaling molecules. Among the signaling molecules, jasmonates (JAs) are well characterized as an inducer of glucosinolates (Dombrecht et al. 2007). In response to stress, JA binds to a receptor complex consisting of coronatine insensitive1 (COI1) and jasmonate zim domain (JAZ) repressors (Chini et al. 2007; Thines et al. 2007). Biochemical analysis has shown that MYC3 and MYC4 interact with JAZ repressors and also form homo- and heterodimers with MYC2 and among themselves (Fernandez-Calvo et al. 2011). Under no-stimulus conditions, JAZ repressors bind to the JID domain of bHLH MYC2, MYC3, and MYC4 complex and inhibit the interaction between another positive regulator, such as R2R3-MYB transcription factors. When plants are attacked by herbivores, JAZ repressors are degraded by the SCF^{COI1} complex, and MYC–MYB complexes bind to promoters of glucosinolate-associated genes (Schweizer et al. 2013).

8.5 Transport

In general, defense compounds accumulate to the highest levels in tissues most likely to be attacked (Mckey 1974). Thus, transport processes are important factors determining the distribution pattern of these compounds, including glucosinolates. For example, whereas glucosinolate concentrations in leaves decrease with age until virtually absent at senescence, glucosinolate content in seeds tends to increase (Brown et al. 2003; Petersen et al. 2002). Brudenell et al. (1999) showed that radiolabeled ^{35}S-gluconapin (3-butenylglucosinolate) and ^{35}S-desulphogluconapin, an aliphatic glucosinolate, are transported via the phloem from mature leaves to inflorescences and fruits in *B. napus* (Brudenell et al. 1999). Similarly, one of the aromatic glucosinolates, *p*-hydroxybenzyl-glucosinolate, was transported to flowers from leaves in radiolabeled glucosinolate feeding experiments in *Arabidopsis* (Chen et al. 2001). These results indicate the presence of a molecular basis for a long-distance transport system. By screening for glucoerucin uptake activity in *Xenopus* oocytes expressing 239 *Arabidopsis* transporters, two nitrate/peptide transporters (NTRs/PTRs), GTR1 and GTR2, were identified as transporters for glucoerucin uptake (Nour-Eldin et al. 2006, 2012). GTR1 and GTR2 transported glucosinolates across the plasma membrane into phloem companion cells in *Arabidopsis* leaves (Nour-Eldin et al. 2012). In roots, these transporters are colocalized in areas that appeared to be cortex cells surrounding lateral root branching points (Andersen et al. 2013). A mutant deficient in both *GTR1* and *GTR2* (*gtr1 gtr2*) showed a dramatical decrease in glucosinolate accumulation in seeds and roots but abundant accumulation in mature leaves (Andersen et al. 2013; Nour-Eldin et al. 2012). These findings indicate that the GTR1 and GTR2 transport machinery controls the organ-specific distribution of glucosinolate in *Arabidopsis*. The glucosinolate profile of leaves is dominated by aliphatic glucosinolates, whereas roots contain higher amount of indolic glucosinolate (Brown et al. 2003; Petersen et al. 2002). To address the question of different profiles of glucosinolate among the organs, grafting between rosette and roots of seedlings of wild type and *gtr1gtr2* was demonstrated. Grafting experiments showed that the rosette is the major source and storage site of short-chain (with three to five methylene groups) aliphatic glucosinolates. Moreover, indolic glucosinolates can be synthesized in both rosette and roots (Andersen et al. 2013).

In single leaves, glucosinolates accumulate to high amounts on the leaf margin (Shroff et al. 2008; Sonderby et al. 2010a). A metabolome approach revealed a glucosinolate gradient from base to tip in a fully expanded leaf (Watanabe et al. 2013). In the *gtr1gtr2* double mutant, the concentration of aliphatic glucosinolates in leaf margin was increased six fold compared with that in the wild-type leaf margin. Similarly, concentrations of aliphatic glucosinolate in *gtr1gtr2* leaf tips and bases were increased 11- and fivefold over that in wild-type leaves, respectively (Madsen et al. 2014). These data indicate that GTR1 and GTR2 play roles not only in long-distance transport of glucosinolate but in the formation of concentration gradients within leaves.

8.6 Variation in glucosinolate composition between and within plants

Plants of the genus *Brassica* carry three genomes, those of *B. nigra* (BB; $2n = 16$), *B. oleracea* (CC; $2n = 18$), and *B. rapa* (AA; $2n = 10$), and three of their amphidiploids, *B. carinata* (BBCC; $2n = 34$), *B. juncea* (AABB; $2n = 36$), and *B. napus* (AACC; $2n = 38$). Genome analysis in *Brassica* and allied genera was performed by Mizushima (1980). The cytogenetic relationship among *Brassica* species was presented by Nagaharu (1935). The genome

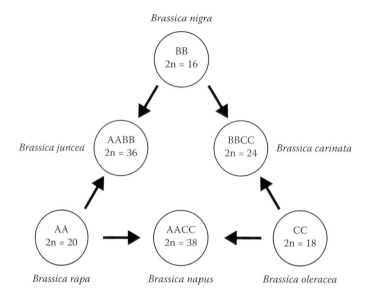

Figure 8.4 Genomic relationships among six cultivated *Brassica* species represented by "U's Triangle." (Adapted from Nagaharu, U. 1935. *Jpn. J. Bot.* 7: 389–452.)

relationships between the three monogenomic and three digenomic species are well known as U's triangle (Figure 8.4).

Great diversity characterizes both the amounts and profiles of glucosinolate in *Brassica* and related species. Most Brassicaceae plants contain one or more aliphatic glucosinolates having an alkyl side chain with three to five carbons (Table 8.3).

B. nigra contains glucosinolates with three-carbon (3C) side chains, derived from a single elongation reaction: *B. oleracea* contains glucosinolates with either 3C or 4C side chains. *B. rapa* contains glucosinolates with either 4C or 5C side chains. Three amphidiploid *Brassica* species, *B. juncea*, *B. napus*, and *B. carinata*, show glucosinolate compositions combining the profiles of two elementary species. Some of the glucosinolates are major glucosinolates of certain plants. For example, glucoraphasatin (4-methylthio-3-butenyl glucosinolate) and glucoerucin (4-methylthiobutyl glucosinolate) and glucoraphanin (4-methylsulfinylbutyl glucosinolate) are glucosinolates specific to radishes and garden rockets, respectively (Table 8.3).

Interestingly, there is variation in glucosinolate profiles within species. Cabbage cultivars have different types of glucosinolates (Poelman et al. 2008). Similarly, 39 accessions of *A. thaliana* have different compositions of 34 glucosinolates (Kliebenstein et al. 2001a). Variation in glucosinolate profiles has resulted from both natural and artificial selection.

Chapter eight: Glucosinolate biosynthesis and functional roles

Table 8.3 Aliphatic glucosinolates in Brassicaceae plants

Species (genome) Representative vegetable	3-Carbon chains				4-Carbon chains					5-Carbon chains		
	Glucoiberverin	Glucoiberin	Sinigrin	Glucoerucin	Glucoraphasatin	Glucoraphanin	Glucoraphenin	Gluconapin	Progoitrin	Glucoalyssin	Glucobrassicanapin	Gluconapoleiferin
Brassica rapa (AA)												
Chinese cabbage		+						++	+		++	+
Turnip		+		+				++	++		++	+
Brassica nigra (BB)			++									
Brassica oleracea (CC)												
Cabbage	+	++	++	+		+		+	+	+	+	
Broccoli		+	+	+		++		++	++	+	+	
Brassica juncea (AABB)			++					+				
Brassica napus (AACC)												
Rapeseed						+			+			+
Brassica carinata (BBCC)			++									
Eruca sativa												
Garden rocket				++		++						
Raphanus sativus												
Radish					++		+					

Source: Adapted from Ishida et al. 2014. *Breed Sci* 64: 48–59.
Note: Dominant glucosinolates in each vegetable are shown with a + symbol.

8.7 For the human diet

Isothiocyanate, a hydrolysis product of glucosinolate, exerts various effects on human health, and broccoli sprouts are a rich source of several isothiocyanates. Sulforaphane (4-methylsulfinylbutyl isothiocyanate) is an organic isothiocyanate derived from glucoraphanin (4-methylsulfinylbutyl glucosinolate), which is abundant in broccoli and broccoli sprouts (Fahey et al. 1997; Zhang et al. 1992). Sulforaphane is already consumed widely as a regular component of the human diet. In this section, aspects of the biological activity of sulforaphane are described.

8.7.1 Anticarcinogenic activity

The first report of anticarcinogenic activity of sulforaphane was by Fahey et al. (1997). Extracts of 3-day-old broccoli sprouts were effective in reducing the incidence of development of mammary tumors in dimethylbenz(a)anthracene (DMBA)-treated rats (Fahey et al. 1997). Today, sulforaphane is known to be protective against carcinogen-induced tumorigenesis in rodents. Sulforaphane also prevents DMBA-induced preneoplastic lesions in mouse mammary glands and rat mammary tumors (Gerhauser et al. 1997; Zhang et al. 1994). Treatment with broccoli sprouts of mice infected with *Helicobacter pylori*, associated with stomach cancer, reduces colonization and attenuates gastritis (Yanaka et al. 2009). Moreover, sulforaphane inhibits breast cancer stem cells and downregulates the Wnt/β-catenin self-renewal pathway (Li et al. 2010).

8.7.2 Induction of phase 2 detoxification enzymes

Sulforaphane functions as a potent inducer of phase 2 detoxification enzymes, such as glutathione *S*-transferase (GST) and NADP(H) quinone oxidoreductase 1 (NQO1) (Dinkova-Kostova et al. 2007; Fahey et al. 1997; Zhang and Callaway 2002; Zhang et al. 2006). Many detoxification enzymes including GST are regulated by the antioxidant response element (ARE) found in their promoter regions. Under basal conditions, nuclear respiratory factor 2 (Nrf2), a member of the NF-E2 family of nuclear basic leucine zipper transcription factors, binds to Kelch-like ECH-associated protein 1 (Keap1), a cytoplasmic protein homologous to the *Drosophila* actin-binding protein Kelch. When inducers such as sulforaphane disrupt the Keap1–Nrf2 complex by reacting with thiol residues in Keap1, Nrf2 migrates to the nucleus, where it forms heterodimers with other transcription factors such as small Maf that bind to ARE and accelerate transcription (Dinkova-Kostova et al. 2002; Watai et al. 2007).

8.7.3 Histone modification

Sulforaphane is implicated in the modification of histone acetylation, which controls gene expression by altering chromatin structure. Histone acetylation depends on a balance between enzymes with histone acetylase activity and histone deacetylase (HDAC) (Struhl 1998). HDAC is recognized as a promising target for the development of anticancer drugs because it is overexpressed in tumor cells (Kim et al. 2003). Myzak et al. (2004) showed that sulforaphane acts as an inhibitor of HDAC. Treatment with sulforaphane reduced HDAC activity and increased acetylated histones H3 and H4 in both human embryonic kidney 293 cells and colon HCT116 cells (Myzak et al. 2004). Activity of HDAC inhibitor and increase in HDAC protein turnover by aliphatic isothiocyanates in human colon cancer cells depended on alkyl chain length (Rajendran et al. 2013).

8.8 For insects

Now that the identity of the genes that controls the diversity of glucosinolates is clear, the question "why are there polymorphisms in these genes?" arises. This question has been discussed on the basis of the relationship between herbivory and the environment. The many natural accessions of *Arabidopsis* survive under diverse environmental conditions and have been successful in adapting to specific environments. Each *Arabidopsis* accession has diverse glucosinolates which is essential for fitness performance (Burow et al. 2010). In particular, diverse aliphatic glucosinolates have been recognized for their ability to adapt to fluctuating biotic pest environments (Bidart-Bouzat and Kliebenstein 2008). Thus, glucosinolates play important roles in plant biotic interactions, making them important determinants of plant fitness in diverse environments (Lankau 2007). Plants often defend themselves against diverse types of herbivore species, including generalists that feed on many different species and specialists that feed on a restricted set of related species. Although generalists are often deterred by secondary compounds such as glucosinolates, many specialists have evolved effective countermeasures to the chemical defenses of their host plants (Siemens and MitchellOlds 1996; Vandam et al. 1995). For example, nitrile, which is synthesized from alkenyl glucosinolates, is effective for strongly attracting parasitic and carnivorous bees, but its toxicity is low for larvae of herbivorous butterflies (Burow et al. 2006; Mumm et al. 2008). Because of this kind of effect, different selective pressures are exerted, depending on the types of herbivorous or carnivorous insects in an environment (Lankau 2007; Lankau and Strauss 2007). The activity shown by each glucosinolate differs among insects. Using the species of insects in the environment surrounding the plant, the optimum glucosinolates are determined. The proportion of occupied plants in an environment varies greatly from year to year. This variation promotes fluctuating selection for glucosinolate composition and variation (Chan et al. 2010).

8.9 Breeding for a glucosinolate profile

The Brassicaceae family contains many economically important plants such as leaf and root vegetables, oilseed and condiment crops. Brassicaceae vegetables are widely cultivated, with many genera, species, and cultivars including *B. rapa* (Chinese cabbage, Chinese mustard, and turnip), *B. oleracea* (cabbage, broccoli, cauliflower, kale, and kohlrabi), *B. napus* (rapeseed and rutabaga), *B. juncea* (mustard greens), and *Raphanus sativus* (radish). Glucosinolate amounts and profiles show great diversity among Brassicaceae, and improvement of glucosinolate composition and content is an important target for breeding these crop species. Here, we introduce examples of research that is focused on glucosinolates in breeding.

8.9.1 Rapeseed

Rapeseed (*B. napus*, AACC, $2n = 38$) is an allopolyploid that originated from spontaneous hybridizations between *B. rapa* (AA, $2n = 20$) and *B. oleracea* (CC, $2n = 18$). Oil purified from rapeseed is as important as soybean oil worldwide. The meal obtained after oil extraction contains 35%–40% of protein and is suitable for animal feed (Dimov et al. 2012; Wanasundara 2011). Although rapeseed meal is excellent as feed, the glucosinolate and their degradation products, when included in feed, cause thyroid gland enlargement and goiter (Fenwick et al. 1983; Walker and Booth 2001). For this reason, the development of varieties with low seed glucosinolate content is desirable. In 1969, the Polish spring rapeseed variety Bronowski was identified as a source of low seed glucosinolates. By backcrossing to introduce this trait into an erucic acid-free cultivar, the first double-low (so-called 00-quality or canola) cultivar Tower was released in 1974. Today the majority of both spring and winter oilseed rape cultivars have double low quality. Given that the trait of low glucosinolate content is controlled by quantitative loci (QTL), identification of traits and DNA markers for responsible genes is desirable for efficient breeding. Hasan et al. (2008) identified markers linked to genes controlling seed glucosinolate content by structure analysis and association studies using 94 gene bank rapeseed accessions and 46 inter oilseed rape genotypes (Hasan et al. 2008). By genomewide association study of 472 rapeseed accessions using the 60 K *Brassica* Infinium SNP array, orthologs of *Arabidopsis high aliphatic glucosinolate 1* were identified as a candidate gene for low glucosinolate content (Li et al. 2014). QTL analysis using a doubled haploid population derived from a cross between two low-glucosinolate oilseed rape parents identified 23 orthologs of a glucosinolate-associated gene of *Arabidopsis* within QTL intervals (Fu et al. 2015). These candidate genes and linked markers will facilitate breeding for low-glucosinolate rapeseed.

8.9.2 Broccoli

Broccoli has a content of glucoraphanin (4-methylsulphinylbutyl glucosinolate) larger than those of other *Brassicaceae* vegetables (Cartea and Velasco 2007; Kushad et al. 1999). This glucosinolate is converted to the isothiocyanate sulforaphane by myrosinase. Sulforaphane has been shown in many cell culture and animal studies to exert health-promoting effects (Juge et al. 2007). The best known among the effects of sulforaphane is its function as a potent inducer of the phase 2 enzyme (Zhang et al. 1992). In particular, broccoli sprouts contains glucoraphanin, a precursor of sulforaphane, at more than twice the level found in mature broccoli (Fahey et al. 1997). Moreover, the inducer potency per miiligram fresh weight decreases with the age of the plant. Thus, to obtain sufficient sulforaphane from mature broccoli, it is necessary to breed broccoli with high sulforaphane concentration. Faulkner et al. (1998) reported increased glucoraphanin in a hybrid F_1 plant of a cross between a commercial broccoli cultivar and the wild species *B. villosa*. QTL mapping in segregating populations derived from F_1 hybrids led to the identification of a major QTL determining the concentrations of methionine-derived glucosinolates (Mithen et al. 2003). All the high-glucoraphanin hybrids harbored an introgressed *B. villosa* genome containing a *B. villosa Myb28* allele. From these studies, two high-glucoraphanin hybrids have been commercialized as Beneforte broccoli (Traka et al. 2013).

8.9.3 Radish

The white radish, also called Japanese daikon, is an important Brassicaceae root vegetable. White radish generally contains glucoraphasatin (4-methylthio-3-butenyl glucosinolate), one of the abundant aliphatic glucosinolates. Glucoraphasatin is contained predominantly in roots (Carlson et al. 1985) and accounts for more than 90% of the total glucosinolates in Japanese cultivars (Ishida et al. 2012). The breakdown of glucoraphasatin by myrosinase and reaction with water produces a yellow pigment and a methanethiol component that is responsible for the color and flavor of pickles (Ozawa et al. 1990a,b; Takahashi et al. 2015). Because variation in the glucosinolate composition of radish is low, breeding targeting glucosinolate content has not been performed to date. Recently, a mutant lacking glucoraphasatin was identified by screening of glucosinolate profiles of 632 Landraces and commercial cultivars (Ishida et al. 2015). Glucoerucin (4-methylthiobutyl glucosinolate) was the dominant component of the mutant instead of glucoraphasatin and accounted for more than 90% of the total glucosinolates in roots. This glucosinolate composition is similar to that of *B. oleracea* plants (Table 8.3). Genetic analysis revealed that the mutant has a lesion in the gene encoding an enzyme functioning in the conversion of glucoerucin to glucoraphasatin (Ishida et al. 2015). Importantly, pickles using the mutant were not shown; amounts of yellow pigment, methanethiol, were extremely low in the product stored for 12 months (Ishida et al. 2015). Use of this mutant for breeding of radish is expected to generate a new cultivar free of odor and yellow pigmentation.

8.10 Conclusion

Glucosinolates are one of the most well-characterized plant compounds. Glucosinolates and their degradation products have been recognized for their distinctive benefits to human nutrition and plant defense. Moreover, metabolic engineering of the customized glucosinolate profile of Brassicaceae vegetables and crops resulted in the rich food culture and health of the livestock. The availability of the *Arabidopsis* genome sequence has enabled functional genomics approaches and facilitated QTL mapping to identify genes involved in glucosinolate biosynthesis. Accumulation of genomic information in Brassicaceae plants will provide information on the relationship between the structure of glucosinolate-related genes and specific glucosinolate profile of the species.

References

Andersen, T.G., H.H. Nour-Eldin, V.L. Fuller, C.E. Olsen, M. Burow, and B.A. Halkier. 2013. Integration of biosynthesis and long-distance transport establish organ-specific glucosinolate profiles in vegetative *Arabidopsis*. *Plant Cell* 25: 3133–3145.

Andreasson, E., L.B. Jorgensen, A.S. Hoglund, L. Rask, and J. Meijer. 2001. Different myrosinase and idioblast distribution in *Arabidopsis* and *Brassica napus*. *Plant Physiol.* 127: 1750–1763.

Bak, S. and R. Feyereisen. 2001. The involvement of two P450 enzymes, CYP83B1 and CYP83A1, in auxin homeostasis and glucosinolate biosynthesis. *Plant Physiology* 127: 108–118.

Beekwilder, J., W. van Leeuwen, N.M. van Dam et al. 2008. The impact of the absence of aliphatic glucosinolates on insect herbivory in *Arabidopsis*. *Plos One* 3: e2068.

Bernardi, R., M.G. Finiguerra, A.A. Rossi, and S. Palmieri. 2003. Isolation and biochemical characterization of a basic myrosinase from ripe Crambe abyssinica seeds, highly specific for epi-progoitrin. *J. Agric. Food Chem.* 51: 2737–2744.

Bidart-Bouzat, M.G. and D.J. Kliebenstein. 2008. Differential levels of insect herbivory in the field associated with genotypic variation in glucosinolates in *Arabidopsis thaliana*. *J. Chem. Ecol.* 34: 1026–1037.

Bones, A.M. and J.T. Rossiter. 1996. The myrosinase-glucosinolate system, its organisation and biochemistry. *Physiol. Plantarum* 97: 194–208.

Brown, P.D., J.G. Tokuhisa, M. Reichelt, and J. Gershenzon. 2003. Variation of glucosinolate accumulation among different organs and developmental stages of *Arabidopsis thaliana*. *Phytochemistry* 62: 471–481.

Brudenell, A.J.P., H. Griffiths, J.T. Rossiter, and D.A. Baker. 1999. The phloem mobility of glucosinolates. *J. Exp. Bot.* 50: 745–756.

Burow, M., B.A. Halkier, and D.J. Kliebenstein. 2010. Regulatory networks of glucosinolates shape *Arabidopsis thaliana* fitness. *Curr. Opin. Plant Biol.* 13: 348–353.

Burow, M., R. Muller, J. Gershenzon, and U. Wittstock. 2006. Altered glucosinolate hydrolysis in genetically engineered *Arabidopsis thaliana* and its influence on the larval development of *Spodoptera littoralis*. *J. Chem. Ecol.* 32: 2333–2349.

Carlson, D.G., M.E. Daxenbichler, C.H. Vanetten, C.B. Hill, and P.H. Williams. 1985. Glucosinolates in radish cultivars. *J. Am. Soc. Hortic. Sci.* 110: 634–638.

Cartea, M.E. and P. Velasco. 2007. Glucosinolates in Brassica foods: Bioavailability in food and significance for human health. *Phytochem. Rev.* 7: 213–229.

Chan, E.K., H.C. Rowe, and D.J. Kliebenstein. 2010. Understanding the evolution of defense metabolites in *Arabidopsis thaliana* using genome-wide association mapping. *Genetics* 185: 991–1007.

Chen, S. and B.A. Halkier. 1999. Functional expression and characterization of the myrosinase MYR1 from *Brassica napus* in *Saccharomyces cerevisiae*. *Protein Expr. Purif.* 17: 414–420.

Chen, S., B.L. Petersen, C.E. Olsen, A. Schulz, and B.A. Halkier. 2001. Long-distance phloem transport of glucosinolates in *Arabidopsis*. *Plant Physiol.* 127: 194–201.

Chen, S.X., E. Glawischnig, K. Jorgensen et al. 2003. CYP79F1 and CYP79F2 have distinct functions in the biosynthesis of aliphatic glucosinolates in *Arabidopsis*. *Plant J.* 33: 923–937.

Chini, A., S. Fonseca, G. Fernandez et al. 2007. The JAZ family of repressors is the missing link in jasmonate signalling. *Nature* 448: 666–U664.

Clarke, D.B. 2010. Glucosinolates, structures and analysis in food. *Anal Methods-Uk* 2: 310–325.

Dimov, Z., E. Suprianto, F. Hermann, and C. Mollers. 2012. Genetic variation for seed hull and fibre content in a collection of European winter oilseed rape material (*Brassica napus* L.) and development of NIRS calibrations. *Plant Breeding* 131: 361–368.

Dinkova-Kostova, A.T., J.W. Fahey, K.L. Wade et al. 2007. Induction of the phase 2 response in mouse and human skin by sulforaphane-containing broccoli sprout extracts. *Cancer Epidemiol. Biomarkers Prev.* 16: 847–851.

Dinkova-Kostova, A.T., W.D. Holtzclaw, R.N. Cole et al. 2002. Direct evidence that sulfhydryl groups of Keap1 are the sensors regulating induction of phase 2 enzymes that protect against carcinogens and oxidants. *Proc. Natl. Acad. Sci. USA* 99: 11908–11913.

Dombrecht, B., G.P. Xue, S.J. Sprague et al. 2007. MYC2 differentially modulates diverse jasmonate-dependent functions in *Arabidopsis*. *Plant Cell* 19: 2225–2245.

Fahey, J.W., A.T. Zalcmann, and P. Talalay. 2001. The chemical diversity and distribution of glucosinolates and isothiocyanates among plants. *Phytochemistry* 56: 5–51.

Fahey, J.W., Y. Zhang, and P. Talalay. 1997. Broccoli sprouts: An exceptionally rich source of inducers of enzymes that protect against chemical carcinogens. *Proc. Natl. Acad. Sci. USA* 94: 10367–10372.

Faulkner, K., R. Mithen, and G. Williamson, 1998. Selective increase of the potential anticarcinogen 4-methylsulphinylbutyl glucosinolate in broccoli. *Carcinogenesis* 19: 605–609.
Fenwick, G.R., R.K. Heaney, and W.J. Mullin. 1983. Glucosinolates and their breakdown products in food and food plants. *CRC Cr. Rev. Food Sci.* 18: 123–201.
Fernandez-Calvo, P., A. Chini, G. Fernandez-Barbero et al. 2011. The *Arabidopsis* bHLH transcription factors MYC3 and MYC4 are targets of JAZ repressors and act additively with MYC2 in the activation of jasmonate responses. *Plant Cell* 23: 701–715.
Field, B., G. Cardon, M. Traka, J. Botterman, G. Vancanneyt, and R. Mithen. 2004. Glucosinolate and amino acid biosynthesis in *Arabidopsis*. *Plant Physiol.* 135: 828–839.
Fu, Y., K. Lu, L. Qian et al. 2015. Development of genic cleavage markers in association with seed glucosinolate content in canola. *Theor. Appl. Genet.* 128: 1029–1037.
Gachon, C.M., M. Langlois-Meurinne, Y. Henry, and P. Saindrenan. 2005. Transcriptional co-regulation of secondary metabolism enzymes in *Arabidopsis*: Functional and evolutionary implications. *Plant Mol. Biol.* 58: 229–245.
Gao, J., X. Yu, F. Ma, and J. Li. 2014. RNA-seq analysis of transcriptome and glucosinolate metabolism in seeds and sprouts of broccoli (*Brassica oleracea* var.). *PLoS One* 9: e88804.
Gerhauser, C., M. You, J.F. Liu et al. 1997. Cancer chemopreventive potential of sulforamate, a novel analogue of sulforaphane that induces phase 2 drug-metabolizing enzymes. *Cancer Res.* 57: 272–278.
Gigolashvili, T., B. Berger, H.P. Mock, C. Muller, B. Weisshaar, and U.I. Flugge. 2007a. The transcription factor HIG1/MYB51 regulates indolic glucosinolate biosynthesis in *Arabidopsis thaliana*. *Plant J.* 50: 886–901.
Gigolashvili, T., M. Engqvist, R. Yatusevich, C. Muller, and U.I. Flugge. 2008. HAG2/MYB76 and HAG3/MYB29 exert a specific and coordinated control on the regulation of aliphatic glucosinolate biosynthesis in *Arabidopsis thaliana*. *New Phytol.* 177: 627–642.
Gigolashvili, T., R. Yatusevich, B. Berger, C. Muller, and U.I. Flugge. 2007b. The R2R3-MYB transcription factor HAG1/MYB28 is a regulator of methionine-derived glucosinolate biosynthesis in *Arabidopsis thaliana*. *Plant J.* 51: 247–261.
Gigolashvili, T., R. Yatusevich, I. Rollwitz, M. Humphry, J. Gershenzon, and U.I. Flugge. 2009. The plastidic bile acid transporter 5 is required for the biosynthesis of methionine-derived glucosinolates in *Arabidopsis thaliana*. *Plant Cell* 21: 1813–1829.
Godoy, M., Franco-Zorrilla, J. Perez-Perez, J.C. Oliveros, O. Lorenzo, and R. Solano. 2011. Improved protein-binding microarrays for the identification of DNA-binding specificities of transcription factors. *Plant J.* 66: 700–711.
Grubb, C.D., B.J. Zipp, J. Ludwig-Muller, M.N. Masuno, T.F. Molinski, and S. Abel. 2004. *Arabidopsis* glucosyltransferase UGT74B1 functions in glucosinolate biosynthesis and auxin homeostasis. *Plant J.* 40: 893–908.
Halkier, B.A. and J. Gershenzon. 2006. Biology and biochemistry of glucosinolates. *Annu. Rev. Plant Biol.* 57: 303–333.
Hansen, B.G., R.E. Kerwin, J.A. Ober et al. 2008. A novel 2-oxoacid-dependent dioxygenase involved in the formation of the goiterogenic 2-hydroxybut-3-enyl glucosinolate and generalist insect resistance in *Arabidopsis*. *Plant Physiol.* 148: 2096–2108.

Hansen, B.G., D.J. Kliebenstein, and B.A. Halkier. 2007. Identification of a flavin-monooxygenase as the S-oxygenating enzyme in aliphatic glucosinolate biosynthesis in *Arabidopsis*. *Plant J.* 50: 902–910.

Hansen, C.H., L.C. Du, P. Naur et al. 2001a. CYP83B1 is the oxime-metabolizing enzyme in the glucosinolate pathway in *Arabidopsis*. *J. Biol. Chem.* 276: 24790–24796.

Hansen, C.H., U. Wittstock, C.E. Olsen, A.J. Hick, J.A. Pickett, and B.A. Halkier. 2001b. Cytochrome p450 CYP79F1 from *arabidopsis* catalyzes the conversion of dihomomethionine and trihomomethionine to the corresponding aldoximes in the biosynthesis of aliphatic glucosinolates. *J. Biol. Chem.* 276: 11078–11085.

Hasan, M., W. Friedt, J. Pons-Kuhnemann, N.M. Freitag, K. Link, and R.J. Snowdon. 2008. Association of gene-linked SSR markers to seed glucosinolate content in oilseed rape (*Brassica napus* ssp. *napus*). *Theor. Appl. Genet.* 116: 1035–1049.

Hemm, M.R., M.O. Ruegger, and C. Chapple. 2003. The *Arabidopsis ref2* mutant is defective in the gene encoding CYP83A1 and shows both phenylpropanoid and glucosinolate phenotypes. *Plant Cell* 15: 179–194.

Hirai, M.Y., T. Fujiwara, M. Awazuhara, T. Kimura, M. Noji, and K. Saito. 2003. Global expression profiling of sulfur-starved *Arabidopsis* by DNA macroarray reveals the role of *O*-acetyl-L-serine as a general regulator of gene expression in response to sulfur nutrition. *Plant J.* 33: 651–663.

Hirai, M.Y., K. Sugiyama, Y. Sawada et al. 2007. Omics-based identification of *Arabidopsis* Myb transcription factors regulating aliphatic glucosinolate biosynthesis. *Proc. Natl. Acad. Sci. USA* 104: 6478–6483.

Husebye, H., S. Chadchawan, P. Winge, O.P. Thangstad, and A.M. Bones. 2002. Guard cell- and phloem idioblast specific expression of thioglucoside glucohydrolase 1 (myrosinase) in *Arabidopsis*. *Plant Physiol.* 128: 1180–1188.

Ishida, M., M. Hara, N. Fukino, T. Kakizaki, and T. Morimitsu. 2014. Glucosinolate metabolism, functionality and breeding for the improvement of Brassicaceae vegetables. *Breed Sci* 64: 48–59.

Ishida, M., T. Kakizaki, Y. Morimitsu et al. 2015. Novel glucosinolate composition lacking 4-methylthio-3-butenyl glucosinolate in Japanese white radish (*Raphanus sativus* L.). *Theor. Appl. Genet.* 128: 2037–2046.

Ishida, M., M. Nagata, T. Ohara, T. Kakizaki, K. Hatakeyama, and T. Nishio. 2012. Small variation of glucosinolate composition in Japanese cultivars of radish (*Raphanus sativus* L.) requires simple quantitative analysis for breeding of glucosinolate component. *Breed. Sci.* 62: 63–70.

Juge, N., R.F. Mithen, and M. Traka. 2007. Molecular basis for chemoprevention by sulforaphane: A comprehensive review. *Cell. Mol. Life Sci.* 64: 1105–1127.

Kim, D.H., M. Kim, and H.J. Kwon. 2003. Histone deacetylase in carcinogenesis and its inhibitors as anti-cancer agents. *J. Biochem. Mol. Biol.* 36: 110–119.

Kitashiba, H., F. Li, H. Hirakawa et al. 2014. Draft sequences of the radish (*Raphanus sativus* L.) genome. *DNA Res.* 21: 481–490.

Kliebenstein, D.J., J. Kroymann, P. Brown et al. 2001a. Genetic control of natural variation in *Arabidopsis* glucosinolate accumulation. *Plant Physiol.* 126: 811–825.

Kliebenstein, D.J., V.M. Lambrix, M. Reichelt, J. Gershenzon, and T. Mitchell-Olds. 2001b. Gene duplication in the diversification of secondary metabolism: Tandem 2-oxoglutarate-dependent dioxygenases control glucosinolate biosynthesis in *Arabidopsis*. *Plant Cell* 13: 681–693.

Knill, T., M. Reichelt, C. Paetz, J. Gershenzon, and S. Binder. 2009. *Arabidopsis thaliana* encodes a bacterial-type heterodimeric isopropylmalate isomerase involved in both Leu biosynthesis and the Met chain elongation pathway of glucosinolate formation. *Plant Mol. Biol.* 71: 227–239.

Knill, T., J. Schuster, M. Reichelt, J. Gershenzon, and S. Binder. 2008. *Arabidopsis* branched-chain aminotransferase 3 functions in both amino acid and glucosinolate biosynthesis. *Plant Physiol.* 146: 1028–1039.

Koroleva, O.A., T.M. Gibson, R. Cramer, and C. Stain. 2010. Glucosinolate-accumulating S-cells in *Arabidopsis* leaves and flower stalks undergo programmed cell death at early stages of differentiation. *Plant J.* 64: 456–469.

Kushad, M.M., A.F. Brown, A.C. Kurilich et al. 1999. Variation of glucosinolates in vegetable crops of *Brassica oleracea*. *J. Agri. Food Chem.* 47: 1541–1548.

Lankau, R.A. 2007. Specialist and generalist herbivores exert opposing selection on a chemical defense. *New Phytol.* 175: 176–184.

Lankau, R.A. and S.Y. Strauss. 2007. Mutual feedbacks maintain both genetic and species diversity in a plant community. *Science* 317: 1561–1563.

Li, F., B. Chen, K. Xu et al. 2014. Genome-wide association study dissects the genetic architecture of seed weight and seed quality in rapeseed (*Brassica napus* L.). *DNA Res.* 21: 355–367.

Li, J., B.G. Hansen, J.A. Ober, D.J. Kliebenstein, and B.A. Halkier. 2008. Subclade of flavin-monooxygenases involved in aliphatic glucosinolate biosynthesis. *Plant Physiol.* 148: 1721–1733.

Li, Y., Y. Sawada, A. Hirai et al. 2013. Novel insights into the function of *Arabidopsis* R2R3-MYB transcription factors regulating aliphatic glucosinolate biosynthesis. *Plant Cell Physiol.* 54: 1335–1344.

Li, Y.Y., T. Zhang, H. Korkaya et al. 2010. Sulforaphane, a dietary component of broccoli/broccoli sprouts, inhibits breast cancer stem cells. *Clin. Cancer Res.* 16: 2580–2590.

Liu, S., Y. Liu, X. Yang et al. 2014. The *Brassica oleracea* genome reveals the asymmetrical evolution of polyploid genomes. *Nature Commun.* 5: 3930.

Madsen, S.R., C.E. Olsen, H.H. Nour-Eldin, and B.A. Halkier. 2014. Elucidating the role of transport processes in leaf glucosinolate distribution. *Plant Physiol.* 166: 1450–1462.

Malitsky, S., E. Blum, H. Less et al. 2008. The transcript and metabolite networks affected by the two clades of *Arabidopsis* glucosinolate biosynthesis regulators. *Plant Physiol.* 148: 2021–2049.

Maruyama-Nakashita, A., E. Inoue, A. Watanabe-Takahashi, T. Yamaya, and H. Takahashi. 2003. Transcriptome profiling of sulfur-responsive genes in *Arabidopsis* reveals global effects of sulfur nutrition on multiple metabolic pathways. *Plant Physiol.* 132: 597–605.

Maruyama-Nakashita, A., Y. Nakamura, T. Tohge, K. Saito, and H. Takahashi. 2006. *Arabidopsis* SLIM1 is a central transcriptional regulator of plant sulfur response and metabolism. *Plant Cell* 18: 3235–3251.

Maruyama-Nakashita, A., Y. Nakamura, A. Watanabe-Takahashi, E. Inoue, T. Yamaya, and H. Takahashi. 2005. Identification of a novel cis-acting element conferring sulfur deficiency response in *Arabidopsis* roots. *Plant J.* 42: 305–314.

Mckey, D. 1974. Adaptive Patterns in Alkaloid Physiology. *Am. Nat.* 108: 305–320.

Mewis, I., J.G. Tokuhisa, J.C. Schultz, H.M. Appel, C. Ulrichs, and J. Gershenzon. 2006. Gene expression and glucosinolate accumulation in *Arabidopsis thaliana* in response to generalist and specialist herbivores of different feeding guilds and the role of defense signaling pathways. *Phytochemistry* 67: 2450–2462.

Mikkelsen, M.D., P. Naur, and B.A. Halkier. 2004. *Arabidopsis* mutants in the C-S lyase of glucosinolate biosynthesis establish a critical role for indole-3-acetaldoxime in auxin homeostasis. *Plant J.* 37: 770–777.

Mikkelsen, M.D., B.L. Petersen, E. Glawischnig, A.B. Jensen, E. Andreasson, and B.A. Halkier. 2003. Modulation of CYP79 genes and glucosinolate profiles in *Arabidopsis* by defense signaling pathways. *Plant Physiol.* 131: 298–308.

Mithen, R., K. Faulkner, R. Magrath, P. Rose, G. Williamson, and J. Marquez. 2003. Development of isothiocyanate-enriched broccoli, and its enhanced ability to induce phase 2 detoxification enzymes in mammalian cells. *Theor. Appl. Genet.* 106: 727–734.

Mitsui, Y., M. Shimomura, K. Komatsu et al. 2015. The radish genome and comprehensive gene expression profile of tuberous root formation and development. *Sci. Rep.* 5: 10835.

Mizushima, U. 1980. *Genome Analysis in Brassica and Allied Genera.* Tokyo: Japan Sci. Soc. Press.

Mumm, R., M. Burow, G. Bukovinszkinékiss et al. 2008. Formation of simple nitriles upon glucosinolate hydrolysis affects direct and indirect defense against the specialist herbivore, *Pieris rapae. J. Chem. Ecol.* 34: 1311–1321.

Myzak, M.C., P.A. Karplus, F.L. Chung, and R.H. Dashwood. 2004. A novel mechanism of chemoprotection by sulforaphane: Inhibition of histone deacetylase. *Cancer Res.* 64: 5767–5774.

Nagaharu, U. 1935. Genome analysis in *Brassica* with special reference to the experimental formation of *B. napus* and peculiar mode of fertilization. *Jpn. J. Bot.* 7: 389–452.

Nikiforova, V., J. Freitag, S. Kempa, M. Adamik, H. Hesse, and R. Hoefgen. 2003. Transcriptome analysis of sulfur depletion in *Arabidopsis thaliana*: Interlacing of biosynthetic pathways provides response specificity. *Plant J.* 33: 633–650.

Nour-Eldin, H.H., T.G. Andersen, M. Burow et al. 2012. NRT/PTR transporters are essential for translocation of glucosinolate defence compounds to seeds. *Nature* 488: 531–534.

Nour-Eldin, H.H., M.H. Norholm, and B.A. Halkier. 2006. Screening for plant transporter function by expressing a normalized *Arabidopsis* full-length cDNA library in *Xenopus oocytes*. *Plant Methods* 2: 17.

Ozawa, Y., S. Kawakishi, Y. Uda, and Y. Maeda. 1990a. Isolation and identification of a novel *beta*-carboline derivative in salted radish roots, Raphanus sativus L. *Agr. Biol. Chem.* 54: 1241–1245.

Ozawa, Y., Y. Uda, and S. Kawakishi. 1990b. Generation of a beta-carboline derivative, the yellowish precursor of processed radish roots, from 4-methylthio-3-butenyl isothiocyanate and L-tryptophan. *Agr. Biol. Chem.* 54: 1849–1851.

Petersen, B.L., S. Chen, C.H. Hansen, C.E. Olsen, and B.A. Halkier. 2002. Composition and content of glucosinolates in developing *Arabidopsis thaliana*. *Planta* 214: 562–571.

Pino Del Carpio, D., R.K. Basnet, D. Arends et al. 2014. Regulatory network of secondary metabolism in Brassica rapa: Insight into the glucosinolate pathway. *PLoS One* 9: e107123.

Piotrowski, M., A. Schemenewitz, A. Lopukhina et al. 2004. Desulfoglucosinolate sulfotransferases from *Arabidopsis thaliana* catalyze the final step in the biosynthesis of the glucosinolate core structure. *J. Biol. Chem.* 279: 50717–50725.

Poelman, E.H., C. Broekgaarden, J.J.A. Van Loon, and M. Dicke. 2008. Early season herbivore differentially affects plant defence responses to subsequently colonizing herbivores and their abundance in the field. *Mol. Ecol.* 17: 3352–3365.

Rajendran, P., A.I. Kidane, T.-W. Yu et al. 2013. HDAC turnover, CtIP acetylation and dysregulated DNA damage signaling in colon cancer cells treated with sulforaphane and related dietary isothiocyanates. *Epigenetics* 8: 612–623.

Reintanz, B., M. Lehnen, M. Reichelt et al. 2001. *bus*, a bushy *Arabidopsis CYP79F1* knockout mutant with abolished synthesis of short-chain aliphatic glucosinolates. *Plant Cell* 13: 351–367.

Sawada, Y., A. Kuwahara, M. Nagano et al. 2009a. Omics-based approaches to methionine side chain elongation in *Arabidopsis*: Characterization of the genes encoding methylthioalkylmalate isomerase and methylthioalkylmalate dehydrogenase. *Plant Cell Physiol.* 50: 1181–1190.

Sawada, Y., K. Toyooka, A. Kuwahara et al. 2009b. *Arabidopsis* bile acid: Sodium symporter family protein 5 is involved in methionine-derived glucosinolate biosynthesis. *Plant Cell Physiol.* 50: 1579–1586.

Schuster, J., T. Knill, M. Reichelt, J. Gershenzon, and S. Binder. 2006. BRANCHED-CHAIN AMINOTRANSFERASE4 is part of the chain elongation pathway in the biosynthesis of methionine-derived glucosinolates in *Arabidopsis*. *Plant Cell* 18: 2664–2679.

Schweizer, F., P. Fernandez-Calvo, M. Zander et al. 2013. *Arabidopsis* basic helix-loop-helix transcription factors MYC2, MYC3, and MYC4 regulate glucosinolate biosynthesis, insect performance, and feeding behavior. *Plant Cell* 25: 3117–3132.

Shroff, R., F. Vergara, A. Muck, A. Svatos, and J. Gershenzon. 2008. Nonuniform distribution of glucosinolates in *Arabidopsis thaliana* leaves has important consequences for plant defense. *Proc. Natl. Acad. Sci. USA* 105: 6196–6201.

Siemens, D.H. and T. MitchellOlds. 1996. Glucosinolates and herbivory by specialists (Coleoptera: Chrysomelidae, Lepidoptera: Plutellidae): Consequences of concentration and induced resistance. *Environ. Entomol.* 25: 1344–1353.

Skirycz, A., M. Reichelt, M. Burow et al. 2006. DOF transcription factor AtDof1.1 (OBP2) is part of a regulatory network controlling glucosinolate biosynthesis in *Arabidopsis*. *Plant J.* 47: 10–24.

Sonderby, I.E., M. Burow, H.C. Rowe, D.J. Kliebenstein, and B.A. Halkier. 2010a. A complex interplay of three R2R3 MYB transcription factors determines the profile of aliphatic glucosinolates in *Arabidopsis*. *Plant Physiol.* 153: 348–363.

Sonderby, I.E., F. Geu-Flores, and B.A. Halkier. 2010b. Biosynthesis of glucosinolates—Gene discovery and beyond. *Trends Plant Sci.* 15: 283–290.

Struhl, K. 1998. Histone acetylation and transcriptional regulatory mechanisms. *Genes Dev.* 12: 599–606.

Takahashi, A., T. Yamada, Y. Uchiyama et al. 2015. Generation of the antioxidant yellow pigment derived from 4-methylthio-3-butenyl isothiocyanate in salted radish roots (*takuan-zuke*). *Biosci. Biotechnol. Biochem.* 79: 1512–1517.

Thines, B., L. Katsir, M. Melotto et al. 2007. JAZ repressor proteins are targets of the SCF(COI1) complex during jasmonate signalling. *Nature* 448: 661–665.

Traka, M.H., S. Saha, S. Huseby et al. 2013. Genetic regulation of glucoraphanin accumulation in Beneforte broccoli. *New Phytol.* 198: 1085–1095.

Vandam, N.M., L.W.M. Vuister, C. Bergshoeff, H. Devos, and E. Vandermeijden. 1995. The "Raison D'être" of pyrrolizidine alkaloids in *Cynoglossum officinale*: Deterrent effects against generalist herbivores. *J. Chem. Ecol.* 21: 507–523.

Walker, K.C. and E.J. Booth. 2001. Agricultural aspects of rape and other Brassica products. *Eur. J. Lipid Sci. Tech.* 103: 441–446.
Wanasundara, J.P. 2011. Proteins of Brassicaceae oilseeds and their potential as a plant protein source. *Crit. Rev. Food Sci. Nutr.* 51: 635–677.
Wang, X., H. Wang, J. Wang et al. 2011. The genome of the mesopolyploid crop species Brassica rapa. *Nat. Genet.* 43: 1035–1039.
Wang, Y., Y. Pan, Z. Liu et al. 2013. De novo transcriptome sequencing of radish (*Raphanus sativus* L.) and analysis of major genes involved in glucosinolate metabolism. *BMC Genom.* 14: 836.
Watai, Y., A. Kobayashi, H. Nagase et al. 2007. Subcellular localization and cytoplasmic complex status of endogenous Keap1. *Genes Cells* 12: 1163–1178.
Watanabe, M., S. Balazadeh, T. Tohge et al. 2013. Comprehensive dissection of spatiotemporal metabolic shifts in primary, secondary, and lipid metabolism during developmental senescence in *Arabidopsis. Plant Physiol.* 162: 1290–1310.
Wentzell, A.M., H.C. Rowe, B.G. Hansen, C. Ticconi, B.A. Halkier, and D.J. Kliebenstein. 2007. Linking metabolic QTLs with network and *cis*-eQTLs controlling biosynthetic pathways. *PLoS Genet.* 3: 1687–1701.
Wittstock, U. and B.A. Halkier. 2002. Glucosinolate research in the *Arabidopsis* era. *Trends Plant Sci.* 7: 263–270.
Yanaka, A., J.W. Fahey, A. Fukumoto et al. 2009. Dietary sulforaphane-rich broccoli sprouts reduce colonization and attenuate gastritis in *Helicobacter pylori*-infected mice and humans. *Cancer Prev. Res.* 2: 353–360.
Zhang, Y. and E.C. Callaway. 2002. High cellular accumulation of sulphoraphane, a dietary anticarcinogen, is followed by rapid transporter-mediated export as a glutathione conjugate. *Biochem. J.* 364: 301–307.
Zhang, Y., R. Munday, H.E. Jobson et al. 2006. Induction of GST and NQO1 in cultured bladder cells and in the urinary bladders of rats by an extract of broccoli (*Brassica oleracea italica*) sprouts. *J. Agric. Food Chem.* 54: 9370–9376.
Zhang, Y.S., T.W. Kensler, C.G. Cho, G.H. Posner, and P. Talalay. 1994. Anticarcinogenic activities of sulforaphane and structurally related synthetic norbornyl isothiocyanates. *Proc. Natl. Acad. Sci. USA* 91: 3147–3150.
Zhang, Y.S., P. Talalay, C.G. Cho, and G.H. Posner. 1992. A major inducer of anticarcinogenic protective enzymes from broccoli: Isolation and elucidation of structure. *Proc. Natl. Acad. Sci. USA* 89: 2399–2403.

chapter nine

Biosynthesis and regulation of plant volatiles and their functional roles in ecosystem interactions and global environmental changes

Gen-ichiro Arimura, Kenji Matsui, Takao Koeduka, and Jarmo K. Holopainen

Contents

9.1	Overview of plant volatiles	186
9.2	Biosynthesis, regulation, and ecological relevance of PVOCs	188
	9.2.1 Terpenes	188
	9.2.1.1 Biosynthesis	188
	9.2.1.2 Regulation	191
	9.2.2 GLVs	194
	9.2.2.1 What are GLVs?	194
	9.2.2.2 Biosynthesis	196
	9.2.2.3 Regulation of GLV formation	198
	9.2.2.4 Physiological and ecological significance	200
	9.2.3 Benzenoid/phenylpropanoid	202
	9.2.3.1 Biosynthesis	202
	9.2.3.2 Regulation	206
9.3	Ecological communications via PVOCs	208
	9.3.1 Plants versus pollinators	208
	9.3.2 Plants versus herbivores	209
	9.3.3 Plants versus herbivore enemies	209
	9.3.4 Plants versus plants	211
9.4	VOCs in global environmental change	212
	9.4.1 Temperature	213
	9.4.2 Droughts and flooding	214

9.4.3 Wind 215
9.4.4 UV radiation 215
9.4.5 Light intensity and quality 216
9.4.6 CO_2 217
9.4.7 Ozone 217
9.4.8 Biotic interactions and ecological functions of VOCs under environmental change 218
References 221

9.1 Overview of plant volatiles

Plant aromas (i.e., volatile organic chemicals, or VOCs, released from plants) are indispensable products that function in mutualism, the environmental stress tolerance, and pest control of plants. Hundreds of millions of tons of VOCs are expelled into the atmosphere annually by living plants (Guenther et al. 1995). As is widely known, "blue haze" is a visual effect resulting from the chemical reactions of plant-derived VOCs (often called PVOCs) generated in the forest atmosphere. Various chemical reactions of PVOCs with hydroxyl radicals and ozone in the atmosphere form aerosol, leading to the nuclei of the cloud. Overall, grasping the global effect of PVOCs on ecosystem and atmosphere conditions is beyond the intelligence of most of us.

Emission of PVOCs is specific to species, cultivar/genotype, organ, environment, and occasion. What is the nature of PVOCs for emitter plants? First, PVOCs are needed for wasting excess carbon in the plant body and/or adjusting their health conditions to adapt to rigorous environmental changes, as homeotherms likely do so by sweating. Otherwise, immobile plants prefer to use volatiles that can release chemicals far away to communicate with organisms including insects, microorganisms, and conspecific plants. In fact, in ancient times, hydrophobic aroma compounds developed in brown algae as pheromones that facilitated the mating process of most brown algae during sexual reproduction (Pohnert and Boland 2002). One of the world's most ancient terrestrial plants, the cosmopolitan moss *Bryum argenteum*, also makes use of a sex-specific blend of PVOCs to attract moss-dwelling microarthropods for sexual reproduction, similarly to the use observed in a scent-based vascular plant-pollinator relationship (Rosenstiel et al. 2012). Nowadays, individual PVOCs and/or combinations of their blend, released from most parts of the plant taxa, special organs (e.g., floral tissues), and circumstances (e.g., stressed foliage and underground roots) provide a broad perspective of ecological communication as it occurs in real time in nature.

In this chapter, we present the biosynthesis, regulation, and evolution of the major class of PVOCs (terpenes, green leaf volatiles [GLVs], and

Chapter nine: Biosynthesis and regulation of plant volatiles

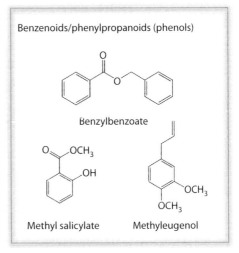

Figure 9.1 Chemical structures of typical examples of volatile terpenes, green leaf volatiles and benzenoids/phenylpropanoids. (TMTT, (*E*,*E*)-4,8,12-trimethyltrideca-1,3,7,11-tetraene.)

phenylpropanoids) in terrestrial plants (Figure 9.1). Central to our theme is the concept that plants are able to induce PVOCs in response to biotic and abiotic stresses; these, in turn, manipulate the behavior of pollinators and members of food webs at the multitrophic level (e.g., herbivores, natural enemies of herbivores) as well as communications with neighboring plants. In this chapter, we therefore explore and discuss how PVOCs are generated and why they matter for the ecological and physiological significance of plants.

9.2 Biosynthesis, regulation, and ecological relevance of PVOCs

9.2.1 Terpenes

9.2.1.1 Biosynthesis

Terpenes (terpenoids or isoprenoids), derived from a common five-carbon building block (isoprene unit), are some of the most structurally diverse natural products. There are two biosynthetic pathways for terpenes: that is, the mevalonate (MVA) pathway in the cytoplasm and the 2-C-methyl-D-erythritol 4-phosphate (MEP) pathway in the plastids. The cytoplasmic formation of sesquiterpenes (C15) relies on isopentenyl diphosphate (IDP) and its isomer dimethylallyl diphosphate, derived specifically from the MVA pathway. In contrast, the plastidial formation of monoterpenes (C10) and diterpenes (C20) relies on IDP derived from the plastidial MEP pathway (Figure 9.2). However, the two pathways are very likely to engage in

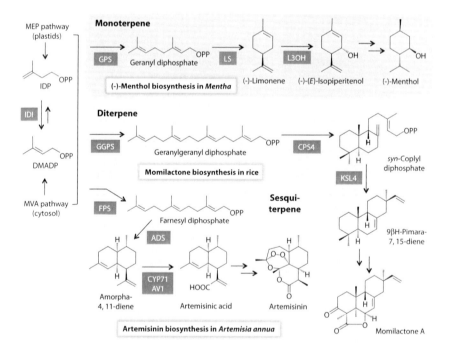

Figure 9.2 Biosynthetic pathway for typical mono-, sesqui-, di-terpenes. (ADS, amorpha-4,11-diene synthase; CPS4, copalyl diphosphate synthase 4; FPS, farnesyl diphosphate synthase; GGPS, geranylgeranyl diphosphate synthase; GPS, geranyl diphosphate synthase; IDI, isopentenyl diphosphate isomerase; KSL4, kaurene synthase-like 4; L3OH, (-)-limonene-3-hydroxylase; LS, (-)-limonene synthase.)

cross-talk by allowing IDP to be shuttled between different compartments (Bartram et al. 2006; Bick and Lange 2003; Piel et al. 1998). It appears that ~80% of IDP from the MEP pathway contributes substantially to sesquiterpene and C11 homoterpene biosynthesis in legume plants in response to herbivory (Arimura et al. 2008a; Bartram et al. 2006), indicating that the MEP pathway also contributes to sesquiterpene biosynthesis. In the case of the floral aroma of snapdragon, only plastidic MEP pathway contributes to the biosynthesis of both (the monoterpene [myrcene] and the sesquiterpene [nerolidol]) (Dudareva et al. 2005). In contrast, Wu et al. (2006) have shown that sesquiterpene biosynthesis occurs predominantly (~80%–90%) in the targeted cytoplasmic (MVA) and plastidic (MEP) pathways, by localizing both patchoulol synthase and farnesyl diphosphate (FDP) synthase to cytoplasmic and plastidic compartments, respectively, in transgenic tobacco (*Nicotiana tabacum*) plants. Curiously, the plastidic formation of sesquiterpenes was artificially manipulated by a chloroplast-targeting signal sequence, and this intracellular targeting resulted in more abundant generation of patchulol (5–30 µg/gram fresh weight), compared to the cases of simply targeting sesquiterpene synthase to the cytosol (average 0.5 µg/gram fresh weight). Thus, the plastidic (MEP) compartment is likely able to manage efficient carbon flux much more efficiently than the cytoplasmic (MVA) compartment in plants. After all, patchoulol biosynthesis occurred predominantly (80%–90%) in the targeted cytosolic or plastidic compartments, and in both cases, with minimal exchange of labeled intermediates between the cellular compartments (Wu et al. 2006).

Given the above conserved pathways, the structural diversity of terpenes is predominantly achieved with the catalysis of a terpene synthase (TPS) family, which uses the different prenyl diphosphates as substrates to synthesize monoterpenes from geranyl diphosphate (GDP), sesquiterpenes from FDP, and diterpenes from geranylgeranyl diphosphate (GGDP). The diversification of a large family of TPSs is believed to originate from repeated duplication and the functional divergence of an ancestral TPS that is involved in plant primary metabolism, for example, hormone biosynthesis (Trapp and Croteau 2001). Evolution has left angiosperms and gymnosperms with a few dozen putative TPS genes in their genomes (Aubourg et al. 2002; Falara et al. 2011; Keeling et al. 2011; Liu et al. 2014), resulting in the structurally different composition of terpenes within or among species: nowadays, at least 40,000 terpene products exist in nature, and many of them are of plant origin (Bohlmann and Keeling 2008). TPSs enable the ionization of the diphosphate group of the substrate to be catalyzed (GDP, FDP, or GGDP) to yield a carbocation that undergoes various rearrangements, with a divalent metal ion (Mg^{2+} or Mn^{2+}) as a cofactor. Three prenyl diphosphates—GDP, FDP, and GGDP—as substrate are specifically accepted by monoterpene synthase (mono-TPS), sesquiterpene synthase (sesqui-TPS), and diterpene synthase (di-TPS), respectively;

notably, there are some TPSs that have the substrate availability with low specificity. For instance, the TPS-g subfamily often converts two or all three sets of prenyl diphosphate substrates to produce linalool (monoterpene), (*E*)-nerolidol (sesquiterpene), and (*E*,*E*)-geranyllinalool (diterpene) from GDP, FDP, and GGDP, respectively (Brillada et al. 2013; Tholl et al. 2011). Phylogenetic analysis based on amino acid sequences has grouped the plant TPSs into seven subfamilies (TPS-a to TPS-g), and most TPSs, except TPS-g, are divided into one of three functional classes, that is, mono-, sesqui-, and di-TPSs (Bohlmann et al. 1998).

Mono-TPSs and di-TPSs are found in the plastid, whereas sesqui-TPSs are found in the cytosol/endoplasmic reticulum (ER). However, there are some exceptions. In the case of Arabidopsis (*Arabidopsis thaliana*) (*E*,*E*)-geranyllinalool synthase, this protein does not target the chloroplast but does target the cytosol/ER (Herde et al. 2008). In another case, when FaNES1, a cytosolic strawberry linalool/nerolidol synthase, is allocated to mitochondria (known as a component of the FDP-derived ubiquinone biosynthesis) in transgenic Arabidopsis plants, unexpectedly, the production of (*E*)-nerolidol and its degradation product (*E*)-4,8-dimethyl-1,3,7-nonatriene (DMNT) are observed (Kappers et al. 2005). Moreover, notably, geraniol, the predominant product of the rose scent, is synthesized not by TPSs but by the Nudix hydrolase RhNUDX1, serving GDP diphosphohydrolase activity in the cytoplasm (Magnard et al. 2015). The discovery of this novel route provided new insight into terpene biosynthesis as the first known case of TPS-independent terpene biosynthesis.

Remember that in our earlier description, "terpene structural diversity is profoundly linked to the functional contribution of TPSs." The TPS enzymes frequently have their catalysts to form not only single but also multiple products. Almost half of all mono-TPSs and sesqui-TPSs generate multiple products, as proven via *in vitro* enzyme assays (Degenhardt et al. 2009). For instance, the recombinant protein sesqui-TPS (MtTPS5) from *Medicago truncatula* enables the plant to produce as many as 27 different products, including cubebol as the major product (38.6% of total), from FDP as a substrate (Garms et al. 2010). However, the product profile from *in vitro* TPS activity is occasionally different from the product profile *in planta*, probably due to the variable protein structure from posttranslational modifications or various enzyme environments (e.g., pH and metal ions) (Fischer et al. 2012; Köllner et al. 2008). Accordingly, the functional properties of a single TPS enzyme depend not only on the enzyme's amino acid sequence but also on the variable, physiological background of those enzymes; the result is a wide range of products found in plants.

In some cases, cytochrome P450 mono-oxygenases (CYP), reductases, dehydrogenases, and transferaseare contribute to the structural diversity of terpene products derived by TPS catabolism. The biosynthesis of

(-)-menthol, a characteristic component of the essential oil of peppermint (*Mentha* x *piperita*), has been intensively studied as a model for the terpene-modifying process (Figure 9.2). The biosynthesis of (-)-menthol is committed by eight enzymatic steps, initially converted from GDP to the monoterpene (-)-limonene by TPS. Following the hydroxylation at C3 of (-)-limonene, a sequence of four redox transformations (one oxidative, three reductive) and an isomerization occur in an "allylic oxidation-conjugate reduction" scheme that installs three chiral centers on the substituted cyclohexanoid ring to form (-)-menthol (Croteau et al. 2005).

In another case, CYP82G1 is able to degrade from the diterpene (*E,E*)-geranyllinalool to the C16 homoterpene volatile (*E,E*)-4,8,12-trimethyltrideca-1,3,7,11-tetraene (TMTT) (Lee et al. 2010). This CYP undergoes similar catalytic activity when it converts from the sesquiterpene alcohol [(*E*)-nerolidol] to the C11 homoterpene [DMNT]. Notably, this homoterpene is also produced by the degradation of the C30 triterpene diol, arabidiol, via the catalysis of the Brassicaceae-specific CYP705A1 in Arabidopsis roots, indicating that there are the two metabolic routes in homoterpene biosynthesis; these routes are differentiated in different tissues of shoots and roots (Sohrabi et al. 2015).

In the context of the terpene-modifying process, diterpene phytoalexines in monocots are especially noteworthy. In rice (*Oryza sativa*), two di-TPSs (copalyl diphosphate synthase) are involved in the first cyclization from GGDP to *ent*- and *syn*-forms of copalyl diphosphate, and the sequential cyclizations are catalyzed by four kaurene synthase-like (KSL) proteins to form diterpene hydrocarbons: *ent*-cassa-12,15-diene, *ent*-sandaracopimaradiene, stemar-13-ene, and 9βH-pimara-7,15-diene. Thanks to the efforts of CYPs and dehydrogenase, four types of phytoalexines (phytocassanes A–E, oryzalexins A–F, oryzalein S, and momilactones A and B) are produced (Figure 9.2) (Miyamoto et al. 2014). All the genes involved in the biosynthetic pathway for momilactone and phytocassane are intensively clustered on the rice chromosome 4, implying that there are sophisticated transcriptional activation systems for the terpene modification process. Similar gene clusters are also observed in the Arabidopsis genomes where six TPSs are arranged in tandem with genes for CYP and glucosyltransferase (Aubourg et al. 2002). The same holds true for the terpene biosynthesis genes that exist in the chromosome 8 cluster in tomato (*Solanum lycopersicum*), as also observed in other Solanaceae species (Matsuba et al. 2013). Altogether, these findings suggest an array of the clustered genes committed for terpene metabolism.

9.2.1.2 Regulation

All aspects of the biosynthesis and emission of terpene compounds, spatial and temporal, are controlled. As is known, floral scents are typical, due to their emissions with diurnal rhythms and developmental-stage and floral-part dependent manners. Three *Lithophragma* species (Saxifragaceae) have been shown to emit species-specific floral scent profiles during the day, probably to attract their day-flying, major pollinator [the floral parasitic moth *Greya politella* (Prodoxidae)] (Friberg et al. 2014). Especially when the divergence of their petal colors is small, local specialization in the pollinator response highly depends on floral scents (Friberg et al. 2014). In turn, (E)-β-ocimene, the major floral scent of *Mirabilis jalapa*, is emitted from the petaloid lobes but not stomata and trichomes on the abaxial flower surface (Effmert et al. 2005). This release occurs in a time-dependent manner, namely, mostly during the evening (5–8 p.m.) (Effmert et al. 2005): this maximum peak of release coincides with the flower opening and activity of crepuscular pollinators, hawk moths such as *Erinnyis ello* and *Hyles lineata* (Martinez del Rio and Búrquez 1986).

In turn, glandular trichomes (oil glands) on the leaf surfaces of the plant are characteristic of sites for terpene biosynthesis and storage. In *Mentha* species, biosynthesis and storage of the (-)-menthol are restricted to the peltate glandular trichomes (oil glands) (Croteau et al. 2005). Likewise, in tomato (*S. lycopersicum*), a linalool synthase gene (*SlTPS5*) appears very locally expressed in glandular trichomes: it has been shown that the transcriptional regulation of this TPS is controlled by a 207 bp fragment containing the minimal promoter and the 5′UTR as well as a glandular trichome–specific EOT1 transcription factor (Spyropoulou et al. 2014). It is also noteworthy to introduce artemisinin, an antimalarial drug derived from the sweet wormwood plant, *Artemisia annua*. Artemisinin, a sesquiterpene lactone, is distilled from the dried leaves (the substance is found mainly in trichomes) or flower clusters of *A. annua*. AsORA, a trichome-specific transcription factor of *A. annua* APETALA2/ethylene-response factor (AP2/ERF) regulates artemisinin biosynthetic genes, amorpha-4,11-diene synthase (ADS) and CYP71AV1 in glandular secretory trichomes and nonglandular T-shaped trichomes of the plants (Lu et al. 2013). It is thus highly likely that trichome-specific terpene biosynthesis is predominantly regulated by transcriptional machinery based on trichome-specific transcription factor(s).

Unlike the constitutive emission of floral and trichome-specific terpenes, plants are able to induce volatiles in response to biotic and abiotic stresses. For instance, plants recruit their bodyguards (predators and parasitoids) by emitting specific blends of volatiles that are induced in response to arthropod herbivory (so-called herbivore-induced plant volatiles: HIPVs) (Arimura et al. 2009). The quality and quantity of HIPVs are very variable among plant species (even among genotype and cultivar) as well as invading herbivore species. For instance, when three different plant species (i.e., tobacco, cotton, and maize) are attacked by two closely related herbivore species (*Heliothis virescens* and *Helicoverpa zea*), plants emit distinct blends of HIPVs, thereby leading the specialist parasitic wasp *Cardiochiles nigriceps* of *H. virescens* to exploit their host *H. virescens* more efficiently than their nonhost *H. zea* (De Moraes et al. 1998).

The physical impact of wounding caused by the foraging performance of herbivores should play a central role in HIPV induction in host plants. Herbivore feeding guilds, in particular—either chewing caterpillars and beetles or sucking aphids and spider mites—manipulate the quantity and quality of HIPVs (Rowen and Kaplan 2016). It is also noteworthy that continuous mechanical damage by herbivores is important to induce certain substance of HIPVs, rather than the mechanical damage that is frequently referred to as plant damage treatment, as the MecWorm device has shown. This device mimics herbivore-caused tissue damage in an absolutely reproducible and quantifiable manner (Mithöfer et al. 2005). Remarkably, the HIPV blend induced by continuous mechanical damage is a qualitatively similar blend to that induced in lima bean (*Phaseolus lunatus*) by real herbivores (*Spodoptera littoralis* and the snail *Cepaea hortensis*) (Arimura et al. 2008b; Mithöfer et al. 2005). However, it appears that not all plant species respond to continuous mechanical damage as they do to real herbivore damage, by emitting an identical blend of HIPVs: for instance, 78%, 60%, and 43% similarities in tobacco, maize, and potato, respectively (Maffei et al. 2007), indicate that additional factors are required for a comprehensive set of HIPV blends. This is why the foraging process is often accompanied by the simultaneous introduction of oral factors (herbivore saliva) into plant tissues.

The first fully characterized herbivore-derived elicitor was volicitin [N-(17-hydroxylinolenoyl)-L-glutamine], a hydroxy fatty acid-amino acid conjugate (FAC) which was purified from the oral secretions of beet armyworm (*Spodoptera exigua*) (Alborn et al. 1997). When applied to wounded corn leaves, pure volicitin compounds elicit the emission of HIPVs which are identical in composition as those induced by *S. exigua* oral secretion. Volicitin is the dominant elicitor component in salivary factors and, moreover, that FACs are ubiquitous in the oral secretions (OS) from several herbivore species, including other lepidopteran species, crickets (*Teleogryllus taiwanemma*), and fruit fly (*Drosophila melanogaster*) larvae (Spiteller et al. 2001; Yoshinaga et al. 2007, 2010). Nowadays, several structures of elicitors have been characterized: "caeliferins," disulfooxy fatty acids from OS of the grasshopper species *Schistocerca americana* (Alborn et al. 2007); "β-glucosidase," from OS of cabbage white butterfly (*Pieris brassicae*) (Mattiacci et al. 1995), "inceptins," the peptide fragments of ATP synthase from OS of fall armyworm (*Spodoptera frugiperda*) (Schmelz et al. 2006); putative β-galactofuranose polysaccharide, from OS of *S. littoralis* (Bricchi et al. 2012); and "benzyl cyanide," from gland secretions released by the female cabbage white butterfly *P. brassicae* during egg deposition (Fatouros et al. 2008). Polyphenol oxidase (PPO) present in the saliva of two cereal aphids, *Sitobion avenae* and *Schizaphis graminum*, induces the expression of the genes for allene oxide synthase (AOS, which is involved in the biosynthesis of jasmonic acid [JA]) and FPS (which is involved in sesqiterpene biosynthesis, as described above) (Ma et al. 2010). On the other hand, aphids can suppress the induced plant responses by delivering salivary proteins inside their hosts when they puncture sieve tubes with their piercing mouthparts (stylets) and ingest phloem sap (Bos et al. 2010). The suppressors aphids use are the so-called "effectors" and these are secreted to host plants to modulate cell processes, repress plant defense systems and promote foraging activity of the aphids. However, only little is known about how ubiquitous effectors work to suppress plant defense responses.

After perceiving oral factors, plants then activate the signaling cascades for defense responses, including HIPV biosynthesis. Once the signaling cascades are activated, JA and its derivatives (jasmonates) play essential roles in mediating signals (Okada et al. 2015). JA signals cross-talks with other phytohormone signaling: for example, salicylic acid (SA) signaling pathways coordinate specific HIPV blends in plants when attacked by sucking arthropods, including aphids (whitefly and spider mite) (Arimura et al. 2009). In other cases, such as that of Colorado potato beetle (*Leptinotarsa decemlineata*) larvae, the antagonistic cross-talk between JA and SA signaling in tomato (*S. lycopersicum*) has been shown to be caused by symbiotic bacteria in the larva's oral secretions (Chung et al. 2013). Since antagonistic interactions often between the SA and JA signaling pathways, herbivores may strategically interfere with the JA-responsive defense pathway by exploiting symbiotic bacteria as a decoy, misleading the host plant's cells into reacting to the perceived threat of microbial pathogenesis (Chung et al. 2013).

Ethylene signaling is also able to mediate HIPV biosynthesis synergistically (Arimura et al. 2008a; Horiuchi et al. 2001). Ethylene, a gaseous phytohormone, contributes to the herbivory-induced terpene biosynthesis at least twice, modulating both early signaling events such as cytoplasmic Ca^{2+}-influx and the JA-dependent biosynthesis of terpenoids (Arimura et al. 2008a). HIPV components are likely orchestrated by a complex array of signaling cascades made up of phytohormones, and their quality may be variable among both plants and herbivores, thereby causing numerous HIPV blends in nature.

9.2.2 GLVs

9.2.2.1 What are GLVs?

"Green leaf odors" are often sensed in day-to-day life. When we mow our lawns, we usually smell these odors. These odors also add qualities to a wide variety of food stuffs. For example, green leaf odor in Japanese green tea is an important determinant of the quality of tea and occurs in most green plant tissues found in the biosphere. Drs. Curtius and Franzen in Germany first identified one of the chemicals having the green leaf odor in 1912 from several woody plants (Curtius and Franzen 1912) as 2-hexenal (the geometry of the double bond had not been determined at that time). Thereafter, other chemicals showing the green leaf odor have been identified, and all were found to have a six-carbon (C6) backbone with the functional group of aldehydes, alcohols, or acetates. Now, these C6 volatile compounds are cumulatively called green leaf volatiles (previously referred to as GLVs) (Figure 9.3).

The physiological and ecological significances of GLVs were first proven by Watanabe (1958). (E)-2-Hexenal emitted from mulberry leaves was shown to play a role in attracting silkworm larvae. Since then, many studies have been conducted to elucidate the roles of GLVs. Because most green land plants in nature have the ability to form GLVs, GLVs are assumed to take a rather general role in the physiology of the plants.

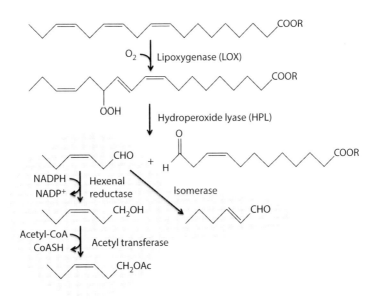

Figure 9.3 GLVs forming pathway in plants. Even though lipoxygenase prefers free fatty acid as substrate *in vitro*, fatty acids esterified as acyl groups in glycerolipids are also substrates, especially *in vivo*. Hydroperoxide lyase also acts on esterified fatty acid hydroperoxide to form volatile six carbon aldehyde and 12 carbon oxo acid esterified with the glycerol backbone of glycerolipids. A portion of volatile aldehyde is reduced to form corresponding alcohol, and subsequently, converted into its acetate. In some plants, isomerase acts on the aldehyde to form α,β-unsaturated carbonyl species. Only linolenic acid is shown in this figure, but linoleic acid and hexadecatrienoic acid are also converted into green leaf volatiles.

9.2.2.2 Biosynthesis

Oxylipins are chemicals formed from lipids through at least one step of oxygenation reaction. GLVs are formed from lipids through an oxygenation reaction by lipoxygenase; thus, they are members of plant oxylipins. Lipoxygenase is the enzyme involved in the oxygenation of fatty acids and acyl groups in lipids. Lipoxygenases need 1Z,4Z-pentadiene moiety in their substrates, and linolenic, linoleic, and hexadecatrienoic acids (and glycerolipids composed with them) are the common substrates found in plant tissues. With linolenic acid, lipoxygenase adds oxygen to its 13-position to yield 13-hydroperoxide of linolenic acid. Some plant lipoxygenases add oxygen to the 9-position to form 9-hydroperoxide, which is further converted into nine carbon (C9) volatile compounds reminiscent of cucumber and melon flavor (Matsui 2006).

Hydroperoxide lyase (HPL) acts on 13-hydroperoxide of linolenic acid and cleaves the C12-C13 bonding to yield two oxo compounds. One is (Z)-3-hexenal and the counterpart is 12-oxo-(Z)-9-dodecenoic acid (Figure 9.3). If 13-hydroperoxide of linoleic acid or 11-hydroperoxide of hexadecatrienoic acid are the substrate, *n*-hexanal and 12-oxo-(Z)-9-dodecenoic acid or (Z)-3-hexenal and 10-oxo-(Z)-7-decenoic acid, respectively, would be formed with the HPL reaction. HPL is a member of the big family of CYP enzymes based on the amino acid sequences and the structure of the active center, and is called CYP74B. However, the catalysis is probably not like that of typical CYPs, because neither molecular oxygen nor a reducing equivalent is needed for HPL. Instead, HPL follows the so-called shunt-pathway where the heme iron is oxidized and reduced during one cycle of its catalysis with the hydroperoxide group in substrate (Hrycay and Bandiera 2012; Lee et al. 2008).

In addition to HPL (CYP74B), plants have several CYP74 enzymes. Allene oxide synthase (AOS, CYP74A) shares a common substrate, that is, 13-hydroperoxide of linolenic acid, with HPL, and converts the substrate into unstable allene oxide through a catalytic pathway almost similar to that conducted by HPL (Lee et al. 2008). The allene oxide is converted into 12-oxophytodienoic acid by allene oxide cyclase, and further converted into JA, a signal molecule regulating plant defense responses against herbivores and pathogens (Wasternack and Hause 2013). CYP74C is HPL-specific to 9-hydroperoxide of fatty acids and forms (Z,Z)-3,6-nonadienal as a volatile product. Because of the cucumber/melon-like flavor properties of most of the nine carbon volatiles including (Z,Z)-3,6-nonadienal, CYP74C was first thought to be limited in Cucurbitaceae; however, its occurrence in a wide variety of plant species other than Cucurbitaceous plants has been reported even though its distribution is still limited (Matsui 2006). CYP74D, divinyl ether synthase, converts hydroperoxides of fatty acids into divinyl ethers, which show bactericidal and fungicidal activities. The distribution of CYP74D is rather limited and found in Solanaceae and several other plants (Mosblech et al. 2009).

CYP74 is also found among the plants belonging to the basal lineages in the evolution of land plants. In the moss *Physcomitrella patens*, three CYP74 genes were identified. One of them was shown to be HPL-specific to 11/12-hydroperoxide of arachidonic acid as well as to 9-hydroperoxide of linoleic acid (Stumpe et al. 2006). CYP74 locates apart from CYP74A and CYP74B in the phylogenetic tree constructed with plant CYP74 genes. In the liverwort, *Marchantia polymorpha*, there are two CYP74 genes, and both encode AOS; however, in the green microalgae, *Klebsormidium flaccidum*, there is only one AOS gene (Koeduka et al. 2015). All the CYP74 genes found so far in nonvascular plants show substantial sequence similarity with each other but are relatively different from CYP74A to D found in vascular plants. Therefore, it seems reasonable to make a clade of nonvascular plant CYP74 in the phylogenetic tree (Koeduka et al. 2015).

Based on phylogenetic relevance, it is assumed that HPL for GLV formation was acquired after the evolution of vascular plants. In fact, a preliminary BLAST search with the genome sequence of the Lycopodiophyta, *Selaginella moellendorffii*, indicated that 17 genes have the sequence feature specific to CYP74. Detailed analysis of the enzymes encoded by the *S. moellendorffii* CYP74-like genes would shed light on the first land plants that emitted GLVs on earth.

In most plants, C6-aldehydes formed by HPL are further converted into their alcohols and, subsequently, into acetate (Figure 9.3). Previously, it was believed that conversion of C6-aldehydes into C6-alcohols was catalyzed by alcohol dehydrogenases (ADHs). For example, one of the tomato ADHs is primarily involved in acetaldehyde metabolism and also involved in GLV metabolism (Bicsak et al. 1982). The gene expression for this tomato alcohol dehydrogenase, *ADH2*, was induced when tomato fruits ripened, and the overexpression of *ADH2* in tomato fruits resulted in increased levels of *n*-hexan-1-ol and (Z)-3-hexen-1-ol but caused little changes in the amount of corresponding aldehydes (Speirs et al. 1998). An ADH mutant of Arabidopsis also resulted in a reduced aldehyde/alcohol ratio (Bate et al. 1998). These lines of evidence support the hypothesis that ADHs are involved in the reduction of C6-aldehydes into C6-alcohols; however, direct evidence supporting this has not been provided, and it is possible that other enzymes are involved. In fact, Matsui et al. (2012) showed that NADPH but not NADH specifically reduced (Z)-3-hexenal into (Z)-3-hexen-1-ol in disrupted Arabidopsis leaves. NADH, a preferable cofactor for most alcohol dehydrogenase, hardly enhanced the reduction. An enzyme having a similarity to NADPH-dependent reductases involved in detoxification of reactive carbonyl species (Yamauchi et al. 2011) might be involved in the reduction of GLVs. In fact, the fumigation of

Arabidopsis plant with (Z)-3-hexenal suppressed photosystem II activity in a concentration-dependent manner, but (Z)-3-hexen-1-ol showed little effect at the same concentration used for (Z)-3-hexenal (Matsui et al. 2012). Therefore, it is assumed that the (Z)-3-hexenal is reduced at least in part because it is detoxified. The purification of the enzyme based on the specific activity assay for NADPH-dependent reduction of (Z)-3-hexenal, and the subsequent gene-knock-out experiment must be done in order to elucidate which enzyme is involved in the reduction step.

The acetylation of (Z)-3-hexen-1-ol to form (Z)-3-hexen-1-yl acetate is catalyzed by a member of the BAHD acyltransferase gene family encoding acetyl CoA-(Z)-3-hexen-1-ol acetyltransferase in Arabidopsis (D'Auria et al. 2007). The recombinant enzyme expressed in bacteria showed wide substrate specificity and accepted several medium-chain-length aliphatic and benzyl-derived alcohols, but showed highest catalytic efficiency with (Z)-3-hexen-1-ol. More important, a loss of function in this gene in Arabidopsis resulted in the loss of the ability to form (Z)-3-hexen-1-yl acetate but not (Z)-3-hexen-1-ol, indicating that this gene is essential for the acetylation step in GLV biosynthesis. However, with a suite of plant species, including lima bean and tea, several alternative products, such as hexenyl butyrate and/or hexenyl hexanoate, are also formed (Boggia et al. 2015; Sun et al. 2014). The substrate specificity of the transferase against the acid component is not likely high, thus, butyric acid and hexanoic acid can be accepted as substrates.

9.2.2.3 Regulation of GLV formation

GLVs do not accumulate highly in intact plant tissues, but their amounts are rapidly enhanced after plant tissues are disrupted. Real-time, highly sensitive volatile analysis carried out with the proton transfer reaction-mass spectrometry (PTR-MS) clearly showed such a so-called GLV-burst after *Arabidopsis* leaves were mechanical wounded (D'Auria et al. 2007). In the experimental system, the emergence of (Z)-3-hexenal was detected within a few seconds after mechanical wounding, and its amount reached a maximum level at 30–45 sec; thereafter, it immediately decreased. (Z)-3-hexen-1-ol and (Z)-3-hexen-1-yl acetate were formed and then peaked at 2.5 and 4.5 min, respectively, which correlates nicely with the order of metabolites in the GLV pathway. The GLV-burst is found in most green tissues of land plants; however, we still do not know how tissue disruption causes the GLV-burst. The GLV-burst can also be quite massive, and in the case of completely disrupted Arabidopsis leaves, the total of GLVs formed accounted for ca. 30% of the all trienoic acid (linolenic acid and hexadecatrienoic acid in either esterified or free form) found in the leaf tissues (Matsui et al. 2012).

Because the GLV-burst occurs a few seconds after cell damage, the reinforcement of the enzyme systems involved in GLV formation through transcriptional control hardly accounts for the extent of the GLV-burst. Substantial amounts of all the components, such as enzymes and substrates, required for the GLV-burst, should preexist in the plant tissues before damage. Furthermore, the metabolic allocation of GLV biosynthesis can be regulated at least by two processes: one is distinctly subcellular and involves the localization of the substrates and enzymes as found in isothiocyanates formed from glucosinolates in Brassicaceae plants (Jørgensen et al. 2015); and the other is the biochemical activation of one of the enzymatic step crucial for regulating GLV metabolism as is found with the activation of 5-lipoxygenase in mammalian cells (Rådmark et al. 2007). However, little is known about the exact rate-limiting process for the GLV-burst because the first committed enzyme in the pathway has not been fully identified.

The accumulation level of free fatty acids in intact, healthy plant tissues is usually kept low because the substantial concentration of free fatty acids would be harmful to cells. On the other hand, enzymatic properties of plant lipoxygenases have been intensively studied with free fatty acids. Lipases are thought to initiate the GLV-burst by sieving free fatty acids through hydrolyzing membrane lipids.

A lipase-like activity, such as lipolytic acyl hydrolase, is likely involved in the GLV-burst (Matos and Pham-Thi 2009), although there is no direct evidence supporting the assumption. With the JA pathway, a lipase-like activity is essential because the β-oxidation needed to truncate the carboxyl terminal in peroxisomes requires a free carboxylic acid moiety. DAD1 phospholipase A1 was identified as a lipase involved in the liberation of free linolenic acid from phospholipids in flower organs of Arabidopsis (Ishiguro et al. 2001). Surprisingly, DAD1 and its isologs in the Arabidopsis genome are not directly involved in the JA pathway in vegetative organs. The ability to form GLVs was also largely unaffected with the gene knockout mutant of DAD1 (K. Matsui et al. unpublished). Another example of a lipase-like activity is found in *Nicotiana attenuata*, GLA1 lipase (Schuck et al. 2014). This enzyme enables to the formation of 9-hydroxide of linoleic acid, but is not involved in the formation of JA, divinyl ethers and GLVs. In short, most attempts to identify the lipase-like activity involved in the formation of GLVs and JAs in vegetative organs have not yet been achieved. It is therefore an open question whether a lipase-like activity is truly involved in the GLV burst in the first place of the octadecanoid pathway.

Importantly, Nakashima et al. (2013) conducted several lines of studies to understand the involvement of a lipase-like activity in the GLV-burst in Arabidopsis leaves. Because the biosynthetic pathway up to the HPL step is intact in the *hpl1* mutant, 13-hydroperoxide of linolenic acid in its free form was expected to accumulate; however, its amount was quite low in comparison to that of GLVs formed in plants in which HPL would be active. Curiously, a substantial accumulation of hydroperoxides of monogalactosyldiacylglycerol (MGDG) was observed instead, leading us to assume that MGDG was directly oxygenated by lipoxygenases, and the resultant MGDG-hydroperoxides might be the substrate for HPL in Arabidopsis. In this case, MGDG containing 12-oxo-(Z)-9-dodecenoic acid as one of the acyl groups should be formed after the reaction of HPL on MGDG hydroperoxides. Through a comprehensive metabolite analysis, MGDG containing 12-oxo-(Z)-9-dodecenoic acid was found in homogenized leaf tissues of Arabidopsis. This novel finding indicates that at least a part of the GLV-burst is supported by the direct reaction of lipoxygenase on galactolipids. Because galactolipids containing 12-oxo-(Z)-9-dodecenoic acid (or its derivatives) were detected in homogenized leaf tissues of several other plant species (Nakashima et al. 2013), the direct reaction of LOX and HPL on galactolipids to form GLVs may be rather common in the plant kingdom.

Accordingly, lipoxygenase might be the key enzyme to commit the GLV-burst, and the initiation of lipoxygenase catalysis on galactolipids might be the trigger of the GLV-burst. Generally, almost all plants examined so far have several to a dozen lipoxygenase isozymes (Andreou and Feussner 2009). The lipoxygenase essential for the formation of GLV has been identified with several plant species. In tomato plants, for example, TomloxC is essential for GLV formation, and it locates in chloroplasts where a lot of galactolipids occur (Chen et al. 2004), raising the question of how the lipoxygenase in chloroplast avoids reacting with galactolipids before initiating the GLV-burst. If lipoxygenase is not found to be harbored in a distinct subdomain in choloroplasts apart from galactolipids, some yet unknown specific biochemical scenario might also be involved. Further studies on the detailed subcellular localization and activation processes of lipoxygenases are essential to elucidate the still-unknown mechanism controlling the GLV-burst.

9.2.2.4 Physiological and ecological significance

As mentioned before, (E)-2-hexenal was found to be a feeding attractant for silkworm larvae (Watanabe 1958); moreover, this GLV classically was known to be a mating stimulant for polyphemus moths (Riddiford 1967). Over the last half-century, an array of GLV functions as attractants, kairomones and pheromones has been discovered for various species of animals as listed in the Pherobase database (http://www.pherobase.com/). A subset of GLVs shows potent antibacterial and antifungal action (Kishimoto et al. 2008; Nakamura and Hatanaka 2002). It may make sense to assume that the ability to form GLVs was, at the beginning, a way for an organism (plant) to cope with microbial invaders that attempted to invade through wounds on plant tissues. This idea is partly supported by the finding that modulating the ability of transgenic Arabidopsis plants to biosynthesize GLVs may affect the susceptibility of the transgenic plants against *Botrytis cinerea* (Shiojiri et al. 2006a).

In the tritrophic system consisting of plants—herbivores—and herbivores' enemies (predators and parasitoids), volatile cues released from plants under attack of herbivores are attractive to the enemies of herbivores (Laothawornkitkul et al. 2009). Within the tritrophic system consisting of Arabidopsis, *Pieris rapae* (an herbivore), and *Cotesia glomerata* (a parasitic wasp), GLVs are one of the volatile cues used to recruit parasites (Shiojiri et al. 2006a). *C. glomerata* is a specialist and parasitizes only *P. rapae*, and *P. rapae* feeds only on Brassicaceae plants. Since GLVs are rather general compounds emitted by almost all land plant taxa, the immediate and transient GLV-burst in damaged plants may provide initial cues for the parasite to determine which herbivores are actively feeding on plants. The parasite might use GLVs to assess the current situation of the plant-herbivore interaction; then, the other volatile compounds such as terpenes might confer a specific command to find its specialized victim. The tritophic ecosystem will be discussed later in this chapter.

GLVs are also important cues for communication between plants: previously, it was shown that Arabidopsis plants exposed to the vapor of (*E*)-2-hexenal or (*Z*)-3-hexenal induced the expression of a subset of defense genes to confer defense traits against *B. cinerea* (Bate and Rothstein 1998; Kishimoto et al. 2005; Yamauchi et al. 2015). JA signaling and ethylene signaling were found to be at least partly involved in the signaling cascades for defense responses in Arabidopsis exposed to the hexenals (Hirao et al. 2012; Kishimoto et al. 2005). Recently, Mirabella et al. (2015) reported that Arabidopsis transcription factors WRKY40 and WRKY6 acted downstream of response of Arabidopsis plants against (*E*)-2-hexeanl. A subset of genes under the regulation of WRKY40/6 is independent of JA-signaling or ethylene signaling. Taken together, (*E*)-2-hexenal and related molecules activate multiple responses in Arabidopsis.

Glutathione (GSH) is highly reactive with α,β-unsaturated carbonyl species, including (*E*)-2-hexenal spontaneously or enzymatically in plant cells. When tomato seedlings were exposed to the vapor of methacrolein, the four carbon α,β-unsaturated carbonyl was efficiently taken up by plants. Essentially, methacrolein was metabolized to form adducts with GSH inside plant tissues (Muramoto et al. 2015). Because the reduced form of GSH reacts quickly with (*E*)-2-hexenal, the concentration of reduced form of GSH in plant cells becomes lower (Muramoto et al. 2015), which in turn causes the disorganized redox status of GSH that would be toxic for plant cells (Gill et al. 2013). However, it must be noticed that (*Z*)-3-hexenal, not α,β-unsaturated carbonyl species, enables the induction of a subset of defense responses in Arabidopsis (Kishimoto et al. 2005). This is probably because a portion of (*Z*)-3-hexenal is converted into (*E*)-2-hexenal in some plant species or by the oral regurgitant of an herbivore (Allmann and Baldwin 2010). Otherwise, (*Z*)-3-hexenal is sensitive to oxygen and easily converted into 4-hydro(pero)xy-(*E*)-2-hexenal, or 4-oxo-(*E*)-2-hexenal, both of which are highly reactive α,β-unsaturated carbonyl (Matsui et al. 2012). Together, (Z)-3-hexenal is considered a potential α,β-unsaturated carbonyl species and thus eligible to activate defense actions after being converted into α,β-unsaturated carbonyl species through the GSH redox regulation system.

(Z)-3-Hexen-1-ol and (Z)-3-hexen-1-yl acetate also induce defense response in several plant taxa including maize, lima bean plants, Arabidopsis, hybrid poplar (Heil and Karban 2010). Because the oxidation of (Z)-3-hexen-1-ol to (Z)-3-hexenal does not naturally occur in plant tissues, the bioactivity of these alcohols and acetates is not in agreement with the above-described GSH-mediated redox regulation. Alternatively, (Z)-3-hexen-1-ol and (Z)-3-hexen-1-yl acetate are perceived by plants in a different way from that employed for their corresponding aldehydes. For instance, lima bean plants enable to perceive various compounds structurally similar to (Z)-3-hexen-1-yl acetate, but there is a relatively narrow specificity in terms of the length of carbon chain, position of double bond, and the terminal functional group (Heil et al. 2008). These findings imply the involvement of a specific receptor-like protein to perceive (Z)-3-hexen-1-yl acetate, albeit no clear evidence has yet supported this possibility. In other cases, plant cells may transport the GLV molecules inside cells. For instance, metabolite analysis using the exposed tomato plants showed the presence of a glycoside of (Z)-3-hexen-1-ol, namely, (Z)-3-hexenyl vicianoside in the cells (Sugimoto et al. 2014). (Z)-3-Hexen-1-ol was one of major components of volatiles emitted by herbivore-damaged tomato plants. The receiver plant takes in the volatile alcohol and further converts it to the glycoside, probably with sequential action on the part of UDP-glucose glycosyltransferase and UDP-arabinose glycosyltransferases, in a similar way as the leaves of tea plants are found to form (Z)-3-hexenyl primeveroside (composed of (Z)-3-hexen-1-ol, glucose, and xylose) (Ohgami et al. 2015). The (Z)-3-hexenyl visianoside showed antifeedant activity against herbivores; thus, the uptake and subsequent glycosylation of (Z)-3-hexen-1-ol is partly accountable for inducing the defense response of the receiver plant to upcoming herbivore threats. In this case, the specific absorption of (Z)-3-hexen-1-ol is likely followed by the specific metabolism of the volatiles with the glycosyltransferases with relatively narrow substrate specificity. In fact, little is known about reception mechanisms for the potentially nonglycosylated GLV molecules, such as (Z)-3-hexen-1-yl acetate, in plant cells.

9.2.3 *Benzenoid/phenylpropanoid*

9.2.3.1 *Biosynthesis*

Volatile benzenoids and phenylpropanoids (VBPs), which contain at least one benzene ring, are originally derived from the shikimate pathway and can be categorized into different subclasses depending on their basic structures: benzenoids with a C_6–C_1 carbon skeleton, phenylpropanoid-related C_6–C_2 compounds, phenylbutanoids with a C_6–C_4 carbon skeleton, and phenylpropenes having a C_6–C_3 building unit. In VBP biosynthesis, the aromatic amino acid phenylalanine (Phe) serves as the initial common precursor for all VBPs and is further converted to individual VBPs via both a direct metabolic pathway and the core phenylpropanoid pathway mediating cinnamic acid and *p*-coumaric acid (Muhlemann et al. 2014).

Phenylpropanoid-related C_6–C_2 compounds, such as phenylacetaldehyde, 2-phenylethanol, and 2-phenethylacetate, are directly formed from Phe by the action of phenylacetaldehyde synthase (PAAS) (Kaminaga et al. 2006) (Figure 9.4). PAAS belongs to pyridoxal-5′-phosphate (PLP)-dependent aromatic amino acid decarboxylase (AADC) family and has

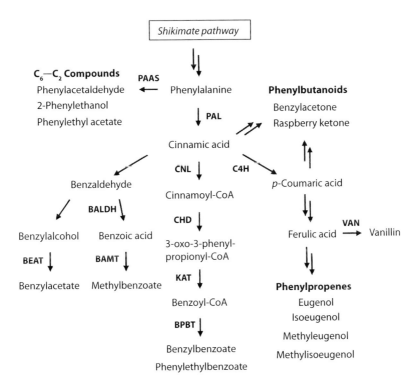

Figure 9.4 Schematic diagram of the proposed biosynthetic pathways leading to the formation of volatile benzenoid/phenylpropanoid compounds in plants. (PAAS, phenylacetaldehyde synthase; BPBT, benzoyl-CoA:benzyl alcohol/phenylethanol benzoyltransferase; PAL, phenylalanine-ammonia lyase; CNL, cinnamate:CoA ligase; CHD, cinnamoyl-CoA hydratase-dehydrogenase; KAT, 3-ketoacyl-CoA thiolase; BALDH, benzaldehyde dehydrogenase; BAMT, benzoic acid carboxyl methyltransferase; BEAT, benzyl alcohol acetyltransferase; VAN, vanillin synthase; C4H, cinnamate-4-hydroxylase.)

been identified only in *Petunia hybrida* and *Rosa hybrida*. Recent study for PAAS homologs resulted in the identification of aromatic aldehyde synthase (AAS) from Arabidopsis, *M. truncatula*, and *Cicer arietinum* (Gutensohn et al. 2011; Torrens-Spence et al. 2014). Both PAAS and AAS are bifunctional enzymes that catalyze phenylalanine decarboxylation and form phenylacetaldehyde, but the latter can also convert L-Dopa to dopaldehyde. In isolated protoplasts of rose petals, an alternative biosynthetic pathway to produce phenylacetaldehyde from Phe via phenylpyruvic acid exists (Hirata et al. 2012). This reaction is catalyzed by aromatic amino acid aminotransferase (AAAT). Thus, rose flowers potentially have two phenylacetaldehyde biosynthetic pathways, the AADC route and the AAAT route. Further reduction of phenylacetaldehyde to 2-phenylethanol is catalyzed by phenylacetaldehyde reductase (PAR) (Chen et al. 2011; Tieman et al. 2007). Phenylethyl acetate is generated from 2-phenylethanol by alcohol acetyltransferase (AAT) (Guterman et al. 2006; Shalit et al. 2003).

The first step in the biosynthetic pathways leading to benzenoids (C_6–C_1), phenylpropenes (C_6–C_3), and phenylbutanoids (C_6–C_4) is the deamination of Phe into cinnamic acid by phenylalanine ammonia lyase (PAL), which shares the common substrate Phe with AADC and AAAT (Figure 9.4). Benzenoid formation from cinnaminc acid requires the shortening of the side chain by two carbons. This shortening can take place by either a CoA-dependent and β-oxidative pathway in peroxisome or a non-β-oxidative pathway in cytosol (Boatright et al. 2004; Orlova et al. 2006). The CoA-dependent β-oxidative pathway involves the activation of cinnamic acid to cinnamoyl-CoA ester and requires the formation of 3-oxo-3-phenylpropionyl-CoA, after which benzoyl CoA is produced in peroxisome. These biochemical reactions have been catalyzed by cinnamoyl CoA ligase (CNL)/acyl-activating enzyme (AAE), cinnamoyl-CoA hydratase-dehydrogenase (CHD), and 3-ketoacyl-CoA thiolase (KAT) (Colquhoun et al. 2012; Klempien et al. 2012; Qualley et al. 2012; Van Moerkercke et al. 2009). Benzoyl CoA is finally converted to benzylbenzoate and phenylethylbenzoate using benzyl alcohol and 2-phenylethanol as cosubstrates, respectively, by benzyl alcohol benzoyl transferase (BEBT)/ benzyl alcohol/phenylethanol benzoyl transferase (BPBT) (Boatright et al. 2004; D'Auria et al. 2002; Dexter et al. 2008). Alternatively, the benzenoid shortening of the propyl side chain can occur via the non-β-oxidative pathway in cytosol with the formation of benzaldehyde as a key intermediate (Long et al. 2009). Benzaldehyde can serve as precursors for benzyl alcohol, benzylacetate, benzoic acid, methylbenzoate by the action of benzaldehyde dehydrogenase (BALDH), benzoic acid methyltransferase (BAMT)/benzoic acid/salicylic acid methyltransferase (BSMT) and benzylalcohol acetyltransferase (BEAT) (Dudareva et al. 1998; Murfitt et al. 2000; Negre et al. 2003; Pott et al. 2004). The side chain-shortening of the cinnamic acid by non-β-oxidative pathway into C_6–C_1 carbon skeleton during benzenoid biosynthesis has long been elusive, and no genes responsible for this reaction step have been discovered. Recently, vanillin synthase (VAN) from *Vanilla planifolia* has for the first time been shown to catalyze the direct conversion of ferulic acid into vanillin (3-methoxy-4-hydroxybenzaldehyde), but not the conversion of *p*-coumaric acid and caffeic acid. Furthermore, a similar enzyme with the same VAN activity was also identified in *Glechoma hederacea*, which is another vanillin-producing plant species (Gallage et al. 2014). The same enzymes probably involve the formation of benzaldehyde from cinnamic acid by a side chain shortening.

Phenylbutanoids include several important plant volatiles such as benzylacetone as attractants for pollinators in wild tobacco *N. attenuata*, raspberry ketone, and zingerone for the characteristic aroma of raspberries and ginger, respectively. In raspberries, the raspberry ketone is biosynthesized from *p*-coumaryl-CoA and malonyl-CoA by a sequential two-step reaction. In the first step, chalcone synthases (CHSs) belonging to type III polyketide synthase generate the intermediate *p*-hydroxybenzalacetone (*p*-hydroxyphenylbut-3-ene-2-one) (Zheng and Hrazdina 2008). In the second step, raspberry ketone/zingerone synthase (RZS) catalyzes the NADPH-dependent reduction of the intermediate *p*-hydroxybenzalacetone to raspberry ketone (Koeduka et al. 2011). The benzalacetone biosynthetic pathway remains elusive in *N. attenuata*; however, the decarboxylative condensation of cinnamoyl-CoA with malonyl-CoA may contribute the biosynthesis of the C_6–C_4 moiety of benzalacetone, as transgenic tobacco with reduced expression *CHS* (*Nachl*) is deficient in benzylacetone emission (Kessler et al. 2008). Benzalacetone synthase (BAS) is also thought to catalyze the production of *p*-hydroxybenzalacetone in rhubarb (*Rheum palmatum*), which accumulates the raspberry glucoside, lindleyin (Abe et al. 2001).

The formation of phenylpropenes such as eugenol, isoeugenol, chavicol and *t*-anol competes with the lignin biosynthetic pathway for the utilization of monolignol including *p*-coumaryl alcohol and coniferyl alcohol (Suzuki et al. 2014). The monolignol are derived from *p*-coumaric acid, and that is produced by the 4-hydroxylation of cinnamic acid and the C_6–C_3 skeletons. These skeletons were enzymatically formed from two enzymatic steps that removed the oxygenated functionality at the C_9 position (Koeduka 2014). The first committed step is the acetylation of monolignol by coniferyl alcohol acetyltransferase (CFAT), belonging to the BEAT, anthocyanin *O*-hydroxylcinnamoyltransferase (AHCT), *N*-hydroxycinnamoyl/benzoyltransferase (HCBT), and deacetylvinodoline 4-*O*-acetyltransferase (DAT) (BHAD) superfamily of plant acyltransferases (Dexter et al. 2007; Koeduka et al. 2009b). Monolignol acetate is then converted to phenylpropenes by the NADPH-dependent reductase eugenol and isoeugenol synthases (EGSs and IGSs), which belong to the pinoresinol-lariciresinol reductase (PLR), isoflavone reductase (IFR), and phenylcoumaran benzylic ether reductase (PCBER) (PIP) family (Koeduka et al. 2006). In grandular trichomes of basil and flower petals of petunia and *Clarkia*, coniferyl acetate is reduced to eugenol by EGS, whereas the same substrate is converted to isoeugenol by IGS. While most EGSs and IGSs belonging to Class I subfamily produce only one product, one EGS (Class II subfamily) from the garden strawberry catalyzes the respective formation of eugenol and isoeugenol from coniferyl acetate (Aragüez et al. 2013; Koeduka 2014). In some cases, O-methyltransferases are committed for further modification of the para-hydroxyl group on the phenylpropene-benzene rings. *Clarkia breweri* (iso)eugenol O-methyltransferase (IEMT) can use both eugenol and isoeugenol as substrates to form methyleugenol and methylisoeugenol, respectively, while basil eugenol O-methyltransferase (EOMT) and anise *t*-anol/isoeugenol O-methyltransferase (AIMT) prefer eugenol and isoeugenol, respectively (Gang et al. 2002; Koeduka et al. 2009a; Wang et al. 1997). These modifications enhance the volatility and diversity of phenylpropene compounds.

9.2.3.2 Regulation

The biosynthesis and emission of VBPs are often restricted to particular tissues and are rhythmically and transcriptionally regulated. The tissue-specific biosynthesis of VBP is a characteristic feature of many plant species. In glandular trichomes on the surface of basil leaves, for example, methyleugenol and chavicol storage is highly restricted. This tissue specificity is regulated at the level of biosynthetic genes, and indeed, basil *EGS*, *EOMT*, and *chavicol O-methyltransfease* (*CVOMT*) genes show a very specific expression in glandular trichomes (Gang et al. 2002; Koeduka et al. 2006). In addition to glandular trichomes, flowers are the primary source of VBP biosynthesis and also contribute to produce a suite of floral VBPs. In petunia, *Nicotiana suaveolens*, and snapdragon, the emission of VBPs is primarily attributed to flower petals (limb, tubes and lobes), although other tissues such as pistils and stamen also biosynthesize floral scent in other plant species (Negre et al. 2003; Pott et al. 2004). Similar tissue-specificity of VBP emissions has also been reported for *C. breweri*, *N. attenuata*, and roses, and is regulated at the levels of VBP biosynthetic genes and enzymes (Baldwin et al. 1997; Chen et al. 2015; Raguso and Pichersky 1995). In the scent-producing part of flowers, VBP formation is often restricted to specific cell layers. In some cases, such as *Stephanotis floribunda* and *N. suaveolens*, BSMT and its products are highly localized in the epidermal and subepidermal cells of petals lobes; these cells show the highest emission of floral scent (Rohrbeck et al. 2006). Similar cell-specific expression of methyl benzoate-synthesizing enzyme is also found in snapdragon flowers (Kolosova et al. 2001).

Different rhythmic pattern VBP emissions have been reported thus far in numerous species and mostly correlate with the respective pollinator activity (van der Niet et al. 2015). Flowers of *N. attenuata*, for example, emit a barely detectable amount of benzylacetone during the daytime, but the amount increases dramatically in the evening (Baldwin et al. 1997). The temporal specificity of benzylacetone emission plays an important role in attracting nocturnal pollinators such as the hawkmoth *Manduca sexta* (Kessler et al. 2008). In contrast, diurnal rhythmicity in methyl benzoate emission is reported in snapdragons pollinated by bumblebees that forage during the day (Negre et al. 2003). Both in nocturnal and diurnal scent-producing flowers, the levels of altered VBP emission are usually highest when flowers bloom, and once pollinated, flowers decrease the level of VBP formation to conserve valuable carbon and energy, and to prevent the potential damage to flowers from more visitors (Dexter et al. 2008; Muhlemann et al. 2006; Negre et al. 2003; Raguso et al. 2003). Although the decline of scent emission after anthesis is different depending on plant species, it is most likely that phytohormone ethylene is involved in the down-regulation of VBP formation and the biosynthetic genes during postpollination process (Dexter et al. 2007; Negre et al. 2003; Underwood et al. 2005). In transgenic ethylene-insensitive petunia, the transcript levels of *BSMT* were unchanged before and after pollination, unlike in the wild-type flowers, in which *BSMT* expression was suppressed after pollination. This difference indicates that ethylene is the signal to cease the expression of VBP biosynthetic genes in pollinated flowers.

In addition to the spatial and temporal regulation of VBP biosynthesis, transcriptional factors control the metabolic flux through the benzenoid/phenylpropanoid network, altering the levels of VBP emission. To date, several transcriptional factors regulating the expression of VBP biosynthetic genes have been identified in petunia. The first such regulator to be characterized was ODORANT1 (ODO1), belonging to the R2R3-type MYB family (Van Moerkercke et al. 2012b; Verdonk et al. 2005). ODO1 regulates the genes (*3-deoxy-D-arabino-heptulosonate-7-phosphate synthase* (*DAHPS*), *5-enol-pyruvyl-shikimate-3-phosphate* (*EPSPS*), and *chorismate mutase* (*CM*)) of the shikimate pathway as well as the phenylpropanoid pathway (*PAL1* and *PAL2*) and also activates the promoter of an ABC transporter *PhABCG1*, which is localizesd in the plasma membrane and is involved in VBP formation (Van Moerkercke et al. 2012a). Next to be characterized was EMISSION OF BENZENOID II (EOBII), which also regulates the promoter of shikimate and phenylpropanoid biosynthetic genes (*chorismate synthase* (*CS*), *CM*, *PAL2*, *CFAT*, *IGS*, and *BPBT*) and ODO1, which is positively regulated by EOBII through a specific MYB binding site (Spitzer-Rimon et al. 2010; Van Moerkercke et al. 2011). Recently, a third flower-specific R2R3-type MYB, EOBI was identified in petunia. EOBI, which directly activates ODO1, is located downstream of EOBII (Spitzer-Rimon et al. 2012). EOBI silencing leads to the down-regulation of numerous genes in both shikimate and phenylpropanoid pathways (*EPSPS*, *CS*, *CM*, *PAL*, *IGS*, and *EGS*).

In contrast to ODO1, EOBI, and EOBII, the recently identified MYB4 in petunia has been found to regulate only the levels of *p*-coumaric acid-derived phenylpropenes such as eugenol and isoeugenol by suppressing C4H (Colquhoun et al. 2011). It therefore appears that the coordinated regulation of biosynthetic genes with several transcriptional factor binding sites controls the orchestrated formation of VBP.

9.3 Ecological communications via PVOCs

PVOCs mediate an array of interactions between plant and arthropods including pollinators, herbivores, and herbivore enemies (predators and parasitoids). In this section, we highlight the behavior of arthropods and ask what components of plant-pollinator interactions and plant-arthropod interactions are based on the standard trophic series (plant–herbivore-natural enemy; Figure 9.5).

9.3.1 Plants versus pollinators

Floral volatiles are often used by pollinators in combination with other signals, such as the shape and color of flowers. Insect pollinators use floral scents to estimate the value of a potential reward obtained from flowers—these scents help insects recognizie specific host flowers—or as infochemicals whose components resemble signals that are valuable for insects in

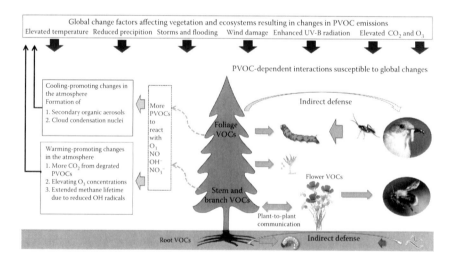

Figure 9.5 (**See color insert.**) Effect of global change factors on PVOC emissions from vegetation, PVOC-based feedback response to the atmosphere, and PVOC-dependent interactions susceptible to global changes in the ecosystem. Cooling-promoting and warming-promoting changes in the atmosphere are listed separately.

other ecological contexts (Schiestl 2015). Here we present typical examples of those phenomena. First, flowers of the herbaceous plant *Scrophularia umbrosa* and the orchid *Epipactis helleborine* release GLVs, which are highly attractive for social wasps. Nonetheless, it is notable that GLVs are emitted mostly from vegetative organs (e.g., leaves) when injured by herbivores (see above). Social wasps use this vegetative signal to hunt caterpillars on plants, and the singular plants, such as *S. umbrosa* and *E. helleborine*, have presumably been co-opted to recruit those wasps as pollinators through sensory exploitation (Schiestl 2015). Second, floral volatiles of the orchid *Epipactis veratrifolia* mimic the components of aphid alarm pheromones—terpenes (α-pinene, β-pinene, β-myrcene and β-phellandrene)—in order to attract hoverflies for pollination (Stökl et al. 2011). From these examples, we see that plants and insects use blends of floral scents as guides to find their partners.

9.3.2 Plants versus herbivores

Herbivore-induced plant volatiles (previously referred to as HIPVs), emitted from vegetative parts of plants, can directly repel herbivores, such as ovipositing butterflies and host-seeking aphids (Unsicker et al. 2009). Diurnal behavior (hiding behavior) of the *Mythimna separate* caterpillar is also affected by HIPVs emitted from *M. separate*-infested maize in the light phase (Shiojiri et al. 2006b). Intriguingly, irrespective of light status, the caterpillars became as active as if they were in the dark when exposed to volatiles emitted from plants in the nighttime. Likewise, irrespective of light status, the caterpillars behaved as if they were in the light when exposed to volatiles emitted from plants in the light. These findings suggest that HIPVs, rather than light cues, are important for confirming the diurnal rhythms of the caterpillars' performance. Probably, *M. separata* larvae perform in this way during the day in order to hide from the specialist parasitoid (*Cotesia kariyai*) that is active in the daytime. In other cases, HIPVs released from tobacco fed on by *H. virescens* larvae at night are highly repellent to the adult females; these females choose fresh vegetative sites for oviposition and avoid the likely presence of particular larval competitors (De Moraes et al. 2001).

9.3.3 Plants versus herbivore enemies

Up to now, numerous studies on HIPV-mediated interactions between herbivore-infested plants and herbivore enemies (predators and parasitoids) have been extensively explored. Such interest focuses mostly on the sophisticated tri-trophic interactions and coevolutions mediated by specific HIPV infochemicals that contribute to plant indirect defense responses. In addition, those interactions may be employed for effective pest control in order to stop or reduce the frequent use of insecticides in agriculture and horticulture. Given broad evidence for the role of HIPVs, we survey some recent examples of behavioral responses that involve the enemies of herbivores; s in these, transgenic plants are used as the powerful toolboxes of modern molecular biology and chemical biology. The use of transgenic plants, especially, represents a novel solution to the challenges of studying the consistent and pronounced effect of a single or a blend of HIPV compound(s) on the nature of plant-arthropod interactions and the potential contribution of HIPVs to biological pest control (Aharoni et al. 2005). For instance, transgenic torenia (*Torenia hybrida*) plants overexpressing a β-ocimene synthase (*PlOS*) gene were assessed for their ability to attract specialist predatory mites (*Phytoseiulus persimilis*) of spider mites (Shimoda et al. 2012). The transgenic plants were highly attractive only when (E)-β-ocimene and a natural blend of HIPVs, comprising linalool (monoterpene), DMNT (C11 homoterpene) and four sesquiterpenes, were blended, in comparison with infested wild-type plants (Figure 9.6). However, the presence of floral aroma abolished the plants' enhanced attractiveness. Although this outcome was unexpected, it was unsurprising. It is thought that floral aroma may inform predator mites of a decline

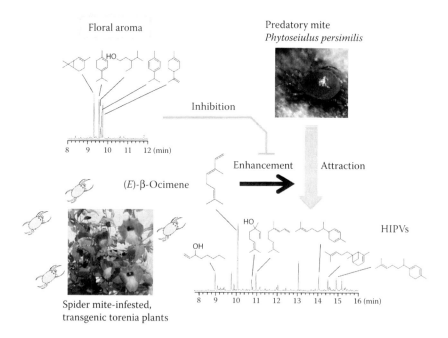

Figure 9.6 Schematic presentation of the effect of transgenic torenia plants that emit both (E)-β-ocimene constitutively and spider mite-induced plant volatiles (HIPVs), on the attraction of predatory mites.

in the quality of prey-inhabiting leaves due to nutrient transmission occurring at the flowering stage. Otherwise, the predatory mites might avoid floral aroma so as to protect themselves against intra-guild predation by flower-visiting predators, such as ants. In both cases, we confirmed the consistent and sophisticated ability of HIPVs to attract predators, depending on the situation of host plants. This is in agreement with behavioral responses of *P. persimilis,* which prefer a blend of HIPVs and the manipulated C16 homoterpene (TMTT) released from the other transgenic *Lotus japonicus* plants infested with spider mites (Brillada et al. 2013). However, unlike this specialist predator, the generalist predator *Neoseiulus californicus* has been shown to be attracted to uninfested, transgenic plants emitting TMTT alone, similarly to how it is attracted to infested, transgenic plants emitting a blend of HIPVs and TMTT. A blend of HIPVs has been shown to be the most powerful attractant for *N. californicus,* beyond individual HIPV cues, with the exception of methyl salicylate (Shimoda 2010). It is, therefore, very likely that TMTT is a strong attractant for *N. californicus*—though not stronger than a blend of HIPVs—and it is unlikely that there is an additive effect when TMTT and HIPVs are blended. Generalist and specialist predators probably have different preferences for volatile blends probably because of their different feeding guilds.

9.3.4   *Plants versus plants*

As described in the section on GLVs, plant volatiles are known to be able to coordinate plant-to-plant communications. Plants can eavesdrop on the cues emitted from their neighboring plants in response to herbivore attack in order to adjust their defenses to suit their current or future risk and to increase their fitness as a result. Plants may have evolved such a communication system—one that allows them to emit and respond to HIPV cues in order to promote within-plant signaling—because volatile signaling within an individual is faster, is independent of the architecture of vascular connections, and expends less water, than vascular plant signaling (Orians 2005). It is therefore likely that within-plant communications can be co-opted to allow communication between emitter plant and receiver plant (Arimura et al. 2010). Moreover, in some cases, receiver plants do not immediately induce defense responses after perceiving HIPVs but, rather, prime defense responses so they are able to respond strongly and rapidly once herbivores begin to feed. It appears that maize receiver plants exposed to HIPVs can maintain a "memory" of such HIPV perception for a long time (~5 days) (Ali et al. 2013).

On the other hand, some PVOCs (GLVs and terpenes) serve as allelochemicals that directly or indirectly cause a plant to negatively affect another (Fischer 1986) (Bate and Rothstein 1998). Chemical communications may lead the emitter to acquire more of the available nutrients, water or light from the environment because of resource competition is reduced. The well-documented example for this is created by *Salvia leucophylla*, a shrub that grows in coastal south California. Camphor and 1,8-cineole of five volatile monoterpenes emitted from *S. leucophylla* was shown to inhibit the growth of cucumber and *Brassica campestris* seedlings (Müller 1970; Nishida et al. 2005). Similarly, several monoterpenes with oxygen-containing functional groups are known to inhibit to the germination of crop and weed seeds (Vaughn and Spencer 1993, 1996).

9.4 VOCs in global environmental change

Plants, particularly trees with long life cycles, use VOCs to adapt to changing environmental conditions, for example, by improving their thermo tolerance to withstand heat stress. Global environmental change caused by human activity is broadly affecting terrestrial vegetation by changing land use; for example, forest removal reduces tree cover and increases the land given to agricultural fields (Crowther et al. 2015; Unger 2014). Regional biogenic PVOC fluxes to atmosphere change substantially as plant species composition changes in response to direct (Rosenkranz et al. 2015; Unger 2014) or indirect (Peñuelas and Staudt 2010; Valolahti et al. 2015) effects of human activity. Global climate change including changes in radiation, temperature, precipitation and concentrations of atmospheric gases as a result of the intensive use of fossil fuels (IPCC 2013) and greenhouse gas emissions is leading to alterations in abiotic growth conditions of plants. These changes will substantially affect plant physiology, plant competition, community structure and, finally, PVOC emissions (Valolahti et al. 2015). Both PVOC emission rates and the composition of PVOCs are species-specific but, typically, depend on temperature and light level (Ghirardo et al. 2010).

The knowledge how the global environmental change affects the biosphere capacity to emit PVOCs to the atmosphere is crucial, because PVOCs are involved to several processes in atmospheric chemistry and physics (Fuentes et al. 2001; Laothawornkitkul et al. 2009; Peñuelas and Staudt 2010). PVOCs have an important role in the control of major drivers of global warming (Figure 9.5). To understand the importance of different abiotic factors for PVOC emissions, experimental studies have assessed mostly one factor at a time in greenhouse, chamber exposures or in field exposures (Blande et al. 2007; Calfapietra et al. 2007). It will be crucial for climate change models to better understand the formation process of secondary aerosols and cloud condensation nuclei (CCN) from PVOCs in the atmosphere and how these atmospheric particles will mitigate warming by increasing the reflection of solar radiation from clouds back to space (Ehn et al. 2014; Paasonen et al. 2013). However, PVOCs in atmosphere may have warming effects by acting directly as greenhouse gases and indirectly by participating in the formation of ozone, the third most important greenhouse gas, and also by lengthening the atmospheric lifetime of methane by competing with methane of hydroxyl radical (OH) sink (Kirschke et al. 2013; Peñuelas and Staudt 2010). Finally, the degradation of PVOC leads to the formation of CO_2 in the atmosphere (Holopainen and Blande 2013).

9.4.1 Temperature

The increase in the global mean surface temperature, a trend observed since the early twentieth century, is explained by radiative forcing from increasing greenhouse gas concentrations (Marotzke and Forster 2015). Temperature is an important driver of plant VOC emissions, due to VOCs high vapor pressures at ambient temperature range; this is why high emission rates are typical in plants grown in elevated temperature (e.g., (Faubert et al. 2010; Ibrahim et al. 2005; Niinemets and Sun 2015). In their review of older literature of warming experiments, Peñuelas and Staudt (2010) indicated that only one report out of 32 published papers did not show an increase of emissions in at least some of the studied PVOC groups.

Plant metabolism and the biosynthesis of VOCs are directly dependent on temperature, but genotypically specific responses have been found. However, in terpene-storing species such as *Pinus sylvestris*, the concentrations of monoterpenes in needle storage of seedlings are not necessarily responding in long-term exposure to warming temperatures in the temperature ramp (Holopainen and Kainulainen 2004) or 2–6°C degrees elevation (Räisänen et al. 2008a), although monoterpene emissions to atmosphere may increase two- to fourfold when grown in 1°C average temperature elevations in the field (Kivimäenpää et al. 2016). The reason for this discrepancy might be that under the stress that accompanies elevated temperature, conifers allocate more carbon to the *de novo* synthesis of PVOCs and immediate emissions instead of storage (Ghirardo et al. 2010). Also, in early spring, when plants are recovering their photosynthetic capacity and when temperatures are cold but solar radiation levels are high, bursts of monoterpene emission are detected from evergreen conifers on warm days that follow cold days (Aalto et al. 2015). Authors have suggested that monoterpene bursts may play a functional role for foliage, protecting it from high radiation levels during this critical transitory state of early season photosynthesis.

Lipophilic VOCs are assumed to help stabilize plant cell membranes when temperatures are stressful and there is a drastic increase in the emission of highly volatile terpenoids, when temperature increases above 30°C (Loreto and Schnitzler 2010). Diffusion (Niinemets et al. 2014; Niinemets and Sun 2015) and other biological transportation mechanisms (Widhalm et al. 2015) have been suggested to explain emission both through stomata and epidermis. According to Widhalm et al. (2015), the passive diffusion from intact cell compartments to the environment cannot be the only system for emission, because VOC emission rates detected in plants when temperatures are high might mean there are phytotoxic VOC concentrations in several subcellular barriers in plants. Authors have suggested that lipid transfer proteins (LTPs) or other types of carrier proteins may contribute to the transport of VOCs through the plastid stroma, cytosol, and/or cell wall. Also, transporters localized in the plasma membrane may have a role in exporting VOCs from the cytosol or membrane into the cell wall. They propose that VOCs may then diffuse through the cuticle before being emitted into the environment (Widhalm et al. 2015).

9.4.2 Droughts and flooding

The availability of water is essential for plant growth, and transpiration (involving water transported from roots to leaves and evaporated into the atmosphere) is the basic physiological mechanism of plants. This process is controlled by stomatal opening, which controls not only the uptake of CO_2, but also the release of VOCs through stomata, such as emissions of methanol, which has high aqueous-phase concentrations (Niinemets et al. 2004). Monoterpenes such as stress-related monoterpene β-ocimene are less dependent on stomatal closure than compounds with high aqueuous-phase concentration (Niinemets et al. 2004).

Drought periods are related to warm climatic periods when plant metabolism is high. Isoprene emissions are known to be strongly temperature and light dependent. The shortage of water limits the metabolisms, and photosynthesis can be constrained by drought. Brilli et al. (2007) studied white poplar (*Populus alba*) and found that ^{13}C incorporation from direct photosynthesis into isoprene decreased when photosynthesis was constrained by drought. Results suggested that *P. alba* can use alternative carbon sources during drought periods to keep isoprene emission high. In general, severe drought stress has been found to reduce monoterpene emissions, while mild to moderate stress does not affect or may even enhance or VOC emissions (Blande et al. 2014). Šimpraga et al. (2011) reported a sudden burst in the emission of a nonidentified, nonmonoterpene BVOCs during acute drought stress in the European beech. In the central Amazonian forest, Jardine et al. (2015) detected an increase in GLV products such as hexanol within the canopy between 11 and 17 m together and an increase in isoprene under warm and dry periods. Above the canopy, 3-methyl furan (an oxidation product of isoprene) was more abundant during drought than during any other season (Jardine et al. 2015).

Flooding is a result of frequent storms in various environments with climate changes, which atypically leads to anoxia in the root system of plants (Rinnan et al. 2014). In *Alnus glutinosa*, *Populus tremula*, and

Quercus rubra, water-logging stress resulted in reduced net assimilation and stomatal conductance, but enhanced emissions of ethanol, acetaldehyde, NO, LOX products and methanol (Copolovici and Niinemets 2010). Isoprene emission capacity was not related to water-logging tolerance (Copolovici and Niinemets 2010), although isoprene emissions are typical for many wetland plants such as sedges, mosses and mangroves (Rinnan et al. 2014). However, emissions of monoterpenes were enhanced in a lemon tree (*Citrus* x *limon*) under winter flooding by salty water (Velikova et al. 2012).

9.4.3 Wind

Windy weather, storms, and hurricanes are expected to become more frequent due to the warming of oceans (Bender et al. 2010). An increase in wind speed directly affects the emission of VOCs from leaves' surface by removing the boundary layer and allowing the improved diffusion to surface. In forests, storms have the capacity to severely damage trees and other vegetation (Seidl et al. 2014). Wind speeds above 15 m s^{-1} may cause damage to Scots pine stands by uprooting and breaking the stems of trees (Talkkari et al. 2000). Conifer stands after normal tree felling (Räisänen et al. 2008b) or after wind damage could be a rich source of mechanically induced emissions of VOCs to the atmosphere. Particularly monoterpenes from resin storage can be emitted in high quantities from the storage of oleoresin in the needles, bark and xylem of fallen trees. Even after forest damage is cleared away, the stumps of cut trees can serve as a significant source of VOC emissions for more than a year (Haapanala et al. 2012; Kivimäenpää et al. 2012).

9.4.4 UV radiation

Stratospheric ozone layer filters out the shortest wavelengths (UV-C, 100–280 nm) from high-energy solar ultraviolet radiation, and only UV-B (280–315 nm) and UV-A (315–400 nm) will reach the atmosphere. The Montreal Protocol was signed in 1989 to cut the emissions of stratospheric ozone-depleting substances to the atmosphere. This has reduced depletion of stratospheric ozone layer and limited the increase in penetration of high-energy, short-wavelength UV-B radiation at the lower atmosphere, (Williamson et al. 2014).

Plants protect themselves with filtering pigments against UV radiation. Therefore, UV-screening phenolic compounds accumulate easily in plant tissues under elevated UV-B radiation (Kaling et al. 2015). Also, many volatile compounds such as isoprene are proposed to have important protective and adaptive functions against several environmental stresses including excess UV-B radiation (Loreto and Schnitzler 2010).

Isoprene has shown increased emissions from peatland plants (Tiiva et al. 2007), alfa grass (Guidolotti et al. 2016) and *Quercus gambelii* (Harley et al. 1996) under elevated UV-B radiation in UV-B exposure or UV-B filtration experiments. Kaling et al. (2015) observed an accumulation of phenolic compounds and a small but clear tendency toward lower isoprene emission rates under UV-B exposure in *Populus* x *canescens*, suggesting possible cross-talk between phenolic and terpenoid biosynthesis pathways. Responses of other PVOCs under supplemented UV-B radiation (Blande et al. 2009; Faubert et al. 2010) or under exclusion of the ambient UV-B radiation (Gil et al. 2014) have been negligible.

9.4.5 Light intensity and quality

The photosynthetic efficiency of plants is controlled by the radiation intensity (irradiance) in the photosynthetically active radiation (PAR) range, namely, 400–700 nm. Particularly the emissions of isoprene, but also of monoterpenes, are strongly dependent on PAR level (Ghirardo et al. 2010; Niinemets et al. 2004). In inner parts of the forest canopy, the ability of leaves adapted shady conditions to assimilate CO_2 is strongly impaired due to high light levels and the abundance of sunflecks, but isoprene emission has been found to help the recovery of CO_2 assimilation (Behnke et al. 2010). Increased cloudiness or increased light diffusion by aerosols may reduce light intensity and photosynthesis in the upper part of the forest canopy, but diffused light may improve full canopy photosynthesis by dispersing light to the more shaded parts of the canopy (Roderick et al. 2001). Williams et al. (2014) reported substantially (17%) improved photosynthesis on cloudy days in the inner canopy of tundra shrubs compared to that on days of direct sunlight.

The quality of visible light and far-red light has a substantial effect on plant growth and chemistry; for example, Siipola et al. (2015) found that the flavonoid composition of pea plants changed more in response to high-energy solar blue light than to solar UV radiation. Studies under LED light have shown that the ratio of blue and red light components has affected plant photosynthesis and terpenoid production. Terpenoid concentration in the form of essential oil can be enhanced significantly in *Mentha* spp. with 70:30 red:blue ratio under LED lighting when compared to solar radiation spectrum (Darko et al. 2014). Similarly, the quantity and quality of PVOC emission from flowers and fruits are affected by specific light treatments (Colquhoun et al. 2013). Also, the ratio of red (R) to far-red (FR) light affects plant growth and plants' secondary chemistry. Total PVOC emissions from barley were reduced under low R: FR conditions compared with under controlled light conditions without FR; however, the quantity and quality of individual volatile compounds varied (Kegge et al. 2015).

9.4.6 CO$_2$

Climate warming is a result of increased greenhouse gas concentrations in the atmosphere, mostly CO$_2$, methane and O$_3$. Carbon dioxide is crucial for plant photosynthesis and growth. Therefore, changes in atmospheric CO$_2$ concentrations directly affect plant physiology and the concentration of carbon-based compounds in plants (Peñuelas and Staudt 2010). The average monthly atmospheric CO$_2$ concentration was approximately 320 ppm in the mid-1960s and reached the level of 400 ppm in March 2015 at the Mauna Loa Observatory in Hawaii (ftp://aftp.cmdl.noaa.gov/products/trends/co2/co2_mm_mlo.txt). Three-quarters of the current CO$_2$ increase is likely the result of fossil fuel use, and the remainder of the increase results from changes in land-use, such as increased deforestation (IPCC 2013).

Most experimental studies assessing short- or long-term effects of elevated CO$_2$ on VOC emissions from plants have demonstrated that such elevated levels significantly reduce emission rates (Peñuelas and Staudt 2010). In a review of 30 experimental studies, isoprenoid emissions, in particular, have been highly responsive to changes in atmospheric CO$_2$ concentrations. Not all studies support this trend. Recently, Sun et al. (2013) found that when hybrid aspen *Populus tremula* x *Populus tremuloides*) saplings were grown under ambient (380 µmol mol^{-1}) or elevated CO$_2$ (780 µmol mol^{-1}), at high temperatures, saplings grown at elevated CO$_2$ levels had higher isoprene emission rates. Plants grown at elevated CO$_2$ levels had thicker leaves (e.g., Sun et al. 2013; Vuorinen et al. 2004b), which may reduce PVOC emissions rates due to the allocation of carbon to thicker cell walls and, consequently, restrict carbon availability for the biosynthesis of volatile compounds (Vuorinen et al. 2004b) or increase emissions as a result of the higher density of chloroplasts per unit leaf surface area (Sun et al. 2013).

9.4.7 Ozone

Stratospheric ozone is a result of high-energy UV radiation, which photolyzes oxygen molecules (O$_2$) into highly reactive oxygen (O) atoms. These atoms rapidly react with O$_2$ to form ozone (O$_3$). Tropospheric O$_3$ in the lower atmosphere is formed as a result of reactions between VOCs from biogenic and anthropogenic sources, and from air pollutants such as NO$_x$ and OH radicals, the result mostly of burning of fuels (Peñuelas and Staudt 2010). Furthermore, stratospheric circulation over the coming century will mix stratospheric and tropospheric O$_3$ and could lead to small but important increases in tropospheric O$_3$ levels (Neu et al. 2014). The exposure of plants to elevated ozone levels has produced highly variable results when the PVOC emission rate has been measured (Peñuelas and

Staudt 2010). This variation is partly caused by the variation of ozone resistance that exists among plant species, cultivars and genotypes. The induction of GLV emission from tobacco was found to be related to the O_3 flux density (mol m^{-2} s^{-1}) into the plants, and not to ozone concentrations or dose-responsive relationships such as AOT_{40} values (Beauchamp et al. 2005). PVOC emission started 30 min to 16 h after the start of ozone exposure (Beauchamp et al. 2005). Earlier, it was found that lima bean plants required exposure to 150 ppb of O_3 for 4 h and to 200 ppb of O_3 for 4 h to achieve slight visible damage and emissions of GLVs and homoterpenes (Vuorinen et al. 2004a).

Major PVOCs, isoprene (Loreto and Velikova 2001), monoterpenes (Pinto et al. 2007b), sesquiterpenes (Hallquist et al. 2009) and GLVs (Hamilton et al. 2009) quench tropospheric ozone. The reactions between isoprene and ozone are considered to protect plants from oxidative stress (Loreto and Velikova 2001; Vickers et al. 2009). External supplementation of large quantities of sesquiterpene (E)-ß-farnesene protected ozone-sensitive tobacco plants from visible cell death symptoms of ozone exposure in reaction chambers, but realistic atmospheric concentrations of (E)-ß-farnesene did not provide protection against ozone (Palmer-Young et al. 2015). These authors suggested that the sesquiterpenes emitted by plants do not provide protection against abiotic stresses but may, instead, function against biotic stresses.

9.4.8 Biotic interactions and ecological functions of VOCs under environmental change

Plant volatiles provide protection against abiotic as well as biotic stresses, such as those caused by herbivores and plant pathogens. PVOCs may act directly against herbivores by repelling attackers (Kos et al. 2013) and deterring feeding or indirectly by attracting natural enemies of herbivores including mostly invertebrate parasitoids and predators (Dicke et al. 2009; Himanen et al. 2009; Pinto et al. 2007a), as well as passerine bird predators (Amo et al. 2013; Mäntylä et al. 2008). Hyperparasitoids of plant feeding larvae may use the same volatile cues and mitigate parasitoids' capacity to control herbivores (Zhu et al. 2015). Specialist herbivores and host-altering species, in particular, use host-specific volatiles to orientate to their main host plant (Björklund et al. 2005) or their secondary winter host (Nissinen et al. 2008), but the specialist herbivores also detect typical volatiles of other plant genus or family which they avoid as host plants (Himanen et al. 2015; Kos et al. 2013). These volatiles provide a potential for development of sustainable applications in plant protection (Guarino et al. 2015; Himanen et al. 2015; Kos et al. 2013). Attack by bacterial (Cardoza and Tumlinson 2006) or fungal (Vuorinen et al. 2007) phytopathogens or colonization by a mycorrhizal fungi (Dicke et al. 2009) may induce emission of PVOC composition which differs qualitatively from herbivore-induced PVOCs (i.e., HIPVs), but is mostly composed of the same PVOCs. Volatile plant monoterpenes and sesquiterpenes, in particular, are known to have antibacterial (Shimada et al. 2014) and antifungal (Kusumoto and Shibutani 2015) properties. As described above, PVOCs also have an important role in plant-to-plant communication and in activating defences in intact neighboring plants, if an individual plant is under attack by herbivores (Arimura et al. 2000; Blande et al. 2010; Heil and Karban 2010).

Elevated temperatures are known to affect the ecosystem and, particularly, the life cycles and performance of biotic organism that interact closely with plants (Bale et al. 2002). Recent analyses have shown that the losses of woody plant foliage to insects follow a dome-shaped latitudinal pattern: they peak in temperate zones, slightly decrease toward the equator and strongly decrease toward the poles, but so far no clear climate warming-related temporal trends in the foliar losses of woody plants to insects within the temperate climate zone have been detected (Kozlov and Zvereva 2015). Recently, it has been evidenced that intensive herbivory may shift to the northern coniferous forest zone. The improved overwintering survival rates of mountain pine bark beetles (MPB) in Canada have been connected to increased winter temperature and precipitation changes, and warming has allowed MPB outbreaks to occur at progressively higher elevations and substantially extend infested areas (Berg et al. 2013). When affecting the distribution of herbivores and their natural enemies, warming intensifies herbivore infestations on vegetation and substantially increases the atmospheric load of HIPVs and other reactive VOCs in the atmosphere (Berg et al. 2013; Holopainen and Gershenzon 2010; Joutsensaari et al. 2015). Local insect populations in the conifer forest zone seem to have limited adaptive capacity in response to increasing mean temperatures. The fecundity of the pine needle aphid *Schizolachnus pineti* had a curvilinear response, reaching an optimum at 4–6°C degrees above the local mean daytime temperature (Holopainen and Kainulainen 2004). It can be concluded that the biotic stress to vegetation by local insect populations can increase to a certain limit. If the biotic stress produced by native populations is combined with the stress produced by new invasive herbivore species extending their geographical range to northern coniferous forests, substantial increases in plant stress and HIPV emissions from conifers by herbivory and warming climate can be expected (Berg et al. 2013; Holopainen 2011; Joutsensaari et al. 2015).

Under natural conditions, plants are exposed to multiple stresses simultaneously. However, relatively few studies have attempted to investigate PVOC emission under multiple stresses, either in the laboratory or in the field (Holopainen and Gershenzon 2010). Exposure to a combination of abiotic factors related to global change and biotic stress simultaneously could have additive or opposing effects on PVOC emissions. As an example, Copolovici et al. (2014) found that drought exposure induced GLV, monoterpene and homoterpene emission in *Alnus incana* more efficiently than did sawfly larvae feeding, but when stresses were combined, the emissions were highest. Gouinguene and Turlings (2002) found drought and elevating temperatures together with elevated light and nutrient availability increased the emission of HIPVs from maize. On the other hand, the exposure to 1.5-fold elevated ambient O_3 levels in the field indicated that only very small changes in PVOC emissions were due to increased ozone levels in hybrid aspen (*Populus tremula* x *P. tremuloides*). The plants showed the induction of some terpenes, particularly the monoterpene (E)-β-ocimene and the homoterpene DMNT, in response to insect feeding, and only marginally interactive effects with ozone were detected (Blande et al. 2007).

If the volatile profile of herbivore-attacked plants changes as a result of external abiotic factors, there may be consequences for the signaling value of these PVOCs on other organisms perceiving the signals. Cabbage plants, for example, grown at doubled CO_2 concentrations had some minor reductions of herbivore-induced emission of homoterpene DMNT and sesquiterpene (E,E)-α-farnesene. In olfactometric tests, the larval parasitoid *Cotesia vestalis* of the diamondback moth (*Plutella xylostella*) did not respond to plants damaged by *P. xylostella* and grown at elevated CO_2 (Vuorinen et al. 2004b). The effect of CO_2 on PVOC emission rates was possibly related to the anatomical changes such as thicker leaves detected in plants grown at elevated CO_2 levels (Vuorinen et al. 2004b). In a field study on yellow star thistle (*Centaurea solstitialis*) with 300 ppm CO_2, addition compared to ambient CO_2 levels (Oster et al. 2015) or open-top exposure of silver birch (*Betula pendula*) to doubled ambient CO_2 (Vuorinen et al. 2005) the authors did not find CO_2 effects on sesquiterpene emissions.

Phytotoxic tropospheric ozone is a greenhouse gas; exposure to it can influence plants and their PVOC emission quality and capacity (Pinto et al. 2007a; Vuorinen et al. 2004a). Furthermore, ozone together with other air pollution-related atmospheric oxidants will rapidly degrade PVOCs emitted to the atmosphere after release (McFrederick et al. 2008, 2009). This degradation has important effects on the ecosystem, as the degradation of signal compounds disturbs communication between plants (Blande et al. 2010) or between plants and parasitic and predatory animals (Himanen et al. 2009; Pinto et al. 2007a). Even the orientation of pollinators may be disturbed, if the volatile signal compounds attracting pollinators react with atmospheric pollutants and lose their signalling value in elevated ozone atmospheres (Farré-Armengol et al. 2016).

VOCs released from biogenic and anthropogenic sources are known to participate in various physical and chemical processes in the atmosphere. These processes include, for example, atmospheric formation as well as the quenching of ozone and the formation of secondary organic aerosol (SOA) particles in reactions with ozone and OH and NO_3 radicals (Figure 9.5). Biogenic VOCs are crucial for the formation and growth of SOA particles in the atmosphere; such particles are a component of atmospheric haze (Paasonen et al. 2013) and grow by forming clusters or by combining with semi-volatile compounds on existing particles. VOCs are also involved in cloud formation: when SOA particles reach a critical size, they may form cloud condensation nuclei (CCN) (Riipinen et al. 2012). Clouds have a crucial role in the climatic control of solar radiation and precipitation, both of which substantially influence ecosystem function. There is some evidence from atmospheric models that PVOC emissions stimulated by temperature and herbivory may participate in secondary aerosol formation and may increase negative radiative forcing by aerosols (Arneth and Niinemets 2010; Joutsensaari et al. 2015). Therefore, it has been suggested that PVOCs and their secondary aerosol products may mitigate plant stress resulting from climate warming (Holopainen 2011). Mechanisms could include the reduction of solar radiative forcing by the development of CCN from biogenic SOA and the increase of photosynthesis rates by the dispersion of light into the canopy (Roderick et al. 2001; Williams et al. 2014).

In the future, PVOCs may have the potential to be exploited to mitigate climate warming, for example, by climate-smart forest management by specific tree species, their genotypes and cultivars (Rosenkranz et al. 2015). It is known that isoprene-emitting trees like *Populus* spp. may inhibit with their isoprene emissions the formation of secondary aerosols from monoterpenes (Kiendler-Scharr et al. 2009). At the same time, isoprene emissions may increase the lifetime of greenhouse gas methane in the atmosphere (Kirschke et al. 2013). Thus, using nonisoprene-emitting *Populus* spp. (Rosenkranz et al. 2015) genotypes in forest plantations may mitigate warming by affecting on two positive radiative forcing factors. Particularly in northern conifer forests, the selection of conifer provenances with high monoterpene emission capacity may mitigate warming by promoting secondary aerosol formation (Ehn et al. 2014; Joutsensaari et al. 2015) and the negative radiative forcing by aerosols.

References

Aalto, J., A. Porcar-Castell, J. Atherton et al. 2015. Onset of photosynthesis in spring speeds up monoterpene synthesis and leads to emission bursts. *Plant Cell Environ.* 38: 2299–2312.

Abe, I., Y. Takahashi, H. Morita, and H. Noguchi. 2001. Benzalacetone synthase. A novel polyketide synthase that plays a crucial role in the biosynthesis of phenylbutanones in *Rheum palmatum. Eur. J. Biochem.* 268: 3354–3359.

Aharoni, A., M.A. Jongsma, and H.J. Bouwmeester. 2005. Volatile science? Metabolic engineering of terpenoids in plants. *Trends Plant Sci.* 10: 594–602.

Alborn, H.T., T.V. Hansen, T.H. Jones et al. 2007. Disulfooxy fatty acids from the American bird grasshopper *Schistocerca americana*, elicitors of plant volatiles. *Proc. Natl. Acad. Sci. USA* 104: 12976–12981.

Alborn, H.T., T.C.J. Turlings, T.H. Jones, G. Stenhagen, J.H. Loughrin, and J.H. Tumlinson. 1997. An elicitor of plant volatiles from beet armyworm oral secretion. *Science* 276: 945–949.

Ali, M., K. Sugimoto, A. Ramadan, and G. Arimura. 2013. Memory of plant communications for priming anti-herbivore responses. *Sci. Rep.* 3: 1872.

Allmann, S. and I.T. Baldwin. 2010. Insects betray themselves in nature to predators by rapid isomerization of green leaf volatiles. *Science* 329: 1075–1078.

Amo, L., J.J. Jansen, N.M. van Dam, M. Dicke, and M.E. Visser. 2013. Birds exploit herbivore-induced plant volatiles to locate herbivorous prey. *Eco. Lett.* 16: 1348–1355.

Andreou, A. and I. Feussner. 2009. Lipoxygenases—Structure and reaction mechanism. *Phytochemistry* 70: 1504–1510.

Aragüez, I., S. Osorio, T. Hoffmann et al. 2013. Eugenol production in achenes and receptacles of strawberry fruits is catalyzed by synthases exhibiting distinct kinetics. *Plant Physiol.* 163: 946–958.

Arimura, G., K. Matsui, and J. Takabayashi. 2009. Chemical and molecular ecology of herbivore-induced plant volatiles: Proximate factors and their ultimate functions. *Plant Cell Physiol.* 50: 911–923.

Arimura, G., S. Garms, M. Maffei et al. 2008a. Herbivore-induced terpenoid emission in *Medicago truncatula*: Concerted action of jasmonate, ethylene and calcium signaling. *Planta* 227: 453–464.

Arimura, G., S. Köpke, M. Kunert et al. 2008b. Effects of feeding *Spodoptera littoralis* on Lima bean leaves: IV. Diurnal and nocturnal damage differentially initiate plant volatile emission. *Plant Physiol.* 146: 965–973.

Arimura, G., R. Ozawa, T. Shimoda, T. Nishioka, W. Boland, and J. Takabayashi. 2000. Herbivory-induced volatiles elicit defence genes in lima bean leaves. *Nature* 406: 512–515.

Arimura, G., K. Shiojiri, and R. Karban. 2010. Acquired immunity to herbivory and allelopathy caused by airborne plant emissions. *Phytochemistry* 71: 1642–1649.

Arneth, A. and U. Niinemets. 2010. Induced BVOCs: How to bug our models? *Trends Plant Sci.* 15: 118–125.

Aubourg, S., A. Lecharny, and J. Bohlmann. 2002. Genomic analysis of the terpenoid synthase (AtTPS) gene family of *Arabidopsis thaliana*. *Mol. Gen. Genomics* 267: 730–745.

Baldwin, I.T., C. Preston, M. Euler, and D. Gorham. 1997. Patterns and consequences of benzyl acetone floral emissions from *Nicotiana attenuata* plants. *J. Chem. Ecol.* 23: 2327–2343.

Bale, J.S., G.J. Masters, I.D. Hodkinson et al. 2002. Herbivory in global climate change research: Direct effects of rising temperature on insect herbivores. *Global Change Biol.* 8: 1–16.

Bartram, S., A. Jux, G. Gleixner, and W. Boland. 2006. Dynamic pathway allocation in early terpenoid biosynthesis of stress-induced lima bean leaves. *Phytochemistry* 67: 1661–1672.

Bate, N.J., J.C.M. Riley, J.E. Thompson, and S.J. Rothstein. 1998. Quantitative and qualitative differences in C_6-volatile production from the lipoxygenase pathway in an alcohol dehydrogenase mutant of *Arabidopsis thaliana*. *Physiol. Plant.* 104: 97–104.

Bate, N.J. and S.J. Rothstein. 1998. C_6-volatiles derived from the lipoxygenase pathway induce a subset of defense-related genes. *Plant J.* 16: 561–569.
Beauchamp, J., A. Wisthaler, A. Hansel et al. 2005. Ozone-induced emissions of biogenic VOC from tobacco: Relationships between ozone uptake and emission of LOX products. *Plant Cell Environ.* 28: 1334–1343.
Behnke, K., M. Loivamäki, I. Zimmer, H. Rennenberg, J.P. Schnitzler, and S. Louis. 2010. Isoprene emission protects photosynthesis in sunfleck exposed Grey poplar. *Photosynth. Res.* 104: 5–17.
Bender, M.A., T.R. Knutson, R.E. Tuleya et al. 2010. Modeled impact of anthropogenic warming on the frequency of intense Atlantic hurricanes. *Science* 327: 454–458.
Berg, A.R., C.L. Heald, K.E.H. Hartz et al. 2013. The impact of bark beetle infestations on monoterpene emissions and secondary organic aerosol formation in western North America. *Atmos. Chem. Phys.* 13: 3149–3161.
Bick, J.A. and M. Lange. 2003. Metabolic cross talk between cytosolic and plastidial pathways of isoprenoid biosynthesis: Unidirectional transport of intermediates across the chloroplast envelope membrane. *Arch. Biochem. Biophys.* 415: 146–154.
Bicsak, T.A., L.R. Kann, A. Reiter, and T. Chase, Jr. 1982. Tomato alcohol dehydrogenase: Purification and substrate specificity. *Arch. Biochem. Biophys.* 216: 605–615.
Björklund, N., G. Nordlander, and H. Bylund. 2005. Olfactory and visual stimuli used in orientation to conifer seedlings by the pine weevil, *Hylobius Abietis*. *Physiol. Entomol.* 30: 225–231.
Blande, J.D., J.K. Holopainen, and T. Li. 2010. Air pollution impedes plant-to-plant communication by volatiles. *Ecol. Lett.* 13: 1172–1181.
Blande, J.D., J.K. Holopainen, and U. Niinemets. 2014. Plant volatiles in polluted atmospheres: Stress responses and signal degradation. *Plant Cell Environ.* 37: 1892–1904.
Blande, J.D., K. Turunen, and J.K. Holopainen. 2009. Pine weevil feeding on Norway spruce bark has a stronger impact on needle VOC emissions than enhanced ultraviolet-B radiation. *Environ. Pollut.* 157: 174–180.
Blande, J.D., P. Tiiva, E. Oksanen, and J.K. Holopainen. 2007. Emission of herbivore-induced volatile terpenoids from two hybrid aspen (*Populus tremula* x *tremuloides*) clones under ambient and elevated ozone concentrations in the field. *Global Change Biol.* 13: 2538–2550.
Boatright, J., F. Negre, X. Chen et al. 2004. Understanding *in vivo* benzenoid metabolism in petunia petal tissue. *Plant Physiol.* 135: 1993–2011.
Boggia, L., B. Sgorbini, C.M. Bertea et al. 2015. Direct contact—Sorptive tape extraction coupled with gas chromatography–mass spectrometry to reveal volatile topographical dynamics of lima bean (*Phaseolus lunatus* L.) upon herbivory by *Spodoptera littoralis* Boisd. *BMC Plant Biol.* 15: 102.
Bohlmann, J. and C.I. Keeling. 2008. Terpenoid biomaterials. *Plant J.* 54: 656–669.
Bohlmann, J., G. Meyer-Gauen, and R. Croteau. 1998. Plant terpenoid synthases: Molecular biology and phylogenetic analysis. *Proc. Natl. Acad. Sci. USA* 95: 4126–4133.
Bos, J.I., D. Prince, M. Pitino, M.E. Maffei, J. Win, and S.A. Hogenhout. 2010. A functional genomics approach identifies candidate effectors from the aphid species *Myzus persicae* (green peach aphid). *PLoS Genet.* 6: e1001216.
Bricchi, I., A. Occhipinti, C.M. Bertea et al. 2012. Separation of early and late responses to herbivory in Arabidopsis by changing plasmodesmal function. *Plant J.* 73: 14–25.

Brillada, C., M. Nishihara, T. Shimoda et al. 2013. Metabolic engineering of the C_{16} homoterpene TMTT in *Lotus japonicus* through overexpression of (E,E)-geranyllinalool synthase attracts generalist and specialist predators in different manners. *New Phytol.* 200: 1200–1211.

Brilli, F., C. Barta, A. Fortunati, M. Lerdau, F. Loreto, and M. Centritto. 2007. Response of isoprene emission and carbon metabolism to drought in white poplar (*Populus alba*) saplings. *New Phytol.* 175: 244–254.

Calfapietra, C., A.E. Wiberley, T.G. Falbel et al. 2007. Isoprene synthase expression and protein levels are reduced under elevated O_3 but not under elevated CO_2 (FACE) in field-grown aspen trees. *Plant Cell Environ.* 30: 654–661.

Cardoza, Y.J. and J.H. Tumlinson. 2006. Compatible and incompatible *Xanthomonas* infections differentially affect herbivore-induced volatile emission by pepper plants. *J. Chem. Ecol.* 32: 1755–1768.

Chen, G., R. Hackett, D. Walker, A. Taylor, Z. Lin, and D. Grierson. 2004. Identification of a specific isoform of tomato lipoxygenase (TomloxC) involved in the generation of fatty acid-derived flavor compounds. *Plant Physiol.* 136: 2641–2651.

Chen, X., S. Baldermann, S. Cao et al. 2015. Developmental patterns of emission of scent compounds and related gene expression in roses of the cultivar *Rosa* x *hybrida* cv. "Yves Piaget." *Plant Physiol. Biochem.* 87: 109–114.

Chen, X.M., H. Kobayashi, M. Sakai et al. 2011. Functional characterization of rose phenylacetaldehyde reductase (PAR), an enzyme involved in the biosynthesis of the scent compound 2-phenylethanol. *J. Plant Physiol.* 168: 88–95.

Chung, S.H., C. Rosa, E.D. Scully et al. 2013. Herbivore exploits orally secreted bacteria to suppress plant defenses. *Proc. Natl. Acad. Sci. USA* 110: 15728–15733.

Colquhoun, T.A., J.Y. Kim, A.E. Wedde et al. 2011. PhMYB4 fine-tunes the floral volatile signature of *Petunia* x *hybrida* through PhC4H. *J. Exp. Bot.* 62: 1133–1143.

Colquhoun, T.A., D.M. Marciniak, A.E. Wedde et al. 2012. A peroxisomally localized acyl-activating enzyme is required for volatile benzenoid formation in a *Petunia* x *hybrida* cv. "Mitchell Diploid" flower. *J. Exp. Bot.* 63: 4821–4833.

Colquhoun, T.A., M.L. Schwieterman, J.L. Gilbert et al. 2013. Light modulation of volatile organic compounds from petunia flowers and select fruits. *Postharvest Biol. Tec.* 86: 37–44.

Copolovici, L., A. Kännaste, T. Remmel, and U. Niinemets. 2014. Volatile organic compound emissions from *Alnus glutinosa* under interacting drought and herbivory stresses. *Environ. Exp. Bot.* 100: 55–63.

Copolovici, L. and U. Niinemets. 2010. Flooding induced emissions of volatile signalling compounds in three tree species with differing waterlogging tolerance. *Plant Cell Environ.* 33: 1582–1594.

Croteau, R.B., E.M. Davis, K.L. Ringer, and M.R. Wildung. 2005. (-)-Menthol biosynthesis and molecular genetics. *Naturwissenschaften* 92: 562–577.

Crowther, T.W., H.B. Glick, K.R. Covey et al. 2015. Mapping tree density at a global scale. *Nature* 525: 201–205.

Curtius, T. and H. Franzen. 1912. Über den chemischen Bestandteile grüner Pflanzen. Über den Blätteraldehyd. *Justus Liebigs Annalen der Chemie* 390: 89–121.

Darko, E., P. Heydarizadeh, B. Schoefs, and M.R. Sabzalian. 2014. Photosynthesis under artificial light: The shift in primary and secondary metabolism. *Philos. Trans. R. Soc. Lond. B Biol. Sci.* 369: 20130243.

D'Auria, J.C., F. Chen, and E. Pichersky. 2002. Characterization of an acyltransferase capable of synthesizing benzylbenzoate and other volatile esters in flowers and damaged leaves of *Clarkia breweri*. *Plant Physiol.* 130: 466–476.

D'Auria, J.C., E. Pichersky, A. Schaub, A. Hansel, and J. Gershenzon. 2007. Characterization of a BAHD acyltransferase responsible for producing the green leaf volatile (Z)-3-hexen-1-yl acetate in *Arabidopsis thaliana*. *Plant J.* 49: 194–207.

De Moraes, C.M., W.J. Lewis, P.W. Paré, H.T. Alborn, and J.H. Tumlinson. 1998. Herbivore-infested plants selectively attract parasitoids. *Nature* 393: 1907–1922.

De Moraes, C.M., M.C. Mescher, and J.H. Tumlinson. 2001. Caterpillar-induced nocturnal plant volatiles repel conspecific females. *Nature* 410: 577–580.

Degenhardt, J., T.G. Köllner, and J. Gershenzon. 2009. Monoterpene and sesquiterpene synthases and the origin of terpene skeletal diversity in plants. *Phytochemistry* 70: 1621–1637.

Dexter, R., A. Qualley, C.M. Kish et al. 2007. Characterization of a petunia acetyltransferase involved in the biosynthesis of the floral volatile isoeugenol. *Plant J.* 49: 265–275.

Dexter, R.J., J.C. Verdonk, B.A. Underwood, K. Shibuya, E.A. Schmelz, and D.G. Clark. 2008. Tissue-specific *PhBPBT* expression is differentially regulated in response to endogenous ethylene. *J. Exp. Bot.* 59: 609–618.

Dicke, M., J.J. van Loon, and R. Soler. 2009. Chemical complexity of volatiles from plants induced by multiple attack. *Nat. Chem. Biol.* 5: 317–324.

Dudareva, N., S. Andersson, I. Orlova et al. 2005. The nonmevalonate pathway supports both monoterpene and sesquiterpene formation in snapdragon flowers. *Proc. Natl. Acad. Sci. USA* 102: 933–938.

Dudareva, N., J.C. D'Auria, K.H. Nam, R.A. Raguso, and E. Pichersky. 1998. Acetyl-CoA: Benzylalcohol acetyltransferase—an enzyme involved in floral scent production in *Clarkia breweri*. *Plant J.* 14: 297–304.

Effmert, U., J. Grosse, U.S. Rose, F. Ehrig, R. Kagi, and B. Piechulla. 2005. Volatile composition, emission pattern, and localization of floral scent emission in *Mirabilis jalapa* (Nyctaginaceae). *Am. J. Bot.* 92: 2–12.

Ehn, M., J.A. Thornton, E. Kleist et al. 2014. A large source of low-volatility secondary organic aerosol. *Nature* 506: 476–479.

Falara, V., T.A. Akhtar, T.T. Nguyen et al. 2011. The tomato terpene synthase gene family. *Plant Physiol.* 157: 770–789.

Farré-Armengol, G., J. Peñuelas, T. Li et al. 2016. Ozone degrades floral scent and reduces pollinator attraction to flowers. *New Phytol.* 209: 152–160.

Fatouros, N.E., C. Broekgaarden, G. Bukovinszkine'Kiss et al. 2008. Male-derived butterfly anti-aphrodisiac mediates induced indirect plant defense. *Proc. Natl. Acad. Sci. USA* 105: 10033–10038.

Faubert, P., P. Tiiva, A. Rinnan et al. 2010. Non-methane biogenic volatile organic compound emissions from a subarctic peatland under enhanced UV-B radiation. *Ecosystems* 13: 860–873.

Fischer, M.J., S. Meyer, P. Claudel et al. 2012. Specificity of *Ocimum basilicum* geraniol synthase modified by its expression in different heterologous systems. *J. Biotechnol.* 163: 24–29.

Fischer, N.H. 1986. The function of mono and sesquiterpenes as plant germination and growth regulators. In *The Science of Allelopathy*, eds. A.R. Putnam and C.S. Tang, 203–218. New York: John Wiley & Sons.

Friberg, M., C. Schwind, L.C. Roark, R.A. Raguso, and J.N. Thompson. 2014. Floral scent contributes to interaction specificity in coevolving plants and their insect pollinators. *J. Chem. Ecol.* 40: 955–965.

Fuentes, J.D., B.P. Hayden, M. Garstang et al. 2001. New Directions: VOCs and biosphere-atmosphere feedbacks. *Atmos. Environ.* 35: 189–191.

Gallage, N.J., E.H. Hansen, R. Kannangara et al. 2014. Vanillin formation from ferulic acid in *Vanilla planifolia* is catalysed by a single enzyme. *Nat. Commun.* 5: 4037.

Gang, D.R., N. Lavid, C. Zubieta et al. 2002. Characterization of phenylpropene O-methyltransferases from sweet basil: Facile change of substrate specificity and convergent evolution within a plant O-methyltransferase family. *Plant Cell* 14: 505–519.

Garms, S., T.G. Köllner, and W. Boland. 2010. A multiproduct terpene synthase from *Medicago truncatula* generates cadalane sesquiterpenes via two different mechanisms. *J. Org. Chem.* 75: 5590–5600.

Ghirardo, A., K. Koch, R. Taipale, I. Zimmer, J.P. Schnitzler, and J. Rinne. 2010. Determination of *de novo* and pool emissions of terpenes from four common boreal/alpine trees by $^{13}CO_2$ labelling and PTR-MS analysis. *Plant Cell Environ.* 33: 781–792.

Gil, M., R. Bottini, M. Pontin, F.J. Berli, M.V. Salomon, and P. Piccoli. 2014. Solar UV-B radiation modifies the proportion of volatile organic compounds in flowers of field-grown grapevine (*Vitis vinifera* L.) cv. Malbec. *Plant Growth Regul.* 74: 193–197.

Gill, S.S., N.A. Anjum, M. Hasanuzzaman et al. 2013. Glutathione and glutathione reductase: A boon in disguise for plant abiotic stress defense operations. *Plant Physiol. Biochem.* 70: 204–212.

Gouinguene, S.P. and T.C.J. Turlings. 2002. The effects of abiotic factors on induced volatile emissions in corn plants. *Plant Physiol.* 129: 1296–1307.

Guarino, S., S. Colazza, E. Peri et al. 2015. Behaviour-modifying compounds for management of the red palm weevil (*Rhynchophorus ferrugineus* Oliver). *Pest Manag. Sci.* 71: 1605–1610.

Guenther, A.B., C.N. Hewitt, D. Erickson et al. 1995. A global model of natural volatile organic compound emissions. *J. Geophys. Res.* 100: 8873–8892.

Guidolotti, G., A. Rey, M. Medori, and C. Calfapietra. 2016. Isoprenoids emission in *Stipa tenacissima* L.: Photosynthetic control and the effect of UV light. *Environ. Pollut.* 208: 336–344.

Gutensohn, M., A. Klempien, Y. Kaminaga et al. 2011. Role of aromatic aldehyde synthase in wounding/herbivory response and flower scent production in different Arabidopsis ecotypes. *Plant J.* 66: 591–602.

Guterman, I., T. Masci, X. Chen et al. 2006. Generation of phenylpropanoid pathway-derived volatiles in transgenic plants: Rose alcohol acetyltransferase produces phenylethyl acetate and benzyl acetate in petunia flowers. *Plant Mol. Biol.* 60: 555–563.

Haapanala, S., H. Hakola, H. Hellén, M. Vestenius, J. Levula, and J. Rinne. 2012. Is forest management a significant source of monoterpenes into the boreal atmosphere? *Biogeosciences* 9: 1291–1300.

Hallquist, M., J.C. Wenger, U. Baltensperger et al. 2009. The formation, properties and impact of secondary organic aerosol: Current and emerging issues. *Atmos. Chem. Phys.* 9: 5155–5236.

Hamilton, J.F., A.C. Lewis, T.J. Carey, J.C. Wenger, E.B.I. Garcia, and A. Muñoz. 2009. Reactive oxidation products promote secondary organic aerosol formation from green leaf volatiles. *Atmos. Chem. Phys.* 9: 3815–3823.

Harley, P., G. Deem, S. Flint, and M. Caldwell. 1996. Effects of growth under elevated UV-B on photosynthesis and isoprene emission in *Quercus gambelii* and *Mucuna pruriens*. *Global Change Biol.* 2: 149–154.

Heil, M. and R. Karban. 2010. Explaining evolution of plant communication by airborne signals. *Trends Ecol. Evol.* 25: 137–144.

Heil, M., U. Lion, and W. Boland. 2008. Defense-inducing volatiles: In search of the active motif. *J. Chem. Ecol.* 34: 601–604.

Herde, M., K. Gärtner, T.G. Köllner et al. 2008. Identification and regulation of TPS04/GES, an *Arabidopsis* geranyllinalool synthase catalyzing the first step in the formation of the insect-induced volatile C_{16}-homoterpene TMTT. *Plant Cell* 20: 1152–1168.

Himanen, S.J., T.N. Bui, M.M. Maja, and J.K. Holopainen. 2015. Utilizing associational resistance for biocontrol: Impacted by temperature, supported by indirect defence. *BMC Ecol.* 15: 16.

Himanen, S.J., A.M. Nerg, A. Nissinen et al. 2009. Effects of elevated carbon dioxide and ozone on volatile terpenoid emissions and multitrophic communication of transgenic insecticidal oilseed rape (*Brassica napus*). *New Phytol.* 181: 174–186.

Hirao, T., A. Okazawa, K. Harada, A. Kobayashi, T. Muranaka, and K. Hirata. 2012. Green leaf volatiles enhance methyl jasmonate response in Arabidopsis. *J. Biosci. Bioeng.* 114: 540–545.

Hirata, H., T. Ohnishi, H. Ishida et al. 2012. Functional characterization of aromatic amino acid aminotransferase involved in 2-phenylethanol biosynthesis in isolated rose petal protoplasts. *J. Plant Physiol.* 169: 444–451.

Holopainen, J.K. 2011. Can forest trees compensate for stress-generated growth losses by induced production of volatile compounds? *Tree Physiol.* 31: 1356–1377.

Holopainen, J.K. and J.D. Blande. 2013. Where do herbivore-induced plant volatiles go? *Front. Plant Sci.* 4: 185.

Holopainen, J.K. and J. Gershenzon. 2010. Multiple stress factors and the emission of plant VOCs. *Trends Plant Sci.* 15: 176–184.

Holopainen, J.K. and P. Kainulainen. 2004. Reproductive capacity of the grey pine aphid and allocation response of Scots pine seedlings across temperature gradients: A test of hypotheses predicting outcomes of global warming. *Can. J. Forest Res.* 34: 94–102.

Horiuchi, J., G. Arimura, R. Ozawa, T. Shimoda, J. Takabayashi, and T. Nishioka. 2001. Exogenous ACC enhances volatiles production mediated by jasmonic acid in lima bean leaves. *FEBS Lett.* 509: 332–336.

Hrycay, E.G. and S.M. Bandiera. 2012. The monooxygenase, peroxidase, and peroxygenase properties of cytochrome P450. *Arch. Biochem. Biophys.* 522: 71–89.

Ibrahim, M.A., A. Nissinen, and J.K. Holopainen. 2005. Response of *Plutella xylostella* and its parasitoid *Cotesia plutellae* to volatile compounds. *J. Chem. Ecol.* 31: 1969–1984.

IPCC. 2013. Climate change 2013: The physical science basis. *Contribution of Working Group I to the Fifth Assessment Report of the Intergovernmental Panel on Climate Change.* Cambridge, UK and New York, NY: Cambridge University Press.

Ishiguro, S., A. Kawai-Oda, J. Ueda, I. Nishida, and K. Okada. 2001. The *DEFECTIVE IN ANTHER DEHISCIENCE1* gene encodes a novel phospholipase A1 catalyzing the initial step of jasmonic acid biosynthesis, which synchronizes pollen maturation, anther dehiscence, and flower opening in Arabidopsis. *Plant Cell* 13: 2191–2209.

Jardine, K., J. Chambers, J. Holm et al. 2015. Green leaf volatile emissions during high temperature and drought stress in a central Amazon rainforest. *Plants* 4: 678–690.

Jørgensen, M.E., H.H. Nour-Eldin, and B.A. Halkier. 2015. Transport of defense compounds from source to sink: Lessons learned from glucosinolates. *Trends Plant Sci.* 20: 508–514.

Joutsensaari, J., P. Yli-Pirilä, H. Korhonen et al. 2015. Biotic stress accelerates formation of climate-relevant aerosols in boreal forests *Atmos. Chem. Phys. Discuss.* 15: 10853–10898.

Kaling, M., B. Kanawati, A. Ghirardo et al. 2015. UV-B mediated metabolic rearrangements in poplar revealed by non-targeted metabolomics. *Plant Cell Environ.* 38: 892–904.

Kaminaga, Y., J. Schnepp, G. Peel et al. 2006. Plant phenylacetaldehyde synthase is a bifunctional homotetrameric enzyme that catalyzes phenylalanine decarboxylation and oxidation. *J. Biol. Chem.* 281: 23357–23366.

Kappers, I.F., A. Aharoni, T.W.J.M. van Herpen, L.L.P. Luckerhoff, M. Dicke, and H.J. Bouwmeester. 2005. Genetic engineering of terpenoid metabolism attracts bodyguards to *Arabidopsis*. *Science* 309: 2070–2072.

Keeling, C.I., S. Weisshaar, S.G. Ralph et al. 2011. Transcriptome mining, functional characterization, and phylogeny of a large terpene synthase gene family in spruce (Picea spp.). *BMC Plant Biol.* 11: 43.

Kegge, W., V. Ninkovic, R. Glinwood, R.A. Welschen, L.A. Voesenek, and R. Pierik. 2015. Red: Far-red light conditions affect the emission of volatile organic compounds from barley (*Hordeum vulgare*), leading to altered biomass allocation in neighbouring plants. *Ann. Bot.* 115: 961–970.

Kessler, D., K. Gase, and I.T. Baldwin. 2008. Field experiments with transformed plants reveal the sense of floral scents. *Science* 321: 1200–1202.

Kiendler-Scharr, A., J. Wildt, M. Dal Maso et al. 2009. New particle formation in forests inhibited by isoprene emissions. *Nature* 461: 381–384.

Kirschke, S., P. Bousquet, P. Ciais et al. 2013. Three decades of global methane sources and sinks. *Nat. Geosci.* 6: 813–823.

Kishimoto, K., K. Matsui, R. Ozawa, and J. Takabayashi. 2005. Volatile C_6-aldehydes and allo-ocimene activate defense genes and induce resistance against *Botrytis cinerea* in *Arabidopsis thaliana*. *Plant Cell Physiol.* 46: 1093–1102.

Kishimoto, K., K. Matsui, R. Ozawa, and J. Takabayashi. 2008. Direct fungicidal activities of C6-aldehydes are important constituents for defense responses in Arabidopsis against *Botrytis cinerea*. *Phytochemistry* 69: 2127–2132.

Kivimäenpää, M., R.P. Ghimire, S. Sutinen et al. 2016. Increases of volatile organic compound emissions of Scots pine in response to elevated ozone and warming are modified by herbivory and soil nitrogen availability. *Eur. J. Forest Res.* 135: 343–360.

Kivimäenpää, M., N. Magsarjav, R. Ghimire et al. 2012. Influence of tree provenance on biogenic VOC emissions of Scots pine (*Pinus sylvestris*) stumps. *Atmos. Environ.* 60: 477–485.

Klempien, A., Y. Kaminaga, A. Qualley et al. 2012. Contribution of CoA ligases to benzenoid biosynthesis in petunia flowers. *Plant Cell* 24: 2015–2030.

Koeduka, T. 2014. The phenylpropene synthase pathway and its applications in the engineering of volatile phenylpropanoids in plants. *Plant Biotechnol.* 31: 401–407.

Koeduka, T., T.J. Baiga, J.P. Noel, and E. Pichersky. 2009a. Biosynthesis of *t*-anethole in anise: Characterization of *t*-anol/isoeugenol synthase and an *O*-methyltransferase specific for a C7-C8 propenyl side chain. *Plant Physiol.* 149: 384–394.

Koeduka, T., E. Fridman, D.R. Gang et al. 2006. Eugenol and isoeugenol, characteristic aromatic constituents of spices, are biosynthesized via reduction of a coniferyl alcohol ester. *Proc. Natl. Acad. Sci. USA* 103: 10128–10133.

Koeduka, T., K. Ishizaki, C.M. Mwenda et al. 2015. Biochemical characterization of allene oxide synthases from the liverwort *Marchantia polymorpha* and green microalgae *Klebsormidium flaccidum* provides insight into the evolutionary divergence of the plant CYP74 family. *Planta* 242: 1175–1186.

Koeduka, T., I. Orlova, T.J. Baiga, J.P. Noel, N. Dudareva, and E. Pichersky. 2009b. The lack of floral synthesis and emission of isoeugenol in *Petunia axillaris* subsp. *parodii* is due to a mutation in the *isoeugenol synthase* gene. *Plant J.* 58: 961–969.

Koeduka, T., B. Watanabe, S. Suzuki, J. Hiratake, J. Mano, and K. Yazaki. 2011. Characterization of raspberry ketone/zingerone synthase, catalyzing the alpha, beta-hydrogenation of phenylbutenones in raspberry fruits. *Biochem. Biophys. Res. Commun.* 412: 104–108.

Köllner, T.G., M. Held, C. Lenk et al. 2008. A maize (E)-β-caryophyllene synthase implicated in indirect defense responses against herbivores is not expressed in most American maize varieties. *Plant Cell* 20: 482–494.

Kolosova, N., D. Sherman, D. Karlson, and N. Dudareva. 2001. Cellular and subcellular localization of *S*-adenosyl-L-methionine: Benzoic acid carboxyl methyltransferase, the enzyme responsible for biosynthesis of the volatile ester methylbenzoate in snapdragon flowers. *Plant Physiol.* 126: 956–964.

Kos, M., B. Houshyani, A.J. Overeem et al. 2013. Genetic engineering of plant volatile terpenoids: Effects on a herbivore, a predator and a parasitoid. *Pest Manag. Sci.* 69: 302–311.

Kozlov, M.V. and E.L. Zvereva. 2015. Changes in the background losses of woody plant foliage to insects during the past 60 years: Are the predictions fulfilled? *Biol. Lett.* 11.

Kusumoto, N. and S. Shibutani. 2015. Evaporation of volatiles from essential oils of Japanese conifers enhances antifungal activity. *J. Essent. Oil Res.* 27: 380–394.

Laothawornkitkul, J., J.E. Taylor, N.D. Paul, and C.N. Hewitt. 2009. Biogenic volatile organic compounds in the Earth system. *New Phytol.* 183: 27–51.

Lee, D.S., P. Nioche, M. Hamberg, and C.S. Raman. 2008. Structural insights into the evolutionary paths of oxylipin biosynthetic enzymes. *Nature* 455: 363–368.

Lee, S., S. Badieyan, D.R. Bevan, M. Herde, C. Gatz, and D. Tholl. 2010. Herbivore-induced and floral homoterpene volatiles are biosynthesized by a single P450 enzyme (CYP82G1) in *Arabidopsis*. *Proc. Natl. Acad. Sci. USA* 107: 21205–21210.

Liu, J., F. Huang, X. Wang et al. 2014. Genome-wide analysis of terpene synthases in soybean: Functional characterization of GmTPS3. *Gene* 544: 83–92.

Long, M.C., D.A. Nagegowda, Y. Kaminaga et al. 2009. Involvement of snapdragon benzaldehyde dehydrogenase in benzoic acid biosynthesis. *Plant J.* 59: 256–265.
Loreto, F. and J.P. Schnitzler. 2010. Abiotic stresses and induced BVOCs. *Trends Plant Sci.* 15: 154–166.
Loreto, F. and V. Velikova. 2001. Isoprene produced by leaves protects the photosynthetic apparatus against ozone damage, quenches ozone products, and reduces lipid peroxidation of cellular membranes. *Plant Physiol.* 127: 1781–1787.
Lu, X., L. Zhang, F. Zhang et al. 2013. AaORA, a trichome-specific AP2/ERF transcription factor of *Artemisia annua*, is a positive regulator in the artemisinin biosynthetic pathway and in disease resistance to *Botrytis cinerea*. *New Phytol.* 198: 1191–1202.
Ma, R., J.L. Chen, D.F. Cheng, and J.R. Sun. 2010. Activation of defense mechanism in wheat by polyphenol oxidase from aphid saliva. *J. Agric. Food Chem.* 58: 2410–2418.
Maffei, M.E., A. Mithöfer, and W. Boland. 2007. Insects feeding on plants: Rapid signals and responses preceding the induction of phytochemical release. *Phytochemistry* 68: 2946–2959.
Magnard, J.L., A. Roccia, J.C. Caissard et al. 2015. Plant volatiles. Biosynthesis of monoterpene scent compounds in roses. *Science* 349: 81–83.
Mäntylä, E., G.A. Alessio, J.D. Blande et al. 2008. From plants to birds: Higher avian predation rates in trees responding to insect herbivory. *PLoS One* 3: e2832.
Marotzke, J. and P.M. Forster. 2015. Forcing, feedback and internal variability in global temperature trends. *Nature* 517: 565–570.
Martinez del Rio, C. and A. Búrquez. 1986. Nectar production and temperature dependent pollination in *Mirabilis jalapa* L. *Biotropica* 18: 28–31.
Matos, A.R. and A.T. Pham-Thi. 2009. Lipid deacylating enzymes in plants: Old activities, new genes. *Plant Physiol. Biochem.* 47: 491–503.
Matsuba, Y., T.T. Nguyen, K. Wiegert et al. 2013. Evolution of a complex locus for terpene biosynthesis in *solanum*. *Plant Cell* 25: 2022–2036.
Matsui, K. 2006. Green leaf volatiles: Hydroperoxide lyase pathway of oxylipin metabolism. *Curr. Opin. Plant Biol.* 9: 274–280.
Matsui, K., K. Sugimoto, J. Mano, R. Ozawa, and J. Takabayashi. 2012. Differential metabolisms of green leaf volatiles in injured and intact parts of a wounded leaf meet distinct ecophysiological requirements. *PloS One* 7: e36433.
Mattiacci, L., M. Dicke, and M.A. Posthumus. 1995. β-Glucosidase: An elicitor of herbivore-induced plant odor that attracts host-searching parasitic wasps. *Proc. Natl. Acad. Sci. USA* 92: 2036–2040.
McFrederick, Q.S., J.D. Fuentes, T. Roulston, J.C. Kathilankal, and M. Lerdau. 2009. Effects of air pollution on biogenic volatiles and ecological interactions. *Oecologia* 160: 411–420.
McFrederick, Q.S., J.C. Kathilankal, and J.D. Fuentes. 2008. Air pollution modifies floral scent trails. *Atmos. Environ.* 42: 2336–2348.
Mirabella, R., H. Rauwerda, S. Allmann et al. 2015. WRKY40 and WRKY6 act downstream of the green leaf volatile E-2-hexenal in Arabidopsis. *Plant J.* 83: 1082–1096.
Mithöfer, A., G. Wanner, and W. Boland. 2005. Effects of feeding *Spodoptera littoralis* on lima bean leaves. II. Continuous mechanical wounding resembling insect feeding is sufficient to elicit herbivory-related volatile emission. *Plant Physiol.* 137: 1160–1168.

Miyamoto, K., T. Shimizu, and K. Okada. 2014. Transcriptional regulation of the biosynthesis of phytoalexin: A lesson from specialized metabolites in rice. *Plant Biotechnol.* 31: 377–388.

Mosblech, A., I. Feussner, and I. Heilmann. 2009. Oxylipins: Structurally diverse metabolites from fatty acid oxidation. *Plant Physiol. Biochem.* 47: 511–517.

Muhlemann, J.K., A. Klempien, and N. Dudareva. 2014. Floral volatiles: From biosynthesis to function. *Plant Cell Environ.* 37: 1936–1949.

Muhlemann, J.K., M.O. Waelti, A. Widmer, and F.P. Schiestl. 2006. Postpollination changes in floral odor in *Silene latifolia*: Adaptive mechanisms for seed-predator avoidance? *J. Chem. Ecol.* 32: 1855–1860.

Müller, C.H. 1970. Phytotoxins as plant habitat variables. *Recent Adv. Phytochem.* 3: 106–121.

Muramoto, S., Y. Matsubara, C.M. Mwenda et al. 2015. Glutathionylation and reduction of methacrolein in tomato plants account for its absorption from the vapor phase. *Plant Physiol.* 169: 1744–1754.

Murfitt, L.M., N. Kolosova, C.J. Mann, and N. Dudareva. 2000. Purification and characterization of S-adenosyl-L-methionine: Benzoic acid carboxyl methyltransferase, the enzyme responsible for biosynthesis of the volatile ester methyl benzoate in flowers of *Antirrhinum majus*. *Arch. Biochem. Biophys.* 382: 145–151.

Nakamura, S. and A. Hatanaka. 2002. Green-leaf-derived C6-aroma compounds with potent antibacterial action that act on both Gram-negative and Gram-positive bacteria. *J. Agric. Food Chem.* 50: 7639–7644.

Nakashima, A., S.H. von Reuss, H. Tasaka et al. 2013. Traumatin- and dinortraumatin-containing galactolipids in *Arabidopsis*: Their formation in tissue-disrupted leaves as counterparts of green leaf volatiles. *J. Biol. Chem.* 288: 26078–26088.

Negre, F., C.M. Kish, J. Boatright et al. 2003. Regulation of methylbenzoate emission after pollination in snapdragon and petunia flowers. *Plant Cell* 15: 2992–3006.

Neu, J.L., T. Flury, G.L. Manney, M.L. Santee, N.J. Livesey, and J. Worden. 2014. Tropospheric ozone variations governed by changes in stratospheric circulation. *Nat. Geosci.* 7: 340–344.

Niinemets, U., S. Fares, P. Harley, and K.J. Jardine. 2014. Bidirectional exchange of biogenic volatiles with vegetation: Emission sources, reactions, breakdown and deposition. *Plant Cell Environ.* 37: 1790–1809.

Niinemets, U., F. Loreto, and M. Reichstein. 2004. Physiological and physicochemical controls on foliar volatile organic compound emissions. *Trends Plant Sci.* 9: 180–186.

Niinemets, U. and Z.H. Sun. 2015. How light, temperature, and measurement and growth [CO_2] interactively control isoprene emission in hybrid aspen. *J. Exp. Bot.* 66: 841–851.

Nishida, N., S. Tamotsu, N. Nagata, C. Saito, and A. Sakai. 2005. Allelopathic effects of volatile monoterpenoids produced by *Salvia leucophylla*: Inhibition of cell proliferation and DNA synthesis in the root apical meristem of *Brassica campestris* seedlings. *J. Chem. Ecol.* 31: 1187–1203.

Nissinen, A., L. Kristoffersen, and O. Anderbrant. 2008. Physiological state of female and light intensity affect the host-plant selection of carrot psyllid, *Trioza apicalis* (Hemiptera: Triozidae). *Eur. J. Entomol.* 105: 227–232.

Ohgami, S., E. Ono, M. Horikawa et al. 2015. Volatile glycosylation in tea plants: Sequential glycosylations for the biosynthesis of aroma beta-primeverosides are catalyzed by two *Camellia sinensis* glycosyltransferases. *Plant Physiol.* 168: 464–477.

Okada, K., H. Abe, and G. Arimura. 2015. Jasmonates induce both defense responses and communication in monocotyledonous and dicotyledonous plants. *Plant Cell Physiol.* 56: 16–27.

Orians, C. 2005. Herbivores, vascular pathways, and systemic induction: Facts and artifacts. *J. Chem. Ecol.* 31: 2231–2242.

Orlova, I., A. Marshall-Colón, J. Schnepp et al. 2006. Reduction of benzenoid synthesis in petunia flowers reveals multiple pathways to benzoic acid and enhancement in auxin transport. *Plant Cell* 18: 3458–3475.

Oster, M., J.J. Beck, R.E. Furrow, K. Yeung, and C.B. Field. 2015. In-field yellow starthistle (*Centaurea solstitialis*) volatile composition under elevated temperature and CO_2 and implications for future control. *Chemoecology* 25: 313–323.

Paasonen, P., A. Asmi, T. Petäjä et al. 2013. Warming-induced increase in aerosol number concentration likely to moderate climate change. *Nature Geosci.* 6: 438–442.

Palmer-Young, E.C., D. Veit, J. Gershenzon, and M.C. Schuman. 2015. The sesquiterpenes(*E*)-beta-farnesene and (*E*)-alpha-bergamotene quench ozone but fail to protect the wild tobacco *Nicotiana attenuata* from ozone, UVB, and drought stresses. *PLoS One* 10: e0127296.

Peñuelas, J. and M. Staudt. 2010. BVOCs and global change. *Trends Plant Sci.* 15: 133–144.

Piel, J., J. Donath, K. Bandemer, and W. Boland. 1998. Mevalonate-independent biosynthesis of terpenoid volatiles in plants: Induced and constitutive emission of volatiles. *Angew. Chem. Int. Ed.* 37: 2478–2481.

Pinto, D.M., A.M. Nerg, and J.K. Holopainen. 2007a. The role of ozone-reactive compounds, terpenes, and green leaf volatiles (glvs), in the orientation of Cotesia plutellae. *J. Chem. Ecol.* 33: 2218–2228.

Pinto, D.M., P. Tiiva, P. Miettinen et al. 2007b. The effects of increasing atmospheric ozone on biogenic monoterpene profiles and the formation of secondary aerosols. *Atmos. Environ.* 41: 4877–4887.

Pohnert, G. and W. Boland. 2002. The oxylipin chemistry of attraction and defense in brown algae and diatoms. *Nat. Prod. Rep.* 19: 108–122.

Pott, M.B., F. Hippauf, S. Saschenbrecker et al. 2004. Biochemical and structural characterization of benzenoid carboxyl methyltransferases involved in floral scent production in *Stephanotis floribunda* and *Nicotiana suaveolens*. *Plant Physiol.* 135: 1946–1955.

Qualley, A.V., J.R. Widhalm, F. Adebesin, C.M. Kish, and N. Dudareva. 2012. Completion of the core beta-oxidative pathway of benzoic acid biosynthesis in plants. *Proc. Natl. Acad. Sci. USA* 109: 16383–16388.

Rådmark, O., O. Werz, D. Steinhilber, and B. Samuelsson. 2007. 5-Lipoxygenase: Regulation of expression and enzyme activity. *Trends Biochem. Sci.* 32: 332–341.

Raguso, R.A., R.A. Levin, S.E. Foose, M.W. Holmberg, and L.A. McDade. 2003. Fragrance chemistry, nocturnal rhythms and pollination "syndromes" in *Nicotiana*. *Phytochemistry* 63: 265–284.

Raguso, R.A. and E. Pichersky. 1995. Floral volatiles from *Clarkia brewevi* and *C. concinna* (Onagraceae): Recent evolution of floral scent and moth pollination. *Plant Syst. Evol.* 194: 55–67.

Räisänen, T., A. Ryyppö, R. Julkunen-Tiitto, and S. Kellomäki. 2008a. Effects of elevated CO_2 and temperature on secondary compounds in the needles of Scots pine (*Pinus sylvestris* L.). *Trees-Struct. Funct.* 22: 121–135.

Räisänen, T., A. Ryyppö, and S. Kellomäki. 2008b. Impact of timber felling on the ambient monoterpene concentration of a Scots pine (*Pinus sylvestris* L.) forest. *Atmos. Environ.* 42: 6759–6766.
Riddiford, L.M. 1967. *Trans*-2-hexenal: Mating stimulant for polyphemus moths. *Science* 158: 139–141.
Riipinen, I., T. Yli-Juuti, J.R. Pierce et al. 2012. The contribution of organics to atmospheric nanoparticle growth. *Nat. Geosci.* 5: 453–458.
Rinnan, R., M. Steinke, T. McGenity, and F. Loreto. 2014. Plant volatiles in extreme terrestrial and marine environments. *Plant Cell Environ.* 37: 1776–1789.
Roderick, M.L., G.D. Farquhar, S.L. Berry, and I.R. Noble. 2001. On the direct effect of clouds and atmospheric particles on the productivity and structure of vegetation. *Oecologia* 129: 21–30.
Rohrbeck, D., D. Buss, U. Effmert, and B. Piechulla. 2006. Localization of methyl benzoate synthesis and emission in *Stephanotis floribunda* and *Nicotiana suaveolens* flowers. *Plant Biol.* 8: 615–626.
Rosenkranz, M., T.A.M. Pugh, J.P. Schnitzler, and A. Arneth. 2015. Effect of land-use change and management on biogenic volatile organic compound emissions—selecting climate-smart cultivars. *Plant Cell Environ.* 38: 1896–1912.
Rosenstiel, T.N., E.E. Shortlidge, A.N. Melnychenko, J.F. Pankow, and S.M. Eppley. 2012. Sex-specific volatile compounds influence microarthropod-mediated fertilization of moss. *Nature* 489: 431–433.
Rowen, E. and I. Kaplan. 2016. Eco-evolutionary factors drive induced plant volatiles: Meta-analysis. *New Phytol.* 210: 284–294.
Schiestl, F.P. 2015. Ecology and evolution of floral volatile-mediated information transfer in plants. *New Phytol.* 206: 571–577.
Schmelz, E.A., M.J. Carroll, S. LeClere et al. 2006. Fragments of ATP synthase mediate plant perception of insect attack. *Proc. Natl. Acad. Sci. USA* 103: 8894–8899.
Schuck, S., M. Kallenbach, I.T. Baldwin, and G. Bonaventure. 2014. The *Nicotiana attenuata* GLA1 lipase controls the accumulation of *Phytophthora parasitica*-induced oxylipins and defensive secondary metabolites. *Plant Cell Environ.* 37: 1703–1715.
Seidl, R., M.J. Schelhaas, W. Rammer, and P.J. Verkerk. 2014. Increasing forest disturbances in Europe and their impact on carbon storage. *Nat. Clim. Change* 4: 806–810.
Shalit, M., I. Guterman, H. Volpin et al. 2003. Volatile ester formation in roses. Identification of an acetyl-coenzyme A. Geraniol/Citronellol acetyltransferase in developing rose petals. *Plant Physiol.* 131: 1868–1876.
Shimada, T., T. Endo, H. Fujii, A. Rodriguez, L. Peña, and M. Omura. 2014. Characterization of three linalool synthase genes from *Citrus unshiu* Marc. and analysis of linalool-mediated resistance against *Xanthomonas citri* subsp. citri and *Penicilium italicum* in citrus leaves and fruits. *Plant Sci.* 229: 154–166.
Shimoda, T. 2010. A key volatile infochemical that elicits a strong olfactory response of the predatory mite *Neoseiulus californicus*, an important natural enemy of the two-spotted spider mite *Tetranychus urticae*. *Exp. Appl. Acarol.* 50: 9–22.
Shimoda, T., M. Nishihara, R. Ozawa, J. Takabayashi, and G. Arimura. 2012. The effect of genetically enriched (*E*)-β-ocimene and the role of floral scent in the attraction of the predatory mite *Phytoseiulus persimilis* to spider mite-induced volatile blends of torenia. *New Phytol.* 193: 1009–1021.

Shiojiri, K., K. Kishimoto, R. Ozawa et al. 2006a. Changing green leaf volatile biosynthesis in plants: An approach for improving plant resistance against both herbivores and pathogens. *Proc. Natl. Acad. Sci. USA* 103: 16672–16676.

Shiojiri, K., R. Ozawa, and J. Takabayashi. 2006b. Plant volatiles, rather than light, determine the nocturnal behavior of a caterpillar. *PLoS Biol.* 4: e164.

Siipola, S.M., T. Kotilainen, N. Sipari et al. 2015. Epidermal UV-A absorbance and whole-leaf flavonoid composition in pea respond more to solar blue light than to solar UV radiation. *Plant Cell Environ.* 38: 941–952.

Šimpraga, M., H. Verbeeck, M. Demarcke et al. 2011. Clear link between drought stress, photosynthesis and biogenic volatile organic compounds in *Fagus sylvatica* L. *Atmos. Environ.* 45: 5254–5259.

Sohrabi, R., J.H. Huh, S. Badieyan et al. 2015. In planta variation of volatile biosynthesis: An alternative biosynthetic route to the formation of the pathogen-induced volatile homoterpene DMNT via triterpene degradation in Arabidopsis roots. *Plant Cell* 27: 874–890.

Speirs, J., E. Lee, K. Holt et al. 1998. Genetic manipulation of alcohol dehydrogenase levels in ripening tomato fruit affects the balance of some flavor aldehydes and alcohols. *Plant Physiol.* 117: 1047–1058.

Spiteller, D., G. Pohnert, and W. Boland. 2001. Absolute configuration of volicitin, an elicitor of plant volatile biosynthesis from lepidopteran larvae. *Tetra. Lett.* 42: 1483–1485.

Spitzer-Rimon, B., M. Farhi, B. Albo et al. 2012. The R2R3-MYB-like regulatory factor EOBI, acting downstream of EOBII, regulates scent production by activating *ODO1* and structural scent-related genes in petunia. *Plant Cell* 24: 5089–5105.

Spitzer-Rimon, B., E. Marhevka, O. Barkai et al. 2010. *EOBII*, a gene encoding a flower-specific regulator of phenylpropanoid volatiles' biosynthesis in petunia. *Plant Cell* 22: 1961–1976.

Spyropoulou, E.A., M.A. Haring, and R.C. Schuurink. 2014. Expression of Terpenoids 1, a glandular trichome-specific transcription factor from tomato that activates the terpene synthase 5 promoter. *Plant Mol. Biol.* 84: 345–357.

Stökl, J., J. Brodmann, A. Dafni, M. Ayasse, and B.S. Hansson. 2011. Smells like aphids: Orchid flowers mimic aphid alarm pheromones to attract hoverflies for pollination. *Proc. Biol. Sci.* 278: 1216–1222.

Stumpe, M., J. Bode, C. Göbel et al. 2006. Biosynthesis of C9-aldehydes in the moss *Physcomitrella patens*. *Biochim. Biophys. Acta* 1761: 301–312.

Sugimoto, K., K. Matsui, Y. Iijima et al. 2014. Intake and transformation to a glycoside of (Z)-3-hexenol from infested neighbors reveals a mode of plant odor reception and defense. *Proc. Natl. Acad. Sci. USA* 111: 7144–7149.

Sun, X.L., G.C. Wang, Y. Gao, X.Z. Zhang, Z.J. Xin, and Z.M. Chen. 2014. Volatiles emitted from tea plants infested by *Ectropis obliqua* larvae are attractive to conspecific moths. *J. Chem. Ecol.* 40: 1080–1089.

Sun, Z., K. Hüve, V. Vislap, and U. Niinemets. 2013. Elevated [CO_2] magnifies isoprene emissions under heat and improves thermal resistance in hybrid aspen. *J. Exp. Bot.* 64: 5509–5523.

Suzuki, S., T. Koeduka, A. Sugiyama, K. Yazaki, and T. Umezawa. 2014. Microbial production of plant specialized metabolites. *Plant Biotechnol.* 31: 465–482.

Talkkari, A., H. Peltola, S. Kellomäki, and H. Strandman. 2000. Integration of component models from the tree, stand and regional levels to assess the risk of wind damage at forest margins. *Forest Ecol. Manag.* 135: 303–313.

Tholl, D., R. Sohrabi, J.H. Huh, and S. Lee. 2011. The biochemistry of homoterpenes—common constituents of floral and herbivore-induced plant volatile bouquets. *Phytochemistry* 72: 1635–1646.
Tieman, D.M., H.M. Loucas, J.Y. Kim, D.G. Clark, and H.J. Klee. 2007. Tomato phenylacetaldehyde reductases catalyze the last step in the synthesis of the aroma volatile 2-phenylethanol. *Phytochemistry* 68: 2660–2669.
Tiiva, P., R. Räsänen, P. Faubert et al. 2007. Isoprene emission from a subarctic peatland under enhanced UV-B radiation. *New Phytol.* 176: 346–355.
Torrens-Spence, M.P., R. von Guggenberg, M. Lazear, H. Ding, and J. Li. 2014. Diverse functional evolution of serine decarboxylases: Identification of two novel acetaldehyde synthases that uses hydrophobic amino acids as substrates. *BMC Plant Biol.* 14: 247.
Trapp, S.C. and R.B. Croteau. 2001. Genomic organization of plant terpene synthases and molecular evolutionary implications. *Genetics* 158: 811–832.
Underwood, B.A., D.M. Tieman, K. Shibuya et al. 2005. Ethylene-regulated floral volatile synthesis in petunia corollas. *Plant Physiol.* 138: 255–266.
Unger, N. 2014. Human land-use-driven reduction of forest volatiles cools global climate. *Nat. Clim. Change* 4: 907–910.
Unsicker, S.B., G. Kunert, and J. Gershenzon. 2009. Protective perfumes: The role of vegetative volatiles in plant defense against herbivores. *Curr. Opin. Plant Biol.* 12: 479–485.
Valolahti, H., M. Kivimäenpää, P. Faubert, A. Michelsen, and R. Rinnan. 2015. Climate change-induced vegetation change as a driver of increased subarctic biogenic volatile organic compound emissions. *Global Change Biol.* 21: 3478–3488.
van der Niet, T., A. Jurgens, and S.D. Johnson. 2015. Is the timing of scent emission correlated with insect visitor activity and pollination in long-spurred *Satyrium* species? *Plant Biol.* 17: 226–237.
Van Moerkercke, A., C.S. Galván-Ampudia, J.C. Verdonk, M.A. Haring, and R.C. Schuurink. 2012a. Regulators of floral fragrance production and their target genes in petunia are not exclusively active in the epidermal cells of petals. *J. Exp. Bot.* 63: 3157–3171.
Van Moerkercke, A., M.A. Haring, and R.C. Schuurink. 2011. The transcription factor EMISSION OF BENZENOIDS II activates the MYB *ODORANT1* promoter at a MYB binding site specific for fragrant petunias. *Plant J.* 67: 917–928.
Van Moerkercke, A., M.A. Haring, and R.C. Schuurink. 2012b. A model for combinatorial regulation of the petunia R2R3-MYB transcription factor *ODORANT1*. *Plant Sig. Behav.* 7: 518–520.
Van Moerkercke, A., I. Schauvinhold, E. Pichersky, M.A. Haring, and R.C. Schuurink. 2009. A plant thiolase involved in benzoic acid biosynthesis and volatile benzenoid production. *Plant J.* 60: 292–302.
Vaughn, S.F. and G.F. Spencer. 1993. Volatile monoterpenes as potential parent structures for new herbicides. *Weed Sci.* 41: 114–119.
Vaughn, S.F. and G.F. Spencer. 1996. Synthesis and herbicidal activity of modified monoterpenes structurally similar to cimmethylin. *Weed Sci.* 44: 7–11.
Velikova, V., T. La Mantia, M. Lauteri, M. Michelozzi, I. Nogues, and F. Loreto. 2012. The impact of winter flooding with saline water on foliar carbon uptake and the volatile fraction of leaves and fruits of lemon (*Citrus* x *limon*) trees. *Funct. Plant Biol.* 39: 199–213.

Verdonk, J.C., M.A. Haring, A.J. van Tunen, and R.C. Schuurink. 2005. *ODORANT1* regulates fragrance biosynthesis in petunia flowers. *Plant Cell* 17: 1612–1624.
Vickers, C.E., M. Possell, C.I. Cojocariu et al. 2009. Isoprene synthesis protects transgenic tobacco plants from oxidative stress. *Plant Cell Environ.* 32: 520–531.
Vuorinen, T., A.M. Nerg, and J.K. Holopainen. 2004a. Ozone exposure triggers the emission of herbivore-induced plant volatiles, but does not disturb tritrophic signalling. *Environ. Pollut.* 131: 305–311.
Vuorinen, T., A.M. Nerg, M.A. Ibrahim, G.V.P. Reddy, and J.K. Holopainen. 2004b. Emission of *Plutella xylostella*-induced compounds from cabbages grown at elevated CO_2 and orientation behavior of the natural enemies. *Plant Physiol.* 135: 1984–1992.
Vuorinen, T., A.M. Nerg, L. Syrjälä, P. Peltonen, and J.K. Holopainen. 2007. *Epirrita autumnata* induced VOC emission of silver birch differ from emission induced by leaf fungal pathogen. *Arthropod-Plant Interact.* 1: 159–165.
Vuorinen, T., A.M. Nerg, E. Vapaavuori, and J.K. Holopainen. 2005. Emission of volatile organic compounds from two silver birch (*Betula pendula* Roth) clones grown under ambient and elevated CO_2 and different O_3 concentrations. *Atmos. Environ.* 39: 1185–1197.
Wang, J., N. Dudareva, S. Bhakta, R.A. Raguso, and E. Pichersky. 1997. Floral scent production in *Clarkia breweri* (Onagraceae). II. Localization and developmental modulation of the enzyme S-adenosyl-L-methionine: (Iso)eugenol O-methyltransferase and phenylpropanoid emission. *Plant Physiol.* 114: 213–221.
Wasternack, C. and B. Hause. 2013. Jasmonates: Biosynthesis, perception, signal transduction and action in plant stress response, growth and development. An update to the 2007 review in *Annals of Botany*. *Ann. Bot.* 111: 1021–1058.
Watanabe, T. 1958. Substances in mulberry leaves which attract silkworm larvæ (*Bombyx mori*). *Nature* 182: 325–326.
Widhalm, J.R., R. Jaini, J.A. Morgan, and N. Dudareva. 2015. Rethinking how volatiles are released from plant cells. *Trends Plant Sci.* 20: 545–550.
Williams, M., E.B. Rastetter, L. Van der Pol, and G.R. Shaver. 2014. Arctic canopy photosynthetic efficiency enhanced under diffuse light, linked to a reduction in the fraction of the canopy in deep shade. *New Phytol.* 202: 1267–1276.
Williamson, C.E., R.G. Zepp, R.M. Lucas et al. 2014. Solar ultraviolet radiation in a changing climate. *Nat. Clim. Change* 4: 434–441.
Wu, S., M. Schalk, A. Clark, R.B. Miles, R. Coates, and J. Chappell. 2006. Redirection of cytosolic or plastidic isoprenoid precursors elevates terpene production in plants. *Nat. Biotechnol.* 24: 1441–1447.
Yamauchi, Y., A. Hasegawa, A. Taninaka, M. Mizutani, and Y. Sugimoto. 2011. NADPH-dependent reductases involved in the detoxification of reactive carbonyls in plants. *J. Biol. Chem.* 286: 6999–7009.
Yamauchi, Y., M. Kunishima, M. Mizutani, and Y. Sugimoto. 2015. Reactive short-chain leaf volatiles act as powerful inducers of abiotic stress-related gene expression. *Sci. Rep.* 5: 8030.
Yoshinaga, N., T. Aboshi, C. Ishikawa et al. 2007. Fatty acid amides, previously identified in caterpillars, found in the cricket *Teleogryllus taiwanemma* and fruit fly *Drosophila melanogaster* larvae. *J. Chem. Ecol.* 33: 1376–1381.
Yoshinaga, N., H.T. Alborn, T. Nakanishi et al. 2010. Fatty acid-amino acid conjugates diversification in Lepidopteran caterpillars. *J. Chem. Ecol.* 36: 319–325.

Zheng, D. and G. Hrazdina. 2008. Molecular and biochemical characterization of benzalacetone synthase and chalcone synthase genes and their proteins from raspberry (*Rubus idaeus* L.). *Arch. Biochem. Biophys.* 470: 139–145.

Zhu, F., C. Broekgaarden, B.T. Weldegergis et al. 2015. Parasitism overrides herbivore identity allowing hyperparasitoids to locate their parasitoid host using herbivore-induced plant volatiles. *Mol. Ecol.* 24: 2886–2899.

chapter ten

Microbial volatiles and their biotechnological applications

Birgit Piechulla and Marie Chantal Lemfack

Contents

10.1 Introduction 239
10.2 Most frequently emitted mVOCs 240
10.3 Habitats of mVOC producers 241
10.4 Applications of mVOCs 242
    10.4.1 mVOCs of foodstuff 243
    10.4.2 mVOCs as indicators of damp buildings and other hardware 246
    10.4.3 mVOCs for the perfume industry 248
    10.4.4 mVOCs as the next generation of biofuel 248
    10.4.5 mVOCs as biocontrol agents in agriculture 249
    10.4.6 mVOCs for chemotyping and diagnostic tools 251
10.5 Conclusion 252
Acknowledgments 252
References 252

10.1 Introduction

The capability of the emission of volatile organic compounds (VOCs) is well known for plants and animals. Such volatile compounds are characterized by their molecular weights of less than 300 Da, high vapor pressures, low boiling points, and low polarities. Due to these features, evaporation and distribution into the atmosphere as well as into the air- and water-filled pores below ground are facilitated. These airborne signals exhibit the potential to act as infochemicals for inter- and intraspecific communication in different habitats even over long distances (Kai et al. 2009; Wenke et al. 2010). A comprehensive source of such volatiles is constituted by the microbial world, which has been overlooked in the past. The microbial VOCs (mVOCs) affecting other organisms play a role in plant/fungi-microbe and animal/human-microbe interactions including microbial pathogens ("the bad") as well as microbes with protecting potential ("the good") (Bailly and Weisskopf 2012; Schenkel et al. 2015; Wenke et al. 2012a). Central questions include: Which volatiles are emitted by microorganisms? Which impact on fitness (health), development and growth do they have on the receiving organism? Are these mVOCs useful for any applications?

At present, headspace analyses of around 490 microbial species resulted in the identification of around 1200 volatiles (for review, see Effmert et al. 2012; Lemfack et al. 2014; Schulz and Dickschat 2007), which are divided into 48 chemical classes dominated by alcohols, alkenes, ketones, and terpenoids (Wenke et al. 2012b). Due to the incredible microbial diversity (10,000 species are known, more than 1 million are expected on Earth) it is foreseen that the actual number of known microbial volatiles represents just the "tip of the iceberg." Furthermore, microorganisms have not been systematically investigated regarding their capabilities of volatile emissions but the present available results indicate that microbes are a good source for novel and unusual volatiles (Lemfack et al. 2014; Von Reuss et al. 2010). The biological and ecological functions of mVOCs are diverse, for example: (i) they play a role in the food chain of the microbial loop since they are assimilated and incorporated into organic matter (bioconversion); (ii) they influence physiological processes in various target organisms (e.g., laccase activity, nitrification, and nitrogen mineralization); (iii) they function as electron acceptors or donors to support metabolic reactions; (iv) they play a role in quorum sensing/quenching; (v) they act as defense compounds against fungi, nematodes/animals, and bacteria; (vi) they act as communication signals; or (vii) their functions remain to be elusive (summarized in Effmert et al. 2012; Kai et al. 2009; Wenke et al. 2010 and 2012). Nevertheless, the detailed reactions and adaptations at the physiological, transcriptional, protein and metabolic levels of the target organisms were only recently investigated (Bailly and Weisskopf 2012; Wenke et al. 2012a).

10.2 Most frequently emitted mVOCs

The emission of microbial volatiles is commonly—often unconsciously—recognized; for example, the typical smell of cheese varieties, the aroma of wine, and the characteristic odor of mushrooms derive from microorganisms. Furthermore, the earthy, muddy smell in a wet forest is due to the production of geosmin and other volatiles released by *Streptomyces* species. Microbiologists are trained to recognize indole characteristically emitted by *Escherichia coli* and butyric acid released by *Clostridium* spp. These examples indicate that microorganisms contribute significantly to the odors present in our environment. The most comprehensive

Chapter ten: Microbial volatiles 241

Table 10.1 Nineteen most cited bacterial volatiles in the mVOC database

	mVOC name	Chemical classification	Number of bacteria emitting the compound
1	2-Phenylethanol	Alcohol	100
2	3-Methylbutan-1-ol	Alcohol	90
3	Dimethyl disulfide	Sulfide	88
4	Dimethyl trisulfide	Sulfide	79
5	Undecan-2-one	Ketone	54
6	Benzyl alcohol	Alcohol	53
7	Geosmin	Terpenoid	52
8	Tetradecanoic acid	Carboxylic acid	52
9	Acetic acid	Carboxylic acid	47
10	2-Aminoacetophenone	Ketone	43
11	Benzaldehyde	Aldehyde	43
12	Nonan-2-one	Ketone	43
13	Acetoin	Ketone	42
14	1-Undecene	Alkene	41
15	2-Methylpropan-1-ol	Alcohol	41
16	3-Methylbutanoic acid	Carboxylic acid	41
17	Dodecanoic acid	Carboxylic acid	41
18	Acetone	Ketone	39
19	1-Heptanol	Alcohol	36

summary of microbial volatiles is found in the mVOC database (Lemfack et al. 2014). By data mining, the 19 most frequently emitted mVOCs of bacteria (Table 10.1) and the 20 most frequently emitted mVOCs of fungi (Table 10.2) were determined. The following compounds are significantly more often released by bacteria: (1) 2-phenylethanol, (2) 3-methylbutan-1-ol, (3) dimethyl disulfide, and (4) dimethyl trisulfide; and 1-octen-3-ol is most frequently emitted by fungi. These most abundant mVOCs derive from three different metabolic pathways: (1) the shikimate pathway synthesizing the amino acid phenylalanine (phenylpropanoid biosynthesis), (2) reduction product of isovaleric acid or caprylic acid (fatty acid biosynthesis), and (3) sulfur metabolism. It is interesting to note that alcohols, ketones, and carboxylic acids are the most frequently emitted mVOCs, while terpenoids and pyrazines seldom appear in the VOC spectra of microorganisms.

10.3 Habitats of mVOC producers

Microorganisms appear ubiquitously, in various—even extreme—habitats and ecological niches (Horikoshi et al. 2011). The mVOC database was used to extract the habitats of which volatile emitting microorganisms

Table 10.2 Twenty most cited fungal volatiles in the mVOC database

	mVOC name	Chemical classification	Number of fungi emitting the compound
1	1-Octen-3-ol	Alcohol	62
2	2-Pentylfuran	Furan	44
3	Hexan-1-ol	Alcohol	43
4	2-Pentanol	Alcohol	42
5	3-Methylbutanal	Aldehyde	41
6	2-Methylbutanal	Aldehyde	39
7	Hexanal	Aldehyde	38
8	1-Heptanol	Alcohol	37
9	2-Pentanone	Ketone	36
10	2-Ethyl-1-hexanol	Alcohol	34
11	Benzaldehyde	Aldehyde	33
12	Nonanal	Aldehyde	33
13	Decanal	Aldehyde	31
14	Heptanal	Aldehyde	31
15	Styrene	Benzenoid	30
16	1,2,4-Trimethylbenzene	Alkane	29
17	1-Nonanol	Alcohol	29
18	Butan-4-olide	Lactone	29
19	Naphtalene	Alkene	29
20	Pentanal	Aldehyde	29

have been isolated thus far. The microorganisms listed in the mVOC database were obtained from 10 distinct habitats: (1) animals, (2) human/clinical sources, (3) food products, (4) fresh water, (5) humans, (6) marine environment, (7) plants, (8) plant waste, (9) rhizosphere, and (10) soil. Most species listed in the database are of plant sources (70), aquatic environment (66), and from the soil (61). We correlated the VOC profiles of the microorganismal species with the habitat where they originated from (multivariate analysis) (Figure 10.1). No habitat-specific VOC spectra became apparent and it is suspected that it is not the location of isolation, but rather the nutritional supply (growth media) and metabolic capabilities that are relevant for the emission profiles (Fiddaman and Rossall 1993).

10.4 Applications of mVOCs

The biological and ecological functions of the microbial volatiles are very diverse and manifold, and often the biological relevances of mVOCs are not known or understood. In addition to this lack of knowledge, mVOCs

Chapter ten: Microbial volatiles 243

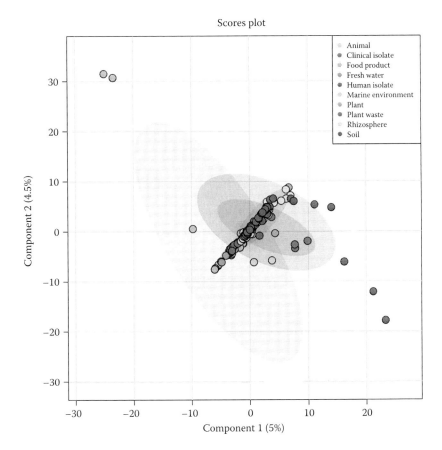

Figure 10.1 (**See color insert.**) Principal component analysis of mVOC-emitting microorganisms of different habitats. Data were extracted from the mVOC database and multivariate analyses were performed using the online comprehensive tool for metabolomic data analysis "MetaboAnalist 3.0." (From Xia, J. et al. 2015. *Nucl. Acid Res.* 43: W251–W257.)

are furthermore important cues for applications. Three major areas of implications are addressed: the agricultural, the medical, and the biotechnological applications (Figure 10.2).

10.4.1 mVOCs of foodstuff

Flavor involves our perception of sugars, organic acids, and of a diverse group of volatile metabolites produced by multiple metabolic pathways. Although the human nose can distinguish many volatiles, there is a given limitation in the detection and proper description of the relevant smell.

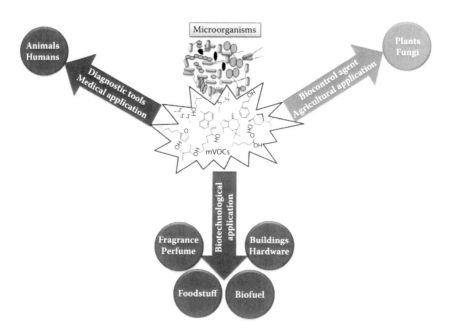

Figure 10.2 Overview of mVOC applications.

Nevertheless, it is still a challenge to improve the flavor of many modern fruit and vegetables. Likewise, the microbial-based volatile aromas of foodstuff such as wine, dairy products, and mushroom (including truffle) are continuously analyzed.

Truffles emit up to 200 volatile metabolites, typical components have a chain length of eight carbon atoms, for example, *trans*-2-octenal, 1-octen-3-ol, and octanol, which are the main components of the mushroom odor (Buzzini et al. 2005; Chitarra et al. 2004) (Table 10.2). The various types of truffles differ in their scent spectra, *Tuber borchii* and *Tuber melanosporum* are distinct to *Tuber indicum* or *Tuber magnatum* (most expensive Piedmont truffle) (Splivallo et al. 2007). New investigations revealed that truffles live in symbiosis with various yeasts, "guest" bacteria, and filamentous fungi, which also contribute to the scent bouquet (Buzzini et al. 2005; Vahdatzadeh et al. 2015). In the case of the white truffle *T. borchii*, it was shown that cyclic sulfur volatiles (thiophene derivatives) actually derive from bacteria inhabiting truffle fruiting bodies (Splivallo et al. 2014). Interestingly, the same thiophene derivatives are also the most important odorants which contribute to human sensed aroma in the latter species (Splivallo and Ebeler 2015). It is believed that truffle fruiting bodies produce their smell to attract mammals and rodents to locate the fungus underground. Actually out of the hundreds of volatiles produced by truffles, only a single one—dimethyl sulfide—has been convincingly implicated in the attraction of mammals (Talou et al. 1990). The role of other truffle volatiles as a location cue is of great interest for the food industry.

Phenolic content, colour and volatile compounds are the most important *wine* quality attributes. The wine aroma is very much influenced by the bacterial and yeast alcoholic fermentation, when sugars are metabolized into acetic acid as well as ethanol. Acetic esters, ethyl esters and higher alcohols are the dominant aroma compounds, for example, ethyl hexanoate and ethyl octanoate contribute to fruity aromas. Defined bacterial starter cultures are applied to grape juice during wine production. The time point is critical and influences significantly the product outcome. Malolactic fermentation, the decarboxlation of L-malic acid to L-lactic acid is conducted by lactic acid bacteria, for example, *Oenococcus oeni*, results in a slight increase in wine pH and provides microbial stability and positively influences wine aroma and flavour (Abrahamse and Bartowsky 2012). A prolonged or delayed malolactic fermentation increases the risk of spoilage by microorganisms such as *Lactobacillus*, *Pediococcus* and *Brettanomyces* species that may produce, for example, biogenic amines or 4-ethyl-phenol (Curtin et al. 2007; Gerbaux et al. 2009).

Microorganisms play an important role in the development of *dairy product* (cheese, yoghurt, curd, etc.) flavor. For example, more than 100 volatiles, sometimes only present in trace amounts, are found in yoghurt, and most flavor compounds are produced from lipolysis of milkfat and microbial transformation of lactose and citrate (Cheng 2010). Most important for the aroma are acetaldehyde, diacetyl, acetoin, acetone, and 2-butanone, while off-flavor compounds appear during extended shelf-lifes due to lipid oxidation. Since flavor is a very important characteristic from the consumer's point of view, new strains isolated from dairy and nondiary environments are tested for their production of odor-active volatile compounds to modulate dairy product flavors (Pogacic et al. 2015). Multivariate analyses are used to evaluate mVOC profiles of typical milk fermenting bacteria such as *Leuconostoc lactis*, *Lactobacillus* spp., *Brachybacterium* spp., *Brevibacterium* sp., and *Propionibacterium* sp. The survey identified 52 mVOCs and certain VOCs such as ethyl esters, sulfur compounds, branched chain alcohols and acids, and diacetyl as well as related carbonyl compounds turned out to be characteristic for each bacterial species (Pogacic et al. 2015). It was also possible to differentiate mVOC profiles of, for example, the New Zealand and the Italian parmesan cheese based on butanoic acid, phenylacetaldehyde, ethyl butyrate, acetaldehyde and methylbutanals (Langford et al. 2012). Thus, microbial volatile aroma compounds are important tools to assess dairy product quality and flavor although much work is still necessary to understand the complete formation of aromas and flavors. Qualitative and quantitative analyses are surely a first step to achieve this goal, but the volatile compound-matrix interaction, release mechanism of VOCs and synergistic actions need to be addressed to assess their aroma contributions (Cheng 2010).

The most well-known *"off-flavor"* volatile geosmin is emitted by soil bacteria such as *Streptomyces* spp., *Anabena* spp., *Oscillatoria* spp., myxobacteria, soil-dwelling, aquatic and airborne fungi. Geosmin has the characteristic earthy and moldy odor and is recognized as a volatile of contaminated food, wine and water (Darriet et al. 2000). The repugnant smell of rotting organic matter results from the release of bacterial and fungal volatiles, consequently mVOCs are good markers for spoiled foodstuff, for example, meat, bread, vegetables and fish. mVOCs used as indicators in food industry will be exemplified here for fish processing. The European sea bass with its white flesh and low fat content is a popular farmed fish. It is very perishable due to microbial spoilage (Gram and Hass 1996). Typical spoilage microorganisms reach high population densities and produce several metabolites (CSIs: chemical spoilage indices), which are responsible for the off-odor resulting in their organoleptic rejection (in Parlapani et al. 2015). During fish deterioration, the production of microbial volatile compounds such as trimethylamine, dimethylamine, and ammonia is measured with a hydrogel-pH-electrode based near-field passive volatile sensor (Bhadra et al. 2015). Other mVOCs have also been studied as potential indicators for CSIs for spoilage/freshness evaluation and became important measures for food quality when the levels vary between the initial and rejection day of seafood. Most of the 40 mVOCs determined on sea bass appeared sporadically or fluctuated, while, for example, 3-methyl-1-butanol, 2-methyl-1-butanol and the ethyl esters acetate, propionate and isobutyrate increased during storage. Various other mVOCs have been reported as metabolites released by *Pseudomonas* spp., *Schewanella* spp., *Enterobacteriaceae*, and *Brochothrix thermosophacta* during fish and/or meat spoilage and specifically ethyl esters were related to *Pseudomonas* activity (Casaburi et al. 2014). Indeed, esters were found only in gutted sea bass stored under air where the aerobic conditions enhance the growth of pseudomonads (Parlapani et al. 2015), while ethyl acetate increased in cod (Olafsdottir et al. 2005) and pangasious filets (Noseda et al. 2012). Acetic acid was mainly attributed to the growth of *B. thermosophacta*. These examples indicate that the characteristic mVOC profiles (= fingerprints) depend on storage conditions (air, vaccum, temperature), fish or meat batches, and on the microorganisms, which preferentially colonize the fish species or the kind of meat. The development of biosensors based on mVOCs is a reachable goal assessing freshness and shelf-life of foodstuff.

10.4.2 mVOCs as indicators of damp buildings and other hardware

In the 1990s, mVOCs were used as indicators for indoor air environment (Bayer and Crow 1994; Wessen et al. 1995). With this technique, it was possible to detect hidden microbial growth behind interior surfaces without destructing the building because it was assumed that gases (mVOCs) may enter the indoor air more easily than fungal or bacterial spores (Lorenz et al. 2002; Wessen et al. 1999). The term "sick building syndrome" (SBS) was coined in the last decade to refer to a set of symptoms that are experienced by the occupants of a building with poor air quality (Polizzi et al. 2012). Buildings with moisture and mold damage were considered risky for health, particular for eyes and the upper respiratory tract (cough and wheeze). Korpi et al. (2009) summarized 96 typical "indoor" mVOCs, of which 15 compounds were toxicologically evaluated (e.g., inhalation studies, lowest administered doses). From the human experimental exposure studies it turned out that symptoms of irritation appeared at mVOC concentrations several orders of magnitude higher than those measured indoors and it was concluded that mVOC concentrations are too low to provoke a nuisance effect on the building occupants. However, a drawback of the present analysis is that the toxicological database is poor, and there may be more potent compounds and endpoints yet not evaluated. Furthermore, in the environment mVOCs may come from various sources, such as building materials, human activities, traffic, foodstuff, smoking and may overlap and act additively in mixtures. The majority of mVOC producers present on buildings are fungi, for example, *Alternaria, Aspergillus, Botrytis, Candida, Fusarium, Penicillium, Trichoderma*, but also bacteria *Streptomyces* and *Pseudomonas* appear frequently (Claeson et al. 2002; Korpi et al. 1998). The main mVOCs produced on building materials were 3-methyl-1-butanol, 1-pentanol, 1-hexanol and 1-octen-3-ol, and Korpi et al. (1998) concluded that no single VOC is a reliable indicator for biocontamination. In contrast, Bennett and Inamdar (2015) reported on some mVOCs that have toxic properties. Experimental tests with tissue cultures and *Drosophila melanogaster* have shown that many single mVOCs as well as mixtures emitted by fungi have toxic effects. Subsequently, they are referred to mycotoxins. Inamdar et al. (2010, 2013) tested low vapor concentrations of C-8 compounds, including 1-octen-3-ol, and showed toxicity to larvae and adult flies, selectively affecting dopaminergic neurons in adult *Drosophila* brains and induced Parkinson like behavioral alterations. The toxicity data on 1-octen-3-ol are of particular concern, because this fungal VOC appears ubiquitously (it is the most prominently released fungal mVOC, Table 10.2) and is largely responsible for the musty odor commonly associated with mold-contaminated damp indoor spaces (Bennet and Inamdar 2015). In accordance, the results of a recent study concerning 6-pentyl-2-pyrone production by *Trichoderma atroviridae* growing on buildings showing that fungi can support SBS symptoms by irritating and damaging mucosal membranes (Polizzi et al. 2011), even when the production of fungal VOCs varies and is suboptimal due to the dependence on temperature and humidity (Polizzi et al. 2012).

Microbial biofilms are also formed on much other hardware, for example, clinical tubings, hulks, and air conditioners of houses and automobiles, which may result in malodors. Alpha-proteobacteria, methylobacteria, Shingomonadales, Burkholderiales, Bacillales, *Alcanovorax* spp., and *Stenotrophomonas* spp. were found on contaminated heat exchanger fins of evaporators from cars, which produced di-, tri-, and multiple sulfides, acetylthiazole, aromatic compounds and diverse pyrazines (Diekmann et al. 2013). Interestingly, a close relationship of the VOC profile and microbial community to the climate and air quality where the car was operating (European, American, Arabic, and Asian) was determined.

10.4.3 mVOCs for the perfume industry

Volatile organic compounds are predestinated and prestige flavor, fragrance, and aroma compounds. The perfume industry has an increasing demand for new VOCs. Although traditionally plants were the sources for aromas and fragrances, many other sources including microorganisms are presently tapped (Rutkin 2015).

10.4.4 mVOCs as the next generation of biofuel

The limitations of fossil fuels are foreseen and contrast steadily increasing fuel requirements (reviewed in Rude and Schirmer 2009). Furthermore to combat climate change and to reach the goal of energy independence, search for alternative and renewable energy sources is ongoing. Three routes to convert such resources into energy-rich, fuel-like molecules or fuel precursors are considered: (i) photosynthesis related production by plants, alga and cyanobacteria, (ii) fermentative and nonfermentative production by heterotrophic microorganisms, and (iii) chemical conversion of biomass. Biofuel platforms based on food and nonfood biomass conversion have several limitations including the energy output per land area, the compatibility with current fuel infrastructures, and insufficient capacity the meet Renewable Fuel Standard (RFS) (Wang et al. 2015). mVOCs are considered as potential alternatives of biofuels from renewable resources. Microbial fuels are fermentative short–chain alcohols (e.g., ethanol and butanol), nonfermentative short-chain alcohols (1-propanol, 1-butanol, isoforms and derivatives of butanol, pentanol, and 1-hexanol), fatty acid-derived hydrocarbons (fatty acid alcohols of C12 to C18 are useful for fuels, while very-long-chain fatty acid alcohols C24–C26 are unsuitable), and isoprenoid-derived hydrocarbons. To overcome the disadvantages of ethanol, fuel research enquired the production of isoprenoid derived biofuels. Sesquiterpenes (e.g., farnesol, farnesene, and farnesane bisabolene) are being developed as precursors for fuel and monoterpenes (e.g., α-pinene, camphene, limonene, and sabinene and terpinene) are discussed as potential next generation jet fuel components (reviewed in Gupta and Phulara 2015; Rude and Schirmer 2009). Isoprenoides are considered as good diesel alternatives because of their low hygroscopy, high energy content and good fluidity at low temperatures. Limitations are the low yields, toxicity and their stereochemical complexities. Presently attempts are undertaken to use genetically engineered microbes that produce a variety of infrastructure-compatible drop-in fuel molecules and to boost up terpene synthesis. In recent years, particularly *Escherichia coli* and *Saccharomyces cerevisiae* as well as less commonly used microbial strains were used as hosts to tune expression of endogenous enzymes of the isoprenoid pathway or to introduce heterologous enzymes. Several obstacles have been addressed and overcome, and advances in isoprenoid-based biofuels have been made using synthetic biology tools, for example, modulating the MEP pathway because it is stoichiometrically more competent than the MVA pathway, introducing the second terpene pathway into the microorganism, reprogramming metabolic nodes of pathways via mutations of transcription factors, improving terpene synthase catalysis, and installing efficient storage and excretion strategies are essential keys for progress (summarized in Gupta and Phulara 2015). Cyanobacteria, due to their fast growth, high photosynthetic rates, ability to grow in nonarable areas, availability of genome sequences, and production from CO_2 and solar energy make them a useful platform for biofuel production (Gupta and Phulara 2015; Wang et al. 2015). Although isoprenoid-based biofuel titers have been significantly improved in microorganisms in recent years, they still do not match with those of ethanol.

10.4.5 mVOCs as biocontrol agents in agriculture

Microbiota are attracted by suitable microenvironments in the soil to colonize and create microecosystems. Consequently, the "networking" communities are characterized by mutualism, commensalism, cooperation, antagonism, competition, and coexistence. In this arena, interactions between bacteria and fungi could have a positive or negative impact on third parties, which is useful if the weakened party is a pathogen and the strengthened party is a valuable member of the community (plant growth–promoting bacteria or fungi). An example of such a kind was shown recently (Kottb et al. 2015). The dominant volatile 6-pentyl-pyrone (coconut-like aroma, 6PP) of *Trichoderma asperellum* was perceived as a stress compound by *A. thaliana* and subsequently initiated multilayered defence adaptations including morphological and physiological alterations as well as activation of signaling cascades to withstand this environmental influence. Most noticeable is that *A. thaliana* preexposed to 6PP showed significantly reduced symptoms when challenged with *Botrytis cinerea* and *Alternaria brassicicola*, indicating that defense-activated plants subsequently became more resistant to pathogen attack. Together, these results support that products that are based on *Trichoderma* volatiles have the potential of being a useful biocontrol agent in agriculture.

Another example is a GFP-tagged *Bacillus subtilis* strain which was able to successfully suppress cucumber wilt by confining growth of *Fusarium oxysporum* f. sp. *cucumerinum* by colonizing the root and persisting on the rhizoplane (Cao et al. 2011). The authors proposed antibiosis as one mode of action caused by diffusible agents. Other experiments with a *B. subtilis* strain isolated from the rhizosphere of wheat and soybean showed bacterial volatiles being involved in the biocontrol of *Botrytis mali* and *Phytophthera sojae*, respectively (Jamalizadeh et al. 2010; Sharifi et al. 2002).

Many *Bacillus* species have been reported to synthesize plant growth-promoting VOCs. Ryu et al. (2004) reported that 2,3-butandiol and acetoin emitted by *Bacillus subtilis* GB013 and *Bacillus amyloliquefaciens* IN937a triggered induced systemic resistance against *Erwinia carotovora* in *Arabidopsis thaliana*. Volatiles of *Bacillus badius* M12 were documented to possess organogenetic potential on callus cultures of *Sesamum indicum* and introduce antioxidative activity in tobacco cultures (Gopinath et al. 2015).

Microorganisms of the soil and rhizosphere are able to protect plants from infections by specific root pathogens, a phenomenon called disease-suppression of soils (Hornby 1983). Beside other antagonistic modes of actions, only in the last decade it appeared that mVOCs play a key role in this pathogen inhibition (Insam and Seewald 2010; McNeal and Herbert 2009). The biological activity of such mVOCs might be an alternative regarding the usage of the methyl bromide for fumigation of soils infected by soil-borne fungal pathogens; in fact dimethyl disulfide, which is frequently emitted by bacteria (Table 10.1), is already used as a novel soil fumigant PALADIN®. Fungistasis results from the presence of mVOCs, for example, blocking germination of spores and inhibiting mycel growth (summarized in Kai et al. 2009). Minerdi et al. (2009) also showed that virulence genes are repressed in *Fusarium oxysporum*. Only a few studies indentified single compounds being responsible for the antifungal activity. Fernando et al. (2005) and Cordovez et al. (2015) demonstrated that six VOCs (cyclohexanal, decanal, 2ethyl-1-hexanol, nonanal, benzothiazole, and dimethyl trisulfide) of *Pseudomonas* and 1,3,5-trichloro-2-methoxy benzene and methyl 2-methylpentanoate of *Streptomyces* species possessed antifungal activity. However, the concentrations tested were quite high and the ecologically relevant doses remain to be determined.

mVOCs are also nematicidial. Gu et al. (2007) showed that mVOCs of bacilli isolates reduced movement of the nematodes *Panagrellus redivivus* and *Bursaphelenchus xylophilus*. Microbial volatiles also influence the tritrophic interactions comprising bacteria, fungi, and nematodes. *Paenibacillus polymyxa* and *Paenibacillus lentimorbus* exhibited strong antifungal activities, thereby interfering with the nematode-fungus interaction (*Meloidogyne incognita—Fusarium oxysporum*), which significantly reduced nematode infestation of tomato plants (Son et al. 2009). Additionally, soil bacteria, including one rhizobacterial strain, enhanced the nematophagous activity of the nematode-trapping fungus *Arthrobotrys oligospora* by increasing trap formation and predaceous activity (Duponnois et al. 1998). There are also numerous instances of mVOCs associated with insect feeding behavior, but some mVOCs are also powerful repellants (summarized in Davis et al. 2013). In some ecosystems, bacterial and fungal volatiles incite insect aggregations, or mVOCs can resemble sexual pheromones that elicit mating and oviposition behaviors from responding insects. An interesting example, which is considered a novel approach to control aphids in the field and greenhouse, is due to the release of volatiles by *Staphylococcus sciuri* allowing to locate the prey in the tritrophic aphid–bacteria hoverfly interaction (Leroy et al. 2010).

Considerable progress has been made in understanding the functions of mVOCs in plants; however, implanting this knowledge under field conditions remains in its infancy (summarized in Kanchiswamy et al. 2015). Field trials are needed to prove the value of mVOCs, but there is a realistic chance to develop new sustainable, cheaper, efficient, effective, and eco-friendly alternatives to pesticides and fertilizers (Kanchiswamy et al. 2015).

10.4.6 *mVOCs for chemotyping and diagnostic tools*

In addition to considering the biological and ecological roles of mVOCs being important for the homeostasis (fitness and health) of humans and plants, these compounds can be used as noninvasive markers. Individual mVOCs as well as clusters of volatiles are useful for phenotyping fungi (Müller et al. 2013) and bacteria (Peñuelas et al. 2014). Even phylogenetically closely related but phenotypically different species of *Streptomyces* isolates or *Clostridum difficile* ribotypes could be differentiated (Cordovez et al. 2015; Kuppusami et al. 2015). Noninvasive markers are particularly demanded and already used as medical diagnostic tools, for example, for the recognition of methyl nicotinate as an indicator for tuberculosis caused by *Mycobacterium tuberculosis* (Syhre and Chambers 2008; Mgode et al. 2012). *Pseudomonas aeruginosa* infections are associated with declining lung function in cystic fibrosis and high mortality rates. 2-aminoacetophenone is a small molecule and an intermediate of the quinazoline biosynthesis and was shown to be significantly higher in *P. aeruginosa* colonized subjects than control patients and is therefore a promising breath biomarker (Scott-Thomas et al. 2010). A noninvasive VOC-based detection was proposed as an alternative technique suitable for surveillance and as diagnostic tool applicable for urological infections by *Proteus* spp. (Aarthi et al.

2014). These examples stand for a number of other cases described in the literature; this chapter was not initiated to present a comprehensive review of this topic.

10.5 Conclusion

Microbial VOCs have been overlooked in the past, but due to the ubiquitous appearance of microorganisms and their broad metabolic capabilities they have to be considered to understand organismal interactions in ecosystems. Besides these semiochemical functions and biological relevances, mVOCs offer chances to provide sustainable and eco-friendly alternatives to pesticides and fertilizers in the agriculture and the environment, and noninvasive diagnostic tools for animal and human health and fitness. mVOCs are already used in biotechnological processes to control, for example, foodstuff, buildings, and other hardware. It is expected that techniques with better sensitivity and wider applications will be used routinely to make our lives safer. Furthermore, we expect to find novel (lead) structures of natural products that are good sources for new applications.

Acknowledgments

We thank Dr. Marco Kai for his help in preparing Figure 10.2 and the University of Rostock for their financial support.

References

Aarthi, R., R. Saranya, and K. Sankaran. 2014. 2-Methylbutanal, a volatile biomarker for non-invasive surveillance of Proteus. *Appl. Microbiol. Biotechnol.* 98: 445–454.

Abrahamse, C.E. and E.J. Bartowsky. 2012. Timing of malolactic fermentation inoculation in shiraz grape must and wine: Influence on chemical composition. *World J. Microbiol. Biotechnol.* 28: 255–265.

Bailly, A. and L. Weisskopf. 2012. The modulating effect of bacterial volatiles on plant growth: Current knowledge and future challenges. *Plant Signal. Behav.* 7: 79–85.

Bayer, C.W. and S. Crow. 1994. Odorous volatile emissions from fungal contamination. In *Proceedings from IAQ'93: Operating and Maintaining Buildings for Health, Comfort and Productivity* ed. K.Y. Teichman, 165–170. Atlanta, GA: American Society of Heating, Refrigerating and Air-Conditioning Engineers.

Bennett, J.W. and A.A. Inamdar. 2015. Are some fungal volatile organic compounds (VOCs) mycotoxins? *Toxins* 7: 3785–3804.

Bhadra S., C. Narvaez, D.J. Thomson, and G.E. Bridges. 2015. Non-destructive detection of fish spoilage using a wireless basic volatile sensor. *Talanta* 134: 718–723.

Buzzini, P., S. Romano, B. Turchetti, A. Vaughan, U.M. Pagnoni, and P. Davoli. 2005. Production of volatile organic sulfur compounds (VOSCs) by basidiomycetous yeasts. *FEMS Yeast Res.* 5: 379–385.

Cao, Y., Z. Zhang, N. Ling, Y. Yuan, X. Zheng, B. Shen, and Q. Shen. 2011. *Bacillus subtilis* SQR9 can control *Fusarium* wilt in cucumber by colonizing plant roots. *Biol. Fertil. Soils* 47: 495–506.
Casaburi, A., P. Piombino, G.J. Nychas, F. Villani, and D. Ercolini. 2014. Bacterial populations and the volatilome associated to meat spoilage. *Food Microbiol.* 45: 83–102.
Cheng, H. 2010. Volatile flavor compounds in yogurt: A review. *Crit. Rev. Food Sci. Nut.* 50: 938–950.
Chitarra, G.S., T. Abee, F.M. Rombouts, M.A. Posthumus, and J. Dijksterhuis. 2004. Germination of *Penicillium paneum* Conidia is regulated by 1-octen-3-ol, a volatile self-inhibitor. *Appl. Environ. Microbiol.* 70: 2823–2829.
Claeson, A.S., J.O. Levin, G. Blomquist, and A.L. Sunesson. 2002. Volatile metabolites from microorganisms grown on humid building materials and synthetic media. *J. Environ. Monit.* 4: 667–672.
Cordovez, V., V.J. Carrion, D.W. Etalo et al. 2015. Diversity and functions of volatile organic compounds produced by *Streptomyces* from a disease-suppressive soil. *Front. Microbiol.* 6: 1081.
Curtin, C.D., J.R. Bellon, P.A. Henschke, P.W. Goddende, and M.A. Barros Lopes. 2007. Genetic diversity of *Dekkera bruxellensis* yeasts isolated from Australian wineries. *FEMS Yeast Res.* 7: 471–481.
Darriet, P., M. Pons, S. Lamy, and D. Dubourdieu. 2000. Identification and quantification of geosmin, an earthy odorant contaminating wines. *J. Agri. Food Chem.* 48: 4835–4838.
Davis T. S., T.L. Crippen, R.W. Hofstetter, and J.K. Tomberlin. 2013. Microbial volatile emissions as insect semiochemicals. *J. Chem. Ecol.* 39: 840–859.
Diekmann, N., M. Burghartz, L. Remus et al. 2013. Microbial communities related to volatile organic compound emission in automobile air conditioning units. *Appl. Microbiol. Biotechnol.* 97: 8777–8793.
Duponnois, R., A.M. Ba, and T. Mateille. 1998. Effects of soil rhizosphere bacteria for the biocontrol of nematodes of the genus *Meloidogyne* with *Arthrobotrys oligospora*. *Fundam. Appl. Nematol.* 21: 157–163.
Effmert, U., J. Kalderas, R. Warnke, and B. Piechulla. 2012. Volatile-mediated interactions between bacteria and fungi in the soil. *J. Chem. Ecol.* 38: 665–703.
Fernando, W.G.D., R. Ramarathnam, A.S. Krishnamoorthy, and S.C. Savchuk. 2005. Identification and use of potential bacterial organic antifungal volatiles in biocontrol. *Soil Biol. Biochem.* 37: 955–964.
Fiddaman, P.J. and S. Rossall. 1993. The production of antifungal volatiles by *Bacillus subtilis*. *J. Appl. Bacteriol.* 74: 119–126.
Gerbaux, V., C. Briffox, A. Dumont, and K. Sibylle. 2009. Influence of inoculation with malolactic bacteria on volatile phenols in wines. *Am. J. Enol. Vitic.* 60: 233–235.
Gopinath, S., K.S. Kumaran, and M. Sundararaman. 2015. A New initiative in micropropagation: Airborne bacterial volatiles modulate organogenesis and antioxidant activity in tobacco (*Nicotiana tabacum* L.) callus. *In Vitro Cell. Dev. Biol. Plant.* 5: 514–523.
Gram, L. and H.H. Huss. 1996. Microbiological spoilage of fish and fish products. *Int. J. Food Microbiol.* 33: 121–137.
Gu, Y.Q., M.H. Mo, J.P. Zhou, C.S. Zou, and K.Q. Zhang. 2007. Evaluation and identification of potential organic nematicidal volatiles from soil bacteria. *Soil Biol. Biochem.* 39: 2567–2575.

Gupta, P. and S.C. Phulara. 2015. Metabolic Engineering for isoprenoid-based biofuel production. *J. Appl. Microbiol.* 119: 605–619.

Horikoshi, K., G. Antranikian, A.T. Bull, F.T. Robb, and K.O. Stetter. 2011. *Extremophiles Handbook.* Berlin: Springer.

Hornby, D. 1983. Suppressive soils. *Ann. Rev. Phytopath.* 2: 65–85.

Inamdar A.A., M.M. Hossain, A.I. Bernstein, G.W. Miller, J.R. Richardson, and J.W. Bennett. 2013. The fungal derived semiochemical 1-octen-3-ol disrupts dopamine packaging and causes neurodegeneration. *Proc. Natl. Acad. Sci. USA* 110: 19561–19566.

Inamdar, A.A., P. Masurekar, and J.W. Bennett. 2010. Neurotoxicity of fungal volatile organic compounds in *Drosophila melanogaster. Toxicol. Sci.* 117: 418–426.

Insam, H. and M.S.A. Seewald. 2010. Volatile organic compounds (VOCs) in soils. *Biol. Fertil. Soils* 46: 199–213.

Jamalizadeh, M., H.R. Etebarian, H. Aminian, and A. Alizadeh. 2010. Biological control of *Botrytis mali* on apple fruit by use of *Bacillus* bacteria, isolated from the rhizosphere of wheat. *Arch. Phytopath. Plant Protect.* 43: 1836–1845.

Kai, M., M. Haustein, F. Molina, A. Petri, B. Scholz, and B. Piechulla. 2009. Bacterial volatiles and their action potential. *Appl Microbiol. Biotechnol.* 81: 1001–1012.

Kanchiswamy, C.N., M. Malnoy, M.E. Maffei. 2015. Bioprospecting bacterial and fungal volatiles for sustainable agriculture. *Trends Plant Sci.* 20: 206–211.

Korpi, A., A.L. Pasanen, and P. Pasanen. 1998. Volatile compounds originating from mixed microbial cultures on building materials under various humidity conditions. *Appl. Environ. Microbiol.* 64: 2914–2919.

Korpi, A., J. Järnberg, and A.L. Pasanen. 2009. Microbial volatile organic compounds. *Crit. Rev. Toxicol.* 39: 139–193.

Kottb, M., T. Gigolashvili., D.K. Großkinsky, and B. Piechulla. 2015. *Trichoderma* volatiles effecting *Arabidopsis*: From inhibition to protection against phytopathogenic fungi. *Front. Microbiol.* 6: 995.

Kuppusami, S., M.R.J. Clokie, T. Panayi, A.M. Ellis, and P.S. Monks. 2015. Metabolite profiling of *Clostridium difficile* ribotypes using small molecular weight volatile organic compounds. *Metabolomics* 11: 251–260.

Langford, V.S., C.J. Reed, D.B. Milligan, M.J. McEwan, S.A. Barringer, and W.J. Harper. 2012. Headspace analysis of Italian and New Zealand parmesan cheeses. *J. Food Sci.* 77: C719–26.

Lemfack, M.C., J. Nickel, M. Dunkel, R. Preissner, and B. Piechulla. 2014. mVOC: A database of microbial volatiles. *Nucl. Acids Res.* 42: D744–D748.

Leroy, S., P. Giammarinaro, J. Chacornac, I. Lebert, and R. Talon. 2010. Biodiversity of indigenous *Staphylococci* of naturally fermented dry sausages and manufacturing environments of small-scale processing units. *Food Microbiol.* 27: 294–301.

Lorenz, W., T. Diederich, and M. Conrad. 2002. Practical Experiences with mVOC as an indicator for microbial growth. *Proc. Indoor Air* 4: 341–346.

McNeal, K.S. and B.E. Herbert. 2009. Volatile organic metabolites as indicators of soil microbial activity and community composition shifts. *Soil Sci. Soc. Am. J.* 73: 579–588.

Mgode, G.F., B.J. Weetjens, T. Nawrath et al. 2012. Diagnosis of tuberculosis by trained African giant pouched rats and confounding impact of pathogens and microflora of the respiratory tract. *J. Clin. Microbiol.* 50: 274–280.

Minerdi, D., S. Bossi, M.L. Gullino, and A. Garibaldi. 2009. Volatile organic compounds: A potential direct long distance mechanism for antagonistic action of *Fusarium oxysporum* strain MSA 35. *Environ. Microbiol.* 11: 844–854.
Müller, A., P. Faubert, M. Hagen et al. 2013. Volatile profiles of fungi- chemotyping of species and ecological functions. *Fungal Genet. Biol.* 54: 25–33.
Noseda, B., M.T. Islam, M. Eriksson et al. 2012. Microbiological spoilage of vacuum and modified atmosphere packaged Vietnamese *Pangasius hypophthalmus* fillets. *Food Microbiol.* 30: 408–419.
Olafsdottir G., E. Chanie, F. Westad et al. 2005. Prediction of microbial and sensory quality of cold smoked Atlantic salmon (*Salmo salar*) by electronic nose. *J. Food Sci.* 70: 563–S574.
Parlapani, F.F., S.A. Haroutounian, G.J. Nychas, and I.S. Boziaris. 2015. Microbiological spoilage and volatiles production of gutted European sea bass stored under air and commercial modified atmosphere package at 2°C. *Food Microbiol.* 50: 44–53.
Peñuelas J., D. Asensio, D. Tholl et al. 2014. Biogenic volatile emissions from the soil. *Plant Cell Environ.* 37: 1866–1891.
Pogacic, T., M.B. Maillard, A. Leclerc et al. 2015. A methodological approach to screen diverse cheese-related bacteria for their ability to produce aroma compounds. *Food Microbiol.* 46: 145–153.
Polizzi, V., A. Adams, A.M. Picco et al. 2011. Influence of environmental conditions on production of volatiles by *Trichoderma atroviride* in relation with the sick building syndrome. *Build. Environ.* 46: 945–954.
Polizzi, V., A. Adams, S. De Saeger, C. Van Peteghem, A. Moretti, and N. De Kimpe. 2012. Influence of various growth parameters on fungal growth and volatile metabolite production by indoor molds. *Sci. Total Environ.* 414: 277–286.
Rude, M.A. and A. Schirmer. 2009. New microbial fuels: A biotech perspective. *Curr. Opin. Microbiol.* 12: 274–281.
Rutkin A. 2015. Yeast's heavenly potential. *New Scientist* 225: 9.
Ryu, C.M., M.A. Farag, C.H. Hu, M.S. Reddy, J.W. Kloepper, and P.W. Paré. 2004. Bacterial volatiles induce systemic resistance in *Arabidopsis*. *Plant Physiol.* 134: 1017–1026.
Schenkel, D., M.C. Lemfack, B. Piechulla, and R. Splivallo. 2015. A meta-analysis approach for assessing the diversity and specificity of belowground root and microbial volatiles. *Front. Plant. Sci.* 6: 707.
Schulz, S. and J.S. Dickschat. 2007. Bacterial volatiles: The smell of small organisms. *Nat. Prod. Rep.* 24: 814–842.
Scott-Thomas, A.J., M. Syhre, P.K. Pattemore et al. 2010. 2-Aminoacetophenone as a potential breath biomarker for *Pseudomonas aeruginosa* in the cystic fibrosis lung. *BMC Pulm. Med.* 10: 56.
Sharifi T.A., A. Zebarjad, G.A. Hedjaroud, and M. Mohammad. 2002. Biological control of soybean damping-off by antagonistic rhizobacteria. *Meded. Rijksuniv. Gent. Fak. Landbouwkd. Toegep. Biol. Wet.* 67: 377–380.
Son, S.H., Z. Khan, S.G. Kim, and Y.H. Kim. 2009. Plant growth-promoting rhizobacteria, *Paenibacillus polymyxa* and *Paenibacillus lentimorbus* suppress disease complex caused by root-knot nematode and fusarium wilt fungus. *J. Appl. Microbiol.* 107: 524–5232.
Splivallo, R., A. Deveau, N. Valdez, N. Kirchhoff, P. Frey-Klett, and P. Karlovsky. 2014. Bacteria associated with truffle-fruiting bodies contribute to truffle aroma. *Environ. Microbiol.* 17: 2647–2660.

Splivallo, R. and S.E. Ebeler. 2015. Sulfur volatiles of microbial origin are key contributors to human-sensed truffle aroma. *Appl. Microbiol. Biotechnol.* 99: 2583–2592.

Splivallo, R., S. Bossi, M. Maffei, and P. Bonfante. 2007. Discrimination of truffle fruiting body versus mycelial aromas by stir bar sorptive extraction. *Phytochem.* 68: 2584–2598.

Syhre, M. and S.T. Chambers. 2008. The scent of *Mycobacterium tuberculosis*. *Tuberculosis (Edinb).* 88: 317–323.

Talou, T., A. Gaset, M. Delmas, M. Kulifaj, and C. Montant. 1990. Dimethyl sulphide: The secret for black truffle hunting by animals? *Mycol. Res.* 94: 277–278.

Vahdatzadeh, M., A. Deveau, and R. Splivallo. 2015. The role of the microbiome of truffles in aroma formation: A meta-analysis approach. *Appl. Environ. Microbiol.* 81: 6946–6952.

Von Reuss, S., M. Kai, B. Piechulla, and W. Francke. 2010. Octamethylbicyclo (3.2.1) octadienes from *Serratia odorifera*. *Angewandte Chemie.* 22: 2053–2054.

Wang, X., D.R. Ort, and J.S. Yuan. 2015. Photosynthetic terpene hydrocarbon production for fuels and chemicals. *Plant Biotechnol. J.* 13: 137–146.

Wenke, K., D. Wanke, J. Kilian, K. Berendzen, K. Harter, and B. Piechulla. 2012a. Volatile-associated molecular pattens of two growth-inhibiting rhizobacteria commonly engage AtWRKY18 function. *Plant J.* 70: 445–459.

Wenke, K., M. Kai, and B. Piechulla. 2010. Belowground volatiles facilitate interactions between plant roots and soil organisms. *Planta* 231: 499–506.

Wenke, K., T. Weise, R. Warnke et al. 2012b. Bacterial volatiles mediating information between bacteria and plants. In *Bio-communication of plants*, ed. G. Witzany and F. Baluska, 327–347. Heidelberg: Springer Verlag.

Wessen, B., G. Stroem, and K.O. Schoeps. 1995. mVOC profiles—A tool for indoor air quality assessment. In *Proceedings of the International Workshop Indoor Air-An Integrated Approach*, Gold Coast Australia, ed. L. Morawska, N.D. Bofinger and M. Maroni, 67–70. Oxford: Elsevier Science & Technology Books.

Wessen, B., G. Stroem, and U. Palmgren. 1999. Microbial problem buildings-analysis and verification. *Proc. Indoor Air* 4: 875–879.

Xia, J., V.I. Sinelnikov, B. Han, and D.S. Wishart. 2015. MetaboAnalyst 3.0—Making metabolomics more meaningful. *Nucl. Acid Res.* 43: W251–W257.

chapter eleven

Volatile glycosylation—A story of glycosyltransferase for volatiles
Glycosylation determining the boundary of volatile and nonvolatile specialized metabolites

Eiichiro Ono and Toshiyuki Ohnishi

Contents

11.1 Introduction	258
11.1.1 General	258
11.1.2 UGT	258
11.1.3 Volatile glycosides	260
11.2 Glycosylation for volatiles	261
11.2.1 General	261
11.2.2 Kiwifruit	263
11.2.3 Grapes	263
11.2.4 Tea	264
11.2.5 Other UGTs for volatiles	265
11.3 Further glycosylation for volatile glycosides	266
11.3.1 General	266
11.3.2 Tea	266
11.3.3 Tomatoes	267
11.4 Perspective	268
11.4.1 Glycoside-specific glycosyltransfearse: Beyond volatiles	268
11.4.2 Ecological and evolutionary landscape of volatile glycosides	273
11.4.3 Promiscuity and diversity	275
References	276

11.1 Introduction

11.1.1 General

Specialized metabolites are lineage-specific phytochemicals that are basically considered to be a consequence of chemical adaptation to their habitat environment. Glycosylation, which involves sugar conjugation, is a key mechanism that regulates the bioactivity and storage of specialized metabolites as well as the detoxification of xenobiotics by increasing water solubility and reducing the reactivity of the metabolites (Bowles et al. 2005). Since it is generally acknowledged that considerable numbers of specialized metabolites are found in a sugar-conjugated form, "glycoside" in nature, a sugar conjugation reaction (glycosylation) is a common for specialized metabolites. Generally, glucose is the most commonly observed sugar moiety in naturally occurring glycosylated metabolites in plants, but other sugar moieties such as rhamnose, galactose, glucuronic acid, arabinose, and xylose are also present.

11.1.2 UGT

A superfamily enzyme, uridine diphosphate (UDP)-sugar:glycosyltransferase (UGT) catalyzes glycosylation by transferring a sugar group from UDP-sugar (sugar donor) to substrates (sugar acceptor). Sugar is usually transferred at the oxygen (O) of a hydroxyl group (OH), but occasionally at a carboxy group (COOH) of sugar acceptors (Richman et al. 2005), both of which are classified as "O-glycosylation." On the contrary, some UGTs are shown to specifically catalyze glycosylation at the carbon (C) or nitrogen (N) of sugar acceptors, classified as "C-glycosylation" and "N-glycosyation," respectively (Brazier-Hicks et al. 2007, 2009; Sasaki et al. 2015a).

Glycosylation is classified based on the UDP-sugar donor that is transferred to sugar acceptable substrates, for example, glucosylation (UDP-*Glu*cose [UDP-Glc]), rhamnosylation (UDP-*Rha*mnose [UDP-Rha], galactosylation (UDP-*Gala*ctose [UDP-Gal]), xylosylation (UDP-*Xyl*ose [UDP-Xyl]), arabinosylation (UDP-*Ara*binose [UDP-Ara]), and glucuronosylation (UDP-*Glu*curonic *A*cid [UDP-GlcA]). UGT basically shows limited specificity for one sugar donor. A glycosylated product (monoglycoside) is often further glycosylated at the sugar moiety, resulting in diglycosides and triglycosides. Thus, UGT significantly impacts the vast structural diversity of specialized metabolites.

From the viewpoint of molecular phylogenetics, a group of structurally similar UGTs possess a conserved enzymatic function, namely similar substrate specificity, beyond species, suggesting that their orthologous gene would have occurred prior to speciation (Caputi et al. 2012). In reverse, a lineage-specific UGT group derived from recent gene multiplication

Chapter eleven: Volatile glycosylation

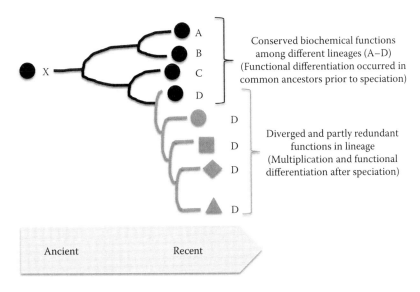

Figure 11.1 A schematic illustration of molecular evolution of a superfamily enzyme gene. The capitalized letter and symbol indicates plant lineage and enzyme function of paralogs, respectively.

suggests its involvement in lineage-specific specialized metabolism which occurs after speciation (Figure 11.1). The similar molecular phylogenetic pattern is also observed in other superfamily enzymes such as cytochrome P450 monooxygenase (P450/CYP) and 2-oxoglutarate-dependent dioxygenase (2OGD/DOX) (Kawai et al. 2014a; Mizutani and Ohta 2010). Increase of these metabolic genes roughly correlates with whole genome duplication (Kawai et al. 2014b). Important to note, is that specialized metabolic enzyme genes are frequently found to be local tandem clusters (Chae et al. 2014). However, the molecular mechanism exerting multiplication for certain genes remains elusive.

UGTs are classified based on their sequences with characteristic UGT numbers given by nomenclature, for example, UGT88D3 in which 88, D, and 3 indicate family number, subfamily (capital alphabet), and gene identifier number, respectively (Mackenzie et al. 1997). Thus far, over 100 and 200 UGT genes have been found in the *Arabidopsis thaliana* genome and *Oryza sativa* genome, respectively (Yonekura-Sakakibara and Hanada 2011). However, most of them have remained biochemically elucidated due to their functional redundancy, high copy numbers, and promiscuity nature, in addition to unavailability of substrates due to rare metabolic intermediates. Here, we describe recent advances in the understanding of UGT functionality specialized for volatiles and volatile glycosides, which enables them to be preserve as water-soluble glycosides *in planta*.

11.1.3 Volatile glycosides

Plants emit volatile organic compounds (VOCs), such as monoterpenoids (C10), sesquiterpenoids (C15), phenylpropanoids (C9), norisoprenoids (C16), aromatic esters, or green leaf volatiles (GLVs) (C6), in response to attacks by insect herbivores, mechanical-wounding or endogenous developmental cues. In general, VOCs are considered not only to be mobile molecules transmitting biological signals from sessile plants to the environment (Arimura et al. 2009) but they are also important commercial products because of their influence on the quality and character of dietary foods and beverages as aromas.

VOCs easily evaporate in the air due to their low molecular and hydrophobic properties. Thus, plants have developed at least two strategies to stably store volatiles in plants during land plant evolution: (1) the development of specific compartments for the accumulation of hydrophobic or high reactive compounds, and (2) the enhancement of water solubility of volatiles by sugar conjugation. The farmer is represented by glandular trichomes, which accumulate large quantities of hydrophobic compounds in the space between their gland cell walls and cuticle. Volatiles quickly emit on physical attack by herbivores since trichomes are located on the surface of plants, which is reasonable for a defense response. For example, glandular trichomes on the leaves of basil (*Ocimum basilicum*); spearmint (*Mentha spicata*); and the female flowers (hop cones) of hop (*Humulus lupulus*), which are aromatic herbs used as flavoring ingredients for food and alcoholic beverages, such as beer; are known to accumulate plenty of specialized volatiles.

The latter are water-soluble volatile glycosides, which are stored stably in aqueous fractions, but instead do not evaporate due to the large and hydrophilic sugar moiety. Thus, glycosylation makes volatile odorless nonvolatile glycosides. In grapes, the high proportion of volatile glycosides is considered wine's "hidden aromatic potential," some of which are converted to free form by hydrolysis of the glycosides during most fermentation and wine maturation (D'Ambrosio et al. 2013; Gunata et al. 1985; Hjelmeland and Ebeler 2015; Strauss et al. 1998). Similarly, the fresh leaves of tea plants (*Camellia sinensis*) barely emit a slightly green note before they are processed for tea or prior to mechanical wounding. This is because aroma volatiles in tea leaves basically accumulate in the water-soluble glycoside form, most of which is a diglycoside, β-primeveroside (6-O-β-D-xylopyranosyl-β-D-glucopyranoside) (Guo et al. 1993; Wang et al. 2000). Floral aroma caused by linalool, geraniol, 2-phenylethanol, and benzyl alcohol are predominant flavor volatile compounds in oolong and black tea, which are deliberately released by hydrolysis of their β-primeveroside by an endogenous diglycoside-specific β-glycosidase of tea plant, β-primeverosidase in the fermentation process of tea manufacturing (Ma et al. 2001; Mizutani et al. 2002). Furthermore, vanillin, one of the most commercially valuable flavors derived from the curled pod of the vanilla orchid (*Vanilla planifolia*), is also known to be synthesized and stored as a nontoxic vanillin glucoside, which is hydrolyzed to form a biologically active defense compound (toxic) on mechanical tissue damage (Gallage et al. 2014). As in these cases of grape, tea, and vanilla, volatile glycosides are stored as aroma precursors which are important components that determine the quality of the crops.

In addition, volatile glycosides are frequently reported in diverse plant species, for example, the apricot (*Prunus armeniaca*), peach (*Prunus persica*), yellow plum (*Prunus domestica*; Krammer et al. 1991), kiwifruit (*Actinidia chinensis*; Young and Paterson 1995), strawberry (Roscher et al. 1996), raspberry (*Rubus* spp.; Pabst et al. 1991), hop (*Humulus lupulus*; Ting et al. 2009), sweet potato (*Ipomoea batatas*; Takamine et al. 2012), and tomato (*Solanum lycopersicum*; Marlatt et al. 1992), suggesting their wider distribution in plants than expected. Importantly, they are often observed in reproductive organs (flower, root tuber, and fruits), which reflects their physiological roles in chemical communication with other organisms such as attraction signals for pollinators and seed dispersers, and repellant signals against unfavorable herbivorous insects or pathogenic microorganisms.

11.2 Glycosylation for volatiles

11.2.1 General

The generality of volatile glycosides implies that the glycosylating machinery responsible for volatile glycosides is widespread and functionally conserved in various seed plants (Aurore et al. 2011; Birtic et al. 2009; Cabrita et al. 2006; Garcia et al. 2011a,b; Loughrin et al. 1992; Oka et al. 1999; Williams et al. 1982). This notion is also supported by the similar observation that transgenic plants ectopically expressing enzyme genes for different volatiles result in the accumulation of its glycosides *in planta* (Aharoni et al. 2003; Boachon et al. 2015; Ginglinger et al. 2013; Koeduka et al. 2013). However, the possibility that biosynthesis of volatile glycosides had occurred independently in various land plants is not excluded.

Recently, structurally similar UGTs, derived from phylogenetically distant edible plants, were identified to catalyze glucosylation for volatiles almost at the same time, for example, in kiwifruit (*Actinidia deliciosa*), grapevine (*Vitis vinifera*), tomato (*Solanum lycopersicum*), and the tea plant (*Camellia sinensis*) (Bönisch et al. 2014a,b; Ohgami et al. 2015; Tikunov et al. 2013; Yauk et al. 2014) (Figure 11.2). These findings support the notion that these UGTs involved in the biosynthesis of volatile glycosides would have been derived from a common ortholog in ancestral seed plants. Here, we summarize the UGTs catalyzing glycosylation for volatiles and volatile glycosides that are stored as water-soluble nonvolatiles in plants.

Figure 11.2 (**See color insert.**) Photographs of dietary crops, which produce volatile glycosides: (a) Tea leaves (*Camellia sinenesis*), (b) Grapes (*Vitis vinifera*), (c) Kiwifruits (*Actinidia deliciosa*), (d) Tomatoes (*Solanum lycopersicum*). Structures: (e) (Z)-3-hexenyl β-primeveroside ((Z)-3-hexenyl 6-O-β-D-xylopyranosyl-β-D-glucopyranoside) found in tea leaves; (f) 8-hydroxyllinaly-acuminoside (8-hydroxylinalool 6-O-α-L-arabinofuranosyl-β-D-glucopyranoside) found in grapes; (g) Eugenol 2-O-β-D-glucopyranosyl-(1 → 2)-[O-β-D-xylopyranosyl-(1 → 6)]-O-β-D-glucopyranoside (triglycoside) found in tomatoes. Gray, blue, and green shading indicates glucose, xylose, and apiose, respectively. Yellow circle shading indicates volatile aglycone.

11.2.2 Kiwifruit

Ripe green-fleshed kiwifruits contain a series of volatiles including methyl and ethyl butanoate esters and the C-6 aldehydes and alcohols (Z)- and (E)-hex-2-enal, hex- anal, (Z)- and (E)-hex-3-enol, and methyl benzoate that are important aroma properties for quality (Garcia et al. 2011a,b). It also contains a variety of volatile glycosides of terpenoids and C-6 alcohols, 3-methylbutanol, benzyl alcohol, and 2-phenylethanol (Yauk et al. 2014).

AdGT4, the encoding a UGT85-related enzyme of kiwifruit, was found to be highly expressed in floral tissues and increased during fruit ripening (Yauk et al. 2014). The recombinant *AdGT4* protein shows glucosylating activity for a range of terpenes and primary alcohols found as glycosides in ripe kiwifruit. Moreover, a series of transgenic experiments with altered expressions of the *AdGT4* gene demonstrates *AdGT4*-overexpressing tomato fruit as having a significantly more "earthy" aroma, and to be globally less "intense" than control fruits, probably due to a drastic reduction of "recognizable" free-volatile alcohols and excessive sequestration of "unrecognizable" volatile glycosides for humans in the transgenic fruits. Thus, volatile UGT is a key enzyme in determining the sequestration of volatiles *in planta* since overexpression of *AdGT4* influences volatile compound release that has a measurable effect on the sensory perception of fruit aroma.

11.2.3 Grapes

As described above, volatile glycosides are regarded as "hidden aromatic potential" since a large fraction of these compounds is present as nonvolatile glycosides in grapes, which can contribute to wine flavor after the hydrolytic release of volatiles during cultivation and fermentation processes from grape to aged wine. A dominant aroma, terpenes are found mainly in the exocarp (skin of grape berries), and the concentration of many terpenes accumulate as the grape ripens (Lund and Bohlmann 2006). Some of the volatile precursors are accumulated in a diglycosylated form.

VvGT14 and VvGT16 were isolated based on the genome database of a Pinot Noir cultivar, and were shown to have glucosylating activity for various terpene volatiles including geraniol, R, S-citronellol, and nerol which are important monoterpenols in traditional grapevine varieties for wine aromas (Bönisch et al. 2014a). VvGT14 and VvGT16 are members of the UGT85-related gene found as multiple genes in the grapevine genome, implying their functional differentiation and redundancy. Spatial and temporal gene expression of these UGTs in combination with metabolite profiles revealed that their enzyme specificity and substrate availability influence the formation of monoterpene glycosides during grape ripening that have an impact on the characteristic aromas of wine (Bönisch et al. 2014a; Luan et al. 2005, 2006).

In addition, a glycosyltransferase, UGT85K14, from the hybrid grapevine cultivar, Muscat Bailey A (*V. labrusca* × *V. vinifera*) was recently found to catalyze glycosylation of furaneol *in vitro* (Sasaki et al. 2015b). The resulting furaneol glucoside is also an important aroma precursor of Muscat Bailey A, since furaneol released via hydrolysis of the glycoside contributes to the characteristic "strawberry-like note" to wines. Considering that wine aromas are significantly influenced by genotypes (cultivar) and environmental factors (terroir), biochemical and environmental responsive properties of UGTs for volatiles would be important for the evaluation of aroma potential in grapes.

11.2.4 Tea

CsGT1 (UGT85K11) was identified to be an enzyme catalyzing the first glucosylation for major volatiles in tea aromas, from tea plants (Ohgami et al. 2015). CsGT1 is highly expressed in young leaves, where volatile glycosides are highly accumulated and show structural similarity to cassava (*Manihot esculenta*) UGT85K4 and UGT85K5, which are involved in the last step of biosynthesis of cyanogenic glucoside (Kannangara et al. 2011). CsGT1 utilizes UDP-glucose as the specific sugar donor in the catalysis, similar to the case of AdGT4 (Yauk et al. 2014). The first attached sugar of naturally occurring volatile glycosides is basically "glucose" regardless of the volatile aglycone's structure, which is consistent with the preferential sugar donor specificity of CsGT1 and AdGT4 for UDP-glucose. In contrast to sugar donor specificity, CsGT1 shows promiscuous specificity for structurally divergent acceptable sugar volatiles including geraniol, eugenol, benzyl alcohol, (Z)-3-hexenol, and 2-phenylethanol. The broad substrate specificities for sugar acceptors might be partly explained by the small molecular size of their substrates (C6-C10), compared to other specialized metabolites such as flavonoids(C15) and sesqui(C15) or diterpenenoids(C20) because amino acid residues for substrate recognition are structurally restricted by substrate size.

Genome and EST data revealed that numerous *UGT* genes structurally similar to *CsGT1* are present throughout the seed plants. Actually, five CsGT1 homologs isolated from the grapevine (Vv_UGT85A33, Vv_UGT85A28, Vv_UGT85A30), sweet potato (*Ipomoea batatas*; Ib_UGT85A32), and snapdragon (*Antirrhinum majus*; Am_UGT85A13) exhibited volatile glycosylating activities similar to CsGT1, including the production of geranyl-glucoside and hexenyl-glucoside. These results show that UGT85-related glycosyltransferase, capable of catalyzing the first glucosylation for volatiles, are widely conserved in various seed plants. In the UGT superfamily genes, the UGT85 family belongs to Orthologous Group 2 (OG2), which is one of the phylogenetic groups outstandingly swelled in angiosperm's genomes (Yonekura-Sakakibara and Hanada 2011). Taken together, glycosylation of volatiles catalyzed by a UGT85-related enzyme is a conserved fashion for sequestration and storage of volatiles in soluble compartments in seed plants.

11.2.5 Other UGTs for volatiles

Other UGTs than UGT85 family for volatiles should be noted herein. A tomato UGT72 family enzyme, SlUGT5 is expressed primarily in ripening fruit and flowers, and is shown to catalyze glycosylation of guaiacol and eugenol, as well as benzyl alcohol and methyl salicylate *in vitro* (Louveau et al. 2011). Similarly, *Vanilla* UGT72U1 is shown specifically to glucosylate vanillin, which is one of the most commercially valuable volatiles in dietary foods, resulting in vanillin glucoside (Gallage et al. 2014). Moreover, *Arabidopsis* UGT72E2 catalyzes glucosylation of 4-*O*-monolignols, such as coniferyl alcohol and sinapyl alcohol (Lanot et al. 2006). Coniferyl alcohol and sinapyl alcohol are further metabolized into other plant volatiles, for example, eugenol and also into nonvolatile lignans and lignins (Koeduka et al. 2006). Therefore, several UGT72 family enzymes take part in volatile glucosylation, especially phenylpropanoid derivatives in seed plants.

In addition to the UGT72 family, different phylogenetic groups of UGT are found to be capable of glucosylating for a variety of terpenoid scaffolds *in vitro* screening of *Arabidopsis* UGTs (Caputi et al. 2008). A grapevine VvGT7, which is most likely classified into a UGT88 family different from UGT85-related VvGT14 and VvGT16 mentioned above, is also capable of catalyzing glucosylation for grape monoterpenoid volatiles including nerol, citronellol, and geraniol *in vitro* that are important aromas in wine (Bönisch et al. 2014b). Volatile glycosylation is another biochemical aspect for the UGT88 family since UGTs in this family have been known to often be involved in flavonoid (C15) glycosylation, for example, chalcone, flavone, isoflavone, and anthocyanidin (Noguchi et al. 2007, 2009; Ogata et al. 2005; Ono et al. 2006). Considering that both UGT88 and UGT72 families belong to OG7, which is also significantly increased in angiosperm's genomes like OG2 (Yonekura-Sakakibara and Hanada 2011), ancestral seed plants have elaborated different UGTs for volatile glycosylation. These findings not only highlight the biochemical plasticity of UGTs but also demonstrate that the metabolic specialization is a consequence of the lineage-specific differentiation of the enzymes. In this context, the volatile UGTs described in this section would not be exceptional examples, but other class volatile UGTs might have occurred in seed plants.

11.3 Further glycosylation for volatile glycosides

11.3.1 General

Volatiles are often found in diglycoside or multiply glycosylated forms in plants, suggesting that sequential glycosylation occurs at sugar moiety of volatile glycosides. For example, a diglycoside, β-primeveroside is an abundant form of volatiles in tea leaves (Wang et al. 2000), suggesting that the first glucosylation is followed by the second xylosylation at the C6-hydroxyl group of the first attached glucose moiety. Since glycosylation for volatile glycosides increases water solubility and structural complexity of the compounds, it is considered that multiple glycosylation for volatile glycosides affects biological properties and ecological roles of volatiles.

11.3.2 Tea

An enzyme corresponding to the second glycosylation was isolated from the tea plant. Tea CsGT2 (UGT94P1) was identified to catalyze the second O-xylosylation at hydroxyl group of C6 position of the glucose moiety of a volatile monoglucoside, geranyl-glucoside, but did not exhibit any activity toward monoterpenols (volatile aglycones) (Ohgami et al. 2015). CsGT2 preferentially utilizes UDP-Xyl as the sugar donor, while weak activity was detected with UDP-Glc and no apparent activity was found for UDP-GlcA or UDP-Gal. The sugar-donor specificity demonstrates that CsGT2 specifically catalyzes the xylosylation for volatile glucosides, leading to the formation of volatile β-primeverosides. In addition, gene expression of *CsGT2* in young leaves is coupled with that of *CsGT1*, which catalyzes the first glucosylation for various volatiles, suggesting a coordinated transcriptional regulation of β-primeveroside biosynthesis *in planta*. Specific signals corresponding to volatile glycosides are enriched in the epidermal layer of tea leaves by imaging MS analysis. The preferential accumulation in the epidermis of juvenile tissues suggests its putative physiological role in defense responses against herbivores. The observation that aroma precursors are enriched in the skin of sweet potato tuber (Takamine et al. 2012) and in the exocarp of grape berries (Lund and Bohlmann 2006) also supports this notion.

On mechanical fermentation in the tea manufacturing process, volatiles (aglycones) are liberated from the preserved primeveroside via hydrolysis by the diglycoside-specific extracellular glycosidase, β-primeverosidase (Mizutani et al. 2002). This volatile release is considered a sophisticated chemical defense system without *de novo* synthesis of volatiles, thereby plants quickly respond to attacks by herbivores. Important to note, β-primverosidase of the tea plant is specific to diglycosides, especially β-primeveroside, but blunt for monoglycosides (Ma et al. 2001). Furthermore, it should also be noted that β-primeverosidase is sensitive to sugar type at the second sugar moiety of diglycoside. It is active to β-primeveroside, but not to β-gentiobioside, which has a glucose moiety instead of a xylose moiety of β-primeveroside (Ma et al. 2001). Therefore, the substrate specificity highlights that the second xylosylation by CsGT2 is a prerequisite for development of the chemical defense and sugar-donor specificity of the second GGT would also be crucial for cleavable or noncleavable properties of volatile glycosides.

11.3.3 Tomatoes

In tomatoes, NON-SMOKY GLYCOSYLTRANSFERASE 1 (NSGT1) was identified to be a UGT enzyme that specifically converts phenylpropanoid diglycoside (mainly primeveroside) to a triglycoside via further 2-O-β-glycosylation for glucose moiety of the diglycosides (Tikunov et al. 2013). Moreover, three phenylpropanoid primeverosides are glucosylated into the triglycosides by the enzyme extract from *Escherichia coli* constitutively expressing the *NSGT1* gene *in vitro*, whereas their aglycones (eugenol, guaiacol, methyl salicylate [MeSA]) were not. The specificity of NSGT1 for volatile glycosides is relevant to that of CsGT2 described above (Ohgami et al. 2015).

A loss-of-function mutation of *NSGT1* (*nsgt1/nsgt1* background) leads to the accumulation of phenylpropanoid β-primeverosides that is rapidly converted into volatiles (smoky phenotypes) via hydrolysis by tomato β-primeverosidase upon physical damage to the fruits. In this respect, diglycoside of volatiles is considered a water-soluble precursor of volatiles that are quickly hydrolyzed enzymatically upon being physically damaged by herbivores, which is similar to the case of tea (Mizutani et al. 2002). Gene expression of *NSGT1* is induced during fruit ripening, which is clearly associated with the formation of triglycosides and nonsmoky phenotypes. These results show that the corresponding β-primeverosidase to the release of phenylpropanoids in the tomato specifically recognizes phenylpropanoid primeverosides (diglycoside), but is blunt to triglycosides (enzymatically non-cleavable). Thus, high diglycosides (cleavable/smoky) and low triglycosides (noncleavable/nonsmoky) of phenylpropanoids in immature fruits might emit repellant signals against seed dispersers, whereas a reverse ratio of low diglycosides and high triglycosides of volatiles might be an ecological signature of seed maturation with red fruit coloration ready to disperse in tomato with the *NSGT1* gene. In terms of tomato breeding processes, human preference for a nonsmoky aroma would also be a strong selection force for *NSGT1* genes since differences of aroma from tomatoes with or without the *NSGT1* gene are detectable by blind sensory analysis (Tikunov et al. 2013). Nonsmoky phenotypes coupled with the accumulation of noncleavable phenylpropanoid triglycosides suggests that bulky triglycosyl-sugar moieties on volatiles would inhibit recognition by substrate binding pockets of diglycoside-specific β-primeverosidase (Saino et al. 2014), resulting in an enzymatically insensitive compound.

11.4 Perspective

11.4.1 Glycoside-specific glycosyltransfearse: Beyond volatiles

Phylogenetically, tea CsGT2 and tomato NSGT1 belong to a structurally conserved UGT cluster designated "glycoside-specific glycosyl*t*ransfearse (GGT)" or "branch-forming UGT," which are commonly known to specifically catalyze glycosylation at sugar moiety of various phytochemical glycosides, and are classified into the OG8 group in plant UGT phylogenetics (Figure 11.3) (Yonekura-Sakakibara and Hanada 2011). Previously characterized GGTs are summarized in Table 11.1. Their acceptable sugar substrate structures are highly diverse, for example, phenylpropanoid derivatives and isoprenoid derivatives, and their sugar donors are also variable, for example, glucose, rhamnose, xylose, and galactose. In contrast, regio-selectivity of GGTs is conserved to be specific, basically for the hydroxyl group of the C2 or C6 position of sugar moiety of the glycosides (Noguchi et al. 2008). These characteristic biochemical properties of GGTs indicate that progenies of ancestral GGT with the established regio-selectivity have independently adapted to specialized glycosides in each plant lineage. The fact, that CsGT2 and NSGT1 are structurally and functionally similar GGTs specialized for volatile glycosides, appears to support the possibility that lineage-specific adaptive specialization of GGTs for endogenous volatile glycosides has often occurred in various plant lineages accumulating multiple sugar conjugates of volatiles.

GGTs are located in the downstream of metabolism, implying that they would be newcomers or new-adaptors, recently participating in the modification of molecules at the end part of specialized metabolism. From this evolutionary view, GGTs are expected to be under purification of their specificity for the substrates if their glycosylated products exert adaptive effects for their host plants. In this way, structural diversity of GGTs is dependent on their "metabolic context" in each plant lineage and it is, therefore, difficult to estimate their substrate specificities for sugar acceptors based on the amino acid sequences only. In sharp contrast, sugar-donor specificity is well characterized using homology structural models of UGT followed by mutational analysis. Thus far, several crucial amino acid residues significantly influence sugar-donor specificity for UDP-glucose, UDP-galactose, UDP-glucuronic acid, or UDP-xylose, which have been determined in several UGTs for specialized metabolites (Noguchi et al. 2009; Ohgami et al. 2015; Ono et al. 2010a; Osmani et al. 2008; Sawada et al. 2005; Sayama et al. 2012). In most cases, a single amino acid change in UGT drastically alters sugar-donor specificity, which is indicative of the evolvable property of UGT. Compared to sugar-acceptor specificity, sugar-donor specificity of UGT has been well established in their catalysis, probably due to definitely fewer types of UDP-sugar

Chapter eleven: Volatile glycosylation

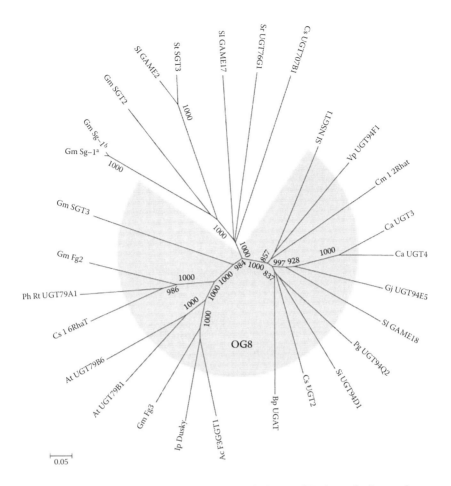

Figure 11.3 A phylogenetic tree of GGTs including UGTs for volatiles and non-volatiles. The amino acid sequences of GGTs were aligned based on codon position using ClustalW ver2.1 in DDBJ website (http://clustalw.ddbj.nig.ac.jp/). All sites containing gaps and missing data were eliminated from the remaining analysis. Unrooted phylogenetic trees were reconstructed by neighbor-joining methods from the translated amino acid sequences. The neighborjoining tree was reconstructed by MEGA6 (Tamura et al. 2013) and the matrix of evolutionary distances was calculated by Poisson correction for multiple substitutions. The reliability of the reconstructed tree was evaluated by a bootstrap test for 1000 replicates. High bootstrap values (greater than 750) are indicated on the branches.

(donors) than structurally diverged sugar-acceptable substrates in plant cells. In other words, amino acids required for the recognition of UDP-sugar donors are highly conserved under the biochemical constraints. Thus, amino acid residues involved in recognition of particular sugar donors are highlighted to be unique residues facing the substrate pocket

Table 11.1 List of of functionally characterized GGTs comprising volatile and nonvolatile UGTs

Cluster	Protein/UGT no.	Plant species	Substrate specificity			Ref.
			Acceptor	Donor	Regio	
OG8	Cs GT2/UGT94P1	Tea (*Camellia sinensis*)	Volatile glucoside	UDP-Xyl	6'	Ohgami et al. (2015)
	Sl NSGT1	Tomato (*Solanum lycopersicum*)	Volatile diglycoside	UDP-Glc	2'	Tikunov et al. (2013)
	Sl GAME18	Tomato (*Solanum lycopersicum*)	Triterpene glycoside	UDP-Glc	2'	Itkin et al. (2013)
	Ac F3GGT1	Kiwifruits (*Actinidia chinensis*)	Flavonoid galactoside	UDP-Xyl	2'	Montefiori et al. (2011)
	At UGT79B1	*Arabidopsis thaliana*	Flavonoid glucoside	UDP-Xyl	2'	Yonekura-Sakakibara et al. (2012)
	At UGT79B6	*Arabidopsis thaliana*	Flavonoid glucoside	UDP-Glc	2'	Yonekura-Sakakibara et al. (2014)
	Si UGT94D1	Sesame (*Sesamum indicum*)	Lignan monoglucoside	UDP-Glc	6'	Noguchi et al. (2008)
	Vp UGT94F1	*Veronica persica*	Anthocyanidin glucoside	UDP-Glc	2'	Ono et al. (2010b)
	Gj UGT9/UGT94E5	*Gardenia jasminoides*	Carotenoid glucoside	UDP-Glc	6'	Nagatoshi et al. (2012)

(*Continued*)

Table 11.1 (Continued) List of functionally characterized GGTs comprising volatile and nonvolatile UGTs

Cluster	Protein/UGT no.	Plant species	Substrate specificity			Regio	Ref.
			Acceptor	Donor			
	Ip Dusky/ UGT79G16	Morning glory (*Ipomoea purpurea*)	Anthocyanidin glucoside	UDP-Glc		2′	Morita et al. (2005)
	Cs 1,6RhaT	Orange (*Citrus sinensis*)	Flavanone glucoside	UDP-Rha		6′	Frydman et al. (2013)
	Cm 1,2RhaT	Pummelo (*Citrus maxima*)	Flavanone glucoside	UDP-Rha		2′	Frydman et al. (2004)
	Ca UGT3	*Catharanthus roseus*	Flavonoid glucoside	UDP-Glc		6′	Masada et al. (2009)
	Ca UGT4	*Catharanthus roseus*	Flavonoid glucoside	UDP-Glc		6′	Masada et al. (2009)
	Bp UGAT/UGT94B1	Red Daisy (*Bellis perennis*)	Anthocyanidin glucoside	UDP-GlcA		2′	Sawada et al. (2005)
	Ph 3RT/UGT79A1	*Petunia hybrida*	Anthocyanidin glucoside	UDP-Rha		6′	Brugliera et al. (1994)
	Pg UGT94Q2	Ginseng (*Panax ginseng*)	Triterpene glucoside	UDP-Glc		2′	Jung et al. (2014)
	Gm Fg2/UGT79A6	Soybean (*Glycine max*)	Flavonoid glycoside	UDP-Rha		6′	Rojas Rodas et al. (2014)

(*Continued*)

Table 11.1 (Continued) List of of functionally characterized GGTs comprising volatile and nonvolatile UGTs

Cluster	Protein/UGT no.	Plant species	Substrate specificity			Ref.
			Acceptor	Donor	Regio	
	Gm Fg3/UGT79B30	Soybean (Glycine max)	Flavonoid glycoside	UDP-Glc	2'	Di et al. (2015)
	Gm SGT3/UGT91H1	Soybean (Glycine max)	Triterpene diglycoside	UDP-Rha	2'	Shibuya et al. (2010)
	Gm SGT2/UGT73P2	Soybean (Glycine max)	Triterpene glycoside	UDP-Gal	2'	Shibuya et al. (2010)
	Gm Sg-1a/UGT73F4	Soybean (Glycine max)	Triterpene glycoside	UDP-Xyl	3'	Sayama et al. (2012)
	Gm Sg-1b/UGT73F2	Soybean (Glycine max)	Triterpene glycoside	UDP-Glc	3'	Sayama et al. (2012)
	St SGT3	Potato (Solanum turberosum)	Steroidal glycoalkaloid	UDP-Rha	2'	McCue et al. (2007)
Non-OG8	Sr UGT76G1	Stevia rebaudiana	Diterpene glucoside	UDP-Glc	3'	Richman et al. (2005)
	Cs UGT707B1	Saffron (Crocus sativus)	Flavonol glucoside	UDP-Glc	2'	Trapero et al. (2012)
	Sl GAME2	Tomato (Solanum lycopersicum)	Triterpene glycoside	UDP-Xyl	3'	Itkin et al. (2013)
	Sl GAME17	Tomato (Solanum lycopersicum)	Triterpene glycoside	UDP-Glc	4'	Itkin et al. (2013)

in comparison with amino acid sequences of other UGTs having different sugar-donor specificity (Noguchi et al. 2009; Ohgami et al. 2015; Ono et al. 2010a; Sayama et al. 2012).

In addition to OG8-class GGTs, exceptional GGTs should be mentioned here. Some non-OG8 class UGTs have also been found to the O-glycosylate sugar moiety of naturally occurring glycosides. For example, *Stevia rebaudiana* UGT76G1 and *Glycine max* UGT73 enzymes (UGT73F2, F4 and P2) are shown to catalyze glycosylation at sugar moiety of glycosides of diterpene and triterpene, respectively (Richman et al. 2005; Sayama et al. 2012; Shibuya et al. 2010), underscoring plasticity in substrate recognition of UGT enzymes.

11.4.2 Ecological and evolutionary landscape of volatile glycosides

As mentioned above, two distinct phylogenetic groups of UGT specialized for volatiles and volatile glycosides are observed among various seed plants. The first glycosylation for volatiles and the second/third glycosylation (sugar–sugar glycosylation) for volatile glycosides are, basically but not absolutely, mediated by two distinct class UGTs separately classified into OG2 and OG8 phylogenetic classes, respectively. From the viewpoints of phylogenomics, paralogs of OG2 and OG8 are commonly observed in seed plants and their gene numbers were outstandingly increased in the genomes of diverse seed plants including *Arabidopsis* and rice (*Oryza sativa*), compared to Lycoppdiophytato (*Selaginella moellendorffii*) (Yonekura-Sakakibara and Hanada 2011). The biased evolutionary trait in extant land plants suggests that evolution of volatile glycosylation would have been temporally associated with multiplication of particular *UGT* genes in an (few) ancestral seed plants, and exerted adaptive effects for their progenies.

During development of ancestral land plants, evolution of UGT catalyzing glycosylation of volatiles and glycosidase catalyzing hydrolysis of volatile glycosides would be an epoch-making event since it enable plants to store volatiles stably as water-soluble compounds prior to unforeseen attacks by herbivores. Volatile glycoside metabolism is more effective than constant emission or *de novo* biosynthesis of volatiles in terms of energy cost and quickness for volatile emission. Volatiles for pollinator attraction is usually *de novo* synthesized on time under the seasonal genetic regulation since induction of volatile biosynthetic genes such as terpene synthases is temporally coincident with volatile production (Boachon et al. 2015), whereas volatile glycosides for an unforeseen herbivore's attack is synthesized in advance and preserved as precursors, enabling a quick response without transcription and translation of biosynthetic genes upon attack (Figure 11.4). The latter system is based on differential compartmentalization of substrates and enzymes although (sub)cellular transport and sequestration of volatile glycosides, namely "metabolic channeling,"

Volatile	Volatile glycoside
Hydrophobic	Water-soluble
Pollinator attraction	Defense response
Seasonal event	Unforeseen attack
Genetically regulated	Quickness required
De novo biosynthesis	Hydrolysis of preserved precursor
Synthesized on time	Synthesized in advance

Figure 11.4 Differences in mode-of-action of volatiles and volatile glycosides.

remains to be elucidated. Glycosylation is exploited for subcellular compartmentalization of nonvolatile precursors (e.g., β-primeveroside) and its activating enzymes (e.g., β-primeverosidase) in plant tissues, but is not the only way for metabolic compartmentalization. For example, wound-induced rapid GLV bursts in tea leaves are considered to be exerted by encounters of preexisting components: endogenous nonglycosidic substrates (fatty acids) and the GLV biosynthetic enzymes, in disrupted tissues (Matsui 2006; Ono et al. 2016).

Conjugation of sugar chain (sugar–sugar) to volatiles is regarded as an anchor for volatiles in aqueous fraction since it not only enhances water-solubility of molecules but also determines sensitivity against hydrolysis by glycosidase as described above. Namely, molecular machineries for volatile glycosylation and hydrolysis determine the molecular state "volatile" or "nonvolatile" *in planta* (Figure 11.5). It is of evolutionary and ecologically interest to unravel how the antagonistic enzymes, diglycoside-specific glycosidase such as β-primeverosidase and UGT for volatile glycosides, evolved and participated in sophisticated volatile metabolism. Speculating on their evolutionary path from the "present" biochemical properties that β-primeverosidase activity is blunt for monoglycoside in tea plants and blunt for triglycoside in tomatoes (Ma et al. 2001; Mizutani et al. 2002; Tikunov et al. 2013), β-primeverosidase might have emerged between occurrence of the second glycosylating UGT represented by CsGT2 and of the third glycosylating UGT represented by NSGT1.

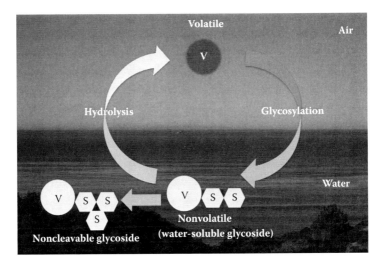

Figure 11.5 Glycosylation-hydrolysis cycle for volatile metabolism in plants. V and S indicate volatiles and sugar moiety, respectively.

From the viewpoint of chemical ecology, it is also worth considering that tomato plants can directly diglycosylate an airborne green leaf volatile, (Z)-3-hexenol which is released by their damaged neighbors, into (Z)-3-hexenylvicianoside, an effective defense compound that reduces the growth and survival of the herbivores (Sugimoto et al. 2014). Thus, glycosylation system is functional for certain exogenous volatiles (derived from other organisms) in atmosphere, and is exploited for chemical defense, thereby increasing environmental fitness of the plants (Mescher and De Moraes 2014). This is another important aspect of volatile glycosylation in sensing biological information of VOCs in proximal gaseous environment where plant faces, although molecular mechanisms underlying the uptake of VOCs and being toxic to herbivores remain to be clarified.

11.4.3 *Promiscuity and diversity*

It is well known that some UGTs catalyze glycosylation for various xenobiotic compounds (Bowles et al. 2005; Taguchi et al. 2000, 2003; Weis et al. 2006). For example, *Arabidopsis* UGT72B1 plays a central role in metabolizing chloroanilines (exogenous pollutants) by glycosylation (Brazier-Hicks and Edwards 2005). In order to detoxify against unexpected structurally diverged xenobiotic compounds, broad substrate specificity would be necessary. In other words, enzymatic promiscuity would be an adaptive property against vast structural diversity of environmental compounds.

This notion would be applied to evolution of UGTs with relatively promiscuous (or sluggish) substrate specificity toward even endogenous metabolites. In the case of terpenes, terpene synthase (TPS), which forms a mid-size family gene in seed plants, is known to produce multiple products from a single substrate, enlarging structural diversity of specialized metabolites (Chen et al. 2011). It is plausible that the promiscuity of substrate specificity of volatile UGTs would be coupled with emergence of multiple substrates during specialized metabolic evolution since excess volatiles is often harmful for plants (Izumi et al. 1999; Matsui et al. 2012). Promiscuity specificity of volatile metabolic enzymes might provide clues for understandings of unexpected metabolic aspects of volatile glycosides abundant in nature.

There are natural variations in UGT function within *Arabidopsis* accessions for xenobiotic drug sensitivity (Zhao et al. 2007), reflecting high genetic diversity in UGTs. Important to note, *Arabidopsis HYR1* encoding UGT71B2 is capable of glucosylation for various xenobiotic compounds exogenously treated, resulting in activation (increasing toxicity) and detoxification of xenobiotic compounds in plants (Zhao et al. 2007). This contrasting responsiveness against the different compounds suggests that effect of glycosylation is dependent on "metabolic context" in preexisting cellular metabolism.

Functional diversity of UGT for specialized metabolites would rely on (1) multiplication of paralogs that serve gene resources sharing redundant function, enabling biochemical differentiation by relaxing genetic constraints from the original function and (2) following metabolic differentiation based on biochemical plasticity promiscuity. Unraveled genome sequences of various plants revealed a significant increase of particular superfamily enzyme genes (e.g., UGT, CYP, and DOX) in land plants (Kawai et al. 2014a; Mizutani and Ohta 2010; Yonekura-Sakakibara and Hanada 2011). For example, gene number explosion in the DOXC clan for specialized metabolism, but not in DOXA and DOXB clans for essential cellular processes, are specifically observed (Kawai et al. 2014b). The vast structural diversity of specialized metabolites would be a consequence of the coordinated metabolic assemblies of superfamily enzyme genes that rapidly evolved in land plant genomes. Then, "drifting metabolites on a border of life."

References

Aharoni, A., A. P. Giri, S. Deuerlein et al. 2003. Terpenoid metabolism in wild-type and transgenic *Arabidopsis* plants. *Plant Cell* 15: 2866–2884.

Arimura, G., K. Matsui, and J. Takabayashi. 2009. Chemical and molecular ecology of herbivore-induced plant volatiles: Proximate factors and their ultimate functions. *Plant Cell Physiol.* 50: 911–923.

Aurore, G., C. Ginies, B. Ganou-parfait, C. M. G. C. Renard, and L. Fahrasmane. 2011. Comparative study of free and glycoconjugated volatile compounds of three banana cultivars from French West Indies: Cavendish, Frayssinette and Plantain. *Food Chem.* 129: 28–34.

Birtic, S., C. Ginies, M. Causse, C. M. G. C. Renard, and D. Page. 2009. Changes in volatiles and glycosides during fruit maturation of two contrasted tomato (*Solanum lycopersicum*) lines. *J. Agric. Food Chem.* 57: 591–598.

Boachon, B., R. R. Junker, L. Miesch et al. 2015. CYP76C1 (Cytochrome P450)-mediated Linalool metabolism and the formation of volatile and soluble linalool oxides in *Arabidopsis* flowers: A strategy for defense against floral antagonists. *Plant Cell* 27: 2972–2990.

Bönisch, F., J. Frotscher, S. Stanitzek et al. 2014a. Activity-based profiling of a physiologic aglycone library reveals sugar acceptor promiscuity of family 1 UDP-glucosyltransferases from grape. *Plant Physiol.* 166: 23–39.

Bönisch, F., J. Frotscher, S. Stanitzek et al. 2014b. A UDP-glucose:monoterpenol glucosyltransferase adds to the chemical diversity of the grapevine metabolome. *Plant Physiol.* 165: 561–581.

Bowles, D., J. Isayenkova, E. K. Lim, and B. Poppenberger. 2005. Glycosyltransferases: Managers of small molecules. *Curr. Opin. Plant Biol.* 8: 254–263.

Brazier-Hicks, M. and R. Edwards. 2005. Functional importance of the family 1 glucosyltransferase UGT72B1 in the metabolism of xenobiotics in *Arabidopsis thaliana*. *Plant J.* 42: 556–566.

Brazier-Hicks, M., K. M. Evans, M. C. Gershater, H. Puschmann, P. G. Steel, and R. Edwards. 2009. The C-glycosylation of flavonoids in cereals. *J. Biol. Chem.* 284: 17926–17934.

Brazier-Hicks, M., W. A. Offen, C. Markus et al. 2007. Characterization and engineering of the bifunctional *N*- and *O*-glucosyltransferase involved in xenobiotic metabolism in plants. *Proc. Natl. Acad. Sci. USA* 104: 20238–20243.

Brugliera, F., T. A. Holton, T.W. Stevenson, E. Farcy, C. Y. Lu, and C. Cornish. 1994. Isolation and characterization of a cDNAclone corresponding to the *Rt* locus of *Petunia hybrida*. *Plant J.* 5: 81–92.

Cabrita, M. J., A. M. C. Freitas, O. Laureano, and R. Di Stefano. 2006. Glycosidic aroma compounds of some Portuguese grape cultivars. *J. Sci. Food Agric.* 86: 922–931.

Caputi, L., E. K. Lim, and D. J. Bowles. 2008. Discovery of new biocatalysts for the glycosylation of terpenoid scaffolds. *Chemistry* 14: 6656–6662.

Caputi, L., M. Malnoy, V. Goremykin, S. Nikiforova, and S. Martens. 2012. A genome-wide phylogenetic reconstruction of family 1 UDP-glycosyltransferases revealed the expansion of the family during the adaptation of plants to life on land. *Plant J.* 69: 1030–1042.

Chae, L., T. Kim, R. Nilo-Poyanco, and S.Y. Rhee. 2014. Genomic signatures of specialized metabolism in plants. *Science* 344: 510–513.

Chen, F., D. Tholl, J. Bohlmann, and E. Pichersky. 2011. The family of terpene synthases in plants: A mid-size family of genes for specialized metabolism that is highly diversified throughout the kingdom. *Plant J.* 66: 212–229.

D'Ambrosio, M., P. Harghel, and V. Guantieri. 2013. Isolation of intact glycosidic aroma precursors from grape juice by hydrophilic interaction liquid chromatography. *Aus. J. Grape Wine Res.* 19: 189–192.

Di, S., F. Yan, F. R. Rodas et al. 2015. Linkage mapping, molecular cloning and functional analysis of soybean gene *Fg3* encoding flavonol 3-*O*-glucoside/galactoside (1 → 2) glucosyltransferase. *BMC Plant Biol.* 15: 126. http://www.biomedcentral.com/content/pdf/s12870-015-0504-7.pdf

Frydman, A., R. Liberman, D. V. Huhman et al. 2013. The molecular and enzymatic basis of bitter/non-bitter flavor of citrus fruit: evolution of branch-forming rhamnosyltransferases under domestication. *Plant J.* 73: 166–178.

Frydman, A., O. Weisshaus, M. Bar-Peled et al. 2004. Citrus fruit bitter flavors: Isolation and functional characterization of the gene *Cm1,2RhaT* encoding a 1,2 rhamnosyltransferase, a key enzyme in the biosynthesis of the bitter flavonoids of citrus. *Plant. J.* 40: 88–100.

Gallage, N. J., E. H. Hansen, R. Kannangara et al. 2014. Vanillin formation from ferulic acid in *Vanilla planifolia* is catalysed by a single enzyme. *Nature Commu.* 5: 4037.

Garcia, C. V., S. Y. Quek, R. J. Stevenson, and R. A. Winz. 2011a. Kiwifruit flavour: A review. *Trends Food Sci. Technol.* 24: 82–91.

Garcia, C. V., S. Y. Quek, R. J. Stevenson, and R. A. Winz. 2011b. Characterization of the bound volatile extract from baby kiwi (*Actinidia arguta*). *J. Agric. Food Chem.* 59: 8358–8365.

Ginglinger, J. F., B. Boachon, R. Höfer et al. 2013. Gene coexpression analysis reveals complex metabolism of the monoterpene alcohol linalool in *Arabidopsis* flowers. *Plant Cell* 25: 4640–4657.

Gunata, Y. Z., C. Bayonove, R. Baumes, and R. Cordonnier. 1985. The aroma of grapes. I. Extraction and determination of free and glycosidically bound fractions of some grape aroma components. *J. Chromatogr. A.* 331: 83–90.

Guo, W., K. Sakata, N. Watanabe et al. 1993. Geranyl 6-*O*-β-D-xylopyranosyl-β-D-glucopyranoside isolated as an aroma precursor from tea leaves for oolong tea. *Phytochemistry* 33: 1373–1375.

Hjelmeland, A. K. and S. E. Ebeler. 2015. Glycosidically bound volatile aroma compounds in grapes and wine. *Am. J. Enol. Vitic.* 66:1–11.

Itkin, M., U. Heinig, O. Tzfadia et al. 2013. Biosynthesis of antinutritional alkaloids in solanaceous crops is mediated by clustered genes. *Science* 341: 175–179.

Izumi, S., O. Takashima, and T. Hirata. 1999. Geraniol is a potent inducer of apoptosis-like cell death in the cultured shoot primordia of *Matricaria chamomilla*. *Biochem. Biophys. Res. Commun.* 259: 519–522.

Jung, S. C., W. Kim, S. C. Park et al. 2014. Two ginseng UDP-glycosyltransferases synthesize ginsenoside Rg3 and Rd. *Plant Cell Physiol.* 55: 2177–2188.

Kannangara, R., M. S. Motawia, N. K. Hansen et al. 2011. Characterization and expression profile of two UDP-glucosyltransferases, UGT85K4 and UGT85K5, catalyzing the last step in cyanogenic glucoside biosynthesis in cassava. *Plant J.* 68: 287–301.

Kawai, Y., E. Ono, and M. Mizutani. 2014a. Expansion of specialized metabolism-related superfamily genes via whole genome duplications during angiosperm evolution. *Plant Biotech.* 31: 579–584.

Kawai, Y., E. Ono, and M. Mizutani. 2014b. Evolution and diversity of the 2-oxoglutarate-dependent dioxygenase superfamily in plants. *Plant J.* 78: 328–343.

Koeduka, T., E. Fridman, D.R. Gang et al. 2006. Eugenol and isoeugenol, characteristic aromatic constituents of spices, are biosynthesized via reduction of a coniferyl alcohol ester. *Proc. Natl. Acad. Sci. USA* 103: 10128–10133.

Koeduka, T., S. Suzuki, Y. Iijima et al. 2013. Enhancement of production of eugenol and its glycosides in transgenic aspen plants via genetic engineering. *Biochem. Biophys. Res. Commun.* 436: 73–78.

Krammer, G., P. Winterhalter, M. Schwab, and P. Schreier. 1991. Glycosidically bound aroma compounds in the fruits of *Prunus* species: apricot (*P. armeniaca*, L.), peach (*P. persica*, L.), yellow plum (*P. domestica*, L. ssp. Syriaca). *J. Agric. Food Chem.* 39: 778–781.

Lanot, A., D. Hodge, and R. G. Jackson. 2006. The glucosyltransferase UGT72E2 is responsible for monolignol 4-*O*-glucoside production in *Arabidopsis thaliana*. *Plant J.* 48: 286–295.

Loughrin, J. H., T. R. Hamiltonkemp, H. R. Burton, R. A. Andersen, and D. F. Hildebrand. 1992. Glycosidically bound volatile components of *Nicotiana sylvestris* and *N. suaveolens* flowers. *Phytochemistry* 31: 1537–1540.

Louveau, T., C. Leitao1, S. Green et al. 2011. Predicting the substrate specificity of a glycosyltransferase implicated in the production of phenolic volatiles in tomato fruit. *FEBS J.* 278: 390–400.

Luan, F., A. Mosandl, M. Gubesch, and M. Wüst. 2006. Enantioselective analysis of monoterpenes in different grape varieties during berry ripening using stir bar sorptive extraction- and solid phase extraction-enantio selective multi dimensional gas chromatography-mass spectrometry. *J. Chromatogr. A.* 1112: 369–374.

Luan, F., A. Mosandl, A. Münch, and M. Wüst. 2005. Metabolism of geraniol in grape berry mesocarp of *Vitis vinifera* L. cv. Scheurebe: demonstration of stereoselective reduction, E/Z-isomerization, oxidation and glycosylation. *Phytochemistry* 66: 295–303.

Lund, S. T. and L. Bohlmann. 2006. The molecular basis for wine grape quality—A volatile subject. *Science* 311: 804–805.

Ma, S. J., M. Mizutani, J. Hiratake et al. 2001. Substrate specificity of β-Primeverosidase, A Key Enzyme in Aroma Formation during Oolong Tea and Black Tea Manufacturing. *Biosci. Biotechnol. Biochem.* 65: 2719–2729.

Mackenzie, P. I., I. S. Owens, B. Burchell et al. 1997. The UDP glycosyltransferase gene superfamily: Recommended nomenclature update based on evolutionary divergence. *Pharmacogenetics* 7: 255–269.

Marlatt, C., C. Ho, and M. J. Chien. 1992. Studies of aroma constituents bound as glycosides in tomato. *J. Agric. Food. Chem.* 40: 249–252.

Masada, S., K. Terasaka, Y. Oguchi, S. Okazaki, T. Mizushima, and H. Mizukami. 2009. Functional and structural characterization of a flavonoid glucoside 1,6-glucosyltransferase from *Catharanthus roseus*. *Plant Cell Physiol.* 50: 1401–1415.

Matsui, K. 2006. Green leaf volatiles: Hydroperoxide lyase pathway of oxylipin metabolism. *Curr. Opin. Plant Biol.* 8: 274–280.

Matsui, K., K. Sugimoto, J. Mano, R. Ozawa, and J. Takabayashi. 2012. Differential metabolisms of green leaf volatiles in injured and intact parts of a wounded leaf meet distinct ecophysiological requirements. *PLoS One* 7: e36433.

McCue, K. F., P. V. Allen, L. V. Shepherd et al. 2007. Potato glycosterol rhamnosyltransferase, the terminal step in triose side-chain biosynthesis. *Phytochemistry* 68: 327–334.

Mescher, M. C. and C. M. De Moraes. 2014. Pass the ammunition. *Nature* 510: 221–222.

Mizutani, M. and D. Ohta. 2010. Diversification of P450 genes during land plant evolution. *Annu. Rev. Plant Biol.* 61: 291–315.

Mizutani, M., H. Nakanishi, J. Ema et al. 2002. Cloning of β-primeverosidase from tea leaves, a key enzyme in tea aroma formation. *Plant Physiol.* 130: 2164–2176.

Montefiori, M., R. V. Espley, D. Stevenson et al. 2011. Identification and characterisation of F3GT1 and F3GGT1, two glycosyltransferases responsible for anthocyanin biosynthesis in red-fleshed kiwifruit (*Actinidia chinensis*). *Plant J.* 65: 106–118.

Morita, Y., A. Hoshino, Y. Kikuchi et al. 2005. Japanese morning glory *dusky* mutants displaying reddish-brown or purplish-gray flowers are deficient in a novel glycosylation enzyme for anthocyanin biosynthesis, UDP-glucose:anthocyanidin 3-O-glucoside- 2"-O-glucosyltransferase, due to 4-bp insertions in the gene. *Plant J.* 42: 353–363.

Nagatoshi, M., K. Terasaka, M. Owaki, M. Sota, T. Inukai, A. Nagatsu, and H. Mizukami. 2012. UGT75L6 and UGT94E5 mediate sequential glucosylation of crocetin to crocin in *Gardenia jasminoides*. *FEBS Lett.* 586: 1055–1061.

Noguchi, A., Y. Fukui, A. Iuchi-Okada et al. 2008. Sequential glucosylation of a furofuran lignan, (+)-sesaminol by *Sesamum indicum* UGT71A9 and UGT94D1 glucosyltransferases. *Plant J.* 54: 415–427.

Noguchi, A., M. Horikawa, Y. Fukui et al. 2009. Local differentiation of sugar donor specificity of flavonoid glycosyltransferase in Lamiales. *Plant Cell* 21: 1556–1572.

Noguchi, A., A. Saito, Y. Homma et al. 2007. A UDP-glucose: Isoflavone 7-O-glucosyltransferase from the roots of soybean (*Glycine max*) seedlings. Purification, gene cloning, phylogenetics, and an implication for an alternative strategy of enzyme catalysis. *J. Biol. Chem.* 282: 23581–23590.

Ohgami, S., E. Ono, M. Horikawa, et al. 2015. Volatile glycosylation in tea plants: Sequential glycosylations for the biosynthesis of aroma β-primeverosides are catalyzed by two *Camellia sinensis* glycosyltransferases. *Plant Physiol.* 168: 464–477.

Ogata, J., Y. Kanno, Y. Itoh, H. Tsugawa, and M. Suzuki. 2005. Plant biochemistry: Anthocyanin biosynthesis in roses. *Nature* 435: 757–758.

Oka, N., H. Ohishi, T. Hatano, M. Hornberger, K. Sakata, and N. Watanabe. 1999. Aroma evolution during flower opening in *Rosa damascena* Mill. *Z. Naturforsch. C.* 54: 889–895.

Ono, E., M. Fukuchi-Mizutani, N. Nakamura et al. 2006. Yellow flowers generated by expression of the aurone biosynthetic pathway. *Proc. Natl. Acad. Sci. USA* 103: 11075–11080.

Ono, E., T. Handa, T. Koeduka et al. 2016. CYP74B24 is the 13-hydroperoxide lyase involved in biosynthesis of green leaf volatiles in tea (*Camellia sinensis*). *Plant Physiol. Biochem.* 98: 112–118.

Ono, E., Y. Homma, M. Horikawa et al. 2010a. Functional differentiation of the glycosyltransferases that contribute to the chemical diversity of bioactive flavonol glycosides in grapevines (*Vitis vinifera*). *Plant Cell* 22: 2856–2871.

Ono, E., M. Ruike, T. Iwashita, K. Nomoto, and Y. Fukui. 2010b. Co-pigmentation and flavonoid glycosyltransferases in blue *Veronica persica* flowers. *Phytochemistry* 71: 726–735.

Osmani, S., S. Bak, A. Imberty, C. E. Olsen, and B. L. Møller. 2008. Catalytic key anmino acids and UDP-sugar donor specificity of a plant glucuronosyltransferase, UGT94B1. *Plant Physiol.* 148: 1295–1308.

Pabst, A., A. Barron, P. Etievant, and P. Schreier. 1991. Studies on the enzymatic hydrolysis of bound aroma constituents from raspberry fruit pulp. *J. Agric. Food Chem.* 39: 173–175.

Richman, A., A. Swanson, T. Humphrey et al. 2005. Functional genomics uncovers three glucosyltransferases involved in the synthesis of the major sweet glucosides of *Stevia rebaudiana*. *Plant J.* 41: 56–67.

Rojas Rodas, F., T. O. Rodriguez, Y. Murai et al. 2014. Linkage mapping, molecular cloning and functional analysis of soybean gene *Fg2* encoding flavonol 3-*O*-glucoside (1 → 6) rhamnosyltransferase. *Plant Mol. Biol.* 84: 287–300.

Roscher, R., M. Herderich, J. P. Steffen, P. Schreier, and W. Schwab. 1996. 2,5-Dimethyl-4-hydroxy-3[2H]-furanone 6′-*O*-malonyl-β-D-glucopyranoside in strawberry fruits. *Phytochemistry* 43: 155–159.

Saino, H., T. Shimizu, J. Hiratake et al. 2014. Crystal structures of β-primeverosidase in complex with disaccharide amidine inhibitors. *J. Biol. Chem.* 289: 16826–16834.

Sasaki, K., H. Takase, H. Kobayashi, H. Matsuo, and R. Takata. 2015b. Molecular cloning and characterization of UDP-glucose: Furaneol glucosyltransferase gene from grapevine cultivar Muscat Bailey A (*Vitis labrusca* × *V. vinifera*). *J. Exp. Bot.* 66: 6167–6174.

Sasaki, N., Y. Nishizaki, E. Yamada et al. 2015a. Identification of the glucosyltransferase that mediates direct flavone C-glucosylation in *Gentiana triflora*. *FEBS Lett.* 589: 182–187.

Sawada, S., H. Suzuki, F. Ichimaida et al. 2005. UDPglucuronic acid:anthocyanin glucuronosyltransferase from red daisy (*Bellis perennis*) flowers. Enzymology and phylogenetics of a novel glucuronosyltransferase involved in flower pigment biosynthesis. *J. Biol. Chem.* 280: 899–906.

Sayama, T., E. Ono, K. Takagi et al. 2012. The *Sg-1* glycosyltransferase locus regulates structural diversity of triterpenoid saponins of soybean. *Plant Cell* 24: 2123–2138.

Shibuya, M., K. Nishimura, N. Yasuyama, and Y. Ebizuka. 2010. Identification and characterization of glycosyltransferases involved in the biosynthesis of soyasaponin I in *Glycine max*. *FEBS Lett.* 584: 2258–2264.

Strauss, C. R., B. Wilson, and P. J. Williams. 1998. Novel monoterpene diols and diol glycosides in *Vitis vinifera* grapes. *J. Agric. Food Chem.* 36: 569–573.

Sugimoto, K., K. Matsui, Y. Iijima et al. 2014. Intake and transformation to a glycoside of (Z)-3-hexenol from infested neighbors reveals a mode of plant odor reception and defense. *Proc. Natl. Acad. Sci. USA* 111: 7144–7149.

Taguchi, G., S. Fujikawa, T. Yazawa et al. 2000. Scopoletin uptake from culture medium andaccumulation in the vacuoles after conversion to scopolin in 2,4-D-treatedtobacco cells. *Plant Sci.* 151: 153–161.

Taguchi, G., M. Nakamura, N. Hayashida, and M. Okazaki. 2003. Exogenously added naphthols induce three glucosyltransferases, and are accu-mulated as glucosides in tobacco cells. *Plant Sci.* 164: 231–240.

Takamine, K., Y. Yoshizaki, Y. Yamamoto et al. 2012. The distribution of monoterpene glycosides in sweet potato. *J. Brew. Soc. Japan* 107: 782–787.

Tamura, K., G. Stecher, D. Peterson, A. Filipski, and S. Kumar. 2013. MEGA6: Molecular evolutionary genetics analysis version 6.0. *Mol. Biol. Evol.* 30: 2725–2729.

Tikunov, Y. M., J. Molthoff, R. C. de Vos et al. 2013. NON-SMOKY GLYCOSYLTRANSFERASE 1 prevents the release of smoky aroma from tomato fruit. *Plant Cell* 25: 3067–3078.

Ting, P. L., S. Kay, and D. Ryder. 2009. The occurrence and nature of kettle hop flavor. Hop flavor and aroma. In *Proceedings of the 1st International Brewers Symposium* ed. T.H. Shellhammer, Master Brewers Association of the Americas, ISBN: 978-0-9770519-8-4: 25–36.

Trapero, A., O. Ahrazem, A. Rubio-Moraga, M. L. Jimeno, M. D. Gómez, and L. Gómez-Gómez. 2012. Characterization of a glucosyltransferase enzyme involved in the formation of kaempferol and quercetin sophorosides in *Crocus sativus*. *Plant Physiol*. 159: 1335–1354.

Wang, D., T. Yoshimura, K. Kubota, and A. Kobayashi. 2000. Analysis of glycosidically bound aroma precursors in tea leaves. 1. Qualitative and quantitative analyses of glycosides with aglycons as aroma compounds. *J. Agric. Food Chem*. 48: 5411–5418.

Weis, M., E. K. Lim, N. Bruce, and D. Bowles. 2006. Regioselective glucosylation of aromatic compounds: Screening of a recombinant glycosyltransferase library to identify biocatalysts. *Angew. Chem. Int. Ed. Engl*. 45: 3534–3538.

Williams, P. J., C. R. Strauss, B. Wilson, and R. A. Massy-Westropp. 1982. Studies on the hydrolysis of *Vitis vinifera* monoterpene precursor compounds and model monoterpene β-D-glucosides rationalizing the monoterpene composition of grapes. *J. Agric. Food Chem*. 30: 1219–1223.

Yauk, Y. K., C. Ged, M. Y. Wang et al. 2014. Manipulation of flavour and aroma compound sequestration and release using a glycosyltransferase with specificity for terpene alcohols. *Plant J*. 80: 317–330.

Yonekura-Sakakibara, K., A. Fukushima, R. Nakabayashi et al. 2012. Two glycosyltransferases involved in anthocyanin modification delineated by transcriptome independent component analysis in *Arabidopsis thaliana*. *Plant J*. 69: 154–167.

Yonekura-Sakakibara, K. and K. Hanada. 2011. An evolutionary view of functional diversity in family 1 glycosyltransferases. *Plant J*. 66: 182–193.

Yonekura-Sakakibara, K., R. Nakabayashi, S. Sugawara et al. 2014. A flavonoid 3-O-glucoside: 2″-O-glucosyltransferase responsible for terminal modification of pollen-specific flavonols in *Arabidopsis thaliana*. *Plant J*. 79: 769–782.

Young, H. and V. J. Paterson. 1995. Characterization of bound flavour components in kiwifruit. *J. Sci. Food Agric*. 68: 257–260.

Zhao, Y., T. F. Chow, R. S. Puckrin et al. 2007. Chemical genetic interrogation of natural variation uncovers a molecule that is glycoactivated. *Nat. Chem. Biol*. 3: 716–721.

chapter twelve

Plant secondary metabolites as an information channel mediating community-wide interactions

André Kessler and Kaori Shiojiri

Contents

12.1 Introduction: Plant volatile organic compounds as
 information channels 283
12.2 The limits of a chemical language 286
   12.2.1 The ecologic niche analogy: Fundamental and realized
    information space 286
   12.2.2 Inherent constraints in secondary metabolite production 289
   12.2.3 Diversity and multifunctionality of plant secondary
    metabolites 291
12.3 VOC-mediated information transfer in biotic interactions 294
   12.3.1 Herbivores and pollinators interacting with plants: The
    mutualist–antagonist continuum 294
   12.3.2 Information-mediated indirect resistance: VOC-
    mediated multitrophic level interactions 300
   12.3.3 VOC-mediated plant–plant communication 302
12.4 Conclusion 305
Acknowledgments 306
References 306

12.1 Introduction: Plant volatile organic compounds as information channels

Plant secondary metabolites in general, and volatile organic compounds (VOCs) in particular, serve a multitude of ecological functions for plants, from protecting them from abiotic stresses, such as frost, drought, and UV radiation (Blande et al. 2014; Loreto and Fineschi 2015; Loreto et al. 2014) to mediating a large diversity of antagonistic and mutualistic biotic interactions (Dicke and Baldwin 2010; Heil and Karban 2010; Turlings and Ton

2006). Plant chemicals are increasingly found as carriers of information that mediate interactions with organisms in the plant's environment. More specifically, it has been hypothesized that the "least common denominator function" of plant secondary metabolites may indeed be information exchange within the plant (e.g., signal transduction between tissues and distant modules) and of the plant with its biotic environment. While this information-mediating function of plant secondary metabolites in general has been discussed elsewhere (Kessler 2015), here we will focus on VOCs, lipophilic liquids with low molecular weight and high vapor pressure at ambient temperatures (Pichersky et al. 2006).

VOCs emitted from all types of plant tissue provide particularly striking examples of the information-transmitting property of plant metabolites. While some of the VOCs can be toxic and directly deterrent when ingested or applied externally to an organism's tissues (Veyrat et al. 2016), most interactions, mediated by VOCs, tend to be a transmission of chemical information. As with plant secondary metabolites in general, the diversity of VOCs is vast, with more than 1700 compounds in 90 angio- and gymnosperm plant families identified thus far (Knudsen et al. 2006). Being just at the beginning of identifying the VOCs produced by plants, we are far from conclusively listing all their biological functions or even general principles associated with their production, transduction, and perception (Wilson et al. 2015). However, it is safe to say that their ecological functions are at least as diverse as the interactions plants can have with distal parts of themselves or other organisms. Some of those VOC-mediated interactions have gotten significant attention and research on them became a driver of the exponential growth in the field of chemical ecology. In the 1980s, VOCs had been found to transmit information between plants. So-called plant–plant communication saw VOCs from herbivore-attacked plants inducing herbivore resistance in undamaged neighboring plants before herbivores arrived (Baldwin and Schultz 1983). At about the same time, insect herbivory-induced VOCs were identified as cues that can attract third trophic level organisms (predators and parasitoids) to herbivore-infested plants, which therefore indirectly help plants cope with herbivore attacks (Dicke and Sabelis 1988; Turlings et al. 1990). Shortly after that, research began to identify so-called chemical elicitors from insect herbivores that specifically induce VOC emission in attacked plants (Alborn et al. 1997; Mattiacci et al. 1995) and explained the frequently, very high specificity in both, the VOC profile emitted and the organisms' responses to those cues (Dicke and Baldwin 2010; Heil and Karban 2010; Kessler and Morrell 2010). The attraction of predacious and parasitic plant "bodyguards" was termed VOC- or information-mediated indirect resistance and has since been the subject of exciting applied and basic research, as well as controversy (Allison and Hare 2009; Kessler and Heil 2011; van der Meijden and Klinkhamer 2000). Herbivores and pollinators also use plant VOCs as part of a complex network of traits to identify their host plants as appropriate foraging or oviposition sites, and conceptually form an interaction continuum from antagonistic to mutualistic, respectively.

The information encoded in plant floral and vegetative VOC emissions can be very specific and have high spatial and temporal resolution (Kessler and Halitschke 2007; McCormick et al. 2012). VOC profiles are correlated with plant ontogeny (Ruíz-Ramón et al. 2014), overall metabolic capability (Halitschke et al. 2011; Kessler and Baldwin 2004), and flowering status (Raguso et al. 2003), as well as with different types of abiotic and biotic physiological stresses of the plant (Blande et al. 2014; Kessler and Halitschke 2007; Loreto et al. 2014). Thus, the information content of VOC emission is a function of the developmental and stress-induced phenotypic plasticity of VOC bouquet composition and intensity (Kessler and Halitschke 2007). Once released into the environment the information is ubiquitously available to whoever is able to perceive the whole or only parts of the bouquet. There are two principles that seem universally associated with chemical communication but specifically apply to VOC-mediated information transfer. First, a large proportion of a plant's secondary metabolite bouquet seems to be functionally redundant. Second, it is generally the specific relative composition of the VOC bouquet that encodes information rather than specific compounds (Bruce et al. 2005) (but see highly specific plant-pollinator interactions, for example, Schiestl 2005). This explains why most compounds seem to be universally detectable by interactors, but it also means that their ecological effects are determined by the environmental and chemical context in which they are presented. If assumed to be adaptive the evolution of the information function of plant secondary metabolites should be driven by (A) the repertoire that's available to a population of plants (inherent, phylogenetically determined metabolic capability) (Knudsen et al. 2006), (B) the sensory (perceptive and processing) systems of the receiver organisms (Bruce et al. 2005) as well as (C) the complexity of the chemical environment in which the information is exchanged. Within this framework species- and genotype-specific differences in constitutive VOC emission can function as cues that facilitate an organism's choice to interact with the plant because it provides information about the plant quality as an interaction environment (e.g., quality of plant tissue as food and reward for herbivores and pollinators, respectively, place to find shelter or prey for predatory organisms). This constitutively emitted bouquet of VOCs is mainly determined by the plants inherent metabolic capabilities. In contrast, VOC emission that is induced by environmental stressors, such as abiotic environmental factors and herbivory, provides information with a much higher time resolution and specificity. For example, VOC emission induced by insect herbivory is very specific to the attacker species (Kessler and Baldwin 2001) and even the herbivore's physiological state (Zhu et al. 2014). Moreover, other environmental factors inducing alterations in the emitted VOC bouquet, will have the potential to provide very precise information about the plant's metabolic status and thus again its suitability as a potential interactor for other organisms. This specificity of information coupled with the ubiquitous availability of the cue allows for the impressive multitude of interactions mediated by VOCs and increases the arena in which plant insect interactions are played out from the plant tissue to the whole community level (Kessler 2015; Kessler and Baldwin 2002). This also means that VOC-mediate plant signaling has the potential to allow plants to incorporate their entire interacting community into their complex strategies to maximize fitness (e.g., herbivore repellence, pollinator and predator attraction, plant–plant communication). In this chapter, we use this chemical information framework to generate hypotheses on VOC cue evolution and function while focusing on the major known biological interactions mediated by plant VOCs, plant herbivore/pollinator interactions, plant- multitrophic level interaction, and plant–plant communication.

12.2 The limits of a chemical language

12.2.1 The ecologic niche analogy: Fundamental and realized information space

The entirety of a plant's chemical information can be visualized as a multidimensional space as is frequently done with multidimensional statistical analyses (e.g., PCA, NMDS) to illustrate differences in VOC bouquets of different plants (Figure 12.1). Each compound is represented by a vector expanding the n-dimensional hyperspace in a certain direction. This fundamental information space, analog to the Hutchinsonian fundamental ecological niche (Hutchinson 1957), characterizes a plant and population of plants as habitat with which other organisms interact. The assumption is that the information encoded in the chemical cues reflects certain habitat qualities such as food quality for an herbivore, presence of a host or prey for a parasitoid or predator, respectively, the availability of pollen and nectar rewards for pollinators, or the risk of being attacked by herbivores or pathogens for a neighboring plant. An interacting organism's spectrum of plant species or genotypes it can interact with is then determined by (and predictably based on) how strongly chemical information spaces differ between plant species/genotypes within the community. This model is rather simplified because it assumes that the interacting organisms use primarily information encoded in VOC emission. However, because other chemical (including defensive secondary metabolism and nutrient-providing primary metabolism) and morphological traits (e.g., leaf surface structure, floral color) are frequently correlated with VOC emission

Chapter twelve: Plant secondary metabolites 287

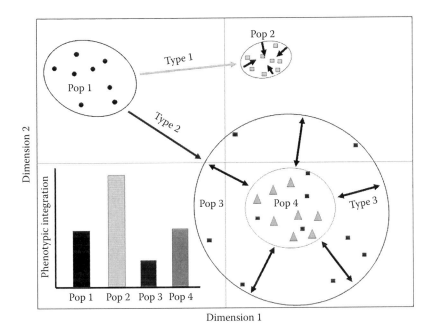

Figure 12.1 The volatile information space. The graph is an illustration of two dimensions for the n-dimensional information spaces filled by individual volatile bouquet phenotypes of four differently shaded plant populations (Pop 1–4). Each volatile organic compound (VOC) of a bouquet is representing a vector in the n-dimensional space, so that the integrated sum of all compound vectors representing a bouquet determine a plant phenotype's position (represented by dots) within the information space. The information space represents all chemical information available to an organism interacting with a plant. Depending on the ecological function, for example, repellency or attraction of an interacting organism, and the fitness effect of the interaction for the plant providing the information, natural selection can result in shifts in information space (e.g., qualitative and quantitative changes in compound composition, Types 1 and 2), as well as in its contraction (stabilizing selection, Type 1) or expansion (diversifying selection, Type 2). Thereby separation and contraction of the information space does not need to occur simultaneously (Type 3). The framework can be applied to predict patterns for natural selection on constitutive VOC production, as well as on patterns of environmentally (e.g., herbivory) induced VOC production. (Inserted Graph: Expected phenotypic integration associated with stabilizing and diversifying selection in the imaginary populations.)

(Kessler and Baldwin 2004; Majetic et al. 2010), it can function as a reliable predictor of the diversity of interactions a plant is likely to be exposed to.

The realized information space, the information actually used by the interacting organism to decide on engagement with the emitting plant, is a function of (A) the plant's metabolic abilities, (B) the interactor's sensory abilities, (C) the presence of biotic and abiotic stressors altering plant metabolism, and, specifically for VOCs, (D) physical environmental factors that affect compound breakdown and VOC plume propagation.

The three later factors (B–D) can be characterized as alterations to the fundamental information that affect perception, generation, and transmission of the VOC cue, respectively. Thus, they can also be viewed as sources of noise affecting the fundamental information space. The noise concept is at the core of information theory (Shannon 1948), and applied to biological systems, is thought to affect ecology and evolution of communication systems (Doyle 2009; McCowan et al. 1999). Noise affects every aspect of the information transfer chain, including signal generation, transmission, and reception, and has been shown to influence the evolution of auditory and visual communication systems (Brumm and Slabbekoorn 2005; Endler 1993). Only very recently has information theory been applied to chemical communication systems; VOC-mediated interactions between plants and third trophic level organisms (Wilson et al. 2015). This exercise aided the discussion on how to identify functional relationships between organisms that are mediated by chemical information. For example, if VOC emission functions as a signal and thus results in fitness benefits for the emitting plant (Allison and Hare 2009; Kessler and Heil 2011), one would expect natural selection to reduce noise, which can be accomplished through multiple mechanisms, such as redundancy and filtering on the emitter side or spectrum spreading and behavioral adaptations on the receivers' side (Wilson et al. 2015). The shrinking or expansion of the information space as a result of natural selection is equivalent to reciprocal changes in phenotypic integration of bouquets of compounds composing the information. Similarly interesting, on an ecological level, environmentally induced changes in VOC emission (e.g., herbivore-induced) can expand and shrink the realized information space and so dramatically influence the outcome of interactions with other organisms. For example, an environmental factor, such as herbivory, inducing an expansion of the VOC-encoded information space of a given plant into the space of another plant species or genotype could increase its attractiveness to another herbivore species that would otherwise not attack this plant. In such cases, herbivore-induced VOC emission would mediate induced susceptibility. The reverse would happen if the information space shifts away from an herbivore's search pattern in response to herbivory (induced resistance/repellency). As a consequence, one can predict natural selection on inducibility in either direction (expansion and shrinkage of the information space), depending on the predominant interactions with other organisms mediating fitness effects in the plant (Figure 12.1). We will discuss some predictive implications of information theories as applied to VOC-mediated information transfer for the major interaction categories.

12.2.2 Inherent constraints in secondary metabolite production

If plant secondary metabolite production in general and VOC emission in particular have primary functions in information transfer, they could be analyzed in analogy to an auditory language. From an ecological function perspective, this then raises the question in how far can a chemical language appropriately and reliably transmit information (specificity, private channel) and how susceptible is the transferred information to noise. This in turn, will determine evolutionary trajectories in chemical signal evolution along the continuum of major types of interactions, from antagonistic over mutualist to deceptive. So, how different is the volatile language of plants when compared at different levels of organization?

Plant species differ dramatically in their secondary metabolism in general and their VOC production in particular, or do they? Humans and certainly other organisms are able to taste and smell difference between species, but are these differences in perception due to differences in the production of specific compounds or different compositions of a common set of compounds? Or, in other words, to what extent do information spaces overlap between species or genotypes?

Each individual plant can produce a large diversity of VOCs, often reaching several hundred compounds (Fraser et al. 2003; Knudsen et al. 1993, 2006; Visser 1986). An analysis of 991 angiosperm and gymnosperm species in 90 families and 38 orders revealed 1719 compounds that are emitted into the headspace of flowers (Knudsen et al. 2006). The seminal study found that most volatile compound classes, such as aliphatics, diterpenes, sesquiterpenes, benzenoids, and phenylpropanoids, are produced by most plant orders with monoterpenes produced by all plant orders. Even the rarer sulfur- and nitrogen-containing compounds were relatively wide-spread and found in 39% and 63% of the plant orders, respectively. Moreover, 12 individual compounds are emitted by more than 50% of the investigated plant families. At the same time, floral scent variation within genera is relatively high (Barkman et al. 1997; Dobson et al. 1997; Jürgens et al. 2002; Knudsen and Mori 1996; Levin et al. 2001; Raguso et al. 2003; Thien et al. 1975) and VOC bouquets may vary as much among species within a genus as among genera within a plant family. Cautioned by the fact that only a subset of species within each plant order/family could be sampled, the data nevertheless suggest that at high taxonomic levels, floral VOC biosynthesis is not phylogenetically constrained and a large number of compounds are produced by a very diverse number of species (Knudsen et al. 2006). However, the large variation between species suggests evolutionary processes shaping VOC patterns and implies significant ecological functions of VOC emission. What factors drive those differences is not yet well understood, however the information transfer framework (Kessler 2015) predicts extrinsic factors, such as the types of pollinators selecting on VOC patterns, for example, associated with florals syndromes (Raguso 2008) or antagonists interacting with flowers (Adler 2000; Theis and Adler 2011) as well as intrinsic factors, such as the mating system (Campbell and Kessler 2013) and phyllogentic constraints (Knudsen et al. 2006) as drivers of between-species variation in VOC emission. In conclusion, the fact that a large number of VOCs are very widely distributed within the plant kingdom as well as the ubiquitously available information encoded in the VOC bouquets indicates them as a universal language aiding the information transfer with diverse organisms. The phylogenetic analysis of floral VOC emissions is spearheading a shift from mechanistic to functional and evolutionary studies in chemical ecology. At this point, no similar studies have been conducted for leaf VOCs or the herbivory-inducibility of floral and leaf VOCs.

Similarly, there are still relatively few studies that compare population level differences in leaf or floral VOC emission (Azuma et al. 2001; Dötterl and Jürgens 2005; Knudsen 1994, 2002; Olesen and Knudsen 1994; Pettersson and Knudsen 2001; Tollsten and Bergstrom 1993; Tollsten and Ovstedal 1994) and even fewer studies measured fitness effects as a function of VOC emission (natural selection) (Parachnowitsch et al. 2012). A comparison of within- and between-population variations of floral scents of the neotropical palm *Geonoma macrostachys* (Arecaceae) revealed a total of 108 VOCs of which only 28 were emitted by all 62 samples in the study. This seems like a relatively small number of common compounds. However, most of the compounds were terpenes with most of the variation between populations sufficiently explained by those few mono- and sesquiterpenes (Knudsen 2002). While these data suggest that there is considerable qualitative variation in floral VOC emissions within a species it poses the question in how far all of the compounds within a headspace are functional in mediating interactions and in how far there is functional redundancy among compounds. The same study also found no differentiation between the study populations but a negative correlation of bouquet similarity with distance, suggesting a clinal variation within the distribution area (Knudsen 2002). This is very likely a typical pattern for outcrossing species with a metapopulation structure and a mix of short and long distance gene flow. Nevertheless, specific pollinator behavior, for example high pollinator constancy, can be predicted to result in selection for similarity and thus strong population separation, in floral chemical cues within a population, even in strictly outcrossing species (Parachnowitsch et al. 2012).

Similarly, if herbivore-induced VOC emission is functional in mediating information between the plant and another organism, with a significant fitness effect for the plant, selection for similarity (e.g., reduction in noise, higher phenotypic integration) of the VOC signal is predicted for cases in which noise disrupts the information transfer (Figure 12.1,

Type 1). This can be the case for interactions like VOC-mediate indirect resistance or induced resistance for which the induced VOC emission signals worsened food quality to herbivores and is avoided in consequence (chemical aposematism) (De Moraes et al. 2001; Kessler and Baldwin 2001). Alternatively, divergent selection on chemical signaling profiles can for example be expected in cases where VOC emission increases the probability of being attacked by a dominant herbivore (Carroll et al. 2006). In such cases, plants can be chemically camouflaged by increasing the noise in the information available within a plant population (Figure 12.1, Type 3).

As on the higher taxonomic level (Campbell and Kessler 2013), the mating system as an intrinsic factor can affect the within-species/population variation of secondary metabolite production. A series of experiments with inbreed and outcrossed progeny of horsenettle (*Solanum carolinense*) revealed that inbreed plants were more attractive and had a much lower resistance to herbivores (Campbell et al. 2013; Kariyat et al. 2014), resulting in inbreeding depression effects that were fully mediated by herbivory (Campbell et al. 2013). These differences in resistance were correlated with a lower inducibility of resistance due to deficiencies in endogenous wound signaling in inbred plants (Campbell et al. 2014) and differences in constitutive and inducible leaf VOC emissions affecting herbivore attraction (Kariyat et al. 2014) and indirect resistance (Kariyat et al. 2012). There is a great need for more studies of VOC emission on a population level to address the predictions underlying the application of information theory to VOC-mediated communication and to document and be able to interpret within and between population variation in floral and constitutive and inducible VOC emission. Moreover, the studies available thus far reflect the high complexity of and potentially high functional redundancy within VOC bouquets (Kessler and Halitschke 2007) and the question remains what proportion of the information is actually used by interacting organisms (Bruce et al. 2005) and if the size and composition of that proportion is correlated with the ecological function of the VOC cue emitted into the environment.

12.2.3 Diversity and multifunctionality of plant secondary metabolites

Seemingly paradox is that the high diversity and potential functional redundancy of plant secondary metabolite production discussed earlier is paired with multifunctionality of individual compounds. Both, multifunctionality as well as high diversity of secondary metabolites, are still at the core of a debate driving research in chemical ecology (Moore et al. 2014). And, even though immense progress has been made in the analytical techniques, we still fall short in assigning biological functions to individual compounds (Firn and Jones 2006; Owen and Peñuelas 2005). One of the widely accepted but rarely tested functional explanations for high plant secondary metabolite production in general is the so-called "screening hypothesis" (Jones et al. 1991). Within this framework, a high-standing (within organism) secondary metabolite diversity increases the probability of resistance activity against a large diversity, including novel herbivores and pathogens. In analogy to an immune system in animals, the metabolic regulatory processes that generate high compound diversity (e.g., mediated by diversification of function and gene duplication) are under positive natural selection in an environment with diverse and unpredictable herbivory (Firn and Jones 2003; Hartmann 2007). Applying this same concept to the production of VOCs with primarily information-mediating functions, rather than direct defensive functions, seems not immediately intuitive. However, if VOCs are viewed as the volatile proportion of the overall metabolic properties of a plant, selection for high compound diversity in general will also result in more diverse VOC bouquets, potentially through the activity of "promiscuous" enzymes able to catalyze the production of compounds with diverse properties (Firn and Jones 2006; Owen and Penuelas 2006). A major mechanistic prediction is than that nonvolatile and volatile secondary metabolite production should be highly correlated, and it provides a reasonable explanation for the potentially high number of volatile and nonvolatile compounds with no obvious biological function.

Alternatively, high standing compound diversity as well as compound diversification in time (environmentally/herbivore-induced changes) have been suggested to be functional in their own right, due to synergistic or additive effects resulting from functional interactions of diverse secondary metabolites (Adler and Karban 1994; Berenbaum et al. 1991; Hay et al. 1994; Richards et al. 2010; Steppuhn and Baldwin 2007). This hypothesis is more readily applied to the diversity of VOCs because compound intrinsic diversity as well as environmentally induced diversification can increase information content and specificity of the VOC bouquet. Thereby information content for an interacting organism can be altered by the plant through additional compounds as well as the variation in compound ratios within the diverse mixture (Bruce et al. 2005), both of which would be in analogy with an expanding or shifting realized information space. Assuming a biological functionality of compound diversity, especially in the information transfer framework, chemodiversity can be hypothesized to be an organizing principle in its own right or an emerging property of the interactions with other organisms (Hilker 2014). Within this later framework plants could be under selection for diversification of the VOC bouquet if information would be added that allows other organisms to interact with higher efficiency. This is a potential mechanism of diversifying selection that drives the evolution of private channels to communicate with mutualists, such as pollinators. For example a study analyzing the floral VOC bouquets of the Australian sex-deceptive orchid *Drakaea glyptodon* (Orchidaceae) revealed a far more complex and diverse chemistry that mediated the interaction with the deceived thynnine wasp pollinator, *Zaspilothynnus trilobatus* (Thynnidae) than previously expected of such a private interaction. The authors suggested that the evolutionary novelty in floral chemical traits is a major driver of the evolutionary dynamics in sex-deceptive plant-pollinator interaction systems. On the other hand, diversification of chemical information can allow plants to escape certain interactions. No explicit tests of this hypothesis have been conducted for constitutive VOC emission phenotypes as far as we know, but many of the differences in herbivore preferences for certain plant genotypes could be driven by differences in the VOC bouquet (Kariyat et al. 2014). Similarly, there are many examples where induced changes in secondary metabolite production result in altered herbivore preferences (Carroll et al. 2006; De Moraes et al. 2001; Kessler and Baldwin 2001). Induced changes in VOC production and thus alteration of the realized information space can be interpreted as a diversification of the information space in time. Thereby the alteration of the scent bouquet or the simple addition of individual compounds can have dramatic effects. The addition of a single green leaf volatile, 1-hexanol, to floral bouquets rendered borage, grapefruit, and alkanet flowers unattractive to two bee species, *Apis mellifera* and *Eucera tetralonia* (Apidae), illustrating the potentially dramatic effect of the expansion or shift of the information space of a plant (Wright and Schiestl 2009).

Because of the ubiquitous availability of VOC information, a genotypic or environmentally induced difference in bouquet composition and diversity that affects a particular interaction is likely to affect many other interactions of the plant with its community. We are only beginning to study the implications of this ecological multifunctionality (Kessler et al. 2011; Poelman et al. 2012; Schiestl et al. 2014). Again, the likely precondition that allows such broad ecological consequences of VOC emission is the universal use of a relatively restricted number of signaling compounds that are common to most plant species (Bruce et al. 2005; Knudsen 2002; Raguso et al. 2015). At the same time, these compounds have to share chemical structural properties that allow them to be detected through receptors (Wicher 2015) or by directly interacting with cell membranes (Bricchi et al. 2010, 2012) by a large number of phylogenetically diverse organisms. This also implies multifunctionality on yet another level.

If a compound has the chemical properties to attach to and be perceived by a receiver's tissues, the same compounds can have signaling functions across multiple organization levels from the ecological levels discussed above to the within plant and between tissue hormonal signal transduction function (Kessler 2015; Raguso et al. 2015). Impressive recent examples come from compounds that were thought to have exclusively herbivore-defensive functions. Benzoxacinoids, potent antiherbivore defenses in maize, *Zea mays* (Poacae), have now been found to also serve as important endogenous defense regulatory signals (Ahmad et al. 2011). Similarly, glucosinolates, the major component of the mustard oil bomb mediating broad herbivore resistance, and their degradation products have now been shown to be involved in microbial pathogen resistance (Clay et al. 2009) as well as cell regulatory functions (Asberg et al. 2015). VOCs such as greenleaf volatiles as well as terpenoids, known to attract predators and parasitoids to herbivore-damaged plants and so functioning as indirect resistance traits, are also found to induce defense and defense priming responses in unattacked neighboring plants or distal branches within the same plant, suggesting a within-plant signaling function (Heil and Bueno 2007).

12.3 VOC-mediated information transfer in biotic interactions

The relatively universal detectability of VOCs and the ubiquitous availability of the information encoded in VOC bouquets predispose them as mediators of a whole spectrum of interactions from antagonism to mutualism and from cellular over whole plant to multitrophic level. However, depending on a particular organism's effect on plant fitness natural selection on VOC traits can be expected to be very different, and may be conflicting. Conflicting natural selection was implied from the beginning of the study of VOC-mediated interactions. For example, predatory or parasitoid organisms may select on intensity and specificity of herbivory-induced VOC emission because the plant can so facilitate the prey/host search behavior of natural enemies, the plant's "bodyguards" (Dicke and Baldwin 2010). On the other hand, the induced changes in VOC emission may make the plant more conspicuous to herbivores (Carroll et al. 2006) or may affect other mutualists, such as pollinators negatively (Kessler et al. 2011; Schiestl et al. 2014), both suggesting negative selection on signal intensity and specificity (Figure 12.1). Similarly, neighboring plants may eavesdrop on their herbivore-attacked neighbors' induced VOC emission and, with that information, gain a competitive advantage (Heil and Karban 2010). Below we develop general predictions for VOC signal construction, mediating different types of interactions and assuming information transfer as the major function of VOC emission.

12.3.1 Herbivores and pollinators interacting with plants: The mutualist–antagonist continuum

Both herbivores and pollinators consume plant tissues or metabolites and both are dependent on efficiently locating an appropriate host plant to meet their nutritional needs. However, they usually represent two opposite ends of the interaction spectrum with pollinators being the mutualists and herbivores the antagonists. Due to their opposite effects on plant fitness they can be expected to select very differently on plant VOC signaling. However, this is assuming that both groups use plant VOCs as major host identification cues or that the VOC emission is strongly covarying with a different host cue (e.g., visual, tactile, nonvolatile metabolites). There are only a few studies measuring natural selection of VOC traits and, to our knowledge, none that explicitly identifies herbivores or pollinators as agents of natural selection on the volatile traits. Nevertheless the vast information on the behavior of herbivores and pollinators toward plant VOCs allows for some prediction on how natural selection may shape those traits.

A recent review (Bruce et al. 2005) analyzed the diverse literature on herbivore use of host VOC cues by evaluating the evidence in support of the two major hypotheses for host recognition that apply for all types for VOC-mediated interactions: (A) species-specific odor recognition (Fraenkel 1959) and (B) compound ratio-specific odor recognition. For herbivores, the use of compounds that are specific to host species or taxonomic groups for host recognition seems to be rather rare (Bruce et al. 2005). For example, specialists on Brassicaceae plants use isothiocyanates, catabolites of the family-specific glucosinolates, to find their hosts (Blight et al. 1995; Nottingham et al. 1991). However, an overwhelming majority of electrophysiological studies found the peripheral receptors of a taxonomically broad range of insects tuned to rather common plant VOCs (Bruce et al. 2005), including most of those greenleaf volatiles, phenylpropanoids and terpenoids that had been found to be common to most plant taxa (Knudsen et al. 2006). The fact that herbivores detect rather common compounds, suggests that they differentiate between host and nonhost plants by the relative ratios of compounds within bouquets rather than specific compounds. Thus, in cases in which herbivores are the major agents of natural selection (e.g., damaged plants with significant fitness consequences), plants should be under selection to alter the relative ratio, the realized information space, of common compounds that are physiologically detectable by herbivores. There are three mechanism that were proposed through which plants can escape herbivory by changing the chemical information space (Kessler 2015). First, compounds or combinations of compounds included in the realized information space are toxic themselves (Nottingham et al. 1991) and so appropriately reflect the nonhost status. Herbivores repelled by such toxic chemical cues are expected to select for novelty and at the same time for the increased production of the resistance-mediating compound(s) relative to all other compounds. Such selection can be expected to result in a shift as well as a shrinking of the information space (increased phenotypic integration) away from the original host-identifying information space used by the herbivore to identify a suitable host plant (Figure 12.1, Type 1). In consequence selection by herbivory under this scenario would result in lower variance of VOC phenotypes within a population. Second, compounds or compound combinations that themselves are not toxic but indicate toxicity or low food quality of a plant are expected to be under a similar selection by herbivores as actual toxic compounds. Such compounds would function as chemical aposematism in the same way herbivore-induced VOC production can signal high predation pressure and low food quality to host-searching herbivores (De Moraes et al. 2001; Kessler and Baldwin 2001) (Figure 12.2a). Accordingly, herbivore-induced VOC emission for both mechanisms should be under selection to significantly differ between the unchallenged and herbivore-attacked state of the plant to advertise induced resistance (Figure 12.1, Type 1, 2). This, however, is only expected if the major agent of natural selection avoids previously damaged plants (De Moraes et al. 2001; Kessler and Baldwin 2001), rather than being attracted to them (Carroll et al. 2006). The third mechanism involves so called "detractant" compounds (Kessler 2015) that are neither toxic nor are they associated with toxicity, but there addition to the bouquet changes the information space in a way so that herbivores will not identify the plant as a host anymore. These could be compounds that throw off the particular compound ratios within a VOC bouquet that are detected by

Figure 12.2 **(See color insert.)** Herbivory-induced volatile organic compound (VOC) emission mediating direct and indirect resistance. (a) Adult females of the major Solanaceae herbivore, *Manduca quinquemaculata,* use VOC information to avoid herbivore-damaged plants for oviposition. (Photo: Danny Kessler.) (b) The generalist predator, *Geocoris pallens,* is attracted to VOCs emitted from herbivore-damaged wild tobacco, *Nicotiana attenata* plants, with positive plant fitness effects. (Photo: André Kessler.)

the herbivores as an appropriate host cue (Wright and Schiestl 2009), or they could be compounds that affect the perception of host-typical cues by the herbivore's olfactory system. A recent study found (E)-4,8-dimethyl-1,3,7-nonatriene (DMNT) as the key compound mediating deterrence to host and mate-searching *Spodoptera litoralis* (Noctuidae) moths on cotton plants. Remarkably that deterrence was a result of DMNT suppressing the moths' sensory responses to the main sex pheromone component, (Z)-9-(E)-11-tetrdecenyl acetate, and the primary host plant attractant, (Z)-3-hexenyl acetate (Hatano et al. 2015). Thus, DMNT is changing the realized (in this case perceived) information space, which moves the overall VOC bouquet out of the herbivores' host search spectrum. Herbivore selection on detracting traits is expected to diversify the leaf VOC bouquet of individual plants as well as to expand the information space describing the plant populations. Similarly, induced changes in the VOC bouquet mediating deterrence through the detractant mechanism are expected to broaden the information space, meaning plants within a population become more dissimilar (lower phenotypic integration) when attacked than plants in the unchallenged state (Figure 12.1, Type 2, 3).

Mutualist pollinators use the same fundamental information space as antagonistic herbivores and thus should follow the same basic assumptions of information theory. However their selection on VOC traits should mostly confound that by herbivores for their usually opposite ecological effects. This becomes most apparent in cases when herbivore-induced changes in whole-plant and floral VOC emission affect the plants' interactions with pollinators. For example in the wild tomato, *Solanum peruvianum* (Solanaceae, Figure 12.3), herbivory induces changes in floral and vegetative VOC emission (Kessler and Halitschke 2009). These changes make pollinators avoid flowers on herbivore-attacked plants with dramatic negative effects on plant fitness ("herbivory-induced pollinator limitation" (Kessler et al. 2011)). A similar conflicting effect of herbivory and pollination, mediated by herbivore-induced changes in VOC emission was found in *Brassica rapa* (Brassicaceae) with lower effects on plant fitness. Nevertheless, both examples illustrate the apparent conflict that derives from the fact that the same chemical language components (Bruce et al. 2005; Knudsen et al. 2006) are used to mediate opposing ecological interactions and thus suggests strong coevolutionary interactions in chemical signaling associated with plant mating system and defense strategies (Andrews et al. 2007; Campbell and Kessler 2013; Kessler and Halitschke 2009).

To avoid this conflict, plants are expected to be under selection to maximize the separation of the signaling associated with pollinator attraction from that mediating herbivore resistance. There are three major ways to accomplish that separation. First, and most obvious, plants tend to separate the chemical signaling in space by separating reproductive tissues

Figure 12.3 (See color insert.) The wild tomato, *Solanum peruvianum* (Solanaceae) with one of its pollinators, a Halictidae bee. Leaf herbivory on *S. peruvianum* induces changes in floral and vegetative volatile organic compound emissions. Pollinators use this alteration in chemical information space to avoid flowers on herbivore-damaged plants with negative fitness effects for the plant. (From Kessler, A., R. Halitschke, and K. Poveda. 2011. *Ecology* 92: 1769–1780. Photo: André Kessler.)

from vegetative tissues and commonly presenting them in different visual and tactile (e.g., floral color and physical structure) environments. Scent production tends to vary on a very small spatial scale (Burdon et al. 2015; Dobson et al. 1999; Dötterl and Jürgens 2005; Piskorski et al. 2011; Raguso 2004) and functionally interacts with other floral cues (Raguso and Willis 2002), which was interpreted as an adaptation that could appropriately guide pollinators to the reproductive structures. However, at the same time, spatial separation can also reduce potential conflicts between different groups of tissue consumers, such as pollinators and herbivores.

Second, floral VOC emission may be restricted in time to reduce potential conflicts with antagonists but still advertise information to mutualists. Floral VOC emission can dramatically change depending on a plant's developmental state or with the time of the day (Ruíz-Ramón et al. 2014; Theis et al. 2007). Temporal changes in VOC emission can indicate if flowers are receptive (Bergström et al. 1995; Raguso et al. 2003; Rodriguez-Saona et al. 2011) or already pollinated (Muhlemann et al. 2006; Theis and

Raguso 2005; Tollsten and Bergstrom 1993) and are often timed to meet the peak activity of the most important pollinators (Borges et al. 2013; Dötterl et al. 2012; Jürgens et al. 2014). With such temporal restriction the advertisement of the plant to potential antagonists can be minimized.

Third, flower VOC emissions can be qualitatively different from leaf emission so that the floral information space minimally overlaps with the leaf information space. In many plant taxa floral VOC emission tends to be significantly different from leaf emissions even in cases in which the same set of compounds appears in both tissues (Kessler and Halitschke 2009; Kessler et al. 2008). When relatively similar sets of compounds mediate the interactions of plants with pollinators and herbivores, the information-mediating cues are hypothesized to be a result of compromising selection (Gomez 2003) and the VOCs expressed in a floral bouquet include repellent and attractive compounds. This hypothesis was supported by a recent study with transgenic *Petunia* x *hybrida* lines silenced in their ability to emit individual components of their floral VOC blend. It demonstrated that certain compounds are repellent, while others are attractive within the same blend (Kessler et al. 2013). A special case of extreme separation of the information space between leaves and flowers are deceptive pollination systems that create private communication channels between plants and their pollinators. In these systems plants mimic sex-pheromones of pollinating bees and wasps (sexual deception) (Schiestl 2005), as well as food and or brood substrates of nonherbivorous organism that may feed on carrion or animal feces (food and brood site deception) (Urru et al. 2011). A recent study even found flowers of *Aristolochia rotunda* (Aristolochiaceae) mimicking host finding cues of kleptoparasitic Chloropidae flies, VOCs normally emitted from Mirid bug bodily fluids when attacked by predators. This new type of deceptive pollinator attraction was named "klepto-myiophily" (Oelschlägel et al. 2015).

Within the information space framework and under the assumption of conflicting selection on chemical signals in general and VOC bouquets by antagonists and mutualists in particular, herbivores using chemical cues for host finding can be expected to select for stronger separation of floral and vegetative information space in time, space and quality. Thereby natural selection for private channel communication should be evident in stronger separation as well as contraction (increased phenotypic integration) of the information space of flowers relative to that of leaves. Accordingly, relief from selection by herbivory should allow for relatively increased overlap between floral and vegetative VOC bouquets. Alternatively it has been proposed that escape from antagonistic organisms could allow for relatively stronger selection of floral VOC cues by pollinators and thus a privatization of the information channels to attract only the most efficient pollinator (Kessler and Halitschke 2009). This discussion again illustrates one of the major problems with identifying generalizable patterns for the evolution of VOC information transfer. The ubiquity of the information encoded in VOC emission makes it difficult to identify the major agents of natural selection in any kind of system.

12.3.2 Information-mediated indirect resistance: VOC-mediated multitrophic level interactions

The attraction of natural enemies of herbivores to insect-attacked plants is a special case within the information transfer framework. Information-mediated indirect resistance only works when the VOC signal is a solid predictor of a prey/host present for a natural enemy and thus is dependent on the herbivore-inducibility of VOCs, rather than their constitutive production (Dicke and Baldwin 2010). Measuring natural selection on inducible traits is experimentally challenging and thus has rarely been done, but it has also been hypothesized that the relatively indirect relationship between plants and the natural enemies of their herbivores makes it less likely that predators and parasitoids function as agents of natural selection on herbivore-induced VOC emission (Kessler and Heil 2011). More directly responding agents of natural selection, such as herbivores and neighboring plants are more likely candidates for primary agents of natural selection on inducible leaf VOC emission. Nevertheless, in some cases significant plant fitness effects have been found to be associated with the plant's ability to induce VOC production in response to herbivory and the consequential attraction of predators. In the wild tobacco *Nicotiana attenuata* (Solanaceae), herbivory by various herbivore species induces specific VOC emission. The induced VOC emission of wildtype plants attracts a generalist predatory bug, *Geocoris pallens* (Geocoridae) (Figure 12.2b), whose actions reduce herbivory on the plant (Kessler and Baldwin 2001, 2004) and so increase plant fitness relative to genetically modified plants that are not able to induce VOC emission or are not able to produce certain components of the complete VOC bouquet (Halitschke et al. 2007; Schuman et al. 2012). In this system, the generalist herbivore seems to use common VOCs indicating herbivore damage, rather than specific information indicating the type of herbivore that is actually on the plant (Kessler and Baldwin 2001, 2004), as well as herbivory-specific shifts in wound-induced isomeric ratios of green leaf volatiles (Allmann and Baldwin 2010) to find herbivore-infected plants and thus its prey. This interaction fits the classic definition of information-mediated indirect resistance (Kessler and Heil 2011). In systems like this, herbivory in the presence of predators should select for increased herbivore-induced production of these common herbivory-indicating compounds. Natural selection on inducible VOC emissions would be similar to the selection predicted with toxic or toxin-indicating compounds mediating interactions with herbivores, which should drive the VOC information space to emphasized difference between the VOC bouquets emitted from undamaged and damaged plants and contraction of the information space (more similar emission patterns) describing the plant population (Figure 12.1, Type 1). Natural selection by herbivores and their natural enemies would point in the same direction towards similar induced VOC phenotypes. Such parallel selection is probably underlying cases in which plant direct and indirect resistance function synergistically (Kessler and Baldwin 2004), it, however, increases the difficulty to experimentally identify the agent of natural selection on herbivore-inducible VOC signals.

Moreover, recent studies on the effects of herbivore-induced VOC emission on entire interaction networks rather than reciprocal interactions between two or three species draw a far more complicated picture (Poelman and Dicke 2014). Plants are imbedded in complex interacting communities and alterations of their VOC phenotype as a result of herbivory will affect the dynamics within the interaction network (Stam et al. 2014). The outcome of the interactions for the plant depends on those dynamics among the members of the interaction network as well as the information landscape in which the interactions are played out (Kessler 2015). Thus, an herbivore-induced change in VOC signaling can have a different fitness effect for a plant, depending on the directly interacting community (Kessler and Baldwin 2004), the composition of which can be determined by the plant chemical phenotype (Broekgaarden et al. 2010; Halitschke et al. 2007; Kessler et al. 2004; Poelman et al. 2009), as well as the phenotypes of their neighboring plants (de Rijk et al. 2013; Schuman et al. 2015). Most striking, and specifically for VOC-mediated information transfer, potentially positive plant fitness effects of indirect resistance, may be negated because the same VOC signals attracting the predators/ and parasitoids of herbivores may attract members of the fourth trophic level or mediate competitive interactions among members of the third trophic level (Harvey et al. 2013). Recent studies in *Brassica* spp. (Brassicaceae) species have provided deep insights into these complications associated with a plant's indirect defense systems. Parasitoid wasps as members of the third trophic level have generally been found to very efficiently and specifically respond to plant VOC emission induced by their hosts and their attraction has been hypothesized to positively affect plant fitness (Kessler and Heil 2011). In the *Brassica oleracea* (Brassicaceae) study system, the herbivore-induced VOC emission does not only attract parasitoid wasps, but adult herbivores (e.g., diamondback moth, *Plutella xylostella* (Plutellidae)) choose their hosts based on whether or not caterpillars already on the plant are parasitized or not. These specific oviposition choices by the adult diamondback moths are mediated by herbivore-induced plant metabolic changes (Poelman et al. 2011). More importantly, plant metabolic changes specifically induced by parasitized and nonparasitized *Pieris* spp. caterpillars affect their attractiveness to hyperparasitoids, wasps that parasitize the third trophic level parasitoid wasp larvae inside the caterpillar. The specific VOC emission induced by parasitized caterpillars function as a cue for the hyperparasitoids to find their hosts, thereby potentially compromising the indirect resistance effect of the third trophic level parasitoids (Poelman et al. 2012). Similarly, induced plant metabolic changes as well as plant genotype affect competitive interactions among the parasitoids of the third trophic level as well (Poelman et al. 2013, 2014). The high specificity of information encoded in this and other systems is impressive and is the basis for the complexity of interactions mediated by herbivore-induced plant VOCs (Kessler and Halitschke 2007; Poelman and Dicke 2014). But what does that mean for plant fitness and thus natural selection on herbivore-inducible information-mediating traits? Although no natural selection data are available the above studies allow for some predictions. As mentioned above, parasitoids and predators as major agents of natural selection can be predicted to select for high specificity of the herbivore-induced VOC signal (there is a clear separation between herbivore-attacked and undamaged plants in the information space) and for similarity of the induced signals within a plant population (contraction of the information space) (Figure 12.1, Type 1). However, any effect compromising the indirect defense function can be predicted to select for lower specificity of the VOC signal as well as for an expanding information space on the population level. This would mean that plants should induce a less specific and less predictable (by plant genotype) VOC emission in response to already parasitized herbivores and relative to the signal induced by non-parasitized herbivores (Figure 12.1, Type 3). Interestingly this hypothesis seems to be supported by the fact that parasitation of two lepidopteran caterpillars, *Pieris rapae* and *Pieris brassicae* (Pieridae), on wild *Brassica oleracea* (Brassicaceae) seems to override the differences in induced transcriptional and metabolic changes (including VOC emission) that are usually observed when the two caterpillar species feed on the plant. This similarity in parasitized herbivore-induced VOC emission results in equal preference for the olfactorial information by the hyperparasitoid wasp *Lysibia nana* (Ichneumonidae) (Zhu et al. 2015). While no population level data able to address these hypotheses are available yet, the community network approach to VOC-mediated information transfer seems certainly promising.

12.3.3 *VOC-mediated plant–plant communication*

The VOC production induced in response to herbivory or plant pathogens can itself induce or prime for transcriptional and metabolic changes in neighboring plants (Arimura et al. 2000a,b, 2002; Engelberth et al. 2004; Kessler et al. 2006). Such plant–plant communication, can mediate resistance in neighboring plants through the expression of direct and indirect resistance traits (Baldwin and Schultz 1983; Heil and Bueno 2007; Karban et al. 2000; Kessler et al. 2006). In some ways the induction of resistance by VOCs from a damaged neighbor can force the receiver plant into investing into costly defense metabolite production without an herbivore present and thus with potentially negative fitness effects for this plant (Karban et al. 2003). This potential allelopathic function of plant–plant communication via VOCs can be overcome by the plant through only readying (priming) resistance responses instead of directly inducing in response to VOCs. This allows plants to use the VOC-mediated information without a large investment in costly defense metabolite production until an herbivore arrives to which the plant can then respond stronger or quicker (Engelberth et al. 2004; Kessler et al. 2006). As with information-mediated indirect resistance, VOC-mediated information transfer only happens when VOC emission is induced in response to environmental stress, such as herbivory.

Many hypotheses for the ecological functions of plant–plant communication have been discussed (Heil and Karban 2010). One major hypothesis is that plants use VOC signals for within-plant information transfer to overcome constraints of vascular signal transduction and so facilitate systemic induced resistance (Heil and Bueno 2007; Park et al. 2007). Neighboring plants then eavesdrop on the herbivore-attacked plant's signal and can so potentially gain a competitive advantage over their signaling neighbor. Thereby for the receiving plant, a direct response to a damaged neighbor's signal is only beneficial if the information is a clear predictor of future herbivory. Emitters, suffering competitive disadvantages from sharing the information, are under natural selection to privatize the emitted information and so reduce the information content for a neighboring plant (Heil and Karban 2010). In this scenario emitter plants are under selection to minimize the separation of the VOC information space between unchallenged and herbivore attacked plants and reduce phenotypic integration of the induced VOC emission so that each genotype response only to emission from itself. Support for this hypothesis comes from recent studies on sagebrush, *Artemisia tridentata* (Asteraceae) (Figure 12.4). In this model system for the study of plant communication, plants respond stronger to induced VOCs from their own genetic clone or close genetic relatives than to VOCs from nonrelated plants (Karban and Shiojiri 2009; Karban et al. 2013). These differences in resistance responses are strongly related to differences in VOC bouquets between kin and nonkin plants (Karban et al. 2014), which strongly supports the hypothesis of diversifying selection on induced VOC emission and the evolution of kin-specific private channels to minimize eavesdropping by neighbors.

As an alternative to the within-plant signal transduction/ eavesdropping hypothesis, herbivore-damaged emitter plants in highly connected

Figure 12.4 Sagebrush, *Artemisia tridentata* (Asteraceae) is a model system for the study of plant–plant communication. Volatile organic compounds from damaged sagebrush can induce herbivore resistance in neighboring plants. Thereby genetically closely related plants respond stronger to the VOC signal than distantly related plants. (From Karban, R. et al. 2013. *Proc. R. Soc. Lond. B Biol. Sci.* 280: 20123062. Photo: Kaori Shiojiri.)

populations could gain a benefit from transmitting information to neighbors if the VOC induction is coupled with strong induced resistance and a resulting movement of herbivores away from the damaged plant. In this scenario induced resistance would spread the risk of damage to the neighbor and the early warning of the neighbor through VOC-mediated information transfer would move the herbivore further away and generate a patch of resistant plants. A recent model using data from tall goldenrod, *Solidago altissima* (Asteraceae), found evidence for this risk-spreading hypothesis. In this system, VOC-mediated information transfer from herbivore-attacked plants to the herbivores, the larvae of the relatively mobile beetle species *Trirhabda virgata* (Chrysomelidae), and to neighboring plants resulted in an overdispersal of herbivory within the plant population, with every individual plant receiving minimum damage (Rubin et al. 2015). With such a strongly positive effect of plant–plant communication on plant performance, the inducibility of the VOC signal can be expected to be under selection by herbivores to be specific to the inducing agent and highly integrated. Whether emitting plants have fitness benefits from transferring information to their neighbors will strongly depend on the mobility of the dominating herbivore and the connectedness of the plant population.

12.4 Conclusion

In this chapter we discussed predictions associated with the information transfer hypothesis in the light of the plant's relationship with specific types of interactors as well as with complex interaction networks. We are well aware that many of the predictions have not yet been directly tested and that population level comparisons as well as studies on natural selection on VOC traits are still very rare. Knowing about the ecological multifunctionality of chemical cues in general and VOC emission in particular (Kessler 2015; Kessler and Halitschke 2007), it may not come as a surprise that the above-discussed prediction of how the information should be structured on the plant and population level can be very similar for different types of interactors. Rather it is more important how the interacting organism is perceiving the information and which consequences the use of that information has for the plant. As a result there are only two basic evolutionary trajectories for (A) increased or decreased separation and (B) increased or decreased integration of the information space. For complex interaction networks the organism using VOC signals to interact with the plant with the highest direct impact on plant fitness (e.g., dominating herbivore, pollinators) or the organism with the highest impact on the network structure (e.g., keystone herbivore, predator) (Poelman and Kessler 2016) will be the major agents of natural selection and are thus predicted to determine the direction of natural selection on the information-mediating traits. However, it is not only the interaction network or the abiotic stresses that affect the evolution of VOC signaling. We have not had enough space to discuss the effects of noise coming from the plant's environment that is not directly involved in the interactions, the information landscape (Kessler 2015; Wilson et al. 2015), but can affect how an interaction network is shaped and how efficiently information can be transmitted between organisms. The interaction landscape can introduce significant noise in VOC-mediated interactions and can potentially affect the signal evolution as a function of the other signaling species surrounding a plant (e.g., biodiversity, plant community composition) (Barbosa et al. 2009; Desurmont et al. 2015; Himanen et al. 2010) or the physical environment that can affect transport and breakdown dynamics of a VOC bouquets (e.g., UV radiation, ozone) (Blande et al. 2014; Farré-Armengol et al. 2016).

There is a significant development within the field of chemical ecology from primarily mechanistic studies to functional and evolutionary studies driven by a more ready availability of modern molecular, chemical and genetic tools (Raguso et al. 2015). Meanwhile, integrative approaches that combine these tools with information theory (van Baalen 2013; Wilson et al. 2015) will build a theoretical framework for chemical information transfer that can conceptually connect it with other types of information processing. The great accomplishments in the mechanistic understanding of VOC signaling (Dudareva et al. 2013) and the identifications of the large diversity of potential functions (Dicke and Baldwin 2010) will fuel this next step into the analysis of micro- and macro evolution of VOC-mediated information transfer and we hope that the framework suggested here can contribute to the design of new experiments.

Acknowledgments

We thank Alexander Chauta, Katja Poveda, Aino Kalske, and Mia Howard for the helpful discussions in preparation of this manuscript. Work associated with this review has been funded by the U.S. National Science Foundation (NSF-IOS 0950225) and a joint research and extension program funded by the Cornell University Agricultural Experiment Station (FFF Hatch Funds) received from the National Institute for Food and Agriculture (NIFA), U.S. Department of Agriculture (USDA). Any opinions, findings, conclusions, or recommendations expressed in this chapter are those of the authors and do not necessarily reflect the views of the USDA.

References

Adler, F.R. and R. Karban. 1994. Defended fortresses or moving targets? Another model of inducible defenses inspired by military metaphors. *Am. Nat.* 144: 813–832.

Adler, L.S. 2000. The ecological significance of toxic nectar. *Oikos* 91: 409–420.

Ahmad, S., N. Veyrat, R. Gordon-Weeks et al. 2011. Benzoxazinoid metabolites regulate innate immunity against aphids and fungi in maize1. *Plant Physiol.* 157: 317–27.

Alborn, H.T., T.C.J. Turlings, T.H. Jones et al. 1997. An elicitor of plant volatiles from beet armyworm oral secretion. *Science* 276: 945–949.

Allison, J.D. and J.D. Hare. 2009. Learned and naïve natural enemy responses and the interpretation of volatile organic compounds as cues or signals. *New Phytol.* 184: 768–782.

Allmann, S. and I.T. Baldwin. 2010. Insects betray themselves in nature to predators by rapid isomerization of green leaf volatiles. *Science* 329: 1075–1078.

Andrews, E.S., N. Theis, and L.S. Adler. 2007. Pollinator and herbivore attraction to Cucurbita floral volatiles. *J. Chem. Ecol.* 33: 1682–1691.

Arimura, G., R. Ozawa, T. Nishioka et al. 2002. Herbivore-induced volatiles induce the emission of ethylene in neighboring lima bean plants. *Plant J.* 29: 87–98.

Arimura, G., R. Ozawa, T. Shimoda, T. Nishioka, W. Boland, and J. Takabayashi. 2000a. Herbivory-induced volatiles elicit defence genes in lima bean leaves. *Nature* 406: 512–515.

Arimura, G., K. Tashiro, S. Kuhara et al. 2000b. Gene responses in bean leaves induced by herbivory and by herbivore-induced volatiles. *Biochem. Biophys. Res. Commun.* 277: 305–310.

Asberg, S.E., A.M. Bones, and A. Overby. 2015. Allyl isothiocyanate affects the cell cycle of *Arabidopsis thaliana*. *Front. Plant Sci.* 6: 364.

Azuma, H., M. Toyota, and Y. Asakawa. 2001. Intraspecific variation of floral scent chemistry in *Magnolia kobus* DC. (Magnoliaceae). *J. Plant Res.* 114: 411–422.

Baldwin, I.T. and J.C. Schultz. 1983. Rapid changes in tree leaf chemistry induced by damage: Evidence for communication between plants. *Science* 221: 277–79.

Barbosa, P., J. Hines, I. Kaplan, H. Martinson, A. Szczepaniec, and Z. Szendrei. 2009. Associational resistance and associational susceptibility: Having right or wrong neighbors. *Annu. Rev. Ecol. Evol. Syst.* 40: 1–20.

Barkman, T.J., J.H. Beaman, and D.A. Gage. 1997. Floral fragrance variation in *Cypripedium*: Implications for evolutionary and ecological studies. *Phytochemistry* 44: 875–82.

Berenbaum, M.R., J.K. Nitao, and A.R. Zangerl. 1991. Adaptive significance of furanocoumarin diversity in *Pastinaca sativa* (Apiaceae). *J. Chem. Ecol.* 17: 207–215.

Bergström, G., H.E.M. Dobson, and I. Groth. 1995. Spatial fragrance patterns within the flowers of *Ranunculus acris* (Ranunculaceae). *Plant Syst. Evol.* 195: 221–242.

Blande, J.D., J.K. Holopainen, and Ü. Niinemets. 2014. Plant volatiles in polluted atmospheres: Stress responses and signal degradation. *Plant Cell Environ.* 37: 1892–1904.

Blight, M.M., J.A. Pickett, L.J. Wadhams, and C.M. Woodcock. 1995. Antennal perception of oilseed rape, *Brassica napus* (Brassicaceae), volatiles by the cabbage seed weevil *Ceutorhynchus assimilis* (Coleoptera: Curculionidae). *J. Chem. Ecol.* 21: 1649–1664.

Borges, R.M., J. Bessière, and Y. Ranganathan. 2013. Diel variation in fig volatiles across syconium development: Making sense of scents. *J. Chem. Ecol.* 39: 630–42.

Bricchi, I., C.M. Bertea, A. Occhipinti, I.A. Paponov, and M.E. Maffei. 2012. Dynamics of membrane potential variation and gene expression induced by *Spodoptera littoralis*, *Myzus persicae*, and *Pseudomonas syringae* in Arabidopsis. *PLoS One* 7: e46673.

Bricchi, I., M. Leitner, M. Foti, A. Mithöfer, W. Boland, and M.E. Maffei. 2010. Robotic mechanical wounding (MecWorm) versus herbivore-induced responses: Early signaling and volatile emission in lima bean (*Phaseolus Lunatus* L.). *Planta* 232: 719–729.

Broekgaarden, C., E.H. Poelman, R.E. Voorrips, M. Dicke, and B. Vosman. 2010. Intraspecific variation in herbivore community composition and transcriptional profiles in field-grown *Brassica oleracea* cultivars. *J. Exp. Bot.* 61: 807–819.

Bruce, T.J.A., L.J. Wadhams, and C.M. Woodcock. 2005. Insect host location: A volatile situation. *Trends Plant Sci.* 10: 269–274.

Brumm, H. and H. Slabbekoorn. 2005. Acoustic communication in noise. *Adv. Stud. Behav.* 35: 151–209.

Burdon, R.C.F., R.A. Raguso, A. Kessler, and A.L. Parachnowitsch. 2015. Spatiotemporal floral scent variation of *Penstemon digitalis*. *J. Chem. Ecol.* 41: 641–650.

Campbell, S.A., R. Halitschke, J.S. Thaler, and A. Kessler. 2014. Plant mating systems affect adaptive plasticity in response to herbivory. *Plant J.* 78: 481–490.

Campbell, S.A., and A. Kessler. 2013. Plant mating system transitions drive the macroevolution of defense strategies. *Proc. Natl. Acad. Sci. USA* 110: 3973–3978.

Campbell, S.A., J.S. Thaler, and A. Kessler. 2013. Plant chemistry underlies herbivore-mediated inbreeding depression in nature. *Ecol. Lett.* 16: 252–260.

Carroll, M.J., E.A. Schmelz, R.L. Meagher, and P.E.A. Teal. 2006. Attraction of *Spodoptera frugiperda* larvae to volatiles from herbivore-damaged maize seedlings. *J. Chem. Ecol.* 32: 1911–1924.

Clay, N.K., A.M. Adio, C. Denoux, G. Jander, and F.M. Ausubel. 2009. Glucosinolate metabolites required for an Arabidopsis innate immune response. *Science* 323: 95–101.

De Moraes, C.M., M.C. Mescher, and J.H. Tumlinson. 2001. Caterpillar-induced nocturnal plant volatiles repel conspecific females. *Nature* 410: 577–580.

De Rijk M., M. Dicke, and E.H. Poelman. 2013. Foraging behaviour by parasitoids in multiherbivore communities. *Anim. Behav.* 85: 1517–1528.

Desurmont, G.A., D. Laplanche, F.P. Schiestl, and T.C.J. Turlings. 2015. Floral volatiles interfere with plant attraction of parasitoids: Ontogeny-dependent infochemical dynamics in *Brassica rapa*. *BMC Ecology* 15: 17.

Dicke, M. and I.T. Baldwin. 2010. The Evolutionary Context for Herbivore-induced plant volatiles: Beyond the cry for Help. *Trends Plant Sci.* 15: 167–175.

Dicke, M. and M.W. Sabelis. 1988. How plants obtain predatory mites as bodyguards. *Neth. J. Zool.* 38: 148–165.

Dobson, H.E.M., J. Arroyo, G. Bergström, and I. Groth. 1997. Interspecific variation in floral fragrances within the genus *Narcissus* (Amaryllidaceae). *Biochem. Syst. Ecol.* 25: 685–706.

Dobson, H.E.M., E.M. Danielson, and I.D. Van Wesep. 1999. Pollen odor chemicals as modulators of bumble bee foraging on *Rosa rugosa* Thunb. (Rosaceae). *Plant Species Biol.* 14: 153–166.

Dötterl, S., A. David, W. Boland, I. Silberbauer-Gottsberger, and G. Gottsberger. 2012. Evidence for behavioral attractiveness of methoxylated aromatics in a *Dynastid scarab* beetle-pollinated Araceae. *J. Chem. Ecol.* 38: 1539–1543.

Dötterl, S. and A. Jürgens. 2005. Spatial fragrance patterns in flowers of *Silene latifolia*: Lilac compounds as olfactory nectar guides? *Plant Syst. Evol.* 255: 99–109.

Doyle, L.R. 2009. Quantification of information in a one-way plant-to-animal communication system. *Entropy* 11: 431–442.

Dudareva, N., A. Klempien, J.K. Muhlemann, and Kaplan. 2013. Biosynthesis, function and metabolic engineering of plant volatile organic compounds. *New Phytol.* 198: 16–32.

Endler, J.A. 1993. Some general comments on the evolution and design of animal communication systems. *Philos. Trans. R. Soc. Lond. B Biol. Sci.* 340: 215–225.

Engelberth, J., H.T. Alborn, E. A. Schmelz, and J.H. Tumlinson. 2004. Airborne signals prime plants against insect herbivore attack. *Proc. Natl. Acad. Sci. USA* 101: 1781–1785.

Farré-Armengol, G., J. Peñuelas, T. Li, P. Yli-Pirilä, I. Filella, J. Llusia, and J.D. Blande. 2016. Ozone degrades floral scent and reduces pollinator attraction to flowers. *New Phytol.* 209: 152–60.

Firn, R.D. and C.G. Jones. 2003. Natural products—A simple model to explain chemical diversity. *Nat. Prod. Rep.* 20: 382–391.

Firn, R.D. and C.G. Jones. 2006. Do we need a new hypothesis to explain plant VOC emissions? *Trends Plant Sci.* 11: 112–113.

Fraenkel, G.S. 1959. The raison d'être of secondary plant substances; these odd chemicals arose as a means of protecting plants from insects and now guide insects to food. *Science* 129: 1466–1470.

Fraser, A.M., W.L. Mechaber, and J.G. Hildebrand. 2003. Electroantennographic and behavioral responses of the sphinx moth *Manduca sexta* to host plant headspace volatiles. *J. Chem. Ecol.* 29: 1813–33.

Gomez, J.M. 2003. Herbivory reduces the strength of pollinator-mediated selection in the mediterranean herb *Erysimum mediohispanicum*: Consequences for plant specialization. *Am. Nat.* 162: 242–56.

Halitschke, R., J.G. Hamilton, and A. Kessler. 2011. Herbivore-specific elicitation of photosynthesis by mirid bug salivary secretions in the wild tobacco *Nicotiana attenuata*. *New Phytol.* 191: 528–535.

Halitschke, R., J.A. Stenberg, D. Kessler, A. Kessler, and I.T. Baldwin. 2007. Shared signals—"Alarm calls" from plants increase apparency to herbivores and their enemies in nature. *Ecol. Lett.* 11: 24–34.

Hartmann, T. 2007. From waste products to ecochemicals: Fifty years research of plant secondary metabolism. *Phytochemistry* 68: 2831–2846.

Harvey, J.A., E.H. Poelman, and T. Tanaka. 2013. Intrinsic inter- and intraspecific competition in parasitoid wasps. *Annu. Rev. Entomol.* 58: 333–351.

Hatano, E., A.M. Saveer, F. Borrero-Echeverry et al. 2015. A herbivore-induced plant volatile interferes with host plant and mate location in moths through suppression of olfactory signalling pathways. *BMC Biol.* 13: 75.

Hay, M.E., Q.E. Kappel, and W. Fenical. 1994. Synergisms in plant defenses against herbivores: Interactions of chemistry, calcification, and plant quality. *Ecology* 75: 1714–1726.

Heil, M. and J.C.S. Bueno. 2007. Within-plant signaling by volatiles leads to induction and priming of an indirect plant defense in nature. *Proc. Natl. Acad. Sci. USA* 104: 5467–5472.

Heil, M. and R. Karban. 2010. Explaining evolution of plant communication by airborne signals. *Trends Ecol. Evol.* 25: 137–144.

Hilker, M. 2014. New synthesis: Parallels between biodiversity and chemodiversity. *J. Chem. Ecol.*40: 225–226.

Himanen, S.J., J.D. Blande, T. Klemola, J. Pulkkinen, J. Heijari, and J.K. Holopainen. 2010. Birch (*Betula* spp.) leaves adsorb and re-release volatiles specific to neighbouring plants—A mechanism for associational herbivore resistance? *New Phytol.* 186: 722–732.

Hutchinson, G.E. 1957. Concluding remarks. *Cold Spring Harb. Symp. Quant. Biol.* 22: 415–427.

Jones, C.G., R.D. Firn, and S.B. Malcolm. 1991. On the evolution of plant secondary chemical diversity [and discussion]. *Philos. Trans. R. Soc. Lond. B Biol. Sci.* 333: 273–280.

Jürgens, A., U. Glück, G. Aas, and S. Dötterl. 2014. Diel fragrance pattern correlates with olfactory preferences of diurnal and nocturnal flower visitors in *Salix caprea* (Salicaceae). *Bot. J. Linn. Soc.* 175: 624–640.

Jürgens, A., T. Witt, and G. Gottsberger. 2002. Flower scent composition in night-flowering *Silene* species (Caryophyllaceae). *Biochem. Syst. Ecol.* 30: 383–397.

Karban, R., I.T. Baldwin, K.J. Baxter, G. Laue, and G.W. Felton. 2000. Communication between plants: Induced resistance in wild tobacco plants following clipping of neighboring sagebrush. *Oecologia* 125: 66–71.

Karban, R., J. Maron, G.W. Felton, G. Ervin, and H. Eichenseer. 2003. Herbivore damage to sagebrush induces resistance in wild tobacco: Evidence for eavesdropping between plants. *Oikos* 100: 325–332.

Karban, R. and K. Shiojiri. 2009. Self-recognition affects plant communication and defense. *Ecol. Lett.* 12: 502–506.

Karban, R., K. Shiojiri, S. Ishizaki, W.C. Wetzel, and R.Y. Evans. 2013. Kin recognition affects plant communication and defence. *Proc. R. Soc. Lond. B Biol. Sci.* 280: 20123062.

Karban, R., W.C. Wetzel, K. Shiojiri, S. Ishizaki, S.R. Ramirez, and J.D. Blande. 2014. Deciphering the language of plant communication: Volatile chemotypes of sagebrush. *New Phytol.* 204: 380–385.

Kariyat, R.R., K.E. Mauck, C.M. De Moraes, A.G. Stephenson, and M.C. Mescher. 2012. Inbreeding alters volatile signalling phenotypes and Influences tritrophic interactions in horsenettle (*Solanum carolinense* L.). *Ecol. Lett.* 15: 301–309.

Kariyat, R.R., S.R. Scanlon, R.P. Moraski, A.G. Stephenson, M.C. Mescher, and C.M. De Moraes. 2014. Plant inbreeding and prior herbivory influence the attraction of caterpillars (*Manduca sexta*) to odors of the host plant *Solanum carolinense* (Solanaceae). *Am. J. Bot.* 101: 376–380.

Kessler, A. 2015. The information landscape of plant constitutive and induced secondary metabolite production. *Curr. Opin. Insect Sci.* 8: 47–53.

Kessler, A. and I.T. Baldwin. 2001. Defensive function of herbivore-induced plant volatile emissions in nature. *Science* 291: 2141–2144.

Kessler, A. and I.T. Baldwin. 2002. Plant responses to insect herbivory: The emerging molecular analysis. *Annu. Rev. Plant Biol.* 53: 299–328.

Kessler, A. and I.T. Baldwin. 2004. Herbivore-induced plant vaccination. Part I. the orchestration of plant defenses in nature and their fitness consequences in the wild tobacco *Nicotiana attenuata*. *Plant J.* 38: 639–649.

Kessler, A. and R. Halitschke. 2007. Specificity and complexity: The impact of herbivore-induced plant responses on arthropod community structure. *Curr. Opin. Plant Biol.* 10: 409–414.

Kessler, A. and R. Halitschke. 2009. Testing the potential for conflicting selection on floral chemical traits by pollinators and herbivores: Predictions and case study. *Funct. Ecol.* 23: 901–912.

Kessler, A., R. Halitschke and I.T. Baldwin. 2004. Silencing the jasmonate cascade: Induced plant defenses and insect populations. *Science* 305: 665–668.

Kessler, A., R. Halitschke, C. Diezel, and I.T. Baldwin. 2006. Priming of plant defense responses in nature by airborne signaling between *Artemisia tridentata* and *Nicotiana attenuata*. *Oecologia* 148: 280–292.

Kessler, A., R. Halitschke, and K. Poveda. 2011. Herbivory-mediated pollinator limitation: Negative impacts of induced volatiles on plant-pollinator interactions. *Ecology* 92: 1769–1780.

Kessler, A. and M. Heil. 2011. The multiple faces of indirect defences and their agents of natural selection: Multiple faces of indirect defences. *Funct. Ecol.* 25: 348–357.

Kessler, A. and K. Morrell. 2010. Plant volatile signalling: Multitrophic interactions in the headspace. In *The Chemistry and Biology of Volatiles*, ed. Herrmann A. New York: John Wiley & Sons.

Kessler, D., C. Diezel, D.G. Clark, T.A. Colquhoun, and I.T. Baldwin. 2013. Petunia flowers solve the defence/apparency dilemma of pollinator attraction by deploying complex floral blends. *Ecol. Lett.* 16: 299–306.

Kessler, D., K. Gase, and I.T. Baldwin. 2008. Field experiments with transformed plants reveal the sense of floral scents. *Science* 321: 1200–1202.

Knudsen, J.T. 1994. Floral scent variation in the *Pyrola rotundifolia* complex in Scandinavia and Western Greenland. *Nord. J. Bot.* 14: 277–282.
Knudsen, J.T. 2002. Variation in floral scent composition within and between populations of *Geonoma macrostachys* (Arecaceae) in the Western Amazon. *Am. J. Bot.* 89: 1772–1778.
Knudsen, J.T., R. Eriksson, J. Gershenzon, and B. Ståhl. 2006. Diversity and distribution of floral scent. *Bot. Rev.* 72: 1–120.
Knudsen, J.T., L. Tollsten, and L.G. Bergström. 1993. Floral scents—A checklist of volatile compounds isolated by head-space techniques. *Phytochemistry* 33: 253–280.
Knudsen, J.T. and S.A. Mori. 1996. Floral scents and pollination in neotropical Lecythidaceae. *Biotropica* 28: 42–60.
Levin, R.A., R.A. Raguso, and L.A. McDade. 2001. Fragrance chemistry and pollinator affinities in Nyctaginaceae. *Phytochemistry* 58: 429–440.
Loreto, F., M. Dicke, J.-P. Schnitzler, and T.C.J. Turlings. 2014. Plant volatiles and the environment. *Plant Cell Environ.* 37: 1905–1908.
Loreto, F. and S. Fineschi. 2015. Reconciling functions and evolution of isoprene emission in higher plants. *New Phytol.* 206: 578–582.
Majetic, C.J., M.D. Rausher, and R.A. Raguso. 2010. The pigment-scent connection: Do mutations in regulatory vs. structural anthocyanin genes differentially alter floral scent production in Ipomoea purpurea? *South Afr. J. Bot.* 76: 632–642.
Mattiacci, L., M. Dicke, and M.A. Posthumus. 1995. Beta-glucosidase: An elicitor of herbivore-induced plant odor that attracts host-searching parasitic wasps. *Proc. Natl. Acad. Sci. USA* 92: 2036–2040.
McCormick A., S.B. Unsicker, and J. Gershenzon. 2012. The specificity of herbivore-induced plant volatiles in attracting herbivore enemies. *Trends Plant Sci.* 17: 303–310.
McCowan, B., S.F. Hanser, and L.R. Doyle. 1999. Quantitative tools for comparing animal communication systems: Information theory applied to bottlenose dolphin whistle repertoires. *Anim. Behav.* 57: 409–419.
Moore, B.D., R.L. Andrew, C. Külheim, and W.J. Foley. 2014. Explaining intraspecific diversity in plant secondary metabolites in an ecological context. *J. Chem. Ecol.* 201: 733–750.
Muhlemann, J.K., M.O. Waelti, A. Widmer, and F.P. Schiestl. 2006. Postpollination changes in floral odor in *Silene latifolia*: adaptive mechanisms for seed-predator avoidance? *J. Chem. Ecol.* 32: 1855–1860.
Nottingham, S.F., J. Hardie, G.W. Dawson et al. 1991. Behavioral and electrophysiological responses of aphids to host and nonhost plant volatiles. *J. Chem. Ecol.* 17: 1231–1242.
Oelschlägel, B., M. Nuss, M. von Tschirnhaus, C. Pätzold, C. Neinhuis, S. Dötterl, and S. Wanke. 2015. The betrayed thief—The extraordinary strategy of *Aristolochia rotunda* to deceive its pollinators. *New Phytol.* 206: 342–351.
Olesen, J.M. and J.T. Knudsen. 1994. Scent profiles of flower colour morphs of *Corydalis cava* (Fumariaceae) in relation to foraging behaviour of bumblebee queens (*Bombus Terrestris*). *Biochem. System. Ecol.* 22: 231–237.
Owen, S.M. and J. Peñuelas. 2005. Opportunistic emissions of volatile isoprenoids. *Trends Plant Sci.* 10: 420–426.
Owen, S.M. and J. Penuelas. 2006. Response to Firn and Jones: Volatile isoprenoids, a special case of secondary metabolism. *Trends Plant Sci.* 11: 113–114.

Parachnowitsch, A.L., R.A. Raguso, and A. Kessler. 2012. Phenotypic selection to increase floral scent emission, but not flower size or colour in bee-pollinated *Penstemon digitalis*. *New Phytol.* 195: 667–675.
Park, S.W., E. Kaimoyo, D. Kumar, S. Mosher, and D.F. Klessig. 2007. Methyl salicylate is a critical mobile signal for plant systemic acquired resistance. *Science* 318: 113–116.
Pettersson, S. and J.T. Knudsen. 2001. Floral scent and nectar production in *Parkia biglobosa* Jacq. (Leguminosae: Mimosoideae). *Bot. J. Linn. Soc.* 135: 97–106.
Pichersky, E., J.P. Noel, and N. Dudareva. 2006. Biosynthesis of plant volatiles: Nature's diversity and ingenuity. *Science* 311: 808–811.
Piskorski, R., S. Kroder, and S. Dorn. 2011. Can pollen headspace volatiles and pollenkitt lipids serve as reliable chemical cues for bee pollinators? *Trends Plant Sci.* 8: 577–586.
Poelman, E.H., M. Bruinsma, F. Zhu et al. 2012. Hyperparasitoids use herbivore-induced plant volatiles to locate their parasitoid host. *PLoS Biol* 10: e1001435.
Poelman, E.H. and M. Dicke. 2014. Plant-mediated interactions among insects within a community ecological perspective. In *Insect-plant interactions*, eds. C. Voelckel and G. Jander, 47:309–337. Chichester: Wiley-Blackwell.
Poelman, E.H., R. Gols, A.V. Gumovsky, A.M. Cortesero, M. Dicke, and J. A. Harvey. 2014. Food Plant and herbivore host species affect the outcome of intrinsic competition among parasitoid larvae. *Ecol. Entomol.* 39: 693–702.
Poelman, E.H., J.A. Harvey, J.J.A. van Loon, L.E.M. Vet, and M. Dicke. 2013. Variation in herbivore-induced plant volatiles corresponds with spatial heterogeneity in the level of parasitoid competition and parasitoid exposure to hyperparasitism. *Funct. Ecol.* 27: 1107–1116.
Poelman, E.H., N.M. van Dam, J.J.A. van Loon, L.E.M. Vet, and M. Dicke. 2009. Chemical diversity in *Brassica oleracea* affects biodiversity of insect herbivores. *Ecology* 90: 1863–1877.
Poelman, E.H. and A. Kessler. 2016. Keystone herbivores and the evolution of plant defenses. *Trends Plant Sci.* 21: 477–485.
Poelman, E.H., S.J. Zheng, Z. Zhang, N.M. Heemskerk, A.M. Cortesero, and M. Dicke. 2011. Parasitoid-specific induction of plant responses to parasitized herbivores affects colonization by subsequent herbivores. *Proc. Natl. Acad. Sci. USA* 108: 19647–19652.
Raguso, R.A. 2004. Why are some floral nectars scented? *Ecology* 85: 1486–94.
Raguso, R.A. 2008. Wake up and smell the roses: The ecology and evolution of floral scent. *Ann. Rev. Ecol. Evol. Sys.* 39: 549–569.
Raguso, R.A., A.A. Agrawal, A.E. Douglas et al. 2015. The raison d'etre of chemical ecology. *Ecology* 96: 617–630.
Raguso, R.A., R.A. Levin, S.E. Foose, M.W. Holmberg, and L.A. McDade. 2003. Fragrance chemistry, nocturnal rhythms and pollination "syndromes" in. *Nicotiana. Phytochemistry* 63: 265–284.
Raguso, R.A. and M.A. Willis. 2002. Synergy between visual and olfactory cues in nectar feeding by naïve hawkmoths, *Manduca Sexta. Anim. Behav.* 64: 685–695.
Richards, L.A., L.A. Dyer, A.M. Smilanich, and C.D. Dodson. 2010. Synergistic effects of amides from two piper species on generalist and specialist ferbivores. *J. Chem. Ecol.* 36: 1105–1113.

Rodriguez-Saona, C., L. Parra, A. Quiroz, and R. Isaacs. 2011. Variation in highbush blueberry floral volatile profiles as a function of pollination status, cultivar, time of day and flower part: Implications for flower visitation by bees. *Ann. Bot.* 107: 1377–1390.

Rubin, I.N., S.P. Ellner, A. Kessler, and K.A. Morrell. 2015. Informed herbivore movement and interplant communication determine the effects of induced resistance in an individual-based model. *J. Anim. Ecol.* 84: 1273–1285.

Ruíz-Ramón, F., D.J. Águila, M. Egea-Cortines, and J. Weiss. 2014. Optimization of fragrance extraction: Daytime and flower age affect scent emission in simple and double narcissi. *Ind. Crop Prod.* 52: 671–678.

Schiestl, F.P. 2005. On the success of a swindle: Pollination by deception in orchids. *Naturwissenschaften* 92: 255–264.

Schiestl, F.P., H. Kirk, L. Bigler, S. Cozzolino, and G.A. Desurmont. 2014. Herbivory and floral signaling: Phenotypic plasticity and tradeoffs between reproduction and indirect defense. *New Phytol.* 203: 257–266.

Schuman, M.C., S. Allmann, and I.T. Baldwin. 2015. Plant defense phenotypes determine the consequences of volatile emission for individuals and neighbors. *eLife* 4: e04490.

Schuman, M.C., K. Barthel, and I.T. Baldwin. 2012. Herbivory-induced volatiles function as defenses increasing fitness of the native plant nicotiana attenuata in nature. *eLife* 1: e00007.

Shannon, C.E. 1948. A mathematical theory of communication. *Bell Syst. Tech. J.* 27: 623–656.

Stam, J.M., A. Kroes, Y. Li, R. Gols, J.J.A. van Loon, E.H. Poelman, and M. Dicke. 2014. Plant interactions with multiple insect herbivores: From community to genes. *Ann. Rev. Plant Biol.* 65: 689–713.

Steppuhn, A. and I.T. Baldwin. 2007. Resistance management in a native plant: Nicotine prevents herbivores from compensating for plant protease inhibitors. *Ecol. Lett.* 10: 499–511.

Thien, L.B., W.H. Heimermann, and R.T. Holman. 1975. Floral odors and quantitative taxonomy of magnolia and liriodendron. *Taxon* 24: 557–568.

Theis, N. and L.S. Adler. 2011. Advertising to the enemy: Enhanced floral Fragrance increases beetle attraction and reduces plant reproduction. *Ecology* 93: 430–435.

Theis, N., M. Lerdau, and R.A. Raguso. 2007. The challenge of attracting pollinators while evading floral herbivores: Patterns of fragrance emission in *Cirsium arvense* and *Cirsium repandum* (Asteraceae). *Inter. J. Plant Sci.* 168: 587–601.

Theis, N. and R.A. Raguso. 2005. The effect of pollination on floral fragrance in Thistles. *J. Chem. Ecol.* 31: 2581–2600.

Tollsten, L. and L.G. Bergstrom. 1993. Fragrance chemotypes of Platanthera (orchidaceae)—The result of adaptation to pollinating moths. *Nord. J. Bot.* 13: 607–613.

Tollsten, L. and D.O. Ovstedal. 1994. Differentiation in floral scent chemistry among populations of *Conopodium majus* (Apiaceae). *Nord. J. Bot.* 14: 361–367.

Turlings, T.C.J. and J. Ton. 2006. Exploiting scents of distress: The prospect of manipulating herbivore-induced plant odours to enhance the control of agricultural pests. *Curr. Opin. Plant Biol.* 9: 421–427.

Turlings, T.C.J., J.H. Tumlinson, and W.I. Lewis. 1990. Exploitation of herbivore-induced plant odors by host-seeking parasitic wasps. *Science* 250: 1251–1253.

Urru, I., M.C. Stensmyr, and B.S. Hansson. 2011. Pollination by brood-site deception. *Phytochemistry* 72: 1655–1666.
Van Baalen, M. 2013. Biological information: Why we need a good measure and the challenges ahead. *Interface Focus* 3: 20130030.
Van der Meijden, E. and P.G.L. Klinkhamer. 2000. Conflicting interests of plants and the natural enemies of herbivores. *Oikos* 89: 202–208.
Veyrat, N., C.A.M. Robert, T.C.J. Turlings, and M. Erb. 2016. Herbivore intoxication as a potential primary function of an inducible volatile plant signal. *J. Ecol.* 104: 591–600.
Visser, J.H. 1986. Host odor perception in phytophagous insects. *Annu. Rev. Entomol.* 31: 121–144.
Wicher, D. 2015. Olfactory signaling in insects. *Prog. Mol. Biol. Transl. Sci.* 130: 37–54.
Wilson, J.K., A. Kessler, and H.A. Woods. 2015. Noisy communication via airborne infochemicals. *BioScience* 65: 667–677.
Wright, G.A. and F.P. Schiestl. 2009. The evolution of floral scent: The influence of olfactory learning by insect pollinators on the honest signalling of floral rewards. *Funct. Ecol.* 23: 841–851.
Zhu, F., C. Broekgaarden, B.T. Weldegergis et al. 2015. Parasitism overrides herbivore identity allowing hyperparasitoids to locate their parasitoid host using herbivore-induced plant volatiles. *Mol. Ecol.* 24: 2886–2899.
Zhu, F., E.H. Poelman, and M. Dicke. 2014. Insect herbivore-associated organisms affect plant responses to herbivory. *New Phytol.* 204: 315–321.

chapter thirteen

Metabolic engineering and synthetic biology of plant secondary metabolism

Dae-Kyun Ro, Yang Qu, and Moonhyuk Kwon

Contents

13.1 Introduction 316
13.2 Terpenoid metabolism 319
    13.2.1 Terpenoid 319
    13.2.2 Biogenesis of terpene scaffolds 319
    13.2.3 Cytochrome P450 monooxygenase 322
    13.2.4 Isopentenyl diphosphate (IPP) biosynthesis 323
        13.2.4.1 MVA pathway 323
        13.2.4.2 MEP pathway 325
    13.2.5 Metabolic engineering of terpenoids 326
        13.2.5.1 Artemisinin 326
        13.2.5.2 Paclitaxel 329
13.3 Alkaloid metabolism 332
    13.3.1 Monoterpenoid indole alkaloid 332
        13.3.1.1 Overview 332
        13.3.1.2 Biosynthesis and cellular compartmentation
                 of MIAs in *Catharanthus roseus* 335
        13.3.1.3 Metabolic engineering of MIA pathway 339
    13.3.2 Benzylisoquinoline alkaloid 340
        13.3.2.1 Overview 340
        13.3.2.2 Biosynthesis of BIA 341
        13.3.2.3 Microbial production of BIA 342
13.4 Glucosinolate metabolism 345
    13.4.1 Glucosinolate biosynthetic pathway 345
    13.4.2 Glucosinolate production in plants 348
    13.4.3 Microbial production of indole glucosinolate 348

13.5 Gene clusters in plant specialized metabolisms 349
13.6 Conclusion 350
Acknowledgments 351
References.. 351

13.1 Introduction

After over 450 million years of evolutionary adaptation, plants have evolved to cope with various biotic challenges in the places where they are planted. One strategy they have adopted is to produce structurally diverse secondary (or specialized) metabolites, which are not essential for reproduction and daily metabolic processes (e.g., photosynthesis, respiration, and hormones) but significantly enhance their surviving fitness in harsh environments. These specialized metabolites are often toxic or unpleasant to herbivores, animals, neighboring plants, and microbial pathogens (e.g., bacteria, fungi, and oomycetes), thereby serving as effective chemical defense products in immobile plants. It is estimated that 422,000 plant species are present on earth (www.iPlant.org), and each of these has their own history of developing unique specialized metabolites in different parts of the earth. Around 10^6 specialized metabolites have been found from plants (Saito and Matsuda 2010), and undoubtedly plants will continue to be an important reservoir for new chemical discoveries.

Of the known plant specialized metabolites, we understand the mechanism of action (i.e., specific receptor-ligand interaction) in only a few cases. For example, on tissue damage by herbivores, the sugar moiety of cyanogenic glucoside is cleaved in sorghum and cassava to release hydrogen cyanide (HCN) (Zagrobelny and Moller 2011). HCN blocks cellular respiration by acting as an inhibitor of cytochrome c oxidase. In another instance, the terpenoid natural product, thapsigargin, from *Thapsia garganica* binds to sarco/endoplasmic reticulum (ER) Ca^{2+} ATPase to release calcium ions from the ER lumen to the cytosol, which leads to cell death (Quynh Doan and Christensen 2015). However, in a vast majority of cases, specific physiological roles of specialized metabolites remain unknown. It has been a long-standing debate over whether the majority of plant secondary metabolites have undetermined functions or are mere metabolic wastes without function.

The functionality debate aside, humankind has used plants as herbal medicines for thousands of years to cure diseases and to improve well-being, while modern scientists have screened a number of specialized metabolites to discover novel natural product drugs, often followed by structure elucidation and ligand-receptor interaction studies. One excellent example of a traditional herbal medicine is artemisinin from wormwood (*Artemisia annua*) (Tu 2011; White 2008). Artemisinin is a sesquiterpene lactone with an unusual peroxide bridge and is only produced from a single species, *A. annua*. Leaf extract of *A. annua* has been used in Southern China to cure malaria for >2000 years with a written prescription dating back to 200 BC. It was rediscovered in the early 1970s by a team of Chinese scientists led by Dr. Youyou Tu and has become the sole antimalarial drug against drug-resistant *Plasmodium falciparum* strains. As there is no reason to believe that *A. annua* engages with the human parasite *P. falciparum*, artemisinin seems to fortuitously possess antimalarial activity we benefit from. Other known examples of natural product drugs are paclitaxel (Taxol®) from the Pacific yew tree (*Taxus brevifolia*) and vinblastine from Madagascar periwinkle (*Catharanthus roseus*) (De Luca et al. 2012; Koehn and Carter 2005). Both natural products show potent anticancer activities by binding to the microtubule subunit and influencing the stabilities of microtubules (Jordan et al. 1998). Another example is the pain-killing drugs morphine and codeine, from opium poppy (*Papaver somniferum*). Morphine, codeine, and their semi-synthetic derivatives (e.g., oxycodone, hydrocodone, and hydromorphone) are the most effective analgesic drugs for severe pains, and opium poppy is the sole source of various opiates (Ziegler et al. 2009).

In addition to these known natural product drugs, many natural products offer health benefits to humans. A number of phenylpropanoid products from plants, such as resveratrol, catechin, anthocyanin, and anthocyanidins, have antioxidant properties and prolong our life expectancies (Baur and Sinclair 2006; Jang et al. 1997; Yilmaz and Toledo 2004). Some terpenoid products, including stevioside from stevia (*Stevia rebaudiana*), glycyrrhizin from licorice (root of *Glycyrrhiza glabra*), and hernandulcin from Aztec sweet herb (*Lippia dulcis*), are known to be low- or no-calorie sugar substitutes with 10- to 1000-fold higher levels of sweetness (Compadre et al. 1985; Pawar et al. 2013; Philippe et al. 2014). Implementing these natural products in our diets can markedly help decrease obesities, diabetes, and other life-style diseases.

The natural product drugs and health supplements described above can benefit the health and life quality of mankind. However, those natural products are present in a minute quantity in plants, thereby making steady and cost-effective supply of the natural products difficult. For example, a few hundred-year-old Pacific yew trees are needed to prepare a dosage necessary to cure one cancer patient (Suffness and Wall 1995). Total organic synthesis of the majority of these products is not economically viable owing to multiple chiral centers. As an alternative, production of natural product drugs by means of metabolic engineering, or synthetic biology from a modern sense, has been suggested, but our lack of knowledge of their metabolism has hindered the progress of the metabolic engineering approach. Only a few genes involved in the biosynthesis of artemisinin, vinblastine, and morphine were known in the early 2000s. Fortunately, an explosion of new technologies between 2005 and 2015 has dramatically changed the research patterns in plant biology, and truly established new stages for synthetic biology beyond traditional metabolic engineering. Three major breakthroughs that set the era of synthetic biology are: (1) next-generation sequencers, particularly Illumina sequencers, which can sequence billions of reads in a week; (2) cost-effective, rapid, and accurate synthesis and assembly of long nucleotides (thousands of base pairs); and (3) targeted genome editing (deletion and insertion) by clustered regularly interspaced short palindromic repeats (CRISPR)/CAS9 system for eukaryotes (Cong et al. 2013; Horwitz et al. 2015; Jakociunas et al. 2015). Combinatorial uses of these technologies and others have allowed us to rapidly sequence transcripts and genomes of nonmodel plants and discover novel genes, to synthesize genes for codon-optimization and mutant variants, and to engineer eukaryotic genomes to stably knock-out undesirable genes and to insert the genes of interest.

Although there are many examples of metabolic engineering and synthetic biology of plant natural products in literatures, incorporating all these is beyond the scope of this book chapter. Here, we will primarily focus on five representative natural products (artemisinin, paclitaxel, vinblastine, morphine, and indole glucosinolate) from three classes of natural products (terpenoid, alkaloid, and glucosinolate). Their chemical structures are depicted in Figure 13.1. Each section describes the

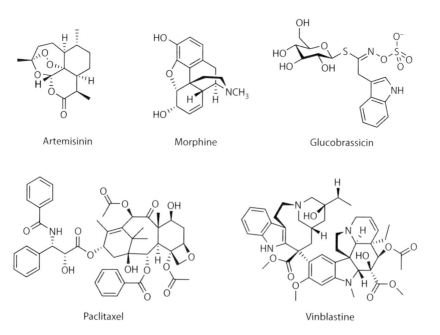

Figure 13.1 The structures of specialized metabolites discussed in this chapter.

specific natural products of interest, biosynthetic pathways, and metabolic engineering attempts.

13.2 Terpenoid metabolism

13.2.1 Terpenoid

Terpenoids (or isoprenoids) are a large and diverse class of natural products synthesized in all living organisms from bacteria to human. More than 50,000 terpenoids have been deposited in the natural product database (Dictionary of Natural Products, dnp.chemnetbase.com). As part of primary metabolism, terpenoids contribute to the structural backbones or moieties of essential metabolites. For example, diverse steroids play roles as hormones (testosterone/estrogen in human and brassinolides in plant) and membrane components; ubiquinone and plastoquinone are electron carriers in mitochondrial oxidative phosphorylation and chloroplastic photosynthesis, respectively; dolichol is a sugar carrier molecule for cell wall biosynthesis in prokaryotes and for posttranslation modifications of protein in eukaryotes; proteins are modified by farnesylation and geranyl geranylation. It is apparent that these terpenoids are indispensable products in prokaryotes or eukaryotes. However, plants have evolved to synthesize a secondary layer of terpenoids that constitute important defense and signaling systems in plants, which in turn benefit humans as aroma, flavor, nutraceuticals, pharmaceuticals, and industrial chemicals (Bohlmann and Keeling 2008; Tholl 2006). Selection and diversification of terpenoids in plants are likely driven by coevolutionary processes, such as plant–herbivore interactions, while applicability of many terpenoids in modern industries has happened by chance owing to the enormous structural diversity of terpenoids that nature created over billions of years. What mechanisms caused the terpenoids to become the most diverse class of natural products in nature? In the next section, we will discuss mechanistic and metabolic details to know more about terpene diversity.

13.2.2 Biogenesis of terpene scaffolds

Despite the structural complexity of terpenoids, the biosynthesis of terpenoids is governed by simple rules (Bohlmann et al. 1998; Tholl 2006). The first step is the preparation of terpenoid precursors of various lengths by *trans*-prenyltransferases. Chemically, C5 isoprene is the building block of terpenoids, but its activated form, isopentenyl diphosphate (IPP), and its isomer dimethylallyl diphosphate (DMAPP) are utilized as biological building blocks in cells (Figure 13.2). As the simplest end product, cleavage of a diphosphate from DMAPP can produce a volatile isoprene product. Isoprene is a natural C5 terpenoid product emitted from some

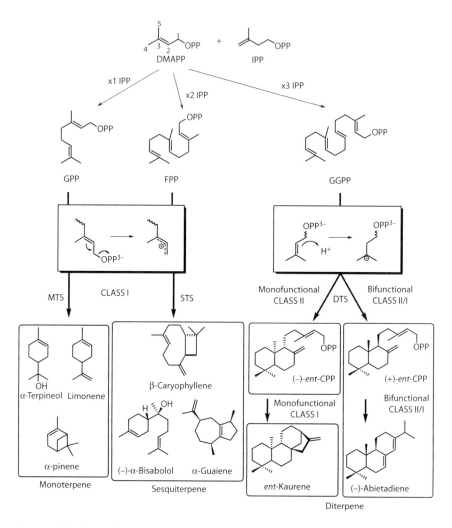

Figure 13.2 Two distinct mechanisms of terpene synthases (class I and class II). The ionization-initiated reaction is shown as the class I reaction, and the protonation-initiated reaction is shown as the class II reaction. (MTS, monoterpene synthase; STS, sesquiterpene synthase; DTS, diterpene synthase.)

trees, such as poplar (Behnke et al. 2007). However, most terpenoids are produced from the elongated versions of DMAPP by *trans*-prenylations. In these reactions, DMAPP serves as a priming molecule and 1, 2, or 3 IPP molecules are sequentially condensed to the dephosphorylated form of DMAPP (Figure 13.2). DMAPP serves as a primer because the double-bond between C2 and C3 of DMAPP can stabilize the C1 carbocation after the cleavage of a diphosphate (see, DMAPP structure in Figure 13.2).

The resulting prenyl diphosphate molecules are C10 geranyl diphosphate (GPP), C15 farnesyl diphosphate (FPP), and C20 geranyl geranyl diphosphate (GGPP), which are synthesized by GPP-, FPP-, and GGPP synthase, respectively. These C10, C15, and C20 prenyl diphosphate molecules are end products of primary metabolism but are starting substrates for various terpenes at the gateway of secondary (specialized) metabolism. The terpenoids synthesized from GPP, FPP, and GGPP are called mono- (C10), sesqui- (C15), and diterpenoids (C20), respectively.

In the second step, terpene synthases (TPSs) utilize these *trans*-prenyl diphosphates to produce structurally diverse terpenoid hydrocarbon skeletons through carbocation cascade reactions. Class I ionization-initiated and class II protonation-initiated TPS reactions have been characterized in TPS reactions (Gao et al. 2012). In the class I TPS reaction, the TPS enzyme initiates the reaction by cleaving a diphosphate moiety of prenyl diphosphate, resulting in the formation of a highly reactive carbocation intermediate (Figure 13.2). Most mono- and sesqui-TPSs, utilizing a GPP or FPP substrate, catalyze terpene-forming reactions by a class I-type reaction, although an atypical sesqui-TPS (drimenol synthase from valerian plant) utilizing both class I and class II reactions was reported (Kwon et al. 2014). Crystal structure of sesqui-TPS showed that hydrophobic tails of prenyl diphosphate molecules are buried inside the enzyme while the negatively charged diphosphate moiety positions at the mouth (entry point) of the substrate-binding pocket (Starks et al. 1997). Since the entry point has multiple basic Arg residues and the binding sites (DDxxD motif) for divalent ions (Mg^{2+} or Mn^{2+}), these features collectively generate a strong positive charge. As a result, the negative charge of the diphosphate and the positive charge of TPS form a strong ionic interaction, inducing the cleavage of a diphosphate from the prenyl diphosphate substrate to yield a carbocation intermediate. The unstable carbocations undergo cascade reactions to rearrange hydrocarbon skeletons typically by double-bond rearrangement, C–C bond rearrangement, hydride-shift, and methyl-shift. Such intramolecular rearrangements are continued until the molecules are stabilized by a deprotonation or a quench by a water molecule. On the other hand, the class II TPS reaction is initiated by a protonation by an acidic residue of TPS to the allylic group, thereby generating an unstable carbocation (Figure 13.2). Subsequently, similar carbocation cascade reactions occur to make different terpene scaffolds. However, it should be noted that a class II reaction is unable to cleave the diphosphate. While mono- and sesqui-terpenoids are mostly formed by the class I reaction, diterpenoid skeletons (C20) are folded by the coordinated reactions of class I and class II (Figure 13.2) (Zerbe and Bohlmann 2015). More specifically, the class II first initiates carbocation reactions by a protonation and then the class II reaction takes products from the class I reaction (terpenoid diphosphates) to reinitiate the class II (ionization–initiation) reaction. The carbocation chain reactions are finalized by a deprotonation and a water quench. Di-TPS enzymes are structurally more diverse and sophisticated, and thus the class I and class II reaction modules can be encoded in a single protein (Bifunctional di-TPS) or in two separate proteins (monofunctional di-TPS) (Figure 13.2) (Zerbe and Bohlmann 2015).

Some TPS enzymes can produce multiple terpene products, but the majority of TPS enzymes are dedicated to the synthesis of a single product (Yoshikuni et al. 2006). It is therefore apparent that the molecular folding patterns are programmed in the polypeptides of each TPS, and each carbocation cascade reaction is finely tuned by a specific (or group) of residues around the substrate-binding pocket. It is not completely understood how the coordinated influence of multiple residues in TPSs direct the sequence of carbocation cascade reactions. However, emergence and disappearance of terpene-forming activities were demonstrated by comprehensive enzyme breeding of *A. annua* β-farnesene/amorpha-4,11-diene synthase (Salmon et al. 2015). It is of particular interest that a mutation of one residue in TPS evokes the emergence of a new terpene; however, a second mutation in another residue (in the background of the first mutant enzyme) can completely silence the effect of the first mutation and a third mutation can reactivate the disappeared activity (Salmon et al. 2015). This result indicates subtle interactions among different residues in TPS can dramatically influence the product profile. Certainly, there is still much to learn about nature's molecular folding skills from TPS reaction mechanisms.

In summary, remarkable diversity of terpene skeletons in nature is attributed to (i) the formation of different prenyl diphosphate substrates (GPP, FPP, and GGPP) and (ii) the intramolecular carbocation reactions molecularly tuned by TPS enzymes. In other words, different types of highly flexible substrates are synthesized, and these are uniquely folded by different TPS enzymes through an unimaginable number of carbocation reactions. Since TPS is the gateway enzyme controlling the flux to the final terpenoid products, from a synthetic biology perspective, it is important to acquire knowledge to manipulate TPS activities (product specificity and turnover number) in a predictable manner.

13.2.3 Cytochrome P450 monooxygenase

The resulting products from TPSs, allylic hydrocarbons, are less prone to additional structural modifications than oxygen-bearing compounds. In terpenoid metabolism, the first modification is often regio- and stereospecific oxygenations mostly by cytochrome P450 monooxygenase (P450) (Pateraki et al. 2015). P450s are a superfamily of the enzymes containing a heme cofactor with highly conserved three-dimensional folding patterns. P450 utilizes the heme cofactor to cleave an oxygen molecule, inserting one oxygen atom to the substrate and the other oxygen atom to water. Although

Chapter thirteen: Metabolic engineering and synthetic biology

hydroxylation is the most common P450 reaction, several other reactions, such as epoxidation, dehydrogenation, C–C bond cleavage, and demethylation, have also been reported (De Montellano and De Voss 2007). Also, multiple oxygenations can be catalyzed by a single P450 as shown in artemisinic acid, abietic acid, and lovastatin biosynthesis (Barriuso et al. 2011; Cochrane and Vederas 2014; Ro et al. 2005, 2006). Eukaryotic P450 is not a self-sufficient enzyme and requires cytochrome P450 reductase (CPR) for reducing equivalents from NADPH. In eukaryotes, P450 and CPR localize on the ER and transiently interact with each other by ionic interactions through protein surface charges (Wang et al. 1997). In metabolic engineering of natural products, P450s are critical enzymes because regio- and stereo-selective oxygenations are difficult and expensive in organic chemical synthesis, thereby making the enzymatic incorporations of oxygen atoms on terpene backbones industrially attractive. However, proper expression of P450s in *E. coli* has been challenging since bacteria do not have native endomembrane systems (Paddon and Keasling 2014). In one study, it was demonstrated that a prokaryote (*E. coli*)-adapted plant P450 can be developed by systemic modifications of the membrane-anchor domain of P450 (Chang et al. 2007). Nonetheless, functional expression of P450 in a prokaryotic background has met tremendous difficulties. Despite some successful cases in *E. coli*, such issues have rendered scientists to use yeast (*Saccharomyces cerevisiae*) for functionalized (oxygenated) terpene production.

13.2.4 Isopentenyl diphosphate (IPP) biosynthesis

In order to increase the production titer of terpenoid products in microbes and other surrogate hosts, cellular levels of the initial precursors (i.e., the end products in primary metabolisms) have to be increased. In terpenoid metabolism, the central precursors of all terpenoids are IPP and DMAPP. Therefore, understanding the IPP biosynthetic pathway is a prerequisite for the synthetic biology of terpenoids. In cells, IPP is synthesized by two distinct routes, mevalonate (MVA) and methylerythritol 4-phosphate (MEP) pathways, depending on the type of organism (Lange et al. 2000).

13.2.4.1 MVA pathway

The MVA pathway was named after the central pathway intermediate, mevalonic acid, and is comprised of six enzymes (Figure 13.3). Acetyl CoA is the starting carbon source for the MVA pathway. Two molecules of acetyl CoA are sequentially condensed onto an acetyl CoA to form 3-hydroxy-3-methyl-glutaryl-CoA (HMGCoA). The key enzyme, HMG CoA reductase (HMGR), then converts HMGCoA to mevalonate, and two sequential phosphorylations of mevalonate activate mevalonate to mevalonate 5-diphosphate. Final decarboxylation completes the process to produce the C5 molecule IPP by consuming ATPs. IPP is converted

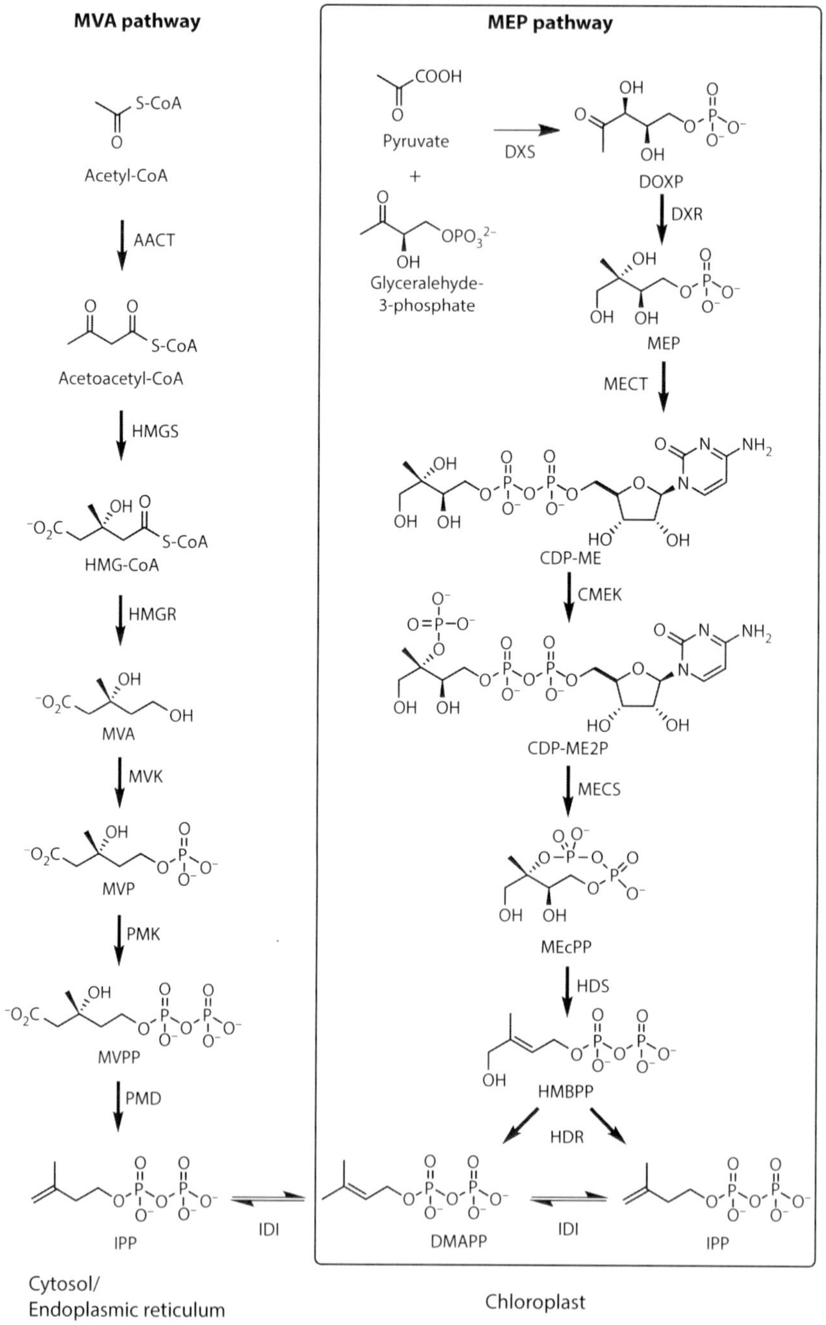

MVA pathway

Acetyl-CoA
↓ AACT
Acetoacetyl-CoA
↓ HMGS
HMG-CoA
↓ HMGR
MVA
↓ MVK
MVP
↓ PMK
MVPP
↓ PMD
IPP

Cytosol/
Endoplasmic reticulum

MEP pathway

Pyruvate + Glyceraldehyde-3-phosphate
→ DXS → DOXP
↓ DXR
MEP
↓ MECT
CDP-ME
↓ CMEK
CDP-ME2P
↓ MECS
MEcPP
↓ HDS
HMBPP
↓ HDR
DMAPP ⇌ IPP (IDI)

IPP ⇌ DMAPP (IDI)

Chloroplast

to DMAPP by a reversible enzyme, IPP isomerase. The MVA pathway is an energy-requiring process since three ATP and one NADPH are consumed to produce one IPP. Although all six enzymes in the MVA pathway contribute to the flux to different degrees, HMGR is the most critical rate-limiting enzyme, and therefore controlling the activity of HMGR at transcription or by feedback inhibition has been a main metabolic point in altering the cellular level of IPP. The MVA pathway operates in the cytosol and ER of animals, fungi, and plants (Chappell 2002). However, prokaryote and algae do not have a MVA pathway, and their supply of IPP depends entirely on the nonmevalonate pathway (see below).

13.2.4.2 MEP pathway

In the early 1990s, stable-isotope labeling studies consistently showed that bacteria do not synthesize IPP via the MVA pathway (Rohmer et al. 1993), implying a hidden, distinct biosynthetic route for IPP supply in bacteria. Further biochemical studies have established that bacteria synthesize the IPP using two glycolytic pathway intermediates, glyceraldehyde 3-phosphate (G3P) and pyruvate. Subsequent biochemical works elucidated a hidden metabolic pathway for the IPP synthesis, evolutionarily conserved in prokaryotes. This IPP-biosynthetic pathway was named methyl erythritol phosphate (MEP) pathway (or Rohmer pathway). The MEP pathway is comprised of seven enzymes as shown in Figure 13.3. At the entry point of the MEP pathway, pyruvate and G3P are condensed by deoxy xylulose phosphate (DXP) synthase to form DXP. DXP is then reduced to methyl erythritol phosphate (MEP) by DXP reductase. Although DXP is the first product synthesized in the MEP pathway, DXP can also be channeled to the vitamin B12 pathway through pyridoxal 5′-phosphate. Due to this reason, this pathway was named after the unique metabolic intermediate, MEP, rather than DXP. The MEP molecule is further activated by CTP and ATP to generate 4-hydroxy-3-methyl-butenyl-2-enyl diphosphate, an unusual

---

Figure 13.3 IPP biosynthetic route by MVA and MEP pathway. (HMG-CoA: 3-hydroxy-3-methylglutaryl-CoA; MVA: mevalonate; MVP: mevalonate-5-phosphate; MVPP: mevalonate-5-diphosphate; IPP: isopentenyl diphosphate; DMAPP: dimethylallyl diphosphate; DOXP: 1-Deoxy-D-xylulose 5-phosphate; MEP: 2-C-methylerythritol 4-phosphate; CDP-ME: 4-diphosphocytidyl-2-C-methylerythritol; CDP-ME2P: 4-diphosphocytidyl-2-C-methyl-D-erythritol 2-phosphate; MEcPP: 2-C-methyl-D-erythritol 2,4-cyclodiphosphate; HMBPP: (E)-4-Hydroxy-3-methyl-but-2-enyl diphosphate; AACT: Acetoacetyl-CoA thiolase; HMGS: HMG-CoA synthase; HMGR: HMG-CoA reductase; MVK: mevalonate-5-kinase; PMK: phosphomevalonate kinase; PMD: mevalonate-5-diphosphate decarboxylase; IDI: isopentenyl diphosphate isomerase; DXS: DOXP synthase; DXR: DOXP reductase; CMS: 2-C-methyl-D-erythritol 4-phosphate cytidylyltransferase; CMK: 4-diphosphocytidyl-2-C-methyl-D-erythritol kinase; MCS: 2-C-methyl-D-erythritol 2,4-cyclodiphosphate synthase; HDS: HMBPP synthase; HDR: HMBPP reductase.)

diphosphate bridged intermediate. This intermediate is converted to either IPP or DMAPP by a reductase. The MEP pathway is also an energy-requiring route as it consumes one ATP, one CTP, and two NADPHs. Comparative studies of the MEP pathway across the kingdoms revealed that the MEP pathway is not present in animal and fungi but is present in prokaryotes and chloroplasts of algae and plants. Therefore, IPPs are supplied by two independent biosynthetic pathways in plants (MVA and MEP pathways), whereas the IPP supply of all other organisms depends on either the MVA or MEP pathway. However, it should be noted that apicomplexan, a phylum of parasitic protists, has a unique organelle called an apicoplast where the MEP pathway operates (Maréchal and Cesbron-Delauw 2001), suggesting the enzymes in the MEP pathway are excellent targets for drug development to cure many tropical parasitic diseases (Ralph et al. 2001).

13.2.5 Metabolic engineering of terpenoids

To produce high-value terpenoids in surrogate hosts, necessary parts (e.g., metabolic genes) have to be discovered, improved, and implemented in the biological chassis (e.g., *E. coli*, yeast, tobacco), and be engineered to produce an improved quantity of metabolic precursor (e.g., IPP in terpenoid metabolism). Synthetic biological production of terpenoids was first attempted in the early 2000s, and here we will focus on discovery and development of the parts and platforms for artemisinin and paclitaxel (Taxol) production.

13.2.5.1 Artemisinin

13.2.5.1.1 Artemisinin biosynthesis Artemisinin is the sole antimalarial drug effective against drug-resistant malarial parasites and is under high demand in Africa. After over 10 years of research, the artemisinin biosynthetic pathway was fully elucidated (Covello 2008). Four enzymes are involved in artemisinin biosynthesis in *A. annua* (Figure 13.4). Amorphadiene synthase (ADS) commits the first reaction to produce amorphadiene from FPP (Bouwmeester et al. 1999; Chang et al. 2000; Mercke et al. 2000). Amorphadiene is sequentially oxidized three times to artemisinic acid by amorphadiene oxidase (AMO) (Ro et al. 2006; Teoh et al. 2006). However, it has been known that a direct precursor for artemisinin is dihydroartemisinic acid, but not artemisinic acid, which prompted the isolation of a C11-C13 double-bond reductase. Enzyme purification from *A. annua* trichomes identified an enoate reductase, named DBR, which can reduce artemisinic aldehyde to dihydroartemisinic aldehyde (Zhang et al. 2008). Dihydroartemisinic aldehyde is further oxidized to dihydroartemisinic acid by aldehyde dehydrogenase (Teoh et al. 2009). Finally, dihydroartemisinic acid can be converted to artemisinin by a nonenzymatic photo-reaction (Brown and Sy 2004).

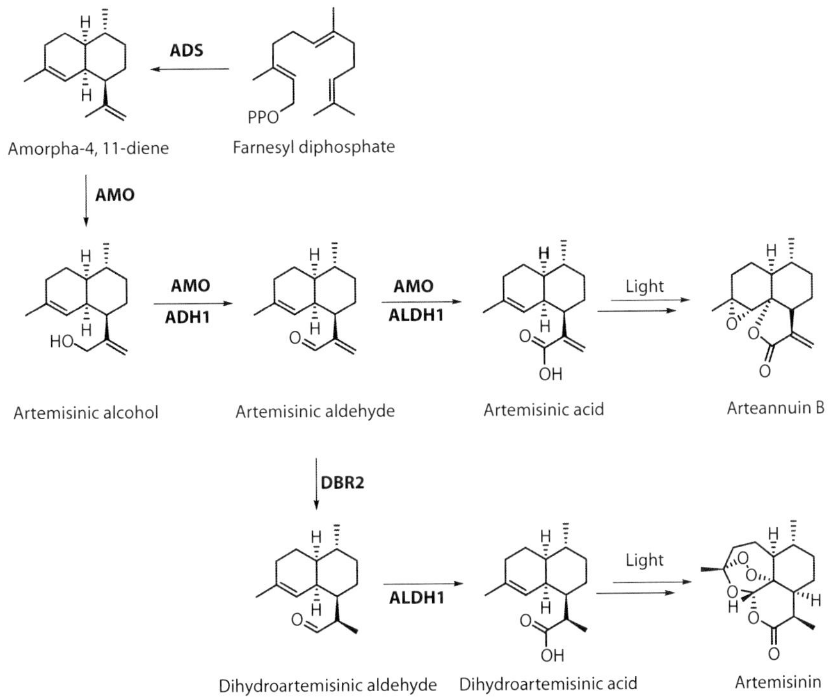

Figure 13.4 Artemisinin biosynthetic pathway. (ADS: amorpha-4,11-diene synthase; ADH1: alcohol dehydrogenase 1; ALDH1: aldehyde dehydrogenase 1; DBR2: artemisinic aldehyde Δ11(13) double bond reductase.)

13.2.5.1.2 Amorphadiene production in E. coli Bacterium *E. coli* was first examined as a workhorse to produce artemisinin precursor. A codon-optimized synthetic *ADS* was expressed in *E. coli*, where yeast MVA pathway genes were overexpressed by a polycistronic expression strategy (note that the MVA pathway is not present in *E. coli*) (Martin et al. 2003). The newly built yeast MVA pathway was free from endogenous metabolic regulations in *E. coli*, thus an unregulated production of an IPP precursor was achieved by this first synthetic MVA pathway in *E. coli*. In shake flask culture, ~100 mg/L amorphadiene could be achieved, and up to 0.5 g/L amorphadiene could be produced in an optimized bioreactor (fermentation) condition (Newman et al. 2006). It was soon realized, however, that the yeast enzymes in the MVA pathway were not optimized for the balanced activities, and therefore toxic metabolites accumulated and interfered with cell growth and other vital metabolic processes. To balance the level of enzyme activities, transcription of multiple metabolic genes in the MVA pathway were tuned by random sequence changes in the intergenic regions in the operon (Pfleger et al. 2006). Several other modifications were also made to improve the production titer. For example, N-terminus truncated version of *HMGR* was expressed to remove the feedback inhibition (Pfleger et al. 2006). Also, unbiased metabolomics and gene expression analyses were conducted to monitor and identify metabolic stresses exerted by the synthetic MVA pathway in *E. coli* (Kizer et al. 2008; Pitera et al. 2007). MVA metabolic genes from organisms other than yeast were also expressed to increase *in vivo* enzyme activities (Tabata and Hashimoto 2004). Such combinatorial metabolic engineering enabled an increased production titer of 25 g/L amorphadiene from the synthetic *E. coli* platform in an optimized bioreactor conditions (Tsuruta et al. 2009), demonstrating a metabolic potential to produce terpenoids in microbes.

13.2.5.1.3 Artemisinic acid production in yeast: Prototype While *E. coli* was the focus for amorphadiene production by a cytosolic ADS enzyme, yeast was developed as an alternative host to harbor membrane-bound P450 enzymes. The multifunctional P450, AMO, catalyzing the synthesis of artemisinic acid from amorphadiene was discovered and expressed in the MVA-pathway-engineered yeast (Ro et al. 2006). In this engineered yeast, the native MVA pathway was manipulated to increase the cellular level of FPP. Some key modifications are as follows. The rate-limiting HMGR was overexpressed in the N-terminus truncated form; a squalene synthase gene was conditionally down-regulated by replacing its promoter with a methionine-repressible promoter (therefore, by adding methionine, flux can be directed to FPP accumulation); a *UPC2-1* transcription factor, regulating expression of multiple MVA metabolic genes, was overexpressed to increase the overall flux in the MVA pathway (*UPC2-1* is a positive transcription factor with a point mutation that appears to stabilize UPC2 protein). Combined expression of *ADS* and *AMO* in this yeast background resulted in an artemisinic acid production titer of ~100 mg/L culture. This yeast strain could synthesize up to 2.5 g/L artemisinic acid by optimizing bioreactor conditions (Lenihan et al. 2008). On the other hand, functional expression of *AMO* in *E. coli* was proven to be recalcitrant and only 0.3 g/L artemisinic acid production could be achieved after extensive AMO protein engineering (Chang et al. 2007). Therefore, this head-to-head comparison of the prototype yeast and *E. coli* suggested that yeast is a superior microbial platform for artemisinic acid production over *E. coli*.

13.2.5.1.4 Development of an industrial yeast strain The prototype yeast for artemisinic acid production was built on yeast S288C, a yeast model strain for genetics, but is not suitable for industrial fermentation. Additionally, this prototype was designed to activate transgenes by a galactose-inducer, an expensive carbon source inadequate for industrial fermentation. In order to generate an elite industrial yeast strain, all metabolic genes in the MVA pathway were chromosomally integrated and constitutively expressed in CEN.PK2, a yeast strain routinely used in industrial fermentation. On bioreactor optimization, the new CEN.PK2 yeast platform could produce 40 g/L amorphadiene, a several-fold higher production titer than the S288C-derived strain, without using costly galactose (Westfall et al. 2012). This result proved the superiority of CEN.PK2 for industrial fermentation. When *AMO* and *CPR* were additionally expressed, it was observed that this strain was under oxidative stress likely due to the production of radical oxygen species released from the uncoupled reactions between P450 and CPR (Ro et al. 2008; Zangar et al. 2004). Additionally, artemisinic aldehyde, a potentially toxic intermediate, accumulated in the CEN.PK2 strain (Paddon et al. 2013). To lessen the oxidative stresses, the cytochrome b5 gene was isolated from *A. annua* because P450 activity is also known to be supported and enhanced by cytochrome b5. The expression of cytochrome b5 in engineered CEN.PK2 reduced the radical oxygen species and increased cell viability. To reduce the accumulation of artemisinic aldehyde and artemisinic alcohol (reaction intermediates of the multifunctional AMO enzyme), artemisinic aldehyde dehydrogenase and putative alcohol dehydrogenase, identified from the trichomes of *A. annua*, were additionally overexpressed in the engineered CEN.PK2. The expression of these two dehydrogenases markedly reduced the accumulation of toxic intermediates and increased the artemisinic acid production to 25 g/L.

The engineered CEN.PK2 strain has been transferred to the Sanofi-Aventis for a large-scale fermentation, artemisinic acid purification, and chemical conversion of artemisinic acid to artemisinin (Turconi et al. 2014). In the chemical process, the double-bond reduction step was chemically achieved by a diastereoselective hydrogenation of artemisinic acid. This semi-synthetic artemisinin represents the first large-scale production of a natural product using synthetic biology. In addition, the knowledge and skills acquired from the metabolic engineering in *E. coli* and yeast, and strain development have become valuable assets for next-generation synthetic biology.

13.2.5.2 Paclitaxel

13.2.5.2.1 Paclitaxel biosynthesis Paclitaxel (commercially known as Taxol) is a structurally complex taxane diterpenoid produced from the bark of the Pacific yew tree, *Taxus brevifolia*, and has been used to treat ovary and breast cancers. Originally, the paclitaxel was purified from the Pacific yew tree; however, its supply from the natural source was extremely limited, and a semisynthetic method was developed from 10-deacetylbaccatin III, a biosynthetic intermediate of paclitaxel found in the needles of various *Taxus* spp. Biosynthesis of paclitaxel is comprised of at least 19 enzymatic steps (Figure 13.5) (Croteau et al. 2006; Jennewein and Croteau 2001; Walker and Croteau 2001). Taxadiene synthase cyclizes GGPP to taxa-4(5),11(12)-diene, a tricyclopentadecane scaffold. On its taxane backbone, eight carbons are oxygenated by regio- and stereo-selective P450s, and two acetyl groups and a benzoyl group are attached to the oxygens at C4, C10 (acetyl-) and C2 (benzoyl-). During the oxygenations, a

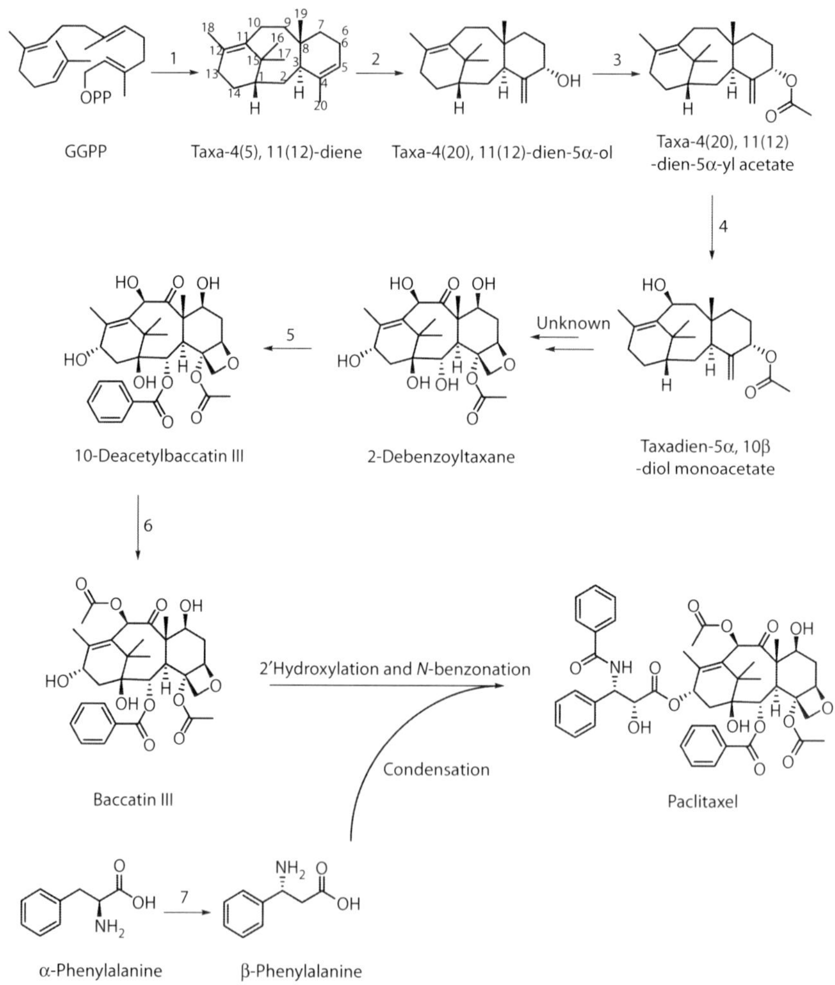

Figure 13.5 Paclitaxel biosynthetic pathway. GGPP, geranylgeranyl diphosphate; (1) taxadiene synthase; (2) cytochrome P450 taxadiene 5-hydroxylase; (3) taxa-4(20), 11(12)-dien-5a-ol-O-acetyltransferase; (4) cytochrome P450 taxane 10-β-hydroxylase; (5) taxane 2a-O-benzoyltransferase; (6) 10-deacetylbaccatin III-10-O-acetyltransferase; (7) phenylalanine aminomutase.

unique oxetane moiety is formed likely by a P450-mediated epoxidation. These reactions synthesize a key precursor, baccatin III or 10-deacetylbaccatin III, to which 2-hydroxy β-phenylalanine (linked with an additional N-benzoyl group) is attached to the oxygen at C13. With the pioneering work by Croteau's group, molecular clones for taxadiene synthase (Wildung and Croteau 1996), five P450s for 2α-, 5α-, 7β-, 10β-, and 13α-hydroxylations (Chau and Croteau 2004; Chau et al. 2004a; Jennewein et al. 2001, 2003, 2004), and four acyl or aroyl transferases for side-carbon modifications (Chau et al. 2004b; Walker and Croteau 2000a,b; Walker et al. 2002) have been identified. It was suggested that taxadiene synthase (di-TPS) and 5α-hydroxylase (P450) catalyze the first two reactions from GGPP since labeled 5α-hydroxytaxadiene fed to the plant stem was readily converted to paclitaxel. An acetylation at C5 and a hydroxylation at C10 have been proposed to follow. However, the exact order of reaction sequences beyond these four steps is unknown. In addition, oxygenations at C1, C4, and C9 of the taxane skeleton remain unknown, and oxetane formation needs to be elucidated as well.

13.2.5.2.2 Microbial production of paclitaxel precursors Despite the early progress in gene discovery in paclitaxel biosynthesis, production of paclitaxel or the semi-synthetic intermediate, 10-deacetylbaccatin III, from any synthetic metabolic platform is in its infancy. Five early biosynthetic genes (GGPP synthase, taxadiene synthase, 5α-hydroxylase, 5α-O-acetyltransferase, and 10β-hydroxylase) were initially expressed in yeast with an aim to produce an oxygenated and acetylated taxadiene precursor, but the transgenic yeast only produced a negligible amount of taxadiene (1 mg/L) and 5α-hydroxytaxadiene (25 μg/L) (Dejong et al. 2006). In another study, to increase the production of taxadiene, metabolic flux of the yeast MVA pathway was enhanced by overexpression of truncated *HMGR, UPC2-1*, and *GGPP synthase* (Engels et al. 2008). By these bioengineering efforts, the yeast was rendered to synthesize 9 mg/L taxadiene, but such a level is certainly at the low end of the metabolic potential of yeast MVA pathway.

In recent years, more comprehensive metabolic engineering was performed in *E. coli* for 5α-hydroxytaxadiene production. Similar to the pathway optimization in yeast for artemisinic acid production, expression levels of the genes in the *E. coli* MEP pathway and the two taxadiene biosynthetic genes (*GGPP* and *taxadiene synthase*) were randomly altered by using different plasmids (copy number variations) and promoter strength (Ajikumar et al. 2010). Combinatorial uses of the metabolic modules identified optimal gene expression levels, resulting in 1 g/L taxadiene production in *E. coli*. Furthermore, a chimeric enzyme fusing *Taxus* CPR to 5α-hydroxylase was generated and expressed in the *E. coli* engineered to produce taxadiene. This *E. coli* strain could synthesize 60 mg/L

5α-hydroxytaxadiene, a significant improvement over the previous result in yeast. Unexpectedly, this transgenic *E. coli* showed an excessive production of 5(12)-oxa-3(11)-cyclotaxane, seemingly an undesirable byproduct of the taxadiene 5α-hydroxylase reaction (Ajikumar et al. 2010). Consistent to this, when the taxadiene synthase and 5α-hydroxylase were overexpressed in tobacco trichome, the same P450 byproduct, 5(12)-oxa-3(11)-cyclotaxane, was also synthesized as a sole oxygenated taxadiene product (Rontein et al. 2008). Enzymatic studies of 5α-hydroxylase detected the formation of an unstable C4–C5 epoxy-taxadiene intermediate (Edgar et al. 2015), as opposed to the previous radical-rebound mechanism (Jennewein et al. 2004). It was suggested that nonselective degradation of the epoxy-taxadiene intermediate is responsible for the unexpected byproduct, explaining the inefficient conversion of taxadiene to 5α-hydroxy taxadiene. Therefore, in order to achieve a high-titer production of oxygenated taxadiene, this initial mechanistic bottleneck has to be resolved in the future.

13.3 Alkaloid metabolism

Alkaloids are low-molecular-weight, nitrogen-containing compounds found in 20% of flowering plants. Based on the core skeletons, alkaloids are categorized into several subfamilies, such as isoquinoline, indole, pyridine, pyrrole, and piperidine. Different from terpenoids, alkaloids are not synthesized from a unifying biosynthetic origin, but diverse amino acids (e.g., ornithine, lysine, tyrosine, phenylalanine, and tryptophan) and other sources (e.g., purine nucleotide) are used as the carbon-nitrogen scaffolds (Figure 13.6). A number of natural alkaloids from plants have been used as medicines. For example, cocaine, caffeine and nicotine are stimulants and recreational drugs; quinine is a traditional antimalarial drug before the wide occurrence of multidrug resistant parasites; morphine and codeine are important analgesics; vinblastine and vincristine are anti-cancer drugs for ovarian and breast cancers; atropine is used for various medical purposes, including pupil dilation and heart-rate increase. Among many important alkaloid natural products from plants, we will focus on the recent progress of two major subclasses of alkaloids, monoterpenoid indole and benzylisoquinoline, with an emphasis on vinblastine and morphine. An overview of their pathways and recent key discoveries, as well as progress in metabolic engineering will be discussed in this section.

13.3.1 Monoterpenoid indole alkaloid

13.3.1.1 Overview

Monoterpenoid indole alkaloid (MIA) is derived from the amino acid tryptophan and a monoterpenoid unit based on secologanin (Figures 13.6 and 13.7). MIAs are commonly found in Apocynaceae, Rubiaceae,

Chapter thirteen: Metabolic engineering and synthetic biology

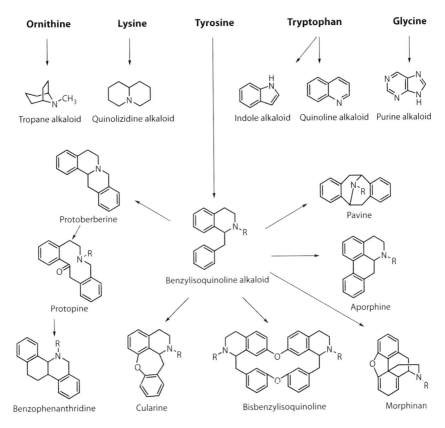

Figure 13.6 Examples of alkaloid natural products in plants with an emphasis on benzyl isoquinoline alkaloids. Main classes of alkaloids biosynthesized from various amino acids, and chemical structures of BIA subgroups derived from the simple benzylisoquinoline alkaloid subunit.

and Loganiaceae families. With approximately 3000 reported structures, MIA makes the largest and most diverse subgroup of alkaloids (Facchini and De Luca 2008). Many MIAs showed antibacterial/antifungal activities and have pharmaceutical properties, such as the anticancer compounds vinblastine and vincristine from *Catharanthus roseus* (Madagascar periwinkle) and the antipsychotic drug reserpine from *Rauwolfia serpentina* (Indian snakeroot).

The enormous diversity of MIAs is of particular scientific interest. Such diversity is a result of the rearrangement of the highly malleable C10 terpene moiety as well as the numerous decoration possibilities. The vast majority of MIAs are biochemically synthesized from the universal MIA precursor strictosidine (Figure 13.7). Tryptophan decarboxylase (TDC) converts tryptophan to tryptamine (De Luca et al. 1989), which is coupled to

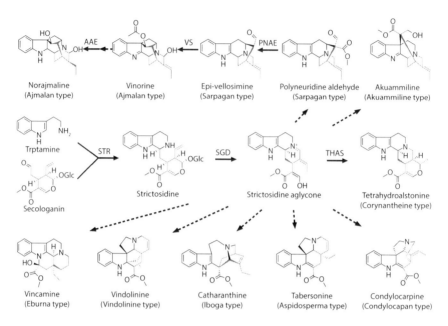

Figure 13.7 **(See color insert.)** Formation of various MIAs from central precursor strictosidine. The C10 monoterpene moiety is labeled in red. Solid arrows show the identified catalytic steps, and the dotted arrows show the unidentified steps. (STR: strictosidine synthase; SGD: strictosidine β-glucosidase; THAS: tetrahydroalstonine synthase; PNAE: polyneuridine aldehyde esterase; VS: vinorine synthase; AAE: acetylnorajmaline acetylesterase.)

secologanin by strictosidine synthase (STR) to give strictosidine (Kutchan et al. 1988). Removal of the glucose moiety by strictosidine β-glucosidase (SGD) yields the highly labile dialdehyde aglycone that forms almost all MIA backbones via various biochemical reactions (Geerlings et al. 2000). During the ring rearrangement, the C10 moiety mainly stays intact with occasional loss of one carbon (O'Connor and Maresh 2006).

Only a few enzymes in this complicated biochemical network for MIA backbone formation are characterized and cloned. These enzymes include polyneuridine aldehyde esterase (PNAE) and vinorine synthase (VS) that form the ajmalan-type backbone in *R. serpentina* (Bayer et al. 2004; Dogru et al. 2000), acetylajmaline acetylesterase (AAE) that forms norajmaline in *R. serpentina* (Ruppert et al. 2005), and tetrahydroalstonine synthase (THAS) that forms a corynantheine-type backbone found in *C. roseus* (Stavrinides et al. 2015) (Figure 13.7). Among these enzymes, THAS is thus far the only identified enzyme that uses strictosidine aglycone as the substrate, representing the class of enzymes in the early part of the pathways.

The monoterpenoid moiety secologanin is only one example of the diverse iridoid family, a series of monoterpenes harboring a methylcyclopentan-[c]-pyran skeleton (Figure 13.7). Cleavage of the cyclopentane ring results in the formation of the subgroup secoiridoid. More than 2500 iridoids have been reported from at least 16 plant families (De Luca et al. 2014). The broad distribution of iridoid/secoiridoid raised the question of their biological significance. $^{14}C-CO_2$ feeding assay in *Antirrhinum majus* (Snapdragon) showed that 47% of the phloem ^{14}C-photoassimilate was the iridoid antirrhinoside that is toxic and deters herbivores (Beninger et al. 2007; Voitsekhovskaja et al. 2006). The aglycone of the secoiridoid oleuropein from *Olea europaea* (Olive) is also implicated in plant defense for its strong protein-crosslinking activities (Konno et al. 1999). Even in the MIA-producing plant *C. roseus*, secologanin content is 10–15 folds higher than that of the major MIAs (Asada et al. 2013; Salim et al. 2013, 2014). It is clear that more studies are required to determine the biological properties, biosynthesis, and transport of iridoid/secoiridoid.

13.3.1.2 Biosynthesis and cellular compartmentation of MIAs in Catharanthus roseus

Among the MIA-producing species, *Catharanthus roseus* (Madagascar periwinkle) is the most extensively studied species for its MIA diversity and being the only commercial source of anticancer drugs vinblastine and vincristine. At least 130 MIAs have been reported in *C. roseus* alone, among which catharanthine and vindoline are the most abundant and unique to *C. roseus* (Facchini and De Luca 2008). Coupling catharanthine and vindoline gives dimeric vinblastine and vincristine, examples of bisindole alkaloids (Figure 13.8).

The majority of MIAs are derived from the coupling of tryptamine and secologanin. Recently, the entire 9-step pathway leading to secologanin from GPP was fully characterized in *C. roseus* (Asada et al. 2013; Collu et al. 2001; Geu-Flores et al. 2012; Irmler et al. 2000; Miettinen et al. 2014; Murata et al. 2008; Salim et al. 2013, 2014) (Figure 13.8). Feeding radiolabeled glucose to *C. roseus* cell culture showed that the terpene moiety of MIA is synthesized from the MEP pathway (Contin et al. 1998). The C10 product geraniol is then transformed through multiple oxidation and cyclization steps to form secologanin. Recent advances in large-scale sequencing greatly accelerated gene discovery. Most of the pathway genes are identified through comparative genomics and metabolite/gene expression coregulation studies, and are verified by virus-induced gene silencing (VIGS) and biochemical analysis. In this pathway, iridoid synthase (IS) is an unusual terpene cyclase that catalyzes the reductive terpene cyclization, distinctive from all other known terpene cyclases (Geu-Flores et al. 2012; Kries et al. 2016). IS belongs to the progesterone 5β-reductase family in phytosterol biosynthesis. Recruited to the iridoid pathway, IS is one example of gene-neofunctionalization during enzyme evolution. Another noteworthy reaction is the final reaction in this pathway that converts

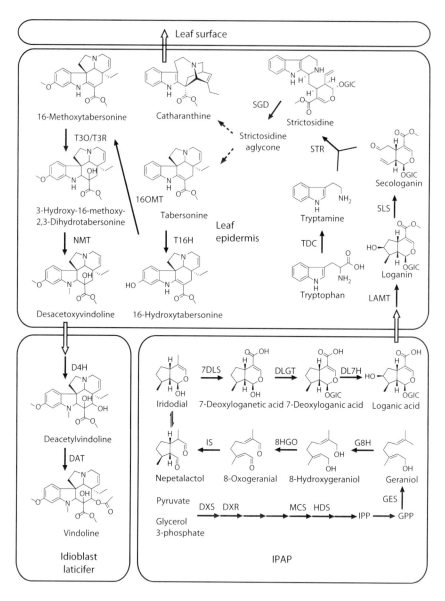

Figure 13.8 An overview of the MIA biosynthetic pathway in *Catharanthus roseus* leaf. The entire MIA pathway is located in four different cell types in the leaf. Hollow arrows show the trafficking of metabolites across various cell types; solid arrows show the identified enzymatic steps; and the dotted arrows show the unidentified catalytic steps. The internal associated parenchyma (IPAP) cells host the early secologanin pathway (geraniol to loganic acid) and the entire MEP pathway. Loganic acid is transported to the leaf epidermis, in which it harbors the late secologanin pathway (loganic acid to secologanin) and the majority of the

loganin to secologanin. Secologanin synthase (SLS) CYP72A1 is a P450 enzyme that catalyzes the oxidative C–C cleavage (Irmler et al. 2000).

Although the formation of various MIA backbones is still elusive, the decorative reactions of the aspidosperma-type MIA tabersonine to vindoline are well characterized at the biochemical level (Besseau et al. 2013; Levac et al. 2008; Liscombe et al. 2010; Qu et al. 2015; St-Pierre et al. 1998; Vazquez-Flota et al. 1997) (Figure 13.8). The seven-enzyme pathway sequentially adds a methoxyl group, a hydroxyl group, an N-methyl group, and another hydroxyl group followed by the O-acetylation. Remarkably, the 3-hydration reaction is catalyzed by the concerted action of a P450 (tabersonine 3-oxygenase, T3O) and a sinapyl alcohol dehydrogenase enzyme (tabersonine 3-reductase, T3R) (Qu et al. 2015). In the absence of T3R, T3O oxidizes tabersonine and 16-methoxytabersonine to a series of epoxides that are no longer the substrates of T3R. T3O and T3R are the last discovered enzymes in this pathway, through gene expression profiling, comparative genomics, and VIGS. The concerted reaction from a membrane-bound P450 and a soluble reductase explains why the traditional enzyme isolation approach was unable to reconstitute such a reaction. Interestingly, the 4-hydroxylation is catalyzed by a dioxygenase, instead of an expected P450 (De Carolis et al. 1990; Vazquez-Flota et al. 1997).

As the model plant for MIA studies, *C. roseus* has become a canonical compartmentation example for plant specialized metabolism. *In situ* hybridization, gene expression studies, and selective metabolite profiling collectively revealed the fascinating logistics of MIA metabolism in *C. roseus* (Figure 13.8). The MEP pathway-generated geraniol is converted to loganic acid inside the internal phloem associated parenchyma (IPAP) cells in the leaf, as supported by the *in situ* hybridization of the pathway genes including DXS, DXR, MCS, and HDS (Burlat et al. 2004; Mahroug et al. 2006; Oudin et al. 2007) in the MEP pathway, and G8H, 8-HGO, IS, 7DLS, DLGT, and DL7H in the secologanin pathway (Burlat et al. 2004; Geu-Flores et al. 2012; Mahroug et al. 2006; Miettinen et al. 2014). On the contrary, the late

---

MIA pathway (strictosidine to desacetoxyvindoline). The last two steps of vindoline formation occur in the leaf mesophyll idioblast and laticifer cells. (DXS: 1-deoxy-D-xylulose 5-phosphate (DOXP) synthase; DXR: DOXP reductase; MCS: 2-C-methyl-D-erythritol 2,4-cyclodiphosphate synthase; HDS: HMB-diphosphate synthase; GES: geraniol synthase; G8H: geraniol 8-hydroxylase; 8HGO: 8-hydroxygeraniol oxidase; IS: iridodial synthase; 7DLS: 7-deoxyloganetic acid synthase; DLGT: 7-deoxyloganetic acid glucosyltransferase; DL7H: 7-deoxyloganic acid 7-hydroxylase; LAMT: loganic acid methyltransferase; SLS: secologanin synthase; TDC: tryptophan decarboxylase; STR: strictosidine synthase; SGD: strictosidine β-glucosidase; T16H: tabersonine 16-hydroxylase; 16OMT: 16-hydroxytabersonine 16-O-methyltransferase; T3O: tabersonine 3-oxidase; T3R: tabersonine 3-reductase; NMT: 3-hydroxy-16-methoxy-2,3-dihydrotabersonine N-methyltransferase; D4H: desacetoxyvindoline 4-hydroxylase; DAT: deacetylvindoline O-acyltransferase.)

secologanin pathway from loganic acid to secologanin, and the early MIA pathway from tryptophan to strictosidine aglycone are entirely located in the leaf epidermis. This is supported by *in situ* hybridization, biochemical localization, and leaf epidermis-enriched transcriptome analysis of LAMT and SLS that finishes the secolognin pathway (Guirimand et al. 2010; Mahroug et al. 2006), TDC that supplies the indole moiety tryptamine (Murata et al. 2008; St-Pierre et al. 1999), and STR and SGD that initiate MIA biosynthesis (Burlat et al. 2004; Mahroug et al. 2006; Miettinen et al. 2014; Murata et al. 2008). Remarkably, the decoration of tabersonine to form vindoline also occurs in two different cell types in the leaf tissue. The early biosynthetic pathway from tabersonine to desacetoxyvindoline is located in the leaf epidermis, as shown by the *in situ* hybridization, biochemical localization, and transcriptome analyses of T16H2, 16-OMT, T3O, T3R, and NMT (Besseau et al. 2013; Levac et al. 2008; Murata and De Luca 2005; Qu et al. 2015). The last two steps of this pathway are found in the leaf mesophyll specialized laticifer and idioblast cells (St-Pierre et al. 1999).

The spatial separation of the biosynthetic pathway will require the transportation of the intermediate across various cell types. Thus far, only one pleiotropic drug-resistance family of ATP-binding cassette transporter CrTPT2 has been characterized. CrTPT2, localized on the plasmamembrane of the leaf epidermis, is responsible for the secretion of catharanthine from the leaf epidermis to the leaf wax exudate (Yu and De Luca 2013). The secretion of MIA to the leaf surface may not be restricted to *C. roseus* but common in other MIA species. Such a mechanism may contribute to the plant defense; however, it requires further investigation. The spatial separation of catharanthine in the leaf wax exudates and vindoline in the leaf mesophyll laticifer and idioblast explains why *C. roseus* only accumulates a trace amount of the dimeric vinblastine and vincristine. It is hypothesized that catharanthine and vindoline are coupled to the potent microtubule assembly inhibitor vinblastine and vincristine by a vacuole peroxidase when the leaf is damaged by herbivores (Yu and De Luca 2013). This mechanism therefore avoids the self-toxicity of the potent dimers (De Luca et al. 2014).

In contrast to the extensive studies of MIA biosynthesis in *C. roseus* leaf, the biosynthetic pathway in the underground root tissue is poorly understood. *C. roseus* root accumulates several MIAs such as horhammericine and minovincinine that are not found in the leaf, suggesting a different set of biosynthetic enzymes. Only one root-specific enzyme in MIA biosynthesis, tabersonine 19-hydroylase (T19H), has been characterized (Giddings et al. 2011). More research is needed to decipher the origin of precursors such as secologanin and tryptamine, cellular localization of biosynthetic genes, and MIA logistics.

Last, the biochemical pathway for the formation of major MIA backbones such as iboga-type catharanthine and aspidosperma-type tabersonine are still entirely unknown. The localization of the early MIA pathway (STR and SGD) and the early tabersonine to vindoline pathway (T16H2 to T3R) in the leaf epidermis strongly suggests that these reactions occur in the same tissue. As a result, exploring the leaf epidermis enriched transcriptome database that was used to identify LAMT and 16-OMT would reveal the most fascinating part of the MIA pathway in *C. roseus*. The complete elucidation of catharanthine and tabersonine biosynthesis will greatly facilitate the gene discovery in other important MIA biosynthesis.

13.3.1.3 Metabolic engineering of MIA pathway

C. roseus cell culture system has been exploited for MIA production over three decades (Zhao and Verpoorte 2007). Despite efforts in optimizing the growth conditions, hormonal treatment, and manipulation of a few transcription factors and biosynthetic genes including tryptophan decarboxylase (TDC) and STR, the cell culture system failed to produce discernable vinblastine and vincristine due to the lack of production of the precursor vindoline. With the elucidation of the entire tabersonine to vindoline pathway, it is now clear that several biosynthetic genes including 16-hydroxytabersonine *O*-methyltransferase (16OMT) and tabersonine 3-oxygenase (T3O) are expressed at very low levels in the cell cultures in comparison to the intact leaves (3% and 10%, respectively) (retrieved from Medicinal Plant Genormics Resource, medicinalplantgenomics.msu. edu). In addition, 16OMT and T3O expresses 6 and 30 fold more in the leaf epidermis where the actual biosynthesis occurs compared to the intact leaves (Qu et al. 2015). As a result, the cell culture is much less efficient in accumulating the valuable precursor vindoline.

Vindoline biosynthesis from tabersonine occurs in different cell types, with the first 5 steps occurring in the leaf epidermis and the last two steps in leaf mesophyll idioblast and laticifer. It is however demonstrated that the entire pathway can be assembled in a single yeast cell (Qu et al. 2015). Overexpression of the pathway genes in yeast in high-copy plasmids could produce vindoline from tabersonine at lower but comparable levels to the live plant (12%–30%, 1.1–2.7 mg/L), together with several fold more of the pathway intermediates. This yeast expression system, however, was able to convert 80% more tabersonine than the live plant (17 mg/L). The result showed great potential in optimizing gene expression to minimize accumulation of the intermediate in favor of vindoline production.

On the upstream of the MIA pathway, *de novo* strictosidine biosynthesis is also demonstrated in the yeast system (Brown et al. 2015). Unlike the plant, yeast does not possess the MEP pathway. Its own MVA pathway produces little geranyl diphosphate (GPP) that would be shuttled to secologanin and eventually strictosidine production. As a "push and pull" strategy, the yeast strain was heavily engineered to produce more GPP.

To enhance the GPP synthesis, a truncated HMGR lacking the product feedback loop, an IPP isomerase, a GPP synthase, and a FPP synthase that has greater GPP synthase activity were introduced. To alleviate IPP and GPP side consumption, the yeast native FPP synthase and two genes metabolizing geraniol to citronellol were knocked out, and a negative regulator of tRNA synthesis that diverts the IPP flux was introduced.

After transferring the entire strictosidine synthetic gene repertoire (11 genes) and cofactor biosynthetic genes (4 genes) to increase the availability of cytochrome P450 reductase, as well as S-adenosyl-L-methionine (SAM) and NADPH, no detectable strictosidine was produced. Further optimization included transferring an alcohol dehydrogenase that appeared to enhance the three-step oxidization of 7-deoxyloganetic acid synthase (7DLS), as well as increasing the copy number of geraniol 8-hydroxylase (G8H), the first committed step in the secologanin/strictosidine pathway. Increasing the G8H transcript by using a high-copy plasmid achieved the most dramatic yield increase for strictosidine, leading to at least a 17-fold increase to 0.5 mg/L.

The results indicated that the key optimization for strictosidine production in yeast is to increase the flux to geraniol and geraniol to 8-hydroxygeraniol. Interestingly, the final yeast strain still accumulated more loganin (0.8 mg/L), the only other detectable intermediate, than strictosidine. However, increasing copy numbers of secologanin synthase (SLS) that converts loganin to secologanin did not substantially increase the strictosidine yield, suggesting further optimization might focus on pulling the strictosidine to downstream MIAs upon the elucidation of the related biosynthetic genes.

Transient expression of the entire strictosidine pathway in tobacco leaves through agrobacterium-mediated coinfiltration was also attempted (Miettinen et al. 2014). Although demonstrating the possibility of such engineering, *de novo* production was only successful up to 7-deoxyloganetic acid, indicating the necessity in restraining the flux only to secologanin. By providing pathway intermediates such as iridodial or 7-deoxyloganetic acid, however, strictosidine was indeed produced.

13.3.2 Benzylisoquinoline alkaloid

13.3.2.1 Overview

Benzylisoquinoline alkaloids (BIAs) are synthesized from the aromatic amino acid, tyrosine (Figure 13.6). BIAs constitute the major subclass of alkaloids with 2500 known natural products. A majority of BIAs are produced from the angiosperm families of Papaveraceae, Berberidaceae, Menispermaceae, Ranunculaceae, and Magnoliaceae. Several plant species synthesizing BIAs with medicinal activities (e.g., analgesic and antimalarial activities) have been traditionally used as herbal medicines in ancient times. It is documented that mankind recognized the analgesic property of opium poppy (*Papaver somniferum*) as early as 4000 BC and used its dried latex to control pains and induce sleep. The active ingredient of the opium poppy latex is morphine synthesized from the morphinan branch of BIA, the most effective and irreplaceable pain-killing drug. Despite the structural complexity of morphine, the entire biosynthetic genes for morphine biosynthesis have recently been identified, followed by the generation of a prototype yeast producing opiates from glucose. Biosynthesis and metabolic engineering of morphine and its related BIA, sanguinarine, will be discussed together with several key gene discoveries.

13.3.2.2 Biosynthesis of BIA

At the entry-point of BIA biosynthesis, L-tyrosine is converted to dopamine and 4-hydroxyphenyl acetaldehyde by decarboxylation, *m*-hydroxylation, and deamination reactions (Figure 13.9). These two metabolites (dopamine and 4-hydroxyphenyl acetaldehyde) are condensed to form (S)-norcoclaurine by norcoclaurine synthase (NCS) via Pictet-Spengler condensation, the first committed step into the BIA metabolic pathway. The first NCS purified from opium poppy belongs to pathogenesis-related

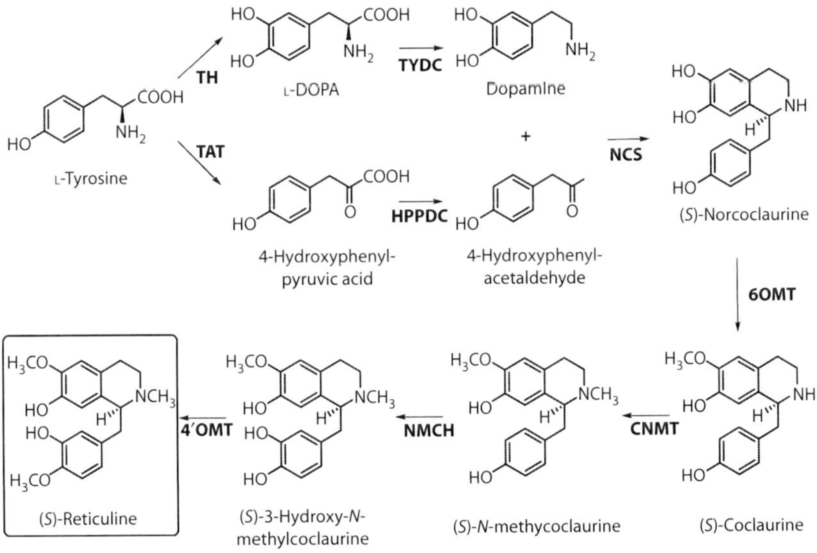

Figure 13.9 (S)-Reticuline biosynthetic pathway from L-tyrosine. (TH: Tyrosine hydroxylase; TAT: L-tyrosine aminotransferase; TYDC: tyrosine/dopa decarboxylase; HPPDC: *p*-hydroxyphenylpyruvate decarboxylase; NCS: (S)-norcoclaurine synthase; 6OMT: norcoclaurine-6-*O*-methyltransferase; CNMT: (S)-coclaurine-*N*-methyltrasnferase; NMCH: (S)-*N*-methylcoclaurine 3′-hydroxylase; 4′OMT: (S)-3′-hydroxy-*N*-methylcoclaurine 4′-*O*-methyltransferase.)

(PR)10/Betv1 protein family (Samanani et al. 2004). Although a distinct NCS belonging to the 2-oxoglutarate-dependent dioxygenase family was independently identified from *Coptis japonica*, its *bona fide* catalytic activity in BIA metabolism has been debated (Lee and Facchini 2010; Minami et al. 2007). (S)-norcoclaurine provides the BIA backbone and is the central intermediate of 2500 BIAs in plants. On the (S)-norcoclaurine backbone, O-methylation, N-methylation, and hydroxylation occur to yield (S)-reticuline (Figure 13.9). (S)-reticuline is the key branching point for the biosynthesis of different BIA subclasses, such as benzophenanthridine, protoberberine, and aporphine. For example, berberine bridge enzyme (BBE) and CYP80G2 act on (S)-reticuline to open new metabolic channels for protoberberine and aporphine, respectively (Figures 13.6 and 13.11).

It should be noted, however, that (S)-reticuline is not a direct precursor for many downstream BIAs, but its epimer (R)-reticuline is used as the precursor, suggesting that a unique epimerase is involved in morphine biosynthesis (Figure 13.10). This epimerization is catalyzed by two sequential reactions of 1,2-dehydroreticuline synthase (DRS) and 1,2-dehydroreticuline reductase (DRR). Notably, the enzyme catalyzing both DRS and DRR reactions was identified as a single polypeptide with the fusion of a P450 and a reductase by two independent research groups (Farrow et al. 2015; Winzer et al. 2015). This enzyme was the last hidden step in the morphinan biosynthesis, and the discovery of this gene in 2015 paved the road for *de novo* synthesis of morphine and related morphinan BIAs.

From (R)-reticuline, six more enzymes are involved to produce morphine (Figure 13.10). It is of particular interest that thebaine 6-O-demethylase (T6OMD; 2-oxoglutarate-dependent dioxygenase family) catalyzes demethylations of thebaine to codeinone and oripavine to morphinone (Hagel and Facchini 2010). Its homolog, codeine 3-O-demethylase (CODM), also similarly catalyzes the demethylation of codeine to morphine and thebaine to oripavine. In the conversion of (R)-reticuline to morphine, all cognate cDNAs encoding specific enzymes have been identified except for thebaine synthase (THS). However, morphine can be enzymatically synthesized without THS because the THS reaction occurs spontaneously, though the enzyme can accelerate the reaction in plant cells. Therefore, all metabolic genes for morphine biosynthesis from tyrosine have become available for synthetic biology in 2015.

13.3.2.3 Microbial production of BIA

At the early stage of BIA pathway reconstitution in microbes, the metabolic genes for BIAs and morphine biosynthesis were not completely identified. Some essential enzymes for morphine biosynthesis, such as salutaridine synthase (SalSyn), thebaine/codeine demethylases (T6ODM/CODM), and epimerase (DRS/DDR fusion enzyme), were only identified after 2009. In addition, enzymatically providing sufficient and balanced

Chapter thirteen: Metabolic engineering and synthetic biology 343

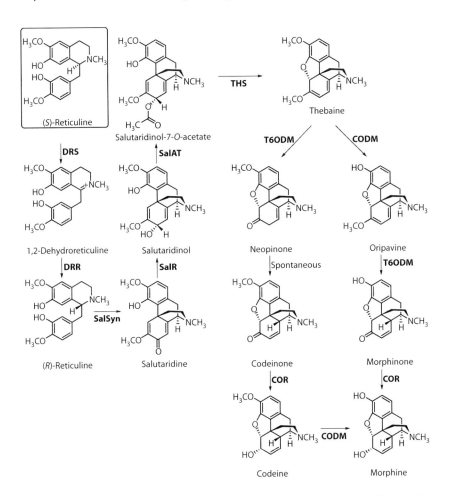

Figure 13.10 Morphine and codeine biosynthetic pathway from (S)-reticuline. (DRS: 1,2-dehydroreticule synthase; DRR: 1,2-dehydroreticuline reductase; SalSyn: salutaridine synthase; SalR: salutaridine reductase; SalAT: salutaridinol 7-O-acetyltransferase; THS: thebaine synthase; T6ODM: thebaine 6-O-demethylase; COR: codeinone reductase; CODM: codeine demethylase.)

amounts of dopamine and 4-hydroxyphenyl acetaldehyde at the entry of BIA pathway has been a difficult task. Indeed, the metabolic genes for this early conversion have yet to be identified.

To overcome these problems, scientists adopted cDNAs from different biological sources (e.g., human and bacteria) or fed intermediates to the microbes implemented with partial BIA biosynthetic pathways. In order to supply (S)-reticuline in *E. coli*, Minami et al. expressed microbial monoamino oxidase (MAO) to deaminate dopamine, thereby producing 3,4-dihydrophenyl acetaldehyde (3,4-DHPAA). By using MAO, a single precursor, dopamine, rather than two precursors, can be added to the culture medium. In addition, the (S)-N-methylcoclaurine 3′-hydroxyase (NMCH) enzymatic step (Figure 13.9) could be omitted since the 3,4-DHPAA has a hydroxyl group at the C3 position at the start. In this MAO-expressing *E. coli*, four more cDNAs necessary for (S)-reticuline biosynthesis were additionally expressed. As a result, after feeding dopamine to the transgenic *E. coli*, a 55 mg/L of (S)-reticuline could be synthesized (Minami et al. 2008). Furthermore, cocultivation of this *E. coli* strain with the yeast expressing either BBE or CYP80G2 allowed for the cell-based production of an aporphine-type alkaloid, magnoflorine and a protoberberine-type alkaloid, scoulerine. In another study, a commercially available compound, (R,S)-norlaudanosoline, was fed to yeast expressing 3 genes [6-OMT, (S)-coclaurine-N-methyltrasnferase (CNMT), 4′-OMT], resulting in the biotransformation of (R,S)-norlaudanosoline to (S)-reticuline (Hawkins and Smolke 2008). In the same report, it was demonstrated that human P450, CYP2D6, can replace SalSyn to produce salutaridine from (R)-reticuline, although the conversion rate was not efficient.

In early 2010, with the arrival of next-generation sequencings, a number of novel genes for morphine and other BIAs were identified and soon implemented in synthetic microbial systems. More specifically, SalSyn, T6ODM/CODM, and other biosynthetic genes for sanguinarine were identified, and their *in vivo* roles in yeast were examined by coexpressing with other BIA-biosynthetic genes. A 10-gene plant pathway was built in yeast to produce the protoberberine BIA, sanguinarine, by feeding (R,S)-norlaudanosoline to the yeast culture (Fossati et al. 2014) (Figure 13.11). Similarly, a 7-gene morphine pathway was implemented in yeast, where about 1.5% of the externally fed substrate, (R)-reticuline, could be converted to codeine but morphine could not be detected (Fossati et al. 2015). The success of these segmental reconstructions of the BIA pathway in *E. coli* and yeast suggested that true *de novo* synthesis of morphine and other BIAs can be achievable. However, all synthetic systems have relied on feeding intermediates, such as dopamine, (R,S)-norlaudanosoline, or (R/S)-reticuline. The two missing links connecting the glucose-derived primary metabolism to morphine or other morphinan BIAs were L-tyrosine *m*-hydroxylation at the entry of the pathway and (S)- to (R)-epimerization in the center of the BIA pathway. These issues were solved by the use of two new genes—L-tyrosine hydroxylase from beet and DRS/DRR-epimerase from opium poppy. By expressing the beet tyrosine hydroxylase in yeast together with all other genes for (S)-reticuline biosynthesis, (S)-reticuline could be *de novo* synthesized in yeast at a titer of 80 µg/L from glucose (DeLoache et al. 2015). Most recently, the last missing link, epimerase, was identified and named as STORR [(S)- to (R)-reticuline] or REPI (reticuline epimerase) from opium poppy (Farrow et al. 2015; Winzer et al. 2015). Finally, the 21- or 23-gene set including the newly identified STORR/REPI was expressed in

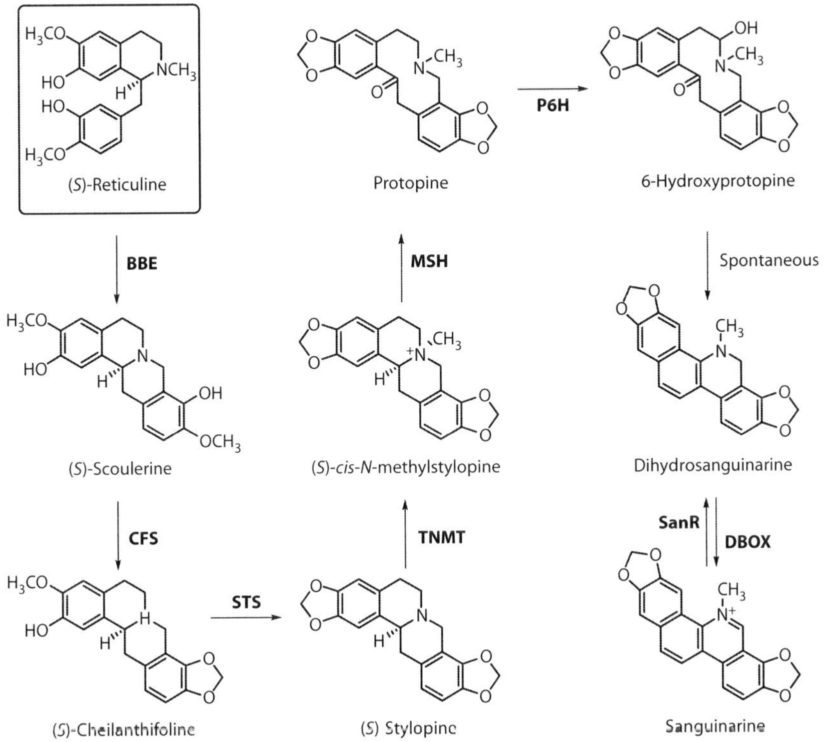

Figure 13.11 Sanguinarine biosynthetic pathway from (S)-reticuline. (BBE: berberine bridge enzyme; CFS: cheilanthifoline synthase; STS: stylopine synthase; TNMT: tetrahydroprotoberberine-N-methyltransferase; MSH: methylstylopine hydroxylase; P6H: protopine 6-hydroxylase; DBOX: dihydrobenzophenanthridine oxidase; SanR: sanguinarine reductase.)

yeast to *de novo* synthesize μg/L-scale of thebaine or hydrocodone (Galanie et al. 2015). In this research, a mutant version of L-tyrosine hydroxylase was used to eliminate feedback inhibition by L-tyrosine.

13.4 Glucosinolate metabolism

13.4.1 Glucosinolate biosynthetic pathway

Glucosinolates (GLSs) are sulfur-rich, nitrogen-bearing natural products stored in glucose-conjugate forms in plant cells. GLSs are primarily synthesized from plants in the Brassicales order, which includes an important crop *Brassica napus* (rapeseed) and the model plant *Arabidopsis thaliana*. The GLSs and their degradation products (isothiocyanate and nitrile) are involved in plant defense against insects and also provide health-related

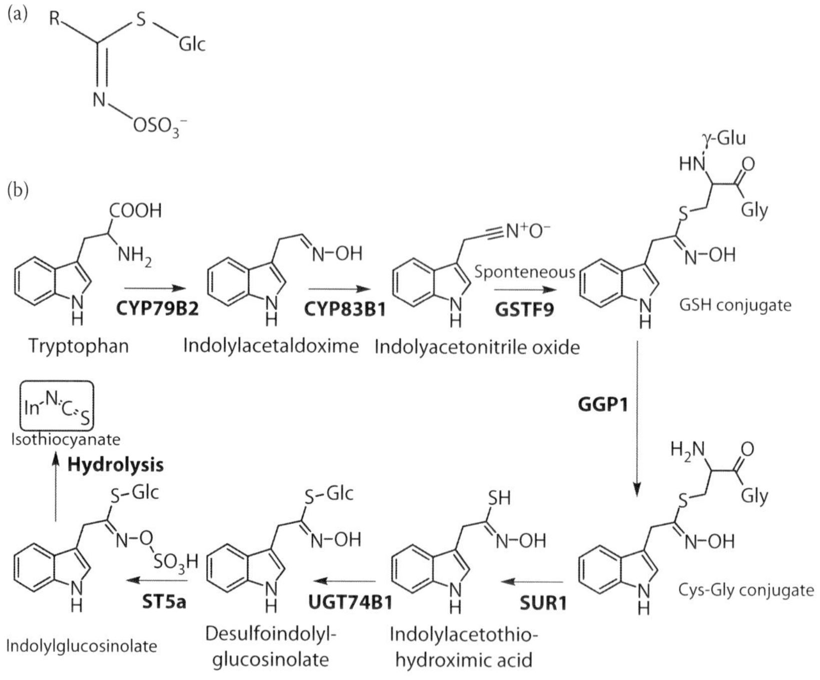

Figure 13.12 Glucosinolate structure and biosynthetic pathway. (a) Core structure of glucosinolate; (b) Enzymes in the glucosinolate biosynthetic pathway. (GSTF9, Glutathione-S-transferase; GGP1, γ-glutamyl peptidase; SUR1, C-S lyase; UGT74B1, glucosyltransferase; ST5a, sulfotransferase.)

benefits to humans (Hopkins et al. 2009; Traka and Mithen 2009). The dietary intake of cruciferous vegetables (Brassicaceae or Cruciferae) is known to reduce the risk of cancer occurrence and to prevent cardiovascular disease (Verkerk et al. 2009). It is believed that such health benefits are attributed to the GLSs in the plants of the Brassicales.

Approximately 120 GLSs, possessing the core structure shown in Figure 13.12a, have been identified (Fahey et al. 2001). From a biogenetic perspective, GLSs are classified into three major groups depending on the types of amino acids used as the building blocks. Aliphatic GLSs are synthesized from Met, Ala, Leu, Ile, and Val; aromatic GLSs from Phe and Tyr; and indole GLSs from Trp (Agerbirk and Olsen 2012). The GLS biosynthesis from these amino acids can be grouped into three enzymatic transformations—elongation of amino acid side-chains, synthesis of GLS core structure, and modification of amino acid R-groups. The mechanism of the GLS core synthesis is conserved in different GLSs, and the diversity of GLS is created by variations in the elongation and R-group modification.

The first elongation step is only relevant to Met and Phe, and all other amino acids skip this step. In this reaction, deamination of amino acids, first, produces 2-oxo acids. This is followed by an enzymatic cycle of condensation by acetyl-CoA, isomerization, and oxidative decarboxylation, which adds one carbon (methylene group) to the side-chain of the amino acids, thus generating a one-carbon elongated 2-oxo acid. This elongation step can either be repeated to add more methylene groups or terminated by transamination. The final products are side-chain elongated amino acids. In the second step of the GLS core synthesis, seven chemical transformations are known to take place, mediated by seven distinct enzymes (Figure 13.12b). This skeletal transformation is best studied for the indole GLS biosynthesis in *Arabidopsis* and thus will be explained using indole GLS as a representative (Figure 13.12b). In this GLS-core framing reaction, Trp is converted to an aldoxime by CYP79 (Hull et al. 2000), and the aldoxime is oxidized to a nitrile oxide or an aci-nitro molecule by CYP83 (Naur et al. 2003). The sulfur atom is yet to be incorporated at this point, and the Cys residue of a γ-Glu-Cys-Gly tripeptide glutathione (GSH) donates a sulfur atom by forming a covalent C–S bond. This GSH-conjugation reaction can happens spontaneously without enzymatic help but can also be facilitated by glutathione *S*-transferase (GST) as shown in yeast bioengineering (Mikkelsen et al. 2012). The Glu residue of the GSH-conjugate is removed by a unique γ-glutamyl peptidase 1 (GGP1), and subsequently the Cys-Gly dipeptide is removed by C-S lyase (SUR1), leaving only a thiol group in the final product (Geu-Flores et al. 2009; Mikkelsen et al. 2004). To this backbone, a glucose is added by UGT74 (glucosyltransferases) family enzyme (Gachon et al. 2005), and finally a sulphonate ($-SO_3^-$) is added by ST5a (sulfotransferase) to complete the indole GLS biosynthesis (Piotrowski et al. 2004). The R-groups of the final GLS structure are further modified to diversify GLSs. For example, two 2-oxoglutarate-dependent dioxygenases (AOP2/3) catalyze an oxidative cleavage of the side-chain in Met-derived GLSs (Kliebenstein et al. 2001). Concerted reactions by P450s from the CYP81F subfamily and *O*-methyltransferases are known to incorporate methoxyl groups to the indole-ring of indole GLSs (Pfalz et al. 2011). However, the metabolic genes involved in the R-group modifications are not fully understood at present. The GLSs and their glucose-hydrolyzing enzyme, called myrosinase, are normally spatially separated in plants. However, upon tissue damage by insects, these two come together to initiate the hydrolysis of a glucose moiety from the GLS. The resulting, unstable aglycone is spontaneously rearranged to form isothiocyanate, nitrile, and other molecules, which can effectively deter herbivores. It is of particular interest that high-affinity GLS transporters, *GTR1* and *GTR2*, were identified that transport the GLS from the apoplasm into the phloem for long distance translocation from leaves to seeds (Nour-Eldin et al. 2012).

13.4.2 Glucosinolate production in plants

The reconstitution of the GLS pathway in non-Brassicales plants can produce health-promoting GLSs for humans, and has also been an effective tool in elucidating and refining GLS biosynthetic pathways. The γ-glutamyl peptidase 1 (GGP1) discovery is one such example. In an effort to bioengineer the GLS pathway in plants, the first two genes for benzylglucosinolate biosynthesis (*CYP83B* and *CYP79A*) were expressed in a translational fusion linked by 2A peptide. As the 2A peptide is cleaved in cells, two genes can be expressed from a single promoter. Similarly, *SUR1*, *UGT74B1*, and *ST5a* were also expressed in a single construct using 2A linkages (Geu-Flores et al. 2009). When these five genes were transiently expressed in tobacco, approximately 8 nmol/mg leaf of GSH-conjugate was produced (Figure 13.12b), but the benzylglucosinolate production was negligible. However, an additional expression of a newly identified *GGP1* resulted in the synthesis of benzylglucosinolate at a titer of 0.6 nmol/mg leaf. Despite the low yield, this was the first demonstration of *de novo* biosynthesis of benzylglucosinolate in tobacco by coexpressing above six genes, and apparently, this unique peptidase is required to remove the Glu moiety from the GSH-conjugate. In a similar study, a seven-gene pathway, which further added *GSTF9* (Figure 13.12b), was constructed in tobacco to produce indole GLS at a yield of 0.2 nmol/mg leaf. This bioengineered plant was used to functionally identify novel P450s from the CYP81F family and O-methyltransferases that are involved in indole-group modifications (Pfalz et al. 2011). In another study, *in planta* conversion from Met to glucoraphanin (Met-derived GLS) was achieved by reconstructing a 13-gene pathway in tobacco. In this case, the first elongation step as well as the core-structure synthesis was also included in the bioengineering, resulting in glucoraphanin production at a titer of 42 nmol/g leaf (Mikkelsen et al. 2000). Although the *de novo* GLS production from these bioengineered tobacco plants were very low, it is clear that multigene pathways for GLSs can be built in plant platforms.

13.4.3 Microbial production of indole glucosinolate

Epidemiological studies showed that the consumption of GLS-producing plants (broccoli and cabbages) prevents cardiovascular disease and cancer development (Verkerk et al. 2009). However, often, multiple GLSs (>30) are synthesized in a single species, making it difficult to purify an individual GLS and to examine its impact on human health. The biosynthesis of GLSs in plants suffers from low yield with endogenous enzymatic activities diverting GLS intermediates into other undefined molecules. In addition, it is difficult to engineer the indole GLS pathway in plants because indole GLS and auxin (plant hormone) share a common precursor, tryptophan.

In order to overcome these problems, microbial production of indole GLS was pursued by stably integrating and expressing *Arabidopsis* genes for indole GLS in yeast (Mikkelsen et al. 2012). In this study, four expression modules, each designed for a simultaneous two-gene expression by a Gal1/10 bidirectional promoter, were sequentially integrated in the yeast genome. The four modules were *CYP79B2/CYP83B1, AtCPR* (Arabidopsis cytochrome P450 reductase)/*GSTF9, GGP1/SUR1*, and *UGT74B1/ST5a*. Expression of *CYP79B2/CYP83B1* in yeast could produce a GSH-conjugate by using endogenous yeast *CPR* and *GST*, but additional expressions of *AtCPR/GSTF9* by the second module increased the GSH-conjugate yield by 2.6-fold. By expressing the third (*GGP1/SUR1*) and the fourth (*UGT74B1/ST5a*) module, indole GLS was *de novo* synthesized in yeast at a scale of 1 mg/L yeast culture. It is a significant improvement over the plant bioengineering, but optimization of bioreactor conditions is necessary to maximize the metabolic capacity of yeast to produce indole GLS.

13.5 Gene clusters in plant specialized metabolisms

One of the most important discoveries in plant metabolism in the past decade is the existence of gene clusters for specialized metabolism. It had been believed that the physically linked gene clusters for specialized metabolisms was an exclusive feature of microbial genomes. However, in 1997, the first gene cluster was discovered in *Zea mays* (Frey et al. 1997). This cluster contains five genes (*Bx1-5*) involved in the biosynthesis of cyclic hydroxamic acid, 2,4-dihydroxy-1,4-benzoxazin-3-one (DIBOA). Since this first report, a number of gene clusters for specialized metabolic biosynthesis have been reported in plants: terpenoid metabolism, thalianol and marneral in *Arabidopsis thaliana* (Field and Osbourn 2008; Field et al. 2011), momilactone A and phytocassanes in *Oryza sativa* (Shimura et al. 2007; Wilderman et al. 2004), and avenacin A1 in *Avena strigosa* (Qi et al. 2004); for glucoside metabolism, Dhurrin in *Sorghum bicolor*, and lotaustralin and linamarin in *Lotus japonicus* (Takos et al. 2011); for alkaloid metabolism, α-tomatine in *Solanum lycopersicum* (Itkin et al. 2013), and noscapine in *Papaver somniferum* (Winzer et al. 2012). Recently, the development of rapid genome sequencings has allowed easier discovery of gene clusters in plants. Such genomic information combined with biochemical knowledge has facilitated rapid elucidation of pathways in entirety since scientists can focus solely on the clustered genes. Recent studies of noscapine and cucurbitacin C biosynthesis are excellent examples of how information on the gene clusters can expedite biochemical research (Dang et al. 2015; Shang et al. 2014). It is apparent that the easier genomic mining from the clustered genes can accelerate the characterizations of gene cluster components, such as promoters, genes, and other regulatory regions (Nutzmann and Osbourn 2014). The characterized gene clusters can be used in synthetic biology by creating minimal gene clusters or synthetic gene clusters.

13.6 Conclusion

Using exemplary studies of artemisinin, morphine, paclitaxel, vinblastine, and indole glucosinolate, this chapter describes the pathway elucidations in plants and the reconstruction of the plant pathways in microbes. Impressive progress was achieved in elucidating the biochemical pathways of these complex plant compounds primarily owing to the affordable high-throughput sequencings, differential expression data, and reverse genetics tools, such as VIGS. In parallel, fine tunings of enzyme/transcript dosages by codon optimization via gene synthesis, mutagenesis, manipulations of *cis*-elementary regions, and rationale pathway design have contributed to the development of massively redesigned metabolic pathways in microbes. To biochemists, a complete pathway elucidation from glucose to morphine is a triumph of modern plant biochemistry. To engineers, tens of grams of artemisinic acid production, followed by semi-synthetic production of artemisinin, testifies that cellular metabolic pathways can surely be engineered for better productivity just as other chemical processes are improved by iterative engineering efforts. Therefore, in principle, any valuable plant natural product can be produced at a large and commercial scale in microbes.

Many valuable lessons were obtained from the metabolic engineering or synthetic biology approaches described here. Repeated findings from various experiments are that balancing catalytic activities in living cells is critically important to reduce unnecessary buildup of toxic intermediates and also to lessen the physiological burden in microbes due to protein overproduction. When generating industrial strains, optimizing the gene expressions of every metabolic gene in the pathway with respect to other enzyme levels is unavoidable. Additionally, it is worth identifying natural allelic variants of metabolic genes or engineering existing enzymes for improved catalytic activities and substrate specificity. Finally, the bioreactor process has to be optimized for the best production of natural products in microbes.

At present, the microbial productions of paclitaxel, morphine, and vinblastine range from µg/L to mg/L scale. However, considering the speed of technology development and microbial metabolic potency, it is likely that orders of magnitude higher levels of production can be achieved in surrogate hosts, promising the sustainable supply of various natural products in the near future.

Acknowledgments

We thank Gillian MacNevin and Eun-Joo Kwon for the critical reading of this chapter. This work was supported by National Science and Engineering Research Council of Canada (NSERC) and the Canada Research Chair Program to D.K. Ro, and is also supported by the Next-Generation BioGreen 21 program (PJ00110552016) to M. Kwon.

References

Agerbirk, N. and C.E. Olsen. 2012. Glucosinolate structures in evolution. *Phytochemistry* 77: 16–45.

Ajikumar, P.K., W.-H. Xiao, K.E.J. Tyo et al. 2010. Isoprenoid pathway optimization for taxol precursor overproduction in *Escherichia coli*. *Science* 330: 70–74.

Asada, K., V. Salim, S. Masada-Atsumi et al. 2013. A 7-deoxyloganetic acid glucosyltransferase contributes a key step in secologanin biosynthesis in Madagascar periwinkle. *Plant Cell* 25: 4123–4134.

Barriuso, J., D.T. Nguyen, J.W.H. Li et al. 2011. Double oxidation of the cyclic nonaketide dihydromonacolin L to monacolin J by a single cytochrome P450 monooxygenase, LovA. *J. Am. Chem. Soc.* 133: 8078–8081.

Baur, J.A. and D.A. Sinclair. 2006. Therapeutic potential of resveratrol: The *in vivo* evidence. *Nat. Rev. Drug Discov.* 5: 493–506.

Bayer, A., X. Ma, and J. Stockigt. 2004. Acetyltransfer in natural product biosynthesis—Functional cloning and molecular analysis of vinorine synthase. *Bioorg. Med. Chem.* 12: 2787–2795.

Behnke, K., B. Ehlting, M. Teuber et al. 2007. Transgenic, non-isoprene emitting poplars don't like it hot. *Plant J.* 51: 485–499.

Beninger, C.W., R.R. Cloutier, M.A. Monteiro, and B. Grodzinski. 2007. The distribution of two major Iridoids in different organs of *Antirrhinum majus* L. at selected stages of development. *J. Chem. Ecol.* 33: 731–747.

Besseau, S., F. Kellner, A. Lanoue et al. 2013. A pair of tabersonine 16-hydroxylases initiates the synthesis of vindoline in an organ-dependent manner in *Catharanthus roseus*. *Plant Physiol.* 163: 1792–1803.

Bohlmann, J. and C.I. Keeling. 2008. Terpenoid biomaterials. *Plant J.* 54: 656–669.

Bohlmann, J., G. Meyer-Gauen, and R. Croteau. 1998. Plant terpenoid synthases: Molecular biology and phylogenetic analysis. *Proc. Natl. Acad. Sci. USA* 95: 4126–4133.

Bouwmeester, H.J., T.E. Wallaart, M.H. Janssen et al. 1999. Amorpha-4,11-diene synthase catalyses the first probable step in artemisinin biosynthesis. *Phytochemistry* 52: 843–854.

Brown, G.D. and L.-K. Sy. 2004. In vivo transformations of dihydroartemisinic acid in *Artemisia annua* plants. *Tetrahedron* 60: 1139–1159.

Brown, S., M. Clastre, V. Courdavault, and S.E. O'Connor. 2015. De novo production of the plant-derived alkaloid strictosidine in yeast. *Proc. Natl. Acad. Sci. USA* 112: 3205–3210.

Burlat, V., A. Oudin, M. Courtois, M. Rideau, and B. St-Pierre. 2004. Co-expression of three MEP pathway genes and *geraniol 10-hydroxylase* in internal phloem parenchyma of *Catharanthus roseus* implicates multicellular translocation of intermediates during the biosynthesis of monoterpene indole alkaloids and isoprenoid-derived primary metabolites. *Plant J.* 38: 131–141.

Chang, M.C.Y., R.A. Eachus, W. Trieu, D.-K. Ro, and J.D. Keasling. 2007. Engineering *Escherichia coli* for production of functionalized terpenoids using plant P450s. *Nat. Chem. Biol.* 3: 274–277.

Chang, Y.J., S.H. Song, S.H. Park, and S.U. Kim. 2000. Amorpha-4,11-diene synthase of *Artemisia annua*: cDNA isolation and bacterial expression of a terpene synthase involved in artemisinin biosynthesis. *Arch. Biochem. Biophys.* 383: 178–184.

Chappell, J. 2002. The genetics and molecular genetics of terpene and sterol origami. *Curr. Opin. Plant Biol.* 5: 151–157.

Chau, M. and R. Croteau. 2004. Molecular cloning and characterization of a cytochrome P450 taxoid 2alpha-hydroxylase involved in Taxol biosynthesis. *Arch. Biochem. Biophys.* 427: 48–57.

Chau, M., S. Jennewein, K. Walker, and R. Croteau. 2004a. Taxol biosynthesis: Molecular cloning and characterization of a cytochrome P450 taxoid 7β-hydroxylase. *Chem. Biol.* 11: 663–672.

Chau, M., K. Walker, R. Long, and R. Croteau. 2004b. Regioselectivity of taxoid-*O*-acetyltransferases: Heterologous expression and characterization of a new taxadien-5α-ol-*O*-acetyltransferase. *Arch. Biochem. Biophys.* 430: 237–246.

Cochrane, R.V. and J.C. Vederas. 2014. Highly selective but multifunctional oxygenases in secondary metabolism. *Acc. Chem. Res.* 47: 3148–3161.

Collu, G., N. Unver, A.M. Peltenburg-Looman, R. van der Heijden, R. Verpoorte, and J. Memelink. 2001. Geraniol 10-hydroxylase, a cytochrome P450 enzyme involved in terpenoid indole alkaloid biosynthesis. *FEBS Lett.* 508: 215–220.

Compadre, C.M., J.M. Pezzuto, A.D. Kinghorn, and S.K. Kamath. 1985. Hernandulcin: An intensely sweet compound discovered by review of ancient literature. *Science* 227: 417–419.

Cong, L., F.A. Ran, D. Cox et al. 2013. Multiplex genome engineering using CRISPR/Cas systems. *Science* 339: 819–823.

Contin, A., R. van der Heijden, A.W. Lefeber, and R. Verpoorte. 1998. The iridoid glucoside secologanin is derived from the novel triose phosphate/pyruvate pathway in a *Catharanthus roseus* cell culture. *FEBS Lett.* 434: 413–416.

Covello, P.S. 2008. Making artemisinin. *Phytochemistry* 69: 2881–2885.

Croteau, R., R.E. Ketchum, R.M. Long, R. Kaspera, and M.R. Wildung. 2006. Taxol biosynthesis and molecular genetics. *Phytochem. Rev.* 5: 75–97.

Dang, T.-T.T., X. Chen, and P.J. Facchini. 2015. Acetylation serves as a protective group in noscapine biosynthesis in opium poppy. *Nat. Chem. Biol.* 11: 104–106.

De Carolis, E., F. Chan, J. Balsevich, and V. De Luca. 1990. Isolation and characterization of a 2-oxoglutarate dependent dioxygenase involved in the second-to-last step in vindoline biosynthesis. *Plant Physiol.* 94: 1323–1329.

Dejong, J.M., Y. Liu, A.P. Bollon et al. 2006. Genetic engineering of taxol biosynthetic genes in *Saccharomyces cerevisiae*. *Biotechnol. Bioeng.* 93: 212–224.

DeLoache, W.C., Z.N. Russ, L. Narcross, A.M. Gonzales, V.J. Martin, and J.E. Dueber. 2015. An enzyme-coupled biosensor enables (*S*)-reticuline production in yeast from glucose. *Nat. Chem. Biol.* 11: 465–471.

De Luca, V., C. Marineau, and N. Brisson. 1989. Molecular cloning and analysis of cDNA encoding a plant tryptophan decarboxylase: Comparison with animal dopa decarboxylases. *Proc. Natl. Acad. Sci. USA* 86: 2582–2586.

De Luca, V., V. Salim, S.M. Atsumi, and F. Yu. 2012. Mining the biodiversity of plants: A revolution in the making. *Science* 336: 1658–1661.

De Luca, V., V. Salim, A. Thamm, S.A. Masada, and F. Yu. 2014. Making iridoids/secoiridoids and monoterpenoid indole alkaloids: Progress on pathway elucidation. *Curr. Opin. Plant Biol.* 19: 35–42.

De Montellano, P.O. and J.J. De Voss. 2007. Substrate oxidation by cytochrome P450 enzymes. In *Cytochrome P450: Structure, Mechanism, and Biochemistry*, ed. P.O. De Montellano, 183–230. New York: Springer Science & Business Media.

Dogru, E., H. Warzecha, F. Seibel, S. Haebel, F. Lottspeich, and J. Stockigt. 2000. The gene encoding polyneuridine aldehyde esterase of monoterpenoid indole alkaloid biosynthesis in plants is an ortholog of the α/β hydrolase super family. *Eur. J. Biochem.* 267: 1397–1406.

Edgar, S., K. Zhou, K. Qiao, J.R. King, J.H. Simpson, and G. Stephanopoulos. 2015. Mechanistic insights into taxadiene epoxidation by taxadiene-5α-hydroxylase. *ACS Chem. Biol.* 11: 460–469.

Engels, B., P. Dahm, and S. Jennewein. 2008. Metabolic engineering of taxadiene biosynthesis in yeast as a first step towards Taxol (Paclitaxel) production. *Metab. Eng.* 10: 201–206.

Facchini, P.J. and V. De Luca. 2008. Opium poppy and Madagascar periwinkle: Model non-model systems to investigate alkaloid biosynthesis in plants. *Plant J.* 54: 763–784.

Fahey, J.W., A.T. Zalcmann, and P. Talalay. 2001. The chemical diversity and distribution of glucosinolates and isothiocyanates among plants. *Phytochemistry* 56: 5–51.

Farrow, S.C., J.M. Hagel, G.A. Beaudoin, D.C. Burns, and P.J. Facchini. 2015. Stereochemical inversion of (S)-reticuline by a cytochrome P450 fusion in opium poppy. *Nat. Chem. Biol.* 11: 728–732.

Field, B., A.-S. Fiston-Lavier, A. Kemen, K. Geisler, H. Quesneville, and A.E. Osbourn. 2011. Formation of plant metabolic gene clusters within dynamic chromosomal regions. *Proc. Natl. Acad. Sci. USA* 108: 16116–16121.

Field, B. and A.E. Osbourn. 2008. Metabolic diversification—Independent assembly of operon-like gene clusters in different plants. *Science* 320: 543–547.

Fossati, E., A. Ekins, L. Narcross et al. 2014. Reconstitution of a 10-gene pathway for synthesis of the plant alkaloid dihydrosanguinarine in *Saccharomyces cerevisiae*. *Nat. Commun.* 5: 3283.

Fossati, E., L. Narcross, A. Ekins, J.P. Falgueyret, and V.J. Martin. 2015. Synthesis of morphinan alkaloids in *Saccharomyces cerevisiae*. *PloS One* 10: e0124459.

Frey, M., P. Chomet, E. Glawischnig et al. 1997. Analysis of a chemical plant defense mechanism in grasses. *Science* 277: 696–699.

Gachon, C.M., M. Langlois-Meurinne, Y. Henry, and P. Saindrenan. 2005. Transcriptional co-regulation of secondary metabolism enzymes in *Arabidopsis*: Functional and evolutionary implications. *Plant Mol. Biol.* 58: 229–245.

Galanie, S., K. Thodey, I.J. Trenchard, M. Filsinger Interrante, and C.D. Smolke. 2015. Complete biosynthesis of opioids in yeast. *Science* 349: 1095–1100.

Gao, Y., R.B. Honzatko, and R.J. Peters. 2012. Terpenoid synthase structures: A so far incomplete view of complex catalysis. *Nat. Prod. Rep.* 29: 1153–1175.

Geerlings, A., M.M. Ibanez, J. Memelink, R. van Der Heijden, and R. Verpoorte. 2000. Molecular cloning and analysis of strictosidine β-D-glucosidase, an enzyme in terpenoid indole alkaloid biosynthesis in *Catharanthus roseus*. *J. Biol. Chem.* 275: 3051–3056.

Geu-Flores, F., M.T. Nielsen, M. Nafisi et al. 2009. Glucosinolate engineering identifies a γ-glutamyl peptidase. *Nat. Chem. Biol.* 5: 575–577.

Geu-Flores, F., N.H. Sherden, V. Courdavault et al. 2012. An alternative route to cyclic terpenes by reductive cyclization in iridoid biosynthesis. *Nature* 492: 138–142.

Giddings, L.A., D.K. Liscombe, J.P. Hamilton et al. 2011. A stereoselective hydroxylation step of alkaloid biosynthesis by a unique cytochrome P450 in *Catharanthus roseus*. *J. Biol. Chem.* 286: 16751–16757.

Guirimand, G., V. Courdavault, A. Lanoue et al. 2010. Strictosidine activation in Apocynaceae: Towards a "nuclear time bomb"? *BMC Plant Biol.* 10: 182.

Hagel, J.M. and P.J. Facchini. 2010. Dioxygenases catalyze the O-demethylation steps of morphine biosynthesis in opium poppy. *Nat. Chem. Biol.* 6: 273–275.

Hawkins, K.M. and C.D. Smolke. 2008. Production of benzylisoquinoline alkaloids in *Saccharomyces cerevisiae*. *Nat. Chem. Biol.* 4: 564–573.

Hopkins, R.J., N.M. van Dam, and J.J. van Loon. 2009. Role of glucosinolates in insect-plant relationships and multitrophic interactions. *Annu. Rev. Entomol.* 54: 57–83.

Horwitz, A.A., J.M. Walter, M.G. Schubert et al. 2015. Efficient multiplexed integration of synergistic alleles and metabolic pathways in yeasts via CRISPR-Cas. *Cell Systems* 1: 88–96.

Hull, A.K., R. Vij, and J.L. Celenza. 2000. *Arabidopsis* cytochrome P450s that catalyze the first step of tryptophan-dependent indole-3-acetic acid biosynthesis. *Proc. Natl. Acad. Sci. USA* 97: 2379–2384.

Irmler, S., G. Schroder, B. St-Pierre et al. 2000. Indole alkaloid biosynthesis in *Catharanthus roseus*: New enzyme activities and identification of cytochrome P450 CYP72A1 as secologanin synthase. *Plant J.* 24: 797–804.

Itkin, M., U. Heinig, O. Tzfadia et al. 2013. Biosynthesis of antinutritional alkaloids in solanaceous crops is mediated by clustered genes. *Science* 341: 175–179.

Jakociunas, T., I. Bonde, M. Herrgard et al. 2015. Multiplex metabolic pathway engineering using CRISPR/Cas9 in *Saccharomyces cerevisiae*. *Metab. Eng.* 28: 213–222.

Jang, M., L. Cai, G.O. Udeani et al. 1997. cancer chemopreventive activity of resveratrol, a natural product derived from grapes. *Science* 275: 218–220.

Jennewein, S. and R. Croteau. 2001. Taxol: Biosynthesis, molecular genetics, and biotechnological applications. *Appl. Microbiol. Biotechnol.* 57: 13–19.

Jennewein, S., R.M. Long, R.M. Williams, and R. Croteau. 2004. Cytochrome p450 taxadiene 5alpha-hydroxylase, a mechanistically unusual monooxygenase catalyzing the first oxygenation step of taxol biosynthesis. *Chem. Biol.* 11: 379–387.

Jennewein, S., C.D. Rithner, R.M. Williams, and R.B. Croteau. 2001. Taxol biosynthesis: Taxane 13 α-hydroxylase is a cytochrome P450-dependent monooxygenase. *Proc. Natl. Acad. Sci. USA* 98: 13595–13600.

Jennewein, S., C.D. Rithner, R.M. Williams, and R. Croteau. 2003. Taxoid metabolism: Taxoid 14beta-hydroxylase is a cytochrome P450-dependent monooxygenase. *Arch. Biochem. Biophys.* 413: 262–270.

Jordan, A., J.A. Hadfield, N.J. Lawrence, and A.T. McGown. 1998. Tubulin as a target for anticancer drugs: Agents which interact with the mitotic spindle. *Med. Res. Rev.* 18: 259–296.

Kizer, L., D.J. Pitera, B.F. Pfleger, and J.D. Keasling. 2008. Application of functional genomics to pathway optimization for increased isoprenoid production. *Appl. Environ. Microbiol.* 74: 3229–3241.

Kliebenstein, D.J., V.M. Lambrix, M. Reichelt, J. Gershenzon, and T. Mitchell-Olds. 2001. Gene duplication in the diversification of secondary metabolism: Tandem 2-oxoglutarate-dependent dioxygenases control glucosinolate biosynthesis in Arabidopsis. *Plant Cell* 13: 681–693.

Koehn, F.E. and G.T. Carter. 2005. The evolving role of natural products in drug discovery. *Nat. Rev. Drug Discov.* 4: 206–220.

Konno, K., C. Hirayama, H. Yasui, and M. Nakamura. 1999. Enzymatic activation of oleuropein: A protein crosslinker used as a chemical defense in the privet tree. *Proc. Natl. Acad. Sci. USA* 96: 9159–9164.

Kries, H., L. Caputi, C.E. Stevenson et al. 2016. Structural determinants of reductive terpene cyclization in iridoid biosynthesis. *Nat. Chem. Biol.* 12: 6–8.

Kutchan, T.M., N. Hampp, F. Lottspeich, K. Beyreuther, and M.H. Zenk. 1988. The cDNA clone for strictosidine synthase from *Rauvolfia serpentina*. DNA sequence determination and expression in *Escherichia coli*. *FEBS Lett.* 237: 40–44.

Kwon, M., S.A. Cochrane, J.C. Vederas, and D.-K. Ro. 2014. Molecular cloning and characterization of drimenol synthase from valerian plant (*Valeriana officinalis*). *FEBS Lett.* 588: 4597–4603.

Lange, B.M., T. Rujan, W. Martin, and R. Croteau. 2000. Isoprenoid biosynthesis: The evolution of two ancient and distinct pathways across genomes. *Proc. Natl. Acad. Sci. USA* 97: 13172–13177.

Lee, E.J. and P. Facchini. 2010. Norcoclaurine synthase is a member of the pathogenesis-related 10/Bet v1 protein family. *Plant Cell* 22: 3489–3503.

Lenihan, J.R., H. Tsuruta, D. Diola, N.S. Renninger, and R. Regentin. 2008. Developing an industrial artemisinic acid fermentation process to support the cost-effective production of antimalarial artemisinin-based combination therapies. *Biotechnol. Prog.* 24: 1026–1032.

Levac, D., J. Murata, W.S. Kim, and V. De Luca. 2008. Application of carborundum abrasion for investigating the leaf epidermis: Molecular cloning of *Catharanthus roseus* 16-hydroxytabersonine-16-*O*-methyltransferase. *Plant J.* 53: 225–236.

Liscombe, D.K., A.R. Usera, and S.E. O'Connor. 2010. Homolog of tocopherol C methyltransferases catalyzes N methylation in anticancer alkaloid biosynthesis. *Proc. Natl. Acad. Sci. USA* 107: 18793–18798.

Mahroug, S., V. Courdavault, M. Thiersault, B. St-Pierre, and V. Burlat. 2006. Epidermis is a pivotal site of at least four secondary metabolic pathways in *Catharanthus roseus* aerial organs. *Planta* 223: 1191–1200.

Maréchal, E. and M.-F. Cesbron-Delauw. 2001. The apicoplast: A new member of the plastid family. *Trends Plant Sci.* 6: 200–205.

Martin, V.J., D.J. Pitera, S.T. Withers, J.D. Newman, and J.D. Keasling. 2003. Engineering a mevalonate pathway in *Escherichia coli* for production of terpenoids. *Nat. Biotechnol.* 21: 796–802.

Mercke, P., M. Bengtsson, H.J. Bouwmeester, M.A. Posthumus, and P.E. Brodelius. 2000. Molecular cloning, expression, and characterization of amorpha-4,11-diene synthase, a key enzyme of artemisinin biosynthesis in *Artemisia annua* L. *Arch. Biochem. Biophys.* 381: 173–180.

Miettinen, K., L. Dong, N. Navrot et al. 2014. The seco-iridoid pathway from *Catharanthus roseus*. *Nat. Commun.* 5: 3606.

Mikkelsen, M.D., L.D. Buron, B. Salomonsen et al. 2012. Microbial production of indolylglucosinolate through engineering of a multi-gene pathway in a versatile yeast expression platform. *Metab. Eng.* 14: 104–111.

Mikkelsen, M.D., P. Naur, and B.A. Halkier. 2004. *Arabidopsis* mutants in the C–S lyase of glucosinolate biosynthesis establish a critical role for indole-3-acetaldoxime in auxin homeostasis. *Plant J.* 37: 770–777.

Mikkelsen, M.D., C.E. Olsen, and B.A. Halkier. 2000. Production of the cancer-preventive glucoraphanin in tobacco. *Molecular Plant* 3: 751–759.

Minami, H., E. Dubouzet, K. Iwasa, and F. Sato. 2007. Functional analysis of norcoclaurine synthase in *Coptis japonica*. *J. Biol. Chem.* 282: 6274–6282.

Minami, H., J.S. Kim, N. Ikezawa et al. 2008. Microbial production of plant benzylisoquinoline alkaloids. *Proc. Natl. Acad. Sci. USA* 105: 7393–7398.

Murata, J. and V. De Luca. 2005. Localization of tabersonine 16-hydroxylase and 16-OH tabersonine-16-O-methyltransferase to leaf epidermal cells defines them as a major site of precursor biosynthesis in the vindoline pathway in *Catharanthus roseus*. *Plant J.* 44: 581–594.

Murata, J., J. Roepke, H. Gordon, and V. De Luca. 2008. The leaf epidermome of *Catharanthus roseus* reveals its biochemical specialization. *Plant Cell* 20: 524–542.

Naur, P., B.L. Petersen, M.D. Mikkelsen et al. 2003. CYP83A1 and CYP83B1, two nonredundant cytochrome p450 enzymes metabolizing oximes in the biosynthesis of glucosinolates in arabidopsis. *Plant Physiol.* 133: 63–72.

Newman, J.D., J. Marshall, M. Chang et al. 2006. High-level production of amorpha-4,11-diene in a two-phase partitioning bioreactor of metabolically engineered *Escherichia coli*. *Biotechnol. Bioeng.* 95: 684–691.

Nour-Eldin, H.H., T.G. Andersen, M. Burow et al. 2012. NRT/PTR transporters are essential for translocation of glucosinolate defence compounds to seeds. *Nature* 488: 531–534.

Nutzmann, H.W. and A. Osbourn. 2014. Gene clustering in plant specialized metabolism. *Curr. Opin. Biotechnol.* 26: 91–99.

O'Connor, S.E. and J.J. Maresh. 2006. Chemistry and biology of monoterpene indole alkaloid biosynthesis. *Nat. Prod. Rep.* 23: 532–547.

Oudin, A., S. Mahroug, V. Courdavault et al. 2007. Spatial distribution and hormonal regulation of gene products from methyl erythritol phosphate and monoterpene-secoiridoid pathways in *Catharanthus roseus*. *Plant Mol. Biol.* 65: 13–30.

Paddon, C.J. and J.D. Keasling. 2014. Semi-synthetic artemisinin: A model for the use of synthetic biology in pharmaceutical development. *Nat. Rev. Microbiol.* 12: 355–367.

Paddon, C.J., P.J. Westfall, D.J. Pitera et al. 2013. High-level semi-synthetic production of the potent antimalarial artemisinin. *Nature* 496: 528–532.

Pateraki, I., A.M. Heskes, and B. Hamberger. 2015. Cytochromes P450 for terpene functionalisation and metabolic engineering. *Adv. Biochem. Eng. Biotechnol.* 148: 107–139.

Pawar, R.S., A.J. Krynitsky, and J.I. Rader. 2013. Sweeteners from plants—With emphasis on *Stevia rebaudiana* (Bertoni) and *Siraitia grosvenorii* (Swingle). *Anal. Bioanal. Chem.* 405: 4397–4407.

Pfalz, M., M.D. Mikkelsen, P. Bednarek, C.E. Olsen, B.A. Halkier, and J. Kroymann. 2011. Metabolic engineering in *Nicotiana benthamiana* reveals key enzyme functions in *Arabidopsis* indole glucosinolate modification. *Plant Cell* 23: 716–729.

Pfleger, B.F., D.J. Pitera, C.D. Smolke, and J.D. Keasling. 2006. Combinatorial engineering of intergenic regions in operons tunes expression of multiple genes. *Nat. Biotechnol.* 24: 1027–1032.
Philippe, R.N., M. De Mey, J. Anderson, and P.K. Ajikumar. 2014. Biotechnological production of natural zero-calorie sweeteners. *Curr. Opin. Biotechnol.* 26: 155–161.
Piotrowski, M., A. Schemenewitz, A. Lopukhina et al. 2004. Desulfoglucosinolate sulfotransferases from *Arabidopsis thaliana* catalyze the final step in the biosynthesis of the glucosinolate core structure. *J. Biol. Chem.* 279: 50717–50725.
Pitera, D.J., C.J. Paddon, J.D. Newman, and J.D. Keasling. 2007. Balancing a heterologous mevalonate pathway for improved isoprenoid production in *Escherichia coli*. *Metab. Eng.* 9: 193–207.
Qi, X., S. Bakht, M. Leggett, C. Maxwell, R. Melton, and A. Osbourn. 2004. A gene cluster for secondary metabolism in oat: Implications for the evolution of metabolic diversity in plants. *Proc. Natl. Acad. Sci. USA* 101: 8233–8238.
Qu, Y., M.L. Easson, J. Froese, R. Simionescu, T. Hudlicky, and V. De Luca. 2015. Completion of the seven-step pathway from tabersonine to the anticancer drug precursor vindoline and its assembly in yeast. *Proc. Natl. Acad. Sci. USA* 112: 6224–6229.
Quynh Doan, N.T. and S.B. Christensen. 2015. Thapsigargin, origin, chemistry, structure-activity relationships and prodrug development. *Curr. Pharm. Des.* 21: 5501–5517.
Ralph, S.A., M.C. D'Ombrain, and G.I. McFadden. 2001. The apicoplast as an antimalarial drug target. *Drug. Resist. Update.* 4: 145–151.
Ro, D.-K., G.-I. Arimura, S.Y.W. Lau, E. Piers, and J. Bohlmann. 2005. Loblolly pine abietadienol/abietadienal oxidase *PtAO* (CYP720B1) is a multifunctional, multisubstrate cytochrome P450 monooxygenase. *Proc. Natl. Acad. Sci. USA* 102: 8060–8065.
Ro, D.K., M. Ouellet, E.M. Paradise et al. 2008. Induction of multiple pleiotropic drug resistance genes in yeast engineered to produce an increased level of anti-malarial drug precursor, artemisinic acid. *BMC Biotechnol.* 8: 83.
Ro, D.K., E.M. Paradise, M. Ouellet et al. 2006. Production of the antimalarial drug precursor artemisinic acid in engineered yeast. *Nature* 440: 940–943.
Rohmer, M., M. Knani, P. Simonin, B. Sutter, and H. Sahm. 1993. Isoprenoid biosynthesis in bacteria: A novel pathway for the early steps leading to isopentenyl diphosphate. *Biochem. J.* 295 (Pt 2): 517–524.
Rontein, D., S. Onillon, G. Herbette et al. 2008. CYP725A4 from yew catalyzes complex structural rearrangement of taxa-4(5),11(12)-diene into the cyclic ether 5(12)-oxa-3(11)-cyclotaxane. *J. Biol. Chem.* 283: 6067–6075.
Ruppert, M., X. Ma, and J. Stockigt. 2005. Alkaloid biosynthesis in rauvolfia-cDNA cloning of major enzymes of the ajmaline pathway. *Curr. Org. Chem.* 9: 1431–1444.
Saito, K. and F. Matsuda. 2010. Metabolomics for functional genomics, systems biology, and biotechnology. *Annu. Rev. Plant Biol.* 61: 463–489.
Salim, V., B. Wiens, S. Masada-Atsumi, F. Yu, and V. De Luca. 2014. 7-Deoxyloganetic acid synthase catalyzes a key 3 step oxidation to form 7-deoxyloganetic acid in *Catharanthus roseus* iridoid biosynthesis. *Phytochemistry* 101: 23–31.
Salim, V., F. Yu, J. Altarejos, and V. De Luca. 2013. Virus-induced gene silencing identifies *Catharanthus roseus* 7-deoxyloganic acid-7-hydroxylase, a step in iridoid and monoterpene indole alkaloid biosynthesis. *Plant J.* 76: 754–765.

Salmon, M., C. Laurendon, M. Vardakou et al. 2015. Emergence of terpene cyclization in *Artemisia annua*. *Nat. Commun.* 6: 6143.
Samanani, N., D.K. Liscombe, and P.J. Facchini. 2004. Molecular cloning and characterization of norcoclaurine synthase, an enzyme catalyzing the first committed step in benzylisoquinoline alkaloid biosynthesis. *Plant J.* 40: 302–313.
Shang, Y., Y. Ma, Y. Zhou et al. 2014. Biosynthesis, regulation, and domestication of bitterness in cucumber. *Science* 346: 1084–1088.
Shimura, K., A. Okada, K. Okada et al. 2007. Identification of a biosynthetic gene cluster in rice for momilactones. *J. Biol. Chem.* 282: 34013–34018.
Starks, C.M., K. Back, J. Chappell, and J.P. Noel. 1997. Structural basis for cyclic terpene biosynthesis by tobacco 5-epi-aristolochene synthase. *Science* 277: 1815–1820.
Stavrinides, A., E.C. Tatsis, E. Foureau et al. 2015. Unlocking the diversity of alkaloids in *Catharanthus roseus*: Nuclear localization suggests metabolic channeling in secondary metabolism. *Chem. Biol.* 22: 336–341.
St-Pierre, B., P. Laflamme, A.M. Alarco, and V. De Luca. 1998. The terminal O-acetyltransferase involved in vindoline biosynthesis defines a new class of proteins responsible for coenzyme A-dependent acyl transfer. *Plant J.* 14: 703–713.
St-Pierre, B., F.A. Vazquez-Flota, and V. De Luca. 1999. Multicellular compartmentation of *Catharanthus roseus* alkaloid biosynthesis predicts intercellular translocation of a pathway intermediate. *Plant Cell* 11: 887–900.
Suffness, M. and M.E. Wall. 1995. *Taxol: Science and Applications*. Boca Raton, FL: CRC Press.
Tabata, K. and S. Hashimoto. 2004. Production of mevalonate by a metabolically-engineered *Escherichia coli*. *Biotechnol. Lett.* 26: 1487–1491.
Takos, A.M., C. Knudsen, D. Lai et al. 2011. Genomic clustering of cyanogenic glucoside biosynthetic genes aids their identification in *Lotus japonicus* and suggests the repeated evolution of this chemical defence pathway. *Plant J.* 68: 273–286.
Teoh, K.H., D.R. Polichuk, D.W. Reed, and P.S. Covello. 2009. Molecular cloning of an aldehyde dehydrogenase implicated in artemisinin biosynthesis in *Artemisia annua*. *Botany* 87: 635–642.
Teoh, K.H., D.R. Polichuk, D.W. Reed, G. Nowak, and P.S. Covello. 2006. *Artemisia annua* L. (Asteraceae) trichome-specific cDNAs reveal CYP71AV1, a cytochrome P450 with a key role in the biosynthesis of the antimalarial sesquiterpene lactone artemisinin. *FEBS Lett.* 580: 1411–1416.
Tholl, D. 2006. Terpene synthases and the regulation, diversity and biological roles of terpene metabolism. *Curr. Opin. Plant Biol.* 9: 297–304.
Traka, M. and R. Mithen. 2009. Glucosinolates, isothiocyanates and human health. *Phytochem. Rev.* 8: 269–282.
Tsuruta, H., C.J. Paddon, D. Eng et al. 2009. High-level production of amorpha-4,11-diene, a precursor of the antimalarial agent artemisinin, in *Escherichia coli*. *PloS One* 4: e4489.
Tu, Y. 2011. The discovery of artemisinin (qinghaosu) and gifts from Chinese medicine. *Nat. Med.* 17: 1217–1220.
Turconi, J., F. Griolet, R. Guevel et al. 2014. Semisynthetic artemisinin, the chemical path to industrial production. *Org. Process Res. Dev.* 18: 417–422.
Vazquez-Flota, F., E. De Carolis, A.M. Alarco, and V. De Luca. 1997. Molecular cloning and characterization of desacetoxyvindoline-4-hydroxylase, a 2-oxoglutarate dependent-dioxygenase involved in the biosynthesis of vindoline in *Catharanthus roseus* (L.) G. Don. *Plant Mol. Biol.* 34: 935–948.

Verkerk, R., M. Schreiner, A. Krumbein et al. 2009. Glucosinolates in *Brassica* vegetables: The influence of the food supply chain on intake, bioavailability and human health. *Mol. Nutr. Food Res.* 53: S219–S219.

Voitsekhovskaja, O.V., O.A. Koroleva, D.R. Batashev et al. 2006. Phloem loading in two scrophulariaceae species. What can drive symplastic flow via plasmodesmata? *Plant Physiol.* 140: 383–395.

Walker, K. and R. Croteau. 2000a. Molecular cloning of a 10-deacetylbaccatin III-10-*O*-acetyl transferase cDNA from *Taxus* and functional expression in *Escherichia coli*. *Proc. Natl. Acad. Sci. USA* 97: 583–587.

Walker, K. and R. Croteau. 2000b. Taxol biosynthesis: Molecular cloning of a benzoyl-CoA:taxane 2α-*O*-benzoyltransferase cDNA from *Taxus* and functional expression in *Escherichia coli*. *Proc. Natl. Acad. Sci. USA* 97: 13591–13596.

Walker, K. and R. Croteau. 2001. Taxol biosynthetic genes. *Phytochemistry* 58: 1–7.

Walker, K., R. Long, and R. Croteau. 2002. The final acylation step in taxol biosynthesis: Cloning of the taxoid C13-side-chain *N*-benzoyltransferase from *Taxus*. *Proc. Natl. Acad. Sci. USA* 99: 9166–9171.

Wang, M., D.L. Roberts, R. Paschke, T.M. Shea, B.S.S. Masters, and J.-J.P. Kim. 1997. Three-dimensional structure of NADPH–cytochrome P450 reductase: Prototype for FMN- and FAD-containing enzymes. *Proc. Natl. Acad. Sci. USA* 94: 8411–8416.

Westfall, P.J., D.J. Pitera, J.R. Lenihan et al. 2012. Production of amorphadiene in yeast, and its conversion to dihydroartemisinic acid, precursor to the antimalarial agent artemisinin. *Proc. Natl. Acad. Sci. USA* 109: E111–118.

White, N.J. 2008. Qinghaosu (Artemisinin): The price of success. *Science* 320: 330–334.

Wilderman, P.R., M. Xu, Y. Jin, R.M. Coates, and R.J. Peters. 2004. Identification of *syn*-pimara-7,15-diene synthase reveals functional clustering of terpene synthases involved in rice phytoalexin/allelochemical biosynthesis. *Plant Physiol.* 135: 2098–2105.

Wildung, M.R. and R. Croteau. 1996. A cDNA clone for taxadiene synthase, the diterpene cyclase that catalyzes the committed step of taxol biosynthesis. *J. Biol. Chem.* 271: 9201–9204.

Winzer, T., V. Gazda, Z. He et al. 2012. A *Papaver somniferum* 10-gene cluster for synthesis of the anticancer alkaloid noscapine. *Science* 336: 1704–1708.

Winzer, T., M. Kern, A.J. King et al. 2015. Morphinan biosynthesis in opium poppy requires a P450-oxidoreductase fusion protein. *Science* 349: 309–312.

Yilmaz, Y. and R.T. Toledo. 2004. Major flavonoids in grape seeds and skins: Antioxidant capacity of catechin, epicatechin, and gallic acid. *J. Agric. Food Chem.* 52: 255–260.

Yoshikuni, Y., T.E. Ferrin, and J.D. Keasling. 2006. Designed divergent evolution of enzyme function. *Nature* 440: 1078–1082.

Yu, F. and V. De Luca. 2013. ATP-binding cassette transporter controls leaf surface secretion of anticancer drug components in *Catharanthus roseus*. *Proc. Natl. Acad. Sci. USA* 110: 15830–15835.

Zagrobelny, M. and B.L. Moller. 2011. Cyanogenic glucosides in the biological warfare between plants and insects: The Burnet moth-Birdsfoot trefoil model system. *Phytochemistry* 72: 1585–1592.

Zangar, R.C., D.R. Davydov, and S. Verma. 2004. Mechanisms that regulate production of reactive oxygen species by cytochrome P450. *Toxicol. Appl. Pharmacol.* 199: 316–331.

Zerbe, P. and J. Bohlmann. 2015. Plant diterpene synthases: Exploring modularity and metabolic diversity for bioengineering. *Trends Biotechnol.* 33: 419–428.

Zhang, Y., K.H. Teoh, D.W. Reed et al. 2008. The molecular cloning of artemisinic aldehyde Δ11(13) reductase and its role in glandular trichome-dependent biosynthesis of artemisinin in *Artemisia annua*. *J. Biol. Chem.* 283: 21501–21508.

Zhao, J. and R. Verpoorte. 2007. Manipulating indole alkaloid production by *Catharanthus roseus* cell cultures in bioreactors: From biochemical processing to metabolic engineering. *Phytochem. Rev.* 6: 435–457.

Ziegler, J., P.J. Facchini, R. Geissler et al. 2009. Evolution of morphine biosynthesis in opium poppy. *Phytochemistry* 70: 1696–1707.

Index

A

AAAT, see Aromatic amino acid aminotransferase
AADC, see Aromatic amino acid decarboxylase
AAE, see Acetylajmalan esterase; Acetylajmaline acetylesterase; Acyl-activating enzyme
AAS, see Aromatic aldehyde synthase
AAT, see Alcohol acetyltransferase
ABC, see ATP-binding cassette
Abietane 12-hydroxy-11,14-diketo-6,8,12-abietatrien-19,20-olide (HABTO), 38
Abiotic stresses, 283
ACC, see 1-Aminocyclopropane-1-carboxylic acid
Acetic acid, 246; see also Microbial volatiles (mVOCs)
7α-Acetoxyroyleanone, 38, 43
Acetylajmalan esterase (AAE), 97
Acetylajmaline acetylesterase (AAE), 334
Acide salicylic, 75; see also Salicaceae salicylates
Acyl-activating enzyme (AAE), 204
ADHs, see Alcohol dehydrogenases
ADS, see Amorpha-4,11-diene synthase; Amorphadiene synthase
AHCT, see Anthocyanin O-hydroxylcinnamoyltransferase
AIMT, see Anise t-anol/isoeugenol O-methyltransferase
Ainslioside, 38, 39
Air pollutants, 217
Ajmalicine, 44, 49
Ajugasterone, 40, 43
AKR, see Aldo–keto reductase
Akuammiline, 334; see also Alkaloid

Alcohol acetyltransferase (AAT), 204
Alcohol dehydrogenases (ADHs), 197
Aldo–keto reductase (AKR), 89
Alkaloid, 42, 44–46; see also Benzylisoquinoline alkaloid (BIA); Chemotaxonomy; Monoterpenoid indole alkaloid (MIA); Nicotine; Pinaceae alkaloids; Plant secondary metabolism
    biosynthesis and regulation in plants, 85
    chemical synthesis of complex, 107
    metabolism, 332
    natural products, 333
    perspectives, 106–108
    purine, 104–106
Allene oxide synthase (AOS), 194, 196
Aloesaponarin II, 32, 33
2-Aminoacetophenone, 251; see also Microbial volatiles (mVOCs)
1-Aminocyclopropane-1-carboxylic acid (ACC), 136
AMO, see Amorphadiene oxidase
Amorpha-4,11-diene synthase (ADS), 192
Amorphadiene oxidase (AMO), 326
Amorphadiene synthase (ADS), 326
β-Amyrenyl acetate, 41, 43
α-Amyrin, 40, 43
Anagyrine, 46, 49
Anise t-anol/isoeugenol O-methyltransferase (AIMT), 206
Anthocyanidins, 29; see also Chemotaxonomy
    Asteraceae, 28
    Lamiaceae, 30
    Leguminosae, 31
Anthocyanin O-hydroxylcinnamoyltransferase (AHCT), 205

Antifungal compounds, 141
Antioxidant response element (ARE), 172
AOS, see Allene oxide synthase
Apicoplast, 326
Apigenin, 30, 33
Aporphine, 333; see also Alkaloid
Arabidopsis transporters, 169
ARE, see Antioxidant response element
Aroma compounds, 245; see also Microbial volatiles (mVOCs)
Aromadendrene, 37, 39
Aromatic aldehyde synthase (AAS), 204
Aromatic amino acid aminotransferase (AAAT), 204
Aromatic amino acid decarboxylase (AADC), 203
Artemisia alcohol, 35, 36
Artemisinin, 316, 318, 326; see also Plant secondary metabolism; Terpenoid metabolism
  amorphadiene production in E. coli, 327–328
  artemisinic acid production in yeast, 328
  biosynthesis, 326
  biosynthetic pathway, 327
  development of industrial yeast strain, 328–329
Asteraceae, 28
ATP-binding cassette (ABC), 92
Azetidine-2-carboxylic acid, 47, 49

B

Bakuchiol, 35, 36
BALDH, see Benzaldehyde dehydrogenase
BAMT, see Benzoic acid methyltransferase
BAS, see Benzalacetone synthase
BASS5, see Bile acid:sodium symporter family protein 5
BAT5, see Bile acid transporter 5
BBE, see Berberine bridge enzyme
BBL, see Berberine bridge enzyme-like protein
BCAT, see Branched-chain amino acid aminotransferase
BEAT, see Benzylalcohol acetyltransferase
BEBT, see Benzyl alcohol benzoyl transferase
Benzalacetone synthase (BAS), 205
Benzaldehyde dehydrogenase (BALDH), 204
Benzoic acid methyltransferase (BAMT), 204

Benzoic acid/salicylic acid methyltransferase (BSMT), 204
Benzophenanthridine, 333; see also Alkaloid
Benzylalcohol acetyltransferase (BEAT), 204
Benzyl alcohol benzoyl transferase (BEBT), 204
Benzyl alcohol/phenylethanol benzoyl transferase (BPBT), 204
Benzylisoquinoline alkaloid (BIA), 86, 333, 340; see also Alkaloid
  biosynthesis of, 341–342
  biosynthetic pathways, 87–88
  canadine, 90
  enzymes for morphine biosynthesis, 342
  epimerase, 342, 344
  microbial production of, 342–345
  morphine and codeine biosynthetic pathway, 343
  pathways and enzymes, 86–91
  (S)-reticuline biosynthetic pathway from L-tyrosine, 341
  sanguinarine biosynthesis, 90, 345
  (S)-scoulerine, 90
  trafficking in opium poppies and C. japonica, 91–92
  transcriptional regulation, 92
Berberine bridge enzyme (BBE), 90, 342
Berberine bridge enzyme-like protein (BBL), 102
cis-α-Bergamotene, 37, 39
β-Cyanoalanine synthase (CAS), 135
β-glucosidase (BGD), 134
BGD, see β-glucosidase
bHLH iridoid synthesis 1 (BIS1), 100
BIA, see Benzylisoquinoline alkaloid
Bicyclogermacrene, 37, 39
Bile acid:sodium symporter family protein 5 (BASS5), 160
Bile acid transporter 5 (BAT5), 160
Bioactive natural products, 132
Biodiversity, 24; see also Chemotaxonomy
  distribution of, 25
  of medicinal plants, 25–26
Biological diversity, see Biodiversity
BIS1, see bHLH iridoid synthesis 1
α-Bisabolol oxide B, 37, 39
Bisbenzylisoquinoline, 333; see also Alkaloid
Black poplar, 142; see also Cyanogenic glucosides (CNglcs)
Blue haze, 186; see also Plant volatiles (PVOCs)

Borneol, 34, 35, 36
Bornyl acetate, 34, 36
BPBT, see Benzyl alcohol/phenylethanol benzoyl transferase
Branched-chain amino acid aminotransferase (BCAT), 160
Brassicaceae, 157–158; see also Glucosinolate
  aliphatic glucosinolates in Brassicaceae plants, 171
  broccoli, 175
  genomic relationships among six cultivated *Brassica* species, 170
  radish, 175–176
  rapeseed, 174–175
Broccoli, 175; see also Brassicaceae
BSMT, see Benzoic acid/salicylic acid methyltransferase

C

C5 isoprene, 319
C6-aldehydes, 197
δ-Cadinene, 36, 39
Caesalfurfuric acid A, 40, 43
Caffeic acid, 29, 33
Caffeine synthase (CS), 106
Camphor, 34, 35, 36
Camptotheca acuminata, 19
Canadine, 90; see also Benzylisoquinoline alkaloid (BIA)
Canadine synthase (CAS), 90
Canavanine, 47, 49
Cannabis sativa, 16
Carnivorous plants, 12
β-carotene, 41, 43
Carvacrol, 35, 36
(E)-β-caryophyllene, 36, 39
Caryophyllene oxide, 36, 39
CAS, see Canadine synthase; β-Cyanoalanine synthase
Catharanthine, 97, 334; see also Alkaloid; Monoterpenoid indole alkaloid (MIA)
Catharanthus roseus, 92
CCN, see Cloud condensation nuclei
CFAT, see Coniferyl alcohol acetyltransferase
CFS, see Cheilanthifoline synthase
Chalcone synthases (CHSs), 205
Chavicol O-methyltransfease (CVOMT), 206
CHD, see Cinnamoyl-CoA hydratase-dehydrogenase
Cheilanthifoline synthase (CFS), 90

Chemical spoilage indices (CSIs), 246
Chemotaxonomy, 26
  alkaloids, 42, 44–46
  Asteraceae, 28–29
  cyanogenic glycosides, 46–47
  fatty acids, 47–48
  flavonoids, 28
  glucosinolates, 46
  Lamiaceae, 29–30
  Leguminosae, 30–32
  nonprotein amino acids, 47
  of phenolic compounds, 27, 33
  plant families, 32
  of specialized compounds, 26
  of specialized products containing nitrogen, 42
  surface alkanes, 50
  of terpenoids, 33
Chorismate mutase (CM), 207
Chorismate synthase (CS), 207
Chrysanthenone, 35, 36
Chrysoeriol, 30, 33
CHSs, see Chalcone synthases
1,8-Cineole, 34, 35, 36
trans-Cinnamic acid, 28, 33
Cinnamoyl-CoA hydratase-dehydrogenase (CHD), 204
Cinnamoyl CoA ligase (CNL), 204
Cirsimaritin, 30, 33
Cislinalool pyran oxide, 35, 36
Climate warming, 217
Cloud condensation nuclei (CCN), 212, 221
Clustered regularly interspaced short palindromic repeats (CRISPR), 318
CM, see Chorismate mutase
CNdglc, see Cyanogenic diglycoside
CNglcs, see Cyanogenic glucosides
CNL, see Cinnamoyl CoA ligase
CNmglc, see Cyanogenic monoglucoside
CNMT, see Coclaurine N-methyltransferase
Coclaurine N-methyltransferase (CNMT), 89, 344
Codeine O-demethylase (CODM), 89, 342
Codeinone reductase (COR), 89
CODM, see Codeine O-demethylase
Coevolutionary theory, 131–132
COI1, see Coronatine insensitive1
Condylocarpine, 334; see also Alkaloid
Coniferyl alcohol acetyltransferase (CFAT), 205
Coniine, 120; see also Pinaceae alkaloids
4-α-Copaenol, 37, 39
COR, see Codeinone reductase

Cordylasins A and B, 31, 33
Coronatine insensitive1 (COI1), 168
Cosmopolitan moss (*Bryum argenteum*), 186
Costunolide, 38
3-*trans*-*p*-Coumaroylglucoside-5-
    malonylglucosides, 30, 33
3-*p*-Coumarylglucoside-5-
    malonylglucoside, 30, 33
CPR, *see* Cytochrome P450 reductase
Crepidiaside B, 38, 39
CRISPR, *see* Clustered regularly interspaced
    short palindromic repeats
CrMYC2, *see* C. *rosues* MYC2
C. rosues MYC2 (CrMYC2), 99
Cryptotanshinone, 39, 43
CS, *see* Caffeine synthase; Chorismate
    synthase
CSIs, *see* Chemical spoilage indices
Cularine, 333; *see also* Alkaloid
CVOMT, *see* Chavicol *O*-methyltransfease
Cyanidin, 29, 33
    derivatives, 31, 33
Cyanogenesis, 46
Cyanogenic diglycoside (CNdglc), 138
Cyanogenic glucosides (CNglcs), 46–47, 131;
        see also Chemotaxonomy
    in *Arabidopsis*, 142
    bioactivation of, 133
    biosynthesis of, 133, 135, 139–140
    biosynthetic pathway for tyrosine-
        derived, 138
    control of cyanogenesis, 136
    crop plants and fruit trees having, 132
    dynamic defense systems, 140–143
    endogenous turnover of, 135
    effectiveness in herbivore defense, 137
    genomic clusters on barley chromosome
        1H, 141
    metabolon, 139
    multifunctional, 140
    phenylalanine-derived cyanogenic
        mono- and diglucosides, 134
    repeated evolution, 140
    resource allocation, 145
    structural diversity of, 137–139
    structure and metabolism of, 137
    toxic effects of, 133
    turnover and metabolic pathways
        involved, 143–145
Cyanogenic monoglucoside (CNmglc), 138
Cyanogenic species, 140
Cyclosativene, 37, 39
p-Cymene, 34, 35, 36

CYP, *see* Cytochrome P450 monooxygenase
CYP74B, 196
CYP74 enzymes, 196; *see also* Green leaf
    volatiles (GLVs)
Cytisine, 46, 49
Cytochrome P450 monooxygenase (CYP;
    P450), 190, 259, 322; *see also*
    Terpenoid metabolism
Cytochrome P450 reductase (CPR), 323

D

D4H, *see* Desacetoxyvindoline
    4-hydroxylase
DAHPS, *see* 3-Deoxy-D-arabino-
    heptulosonate-7-phosphate
    synthase
DAT, *see* Deacetylvindoline
    4-*O*-acetyltransferase
DBOX, *see* Dihydrobenzophenanthridine
    oxidase
Deacetylvinodoline 4-*O*-acetyltransferase
    (DAT), 96, 205
Deciduous trees, 119
3,4-Dehydroneomajucin, 38, 39
1,2-Dehydroreticuline reductase (DRR), 342
1,2-Dehydroreticuline synthase (DRS), 342
Delphinidin, 29, 33
3-Deoxy-D-arabino-heptulosonate-7-
    phosphate synthase (DAHPS), 207
7-Deoxyloganetic acid glucosyltransferase
    (7DLGT), 95
7-Deoxyloganetic acid synthase (7DLS), 340
7-Deoxyloganic acid 7-hydroxylase
    (7DLH), 95
Deoxy xylulose phosphate (DXP), 325
Desacetoxyvindoline 4-hydroxylase
    (D4H), 96
Detractant compounds, 296
3,4-DHPAA, *see* 3,4-Dihydrophenyl
    acetaldehyde
DHR, *see* Dhurrinase
Dhurrinase (DHR), 136
DIBOA, *see* 2,4-Dihydroxy-1,4-benzoxazin-
    3-one
Dietary crops, 262
Dihydrobenzophenanthridine oxidase
    (DBOX), 90
3,4-Dihydrophenyl acetaldehyde (3,4-
    DHPAA), 343
2,4-Dihydroxy-1,4-benzoxazin-3-one
    (DIBOA), 349
Dihydroxyphenylalanine (DOPA), 42

Dimethylallyl diphosphate (DMAPP), 95, 319
Dimethylbenz(a) anthracene (DMBA), 172
(E)-4,8-dimethyl-1,3,7-nonatriene (DMNT), 190, 297
Diterpene, 38–40, 43; see also Terpenoids
Diterpene carnosol, 38, 43
Diterpene manool, 38, 43
Diterpene synthase (di-TPS), 189
di-TPS, see Diterpene synthase
Djenkolic acid, 47, 49
DMAPP, see Dimethylallyl diphosphate
DMBA, see Dimethylbenz(a) anthracene
DMNT, see (E)-4,8-dimethyl-1,3,7-nonatriene
Dolichol, 319
DOPA, see Dihydroxyphenylalanine
Drought periods, 214
DRR, see 1,2-Dehydroreticuline reductase
DRS, see 1,2-Dehydroreticuline synthase
DXP, see Deoxy xylulose phosphate

E

Ecdysone, 40, 43
Echinalide H, 40, 43
Effectors, 194
Eglandine, 44, 49
EGSs, see Eugenol synthases
Embryophytes, 131
Endoplasmic reticulum (ER), 139, 189, 316
5-Enol-pyruvyl-shikimate-3-phosphate (EPSPS), 207
Enzymes for morphine biosynthesis, 342
EOMT, see Eugenol O-methyltransferase
Epidihydropinidine, 120; see also Pinaceae alkaloids
Epimerase, 342, 344
Epi-vellosimine, 334; see also Alkaloid
EPSPS, see 5-Enol-pyruvyl-shikimate-3-phosphate
ER, see Endoplasmic reticulum
Erythrivarine A, 44, 49
α-Erythroidine, 44, 49
Erythroidine alkaloids, 44
Ethylene signaling, 194
Eudesmanolide tanacetin, 38, 39
Eugenol, 35, 36
Eugenol O-methyltransferase (EOMT), 206
Eugenol synthases (EGSs), 205
Euphococcinine, 121; see also Pinaceae alkaloids
Evergreen coniferous trees, 119
Excretion, 10

F

FAC, see Fatty acid-amino acid conjugate
FAD, see Flavin adenine dinucleotide
(E)-β-Farnesene, 36, 37, 39
E-E-α-Farnesene, 36, 39
Farnesyl diphosphate (FPP), 321
Far-red (FR), 216
Fatty acid-amino acid conjugate (FAC), 193
Fatty acids, 47–48; see also Chemotaxonomy
Fenchone, 34, 36
Flavin adenine dinucleotide (FAD), 90
Flavonoids, 28; see also Chemotaxonomy
    exudate, 30
    isoflavonoids, 31
    Leguminosae, 30–32
    in plant families, 32
Flavor, 243
Flooding, 214–215
Floraassamsaponin I, 41, 43
Floral VOC emission, 297–299; see also VOC-mediated information transfer
Floral volatiles; see also Plant volatiles (PVOCs)
4'-O-methyltransferase (4'OMT), 89
4'OMT, see 4'-O-methyltransferase
FPP, see Farnesyl diphosphate
FR, see Far-red

G

G10H, see Geraniol 10-hydroxylase
G3P, see Glyceraldehyde 3-phosphate
G8H, see Geraniol 8-hydroxylase
6'-O-galloyl sambunigrin, 46, 49
γ-glutamyl peptidase 1 (GGP1), 348
GDP, see Geranyl diphosphate
Genistein, 32, 33
Geraniol, 95; see also Monoterpenoid indole alkaloid (MIA)
Geraniol 10-hydroxylase (G10H), 95
Geraniol 8-hydroxylase (G8H), 340
Geraniol synthase (GES), 95
Geranyl diphosphate (GDP; GPP), 95, 189, 321
Geranyl diphosphate synthase (GPPS), 95
Geranylgeranyl diphosphate (GGDP), 189
Germacrene D, 36, 39
GES, see Geraniol synthase
GFP-tagged Bacillus subtilis strain, 250
GGDP, see Geranylgeranyl diphosphate
GGP1, see γ-Glutamyl peptidase 1
GGT, see Glycoside-specific glycosyltransfearse

Glandular trichomes, 12; *see also* Metabolite synthesis site
  in Asteraceae family, 14
  in *Cannabis sativa*, 16
  common form of, 13
  leaves of hops, 16
  in *Mentha piperita*, 14
  pubescence, 13
  secretory tissues in plants, 15
  stem cells, 13
  in tobacco, 16
  VOCs, 12
Glucoalyssin, 46, 49
Glucoberteroin, 46, 49
Glucobrassicanapin, 46, 49
Glucobrassicin, 46, 49, 318
Gluconapin, 46, 49
Gluconasturtiin, 46, 49
β-Glucopyranosyl ester, 38, 39
Glucosinolate (GLS), 46, 157, 176, 345; *see also* Brassicaceae; Chemotaxonomy
  aliphatic and indolic glucosinolate biosynthetic pathways, 161
  aliphatic glucosinolates in Brassicaceae plants, 171
  anticarcinogenic activity, 172
  biosynthesis pathways, 160
  biosynthesis related genes, 163–165
  breeding for glucosinolate profile, 174–176
  chemical name and abbreviation, 162
  chemical structure, 158–159
  conservation of biosynthesis pathway, 166
  core structure formation, 162–165
  degradation products of, 159–160
  gene regulation, 167
  glucosinolate–myrosinase system, 159
  groups in, 346
  histone modification, 173
  for human diet, 172
  hydrolysis and degradation products, 159
  induction of phase 2 detoxification enzymes, 172
  for insects, 173–174
  isothiocyanate, 172
  MYBs, 167–168
  MYCs, 168
  nitrile, 173
  S-cells, 160
  secondary modification of side chain, 165–166
  side-chain elongation, 160–162
  structural classification, 158
  sulforaphane, 172
  transport, 168–169
  U's triangle, 170
  variation in glucosinolate composition, 169
Glucosinolate metabolism, 345; *see also* Plant secondary metabolism
  biosynthetic pathway, 345–348
  glucosinolate structure and biosynthetic pathway, 346
  microbial production of indole glucosinolate, 348–349
  production in plants, 348
Glucotropaeolin, 46, 49
Glutathione (GSH), 201–202; *see also* Green leaf volatiles (GLVs)
Glutathione *S*-transferase (GST), 144, 172, 347
GLVs, *see* Green leaf volatiles
Glyceraldehyde 3-phosphate (G3P), 325
Glycoside-specific glycosyl*t*ransfearse (GGT), 268; *see also* Volatile glycosylation
  exceptional, 273
  functionally characterized, 270–272
  phylogenetic tree of, 269
Glycosylation, 258; *see also* Volatile glycosylation
  -hydrolysis cycle, 275
Glycosyltransferases (GTs), 138
Gossypetin, 31, 33
GPP, *see* Geranyl diphosphate
GPPS, *see* Geranyl diphosphate synthase
Grapes, 263–264
Green leaf odors, *see* Green leaf volatiles
Green leaf volatiles (GLVs), 194; *see also* Plant volatiles (PVOCs)
  biosynthesis, 196–198
  C6-aldehydes, 197
  chemical structures of, 187
  CYP74B, 196
  CYP74 enzymes, 196
  forming pathway in plants, 195
  glutathione, 201–202
  GLV-burst, 198
  (Z)-3-hexen-1-ol, 202
  (Z)-3-hexen-1-yl acetate, 202
  oxylipins, 196
  physiological and ecological significance, 200–202
  regulation of formation, 198–200
  tritrophic system, 201

Index

GSH, see Glutathione
GST, see Glutathione S-transferase
GTs, see Glycosyltransferases
(1β,6α,10α)-guai-4(15)-ene-6,7,10-triol, 37, 39
Guaianolide leucodin, 38, 39

H

H6H, see 6β-Hydroxylase
HABTO, see Abietane 12-hydroxy-11,14-diketo-6,8,12-abietatrien-19,20-olide
Halictidae bee, 298
HCBT, see N-Hydroxycinnamoyl/benzoyltransferase
HCH, see 6-Hydroxy-2-cyclohexen-on-oyl
HCN, see Hydrogen cyanide
HDAC, see Histone deacetylase
Herbivore-derived elicitor, 193
Herbivore-induced plant volatiles (HIPVs), 192–193, 209
Herbivore-induced VOC signal, 302; see also VOC-mediated information transfer
(Z)-3-Hexen-1-ol, 202; see also Green leaf volatiles (GLVs)
(Z)-3-Hexen-1-yl acetate, 202; see also Green leaf volatiles (GLVs)
10HGO, see 10-Hydroxygeraniol oxidoreductase
HIPVs, see Herbivore-induced plant volatiles
Histone deacetylase (HDAC), 173
HIV, see Human immunodeficiency virus
HMGCoA, see 3-Hydroxy-3-methyl-glutaryl-CoA
HMG CoA reductase (HMGR), 323
HMGR, see HMG CoA reductase
HNL, see α-Hydroxynitrile lyase
Holocrine, 10
4-HPAA, see 4-Hydroxyphenylacetaldehyde
HPL, see Hydroperoxide lyase
4-HPP, see 4-Hydroxyphenylpyruvate
Human immunodeficiency virus (HIV), 19
α-Humulene, 36, 39
Hydathodes, 12
Hydrogen cyanide (HCN), 316
Hydroperoxide lyase (HPL), 196
Hydrophilic tissues, 11–12; see also Metabolite synthesis site
N-Hydroxycinnamoyl/benzoyltransferase (HCBT), 205

6-Hydroxy-2-cyclohexen-on-oyl (HCH), 74
12-Hydroxy-11,14-diketo-6,8,12-abietatrien-19,20-olide, 38, 43
10-Hydroxygeraniol oxidoreductase (10HGO), 95
14-Hydroxy-α-humulene, 36, 39
2-Hydroxy-isophytol, 39, 43
6β-Hydroxylase (H6H), 102
12-α-Hydroxylupanine, 46, 49
13-Hydroxylupanine, 46, 49
3-Hydroxy-3-methyl-glutaryl-CoA (HMGCoA), 323
α-Hydroxynitrile lyase (HNL), 134
4-Hydroxyphenylacetaldehyde (4-HPAA), 86
4-Hydroxyphenylpyruvate (4-HPP), 86
16-Hydroxytabersonine O-methyltransferase (16OMT), 96, 339
Hygroline, 126; see also Pinaceae alkaloids
Hyptol, 39, 43

I

Icetexone, 38, 43
IDP, see Isopentenyl diphosphate
IEMT, see (iso) eugenol O-methyltransferase
IFR, see Isoflavone reductase
IGSs, see Isoeugenol synthases
Indole alkaloid, 333; see also Alkaloid
Information-mediated indirect resistance, 300–302; see also VOC-mediated information transfer
Internal phloem associated parenchyma (IPAP), 97, 337
IPAP, see Internal phloem associated parenchyma
IPMDH1, see Isopropylmalate dehydrogenase
IPMI, see Isopropylmalate isomerase
IPP, see Isopentenyl diphosphate
Iridoid oxidase (IRO), 95
Iridoids, 95; see also Monoterpenoid indole alkaloid (MIA)
Iridoid synthase (IRS; IS), 95, 335
IRO, see Iridoid oxidase
IRS, see Iridoid synthase
IS, see Iridoid synthase
(iso) eugenol O-methyltransferase (IEMT), 206
Isoeugenol synthases (IGSs), 205
Isoflavone reductase (IFR), 205
Isoiridomyrmecin, 35, 36
Iso-lyratol, 35, 36

Isopentenyl diphosphate (IDP; IPP), 95, 188, 319, 323; *see also* Terpenoid metabolism
  biosynthesis, 323
  biosynthetic route, 324
  MEP pathway, 325–326
  MVA pathway, 323–325
Isopinocamphone, 34, 36
Isoprene, 319; *see also* Plant volatiles (PVOCs); Terpenoid metabolism
  emissions, 214–215
  emitting trees, 221
Isoprenoids, *see* Terpenes; Terpenoids
Isopropylmalate dehydrogenase (IPMDH1), 160
Isopropylmalate isomerase (IPMI), 160
Isothiocyanate, 172; *see also* Glucosinolate
Isovouacapenol E, 40, 43

J

JA, *see* Jasmonic acid
JA- and elicitor-responsive element (JERE), 99
JA-induced alkaloid transporter (JAT1), 103
Japanese daikon, *see* Radish
JAs, *see* Jasmonates
Jasmonates (JAs), 168
Jasmonate zim domain (JAZ), 168
Jasmonic acid (JA), 194; *see also* Terpenes
JAT1, *see* JA-induced alkaloid transporter
JAZ, *see* Jasmonate zim domain
JERE, *see* JA- and elicitor-responsive element
Jussiaeiine A, 46, 49

K

Kaempferol, 31, 33
KAT, *see* 3-Ketoacyl-CoA thiolase
Kaurene, 37, 39
Kaurene synthase-like (KSL), 191
Keap1, *see* Kelch-like ECH-associated protein 1
Kelch-like ECH-associated protein 1 (Keap1), 172
3-Ketoacyl-CoA thiolase (KAT), 204
Kiwifruit, 263
KSL, *see* Kaurene synthase-like

L

Labdane diterpenoid leoleorin, 39, 43
Ladanein, 30, 33
Lamiaceae, 29–30
LAMT, *see* Loganic acid methyltransferase

Lanthionine, 47, 49
(+)-Lariciresinol 9′-stearate, 37, 39
Latex, 18
Lathyrine, 47, 49
Laticifers, 18–19; *see also* Metabolite synthesis site
Leguminosae, 30–32
Linalool, 34, 35, 36
Linalyl acetate, 34, 36
Linoleic acid, 48, 49
Lipid transfer proteins (LTPs), 214
Lipiferolide, 38, 39
Lipophilic
  molecules, 12
  secretion, 16
Lipoxygenase, 196
Loganic acid, 95; *see also* Monoterpenoid indole alkaloid (MIA)
Loganic acid methyltransferase (LAMT), 95
LTPs, *see* Lipid transfer proteins
Lucumin, 46, 49
Lupanine, 45, 46, 49
Lutein, 41, 43
Luteolin-7-O-β-D-glucoside, 30, 33
Lyratyl butyrate, 35, 36
Lysigeny, 16

M

MAO, *see* Microbial monoamino oxidase
MATE, *see* Multidrug and toxic compound extrusion
Matrine, 45, 49
MDR, *see* Multidrug resistance
Medicinal plants, 25–26
Mentha piperita, 14
Menthofuran, 35, 36
Menthol, 35, 36
MEP, *see* 2-C-Methyl-D-erythritol 4-phosphate; Methylerythritol 4-phosphate
Merocrine, 10
MeSA, *see* Methyl salicylate
Metabolite synthesis site, 9
  excretion, 10
  glandular trichomes, 12–16
  holocrine, 10
  hydathodes, 12
  hydrophilic tissues, 11–12
  laticifers, 18–19
  lysigenous cavities, 17
  merocrine, 10
  nectaries, 12

Index

oil-bearing cells and secretory cells, 17–18
resin ducts, 17
secretion, 9–12
secretion process cytological model, 11
secretory cavities 16–17
secretory tissues, 9
translucent glands, 17
Metabolon, 139; *see also* Cyanogenic glucosides (CNglcs)
2,4-Methanoproline, 47, 49
15-β-Methoxyfasciculatin, 39, 43
Methyl chavicol, 35, 36
γ-Methyleneglutamic acid, 47, 49
Methylerythritol 4-phosphate (MEP), 95, 323 pathway, 325
2-C-Methyl-D-erythritol 4-phosphate (MEP), 188
Methyl eugenol, 35, 36
N-Methylputrescine oxidase (MPO), 102
Methyl salicylate (MeSA), 267
N-Methylsedridine, 126; *see also* Pinaceae alkaloids
(S)-*cis*-N-methylstylopine 14-hydroxylase (MSH), 90
7-Methylxanthosine synthase, 104
Mevalonate (MVA), 188, 323
MGDG, *see* Monogalactosyldiacylglycerol
MIA, *see* Monoterpenoid indole alkaloid
Microbial biofilms, 248; *see also* Microbial volatiles (mVOCs)
Microbial fuels, 248
Microbial monoamino oxidase (MAO), 343
Microbial volatiles (mVOCs), 239, 252
  2-aminoacetophenone, 251
  applications of, 242
  bacterial volatiles in database, 241
  as biocontrol agents in agriculture, 249–251
  biological and ecological functions of, 240
  for chemotyping and diagnostic tools, 251–252
  compounds released by bacteria, 241
  disease-suppression of soils, 250
  emission of, 240
  of foodstuff, 243–246
  frequently emitted, 240–241
  fungal volatiles in database, 242
  habitats of producers, 241
  as indicators of damp buildings and hardware, 246–248
  microbial biofilms, 248
  nematicidal, 250
  nematode-fungus interaction, 251
  as next generation biofuel, 248–249
  overview of, 244
  for perfume industry, 248
  principal component analysis of microorganisms, 243
Microbiota, 249
Mimosine, 47, 49
Minimiflorin, 30, 33
Monogalactosyldiacylglycerol (MGDG), 200
Monoterpenes, 33–36, 214, 248; *see also* Microbial volatiles (mVOCs); Terpenoids; Volatile organic compounds (VOCs)
Monoterpene synthase (mono-TPS), 189
Monoterpenoid indole alkaloid (MIA), 92, 332; *see also* Alkaloid
  biosynthesis and cellular compartmentation, 335–339
  biosynthetic pathways, 93–94, 336
  catharanthine, 97
  diversity, 333
  enzymes for MIA backbone formation, 334
  geraniol, 95
  iridoids, 95
  metabolic engineering, 339–340
  monoterpenoid moiety secologanin, 334
  nuclear time bomb, 97
  ORCA3, 99
  pathways and enzymes, 93–97
  secologanin, 95
  strictosidine, 96
  trafficking in *C. roseus*, 97–99
  transcriptional regulation, 99–100
  vindoline, 96, 97, 339
mono-TPS, *see* Monoterpene synthase
Morphinan, 333; *see also* Alkaloid
Morphine, 318
Mountain pine bark beetles (MPB), 219
MPB, *see* Mountain pine bark beetles
MPO, *see* N-Methylputrescine oxidase
MSH, *see* (S)-*cis*-N-methylstylopine 14-hydroxylase
Multidrug and toxic compound extrusion (MATE), 103
Multidrug resistance (MDR), 92
Multitrophic level interactions, 300–302; *see also* VOC-mediated information transfer
Mundulin, 30, 33
Mutualist–antagonist continuum, 294–300; *see also* VOC-mediated information transfer

Mutualist pollinators, 292, 294, 297
γ-Muurolene, 36, 39
MVA, see Mevalonate
mVOCs, see Microbial volatiles
Myrcene, 35, 36
Myrosin cells, 160

N

NAD, see Nicotinamide adenine
　　dinucleotide
NADP(H) quinone oxidoreductase 1
　　(NQO1), 172
NADPH-dependent cytochrome P450
　　oxidoreductase, 139
Natural product drugs, 316–317; see also
　　Plant secondary metabolism
NCS, see Norcoclaurine synthase
Nectar, 12
Nectaries, 12
Nematode-fungus interaction, 251
(E)-Nerolidol, 37, 39
Nicotinamide adenine dinucleotide
　　(NAD), 102
Nicotine, 5, 100; see also Alkaloid; Plant
　　specialized metabolites (PSMs)
　biosynthesis regulation, 103–104
　nicotinic acid, 102
　pathways and enzymes, 100–102
　putrescine, 102
　tobacco nicotine transporters, 103
Nicotine uptake permease 1 (NUP1), 103
Nicotinic acid, 102
Nitrate/peptide transporters (NTRs/PTRs),
　　169
Nitrilases (NITs), 135
Nitrile, 173; see also Glucosinolate
NITs, see Nitrilases
NMCH, see N-methylcoclaurine
　　3′-hydroxylase
N-methylcoclaurine 3′-hydroxylase
　　(NMCH), 89, 344
N-methyltransferase (NMT), 96
NMT, see N-methyltransferase
Nocturnal pollinators, 207
Nonprotein amino acids (NPAAs), 47; see
　　also Chemotaxonomy
NON-SMOKY GLYCOSYLTRANSFERASE
　　1 (NSGT1), 267
Norajmaline, 334; see also Alkaloid
Norcoclaurine 6-O-methyltransferase
　　(6OMT), 86
Norcoclaurine synthase (NCS), 86, 341

Norpluvine, 44, 49
NOS, see Noscapine synthase
Noscapine synthase (NOS), 91
NPAAs, see Nonprotein amino acids
NQO1, see NADP(H) quinone
　　oxidoreductase 1
Nrf2, see Nuclear respiratory factor 2
NSGT1, see NON-SMOKY
　　GLYCOSYLTRANSFERASE 1
NTRs/PTRs, see Nitrate/peptide
　　transporters
Nuclear respiratory factor 2 (Nrf2), 172
Nuclear time bomb, 97
NUP1, see Nicotine uptake permease 1

O

Obtusifoliol, 40, 43
(Z)-β-Ocimene, 35, 36
Octadecanoid-responsive Catharanthus
　　AP2-domain 2 (ORCA2), 99
ODC, see Ornithine decarboxylase
ODD, see 2-Oxoglutarate-dependent
　　dioxygenase
Off-flavor volatile geosmin, 246; see also
　　Microbial volatiles (mVOCs)
OG2, see Orthologous Group 2
2OGD/DOX, see 2-Oxoglutarate-dependent
　　dioxygenase
Oil-bearing cells, 17; see also Metabolite
　　synthesis site
Ononin, 32, 33
O-oxalylhomoserine, 47, 49
Oplopanone, 37, 39
Oral secretions (OS), 193
ORCA2, see Octadecanoid-responsive
　　Catharanthus AP2-domain 2
Ornithine decarboxylase (ODC), 100
Orobol, 32, 33
Orthologous Group 2 (OG2), 265
OS, see Oral secretions
10-Oxogeranial, 95
2-Oxoglutarate-dependent dioxygenase
　　(ODD; 2OGD/DOX), 86, 259
17-Oxoretamine, 46, 49
Oxylipins, 196; see also Green leaf volatiles
　　(GLVs)

P

P450, see Cytochrome P450 monooxygenase
P6H, see Protopine 6-hydroxylase
PAAS, see Phenylacetaldehyde synthase

Paclitaxel, 318, 329; see also Terpenoid metabolism
  biosynthesis, 329
  biosynthetic pathway, 330
  precursors production of, 331–332
Pain-killing drugs, 317
PAL, see Phenylalanine ammonia lyase
Paleozoic Era, 4
PAR, see Photosynthetically active radiation
Parthenolide, 38
PAs, see Purine alkaloids
Passibiflorin, 46, 49
Patchouli alcohol, 36, 39
Pathogenesis-related (PR), 341
Pathogenesis-related 10 (PR10), 86
Pavine, 333; see also Alkaloid
PCBER, see Phenylcoumaran benzylic ether reductase
Pelargonidin, 29, 33
Pelletierine, 120; see also Pinaceae alkaloids
Peltate trichomes, 16
Penduletin, 30, 33
Peonidin derivatives, 31, 33
Peroxidase (PRX), 96
pGlcPAAc, see p-glucosyloxyphenylacetic acid
p-glucosyloxyphenylacetic acid (pGlcPAAc), 144
Phanginin, 40
Phenolic compounds, 27, 33; see also Chemotaxonomy
Phenylacetaldehyde synthase (PAAS), 203
Phenylalanine ammonia lyase (PAL), 204
Phenylalanine-derived cyanogenic mono- and diglucosides, 134
Phenylbutanoids, 205; see also Volatile benzenoids and phenylpropanoids (VBPs)
Phenylcoumaran benzylic ether reductase (PCBER), 205–206
Phenylpropenes formation, 205; see also Volatile benzenoids and phenylpropanoids (VBPs)
Photosynthesis, 9
Photosynthetically active radiation (PAR), 216
p-hydroxymandelonitrile (pOHMN), 139
(E)-Phydroxyphenylacetaldoxime (pOHPOx), 139
p-hydroxyphenylacetic acid (pOHPAAc), 144

p-hydroxyphenylacetonitrile (pOHPCN), 139
Phytotoxic tropospheric ozone, 220
Pinaceae alkaloids, 119, 128; see also Alkaloid
  alkaloid end products, 125–126
  amounts and localization of, 126–127
  appearance of, 124
  biological roles of, 127
  biosynthesis of, 121–123
  control of biosynthesis, 123–124
  end products of coniferous species, 120
  euphococcinine, 121, 127
  genera and species-specific differences, 124–126
  toxic alkaloid compounds, 120
  trivial names, 120–121
Pinaceae piperidines, 119; see also Pinaceae alkaloids
α-Pinene, 34, 35, 36
β-Pinene, 34, 35, 36
4-oxo-2,6-cis-pinidinol, 126; see also Pinaceae alkaloids
Pinolenic, 48, 49
Pinoresinol-lariciresinol reductase (PLR), 205
PIP family, 205–206; see also Volatile benzenoids and phenylpropanoids (VBPs)
Plant, 132; see also Plant secondary metabolites
  alkaloid natural products, 333
  attraction by volatiles, 132
  cells, 10
  chemicals, 284
  defense mechanisms, 132
  evolution, 4, 316
  feeders, 4
  phenylpropanoid products from, 317
  secretory tissues in, 15
  specialized metabolism, 1
  stomatal opening, 214
  synthetic biology era, 318
  volatile storage, 260–261
Plant's chemical information, 286; see also Plant secondary metabolites
  ecologic niche analogy, 286–288
  herbivore-induced VOC emission, 290
  inherent constraints in metabolite production, 289–291
  PSM diversity and multifunctionality, 291–294
  screening hypothesis, 292
  volatile information space, 286

Plant secondary metabolism, 316, 350–351; see also Alkaloid; Glucosinolate metabolism; Terpenoid metabolism
  artemisinin, 316
  drugs, 316–317
  gene clusters in, 349–350
Plant secondary metabolites, see Plant specialized metabolites
Plant specialized metabolites (PSMs), 1–3, 305–306; see also Plant's chemical information; VOC-mediated information transfer
  application, 6
  characteristics of, 3
  focal point in framing, 1
  as information channels, 283–286
  molecular evolution, 3–5
  needs for human health, 5–6
  nicotine, 5
  qualities of, 1–2
  soy isoflavones, 5
Plant volatiles (PVOCs), 185 186; see also Green leaf volatiles (GLVs); Terpenes; Volatile benzenoids and phenylpropanoids (VBPs); Volatile organic compounds (VOCs)
  biosynthesis, regulation, and ecological relevance of, 188
  blue haze, 186
  effect of CO_2 on, 220
  ecological communications via, 208
  effect of global change factors on PVOC emissions, 208
  effect of transgenic torenia plants, 210
  function, 218
  plants vs. herbivore enemies, 209–211
  plants vs. herbivores, 209
  plants vs. plants, 211–212
  plants vs. pollinators, 208
  potential, 221
  signaling value, 220
Plebeiafuran, 37, 39
Plebeiolide A, 37, 39
PLP, see Pyridoxal-5'-phosphate
PLR, see Pinoresinol-lariciresinol reductase
PMT, see Putrescine N-methyltransferase
PNAE, see Polyneuridine aldehyde esterase
Pogosterol, 40, 43
pOHMN, see p-hydroxymandelonitrile
pOHPAAc, see p-hydroxyphenylacetic acid
pOHPCN, see p-hydroxyphenylacetonitrile

pOHPOx, see (E)-Phydroxyphenylacetaldoxime
Polyneuridine aldehyde, 334; see also Alkaloid
Polyneuridine aldehyde esterase (PNAE), 96, 334
Polyphenol oxidase (PPO), 193
Polyterpenes, 42; see also Terpenoids
PPO, see Polyphenol oxidase
PR, see Pathogenesis-related
PR10, see Pathogenesis-related 10
Progoitrin, 46, 49
Protoberberine, 333; see also Alkaloid
Proton transfer reaction mass spectrometry (PTR-MS), 198
Protopine, 333; see also Alkaloid
Protopine 6-hydroxylase (P6H), 90
Prunasin, 46, 49
PRX, see Peroxidase
Pseudo-alkaloids, 120
PSMs, see Plant specialized metabolites
PTR-MS, see Proton transfer reaction mass spectrometry
Pubescence, 13
Pulegone, 35, 36
PUP, see Purine permease
Purine alkaloids (PAs), 104–106, 333; see also Alkaloid
Purine permease (PUP), 103
Putrescine, 102; see also Nicotine
Putrescine N-methyltransferase (PMT), 102
PVOCs, see Plant volatiles
Pyridoxal-5'-phosphate (PLP), 203

Q

QTL, see Quantitative loci
Quantitative loci (QTL), 174
Quercetagetin, 31, 33
Quercetin, 31, 33
Quinoline alkaloid, 333; see also Alkaloid
Quinolizidine alkaloid, 333; see also Alkaloid

R

Racemic sparteine, 45, 49
Radish, 175–176; see also Brassicaceae
Rapeseed, 174–175; see also Brassicaceae
Raspberry ketone, 205; see also Volatile benzenoids and phenylpropanoids (VBPs)
Raspberry ketone/zingerone synthase (RZS), 205

Rauwolfia serpentina (Apocynaceae), 96
Reactive oxygen species (ROS), 142
Renewable Fuel Standard (RFS), 248
Repeated evolution, 140
Retamine, 46, 49
RFS, *see* Renewable Fuel Standard
Rhoiptelenol, 40, 43
Rohmer pathway, *see* Methylerythritol 4-phosphate (MEP)—pathway
ROS, *see* Reactive oxygen species
Rubixanthin, 41, 43
RZS, *see* Raspberry ketone/zingerone synthase

S

SA, *see* Salicylic acid
Sabinene, 34, 36
S-adenosyl L-methionine (SAM), 136, 340
Safranal, 35, 36
Sagebrush (*Artemisia tridentata*), 304
Salicaceae salicylates, 65, 78
 biological roles of, 75
 biosynthesis of salicylates, 72–74
 composition of, 67–72
 decomposition routes, 76
 as deterrents to rusts, 77
 effect of growing conditions, 71
 to herbivores, 75–77
 in human health, 75
 ontogenetic changes, 70
 in organs of *Salix* species, 68
 production costs of, 69
 Salicaceae species, 65
 salicin, 66
 salicin biosynthesis, 72
 turnover of salicylates, 74–75
Salicaceae species, 65
Salicin, 66
 biosynthesis, 72
Salicylic acid (SA), 194; *see also* Terpenes
SalR, *see* Salutaridine reductase
SalS, *see* Salutaridine synthase
SalSyn, *see* Salutaridine synthase
Salutaridine reductase (SalR), 89
Salutaridine synthase (SalS), 89
Salutaridine synthase (SalSyn), 342
Salvia species, 38
Salvigenin, 30
Salvinorin A, 38, 43
Salviol, 38, 43
SAM, *see* S-adenosyl L-methionine
Sanguinarine reductase (SanR), 90

SanR, *see* Sanguinarine reductase
Santolina alcohol, 35, 36
Santolina triene, 35, 36
SBS, *see* Sick building syndrome
S-cells, 160; *see also* Glucosinolate
Schizogeny, 16
Sciadonic, 48, 49
(S)-Scoulerine, 90; *see also* Benzylisoquinoline alkaloid (BIA)
Scoulerine 9-O-methyltransferase (SOMT), 90
Screening hypothesis, 292
Scutellarein, 30, 33
SDR, *see* Short-chain dehydrogenase/reductase
Secologanin, 44, 49, 95, 334; *see also* Alkaloid; Monoterpenoid indole alkaloid (MIA)
Secologanin synthase (SLS), 95, 337, 340
Secondary organic aerosol (SOA), 221
Secretion process cytological model, 11
Secretory; *see also* Metabolite synthesis site
 cells, 17–18
 ducts, 16–17
 tissues, 9
β-Selinene, 37, 39
Sesquiterpenes, 36–38, 39, 248; *see also* Microbial volatiles (mVOCs); Terpenoids
Sesquiterpene synthase (sesqui-TPS), 189
sesqui-TPS, *see* Sesquiterpene synthase
7DLGT, *see* 7-Deoxyloganetic acid glucosyltransferase
7DLH, *see* 7-Deoxyloganic acid 7-hydroxylase
7DLS, *see* 7-Deoxyloganetic acid synthase
SGD, *see* Strictosidine β-glucosidase
Short-chain dehydrogenase/reductase (SDR), 89
Sick building syndrome (SBS), 247
Single nucleotide polymorphism (SNP), 144
6OMT, *see* Norcoclaurine 6-O-methyltransferase
16OMT, *see* 16-Hydroxytabersonine O-methyltransferase
SLS, *see* Secologanin synthase
SNP, *see* Single nucleotide polymorphism
SOA, *see* Secondary organic aerosol
Solenopsin, 120; *see also* Pinaceae alkaloids
SOMT, *see* Scoulerine 9-O-methyltransferase
SOTs, *see* Sulfotransferases
Soy isoflavones, 5; *see also* Plant specialized metabolites (PSMs)

Spathulenol, 36, 39
Specialized metabolites, 24, 258, 316
SPS, see Stylopine synthase
Stachyspinoside, 30, 33
Stem cells, 13
(S)-tetrahydroberberine oxidase (STOX), 90, 90
Stomatal opening, 214
STORR, see (S)-to (R)-reticuline
(S)-to (R)-reticuline (STORR), 89
STOX, see (S)-tetrahydroberberine oxidase
STR, see Strictosidine synthase
Stratospheric ozone, 217
Strictosidine, 96; see also Alkaloid; Monoterpenoid indole alkaloid (MIA)
  aglycone, 334
Strictosidine synthase (STR), 96, 334
Strictosidine β-glucosidase (SGD), 96, 334
Stylopine synthase (SPS), 90
Sulforaphane, 172; see also Glucosinolate
Sulfotransferases (SOTs), 165
Superfamily enzyme gene, 258, 259
Superfamily enzymes, 259
Surface alkanes, 50; see also Chemotaxonomy
Synthetic biology, 318

T

T16H, see Tabersonine 16-hydroxylase
T19H, see Tabersonine 19-hydroylase
T3O, see Tabersonine 3-oxygenase
T6ODM, see Thebaine 6-O-demethylase
T6ODM/CODM, see Thebaine/codeine demethylases
Tabersonine 16-hydroxylase (T16H), 96
Tabersonine 19-hydroylase (T19H), 338
Tabersonine, 334; see also Alkaloid
Tabersonine 3-oxygenase (T3O), 339
Tanshinone I, 39, 43
Taraxanthin, 41, 43
TAs, see Tropane alkaloids
Taxol, see Paclitaxel
Taxoleic, 48, 49
TCHA, see Tetrahydrocannabinolate
TDC, see Tryptophan decarboxylase
Tea, 264–265, 266–267
Terpenes, 33, 249, 188; see also Chemotaxonomy; Microbial volatiles (mVOCs); Plant volatiles (PVOCs)
  biosynthesis, 188–191
  chemical structures of, 187

  cross-talks with phytohormone signaling, 194
  diterpenes, 38–40, 43
  effectors, 194
  ethylene signaling, 194
  HIPV induction, 192–193
  monoterpenes, 33–36
  polyterpenes, 42
  regulation, 191–194
  sesquiterpenes, 36–38, 39
  tetraterpenes, 41, 43
  triterpenes, 40–41, 43
  volicitin, 193
Terpene synthase (TPS), 189, 276, 321; see also Terpenoid metabolism
Terpenoid metabolism, 319; see also Isopentenyl diphosphate (IDP; IPP); Plant secondary metabolism
  artemisinin, 326–329
  biogenesis of terpene scaffolds, 319–322
  C5 isoprene, 319
  cytochrome P450 monooxygenase, 322–323
  dolichol, 319
  isoprene, 319
  mechanisms of terpene synthases, 320
  metabolic engineering, 326
  paclitaxel, 329–332
  terpenoid, 319
Terpenoids, see Terpenes
Terpinen-4-ol, 35, 36
γ-Terpinene, 35, 36
α-Terpineol, 35, 36
Terrestrialization, 4
Tetrahydroalstonine, 334; see also Alkaloid
Tetrahydroalstonine synthase (THAS), 334
Δ9-Tetrahydrocannabinol (THC), 16
Tetrahydrocannabinolate (TCHA), 16
Tetrahydroprotoberberine cis-N-methyltransferase (TNMT), 90
Tetraterpenes, 41, 43; see also Terpenoids
Teufruintin A, 39, 43
THAS, see Tetrahydroalstonine synthase
THC, see Δ9-Tetrahydrocannabinol
Thebaine, 89; see also Benzylisoquinoline alkaloid (BIA)
Thebaine/codeine demethylases (T6ODM/CODM), 342
Thebaine 6-O-demethylase (T6ODM), 89, 342
Thebaine synthase (THS), 342
Theobromine synthase (TS), 106
3C, see Three-carbon

Index

Three-carbon (3C), 170
THS, see Thebaine synthase
α-Thujone, 34, 36
β-Thujone, 34, 35, 36
Thymol, 35, 36
TMTT, see (E,E)-4,8,12-trimethyltrideca-1,3,7,11-tetraene
TNMT, see Tetrahydroprotoberberine cis-N-methyltransferase
Toxic alkaloid compounds, 120; see also Pinaceae alkaloids
Toxicarol isoflavone, 32
TPS, see Terpene synthase
Translucent glands, 17
(E,E)-4,8,12-trimethyltrideca-1,3,7,11-tetraene (TMTT), 191
Triterpenes, 40–41, 43; see also Terpenoids
Tritrophic system, 201
Tropane alkaloids (TAs), 100, 333; see also Alkaloid; Nicotine
Tropinone reductases (TRs), 102
Trptamine, 334; see also Alkaloid
TRs, see Tropinone reductases
Truffles, 244; see also Microbial volatiles (mVOCs)
Tryptophan decarboxylase (TDC), 93, 333, 339
TS, see Theobromine synthase
TYDC, see Tyrosine decarboxylase
TyrAT, see Tyrosine aminotransferase
Tyrosine aminotransferase (TyrAT), 86
Tyrosine decarboxylase (TYDC), 86

U

UDP, see Uridine diphosphate
UDP-glucose glycosyltransferase (UGT), 95, 139, 258–259; see also Volatile glycosylation
    functional diversity of UGT for specialized metabolites, 276
    for volatiles, 265
UGT, see UDP-glucose glycosyltransferase
Ultraviolet radiation (UVB), 71
Uridine diphosphate (UDP), 258
U's triangle, 170; see also Glucosinolate
UVB, see Ultraviolet radiation

V

Valencene, 37, 39
VAN, see Vanillin synthase
Vanillin synthase (VAN), 205

VBPs, see Volatile benzenoids and phenylpropanoids
VIGS, see Virus-induced gene silencing
Vinblastine, 318
Vincadifformine, 44, 49
Vincamine, 334; see also Alkaloid
Vindoline, 96, 97; see also Monoterpenoid indole alkaloid (MIA) biosynthesis, 339
Vindolinine, 334; see also Alkaloid
Vinorine, 334; see also Alkaloid
Vinorine synthase (VS), 96, 334
Viridiflorene, 37, 39
Virus-induced gene silencing (VIGS), 335
VOC-mediated information transfer, 294; see also Plant secondary metabolites
    allelopathic function of plant–plant communication, 303
    detractant compounds, 296
    floral VOC emission, 297–299
    herbivory-induced VOC emission, 296, 302
    multitrophic level interactions, 300–302
    mutualist–antagonist continuum, 294–300
    mutualist pollinators, 292, 294, 297
    sagebrush, 304
    VOC-mediated plant–plant communication, 302–304
    wild tomato and Halictidae bee, 298
VOC-mediated plant–plant communication, 302–304; see also VOC-mediated information transfer
VOCs, see Volatile organic compounds
Volatile benzenoids and phenylpropanoids (VBPs), 202; see also Plant volatiles (PVOCs)
    biosynthesis, 202–206
    biosynthetic pathways, 203
    chemical structures of, 187
    phenylbutanoids, 205
    phenylpropenes formation, 205
    PIP family, 205–206
    raspberry ketone, 205
    regulation, 206–208
    VAN activity, 205
Volatile glycosylation, 258
    dietary crops producing, 262
    ecological and evolutionary landscape, 273–275
    general, 258

Volatile glycosylation (*Continued*)
   glycoside-specific glycosyltransfearse, 268–273
   glycosylating machinery, 261
   glycosylation-hydrolysis cycle, 275
   grapes, 263–264
   kiwifruit, 263
   mode-of-action of volatiles, 274
   molecular evolution of superfamily enzyme gene, 259
   perspective, 268
   promiscuity and diversity, 275–276
   tea, 264–265, 266–267
   tomatoes, 267
   UGT, 258–259, 265
   volatile glycoside metabolism, 273
   volatile glycosides, 260–261
Volatile organic compounds (VOCs), 12, 13, 260; *see also* Metabolite synthesis site; Plant volatiles (PVOCs)
   in atmospheric formation, 221
   biotic interactions and ecological functions, 218–221
   CO_2, 217
   in global environmental change, 212
   isoprene emission capacity, 214–215
   light intensity and quality, 216
   lipophilic, 213
   ozone, 217–218
   release through stomata, 214
   synthesis, 18
   temperature, 213–214
   UV radiation, 215–216
   vetiver, 17–18
   wind, 215
Volatiles, 266; *see also* Volatile glycosylation; Volatile organic compounds (VOCs)
   compound classes, 289
   glycoside metabolism, 273
Volicitin, 193; *see also* Terpenes
VS, *see* Vinorine synthase

W

Wild tomato (*Solanum peruvianum*), 298
Wine quality attributes, 245
Wistin, 32, 33

X

Xanthosine, 104